S Merritt

CHEMICAL MEDIATION OF COEVOLUTION

Chemical Mediation of Coevolution

EDITED BY
KEVIN C. SPENCER

Department of Medicinal Chemistry and Pharmacognosy
College of Pharmacy, University of Illinois at Chicago
Chicago, Illinois

ACADEMIC PRESS, INC.
Harcourt Brace Jovanovich, Publishers
San Diego New York Berkeley Boston
London Sydney Tokyo Toronto

ACADEMIC PRESS, INC.
San Diego, California 92101

United Kingdom Edition published by
ACADEMIC PRESS, INC. (LONDON) LTD.
24-28 Oval Road, London NW1 7DX

Library of Congress Cataloging-in-Publication Data

Chemical mediation of coevolution.

Includes index.
1. Coevolution. 2. Insect-plant relationships.
3. Chemical evolution. 4. Botanical chemistry.
I. Spencer, Kevin C.
QH372.C44 1988 575 88-3424
ISBN 0-12-656855-3 (hardcover)(alk. paper)
ISBN 0-12-656856-1 (paperback)(alk. paper)

PRINTED IN THE UNITED STATES OF AMERICA
88 89 90 91 9 8 7 6 5 4 3 2 1

CONTENTS

Contributors

May R. Berenbaum Department of Entomology, University of Illinois, Urbana, Illinois 61801

Murray S. Blum Department of Entomology, University of Georgia, Athens, Georgia 30602

Calhoun Bond Department of Biology, University of North Carolina, Chapel Hill, North Carolina 27514

M. Deane Bowers Museum of Comparative Zoology, Harvard University, Cambridge, Massachusetts 02138

Lincoln P. Brower Department of Zoology, University of Florida, Gainesville, Florida 32611

John P. Bryant Institute of Arctic Biology, Irving Building 311-A University of Alaska, Fairbanks, Alaska 99775

Rex G. Cates Department of Botany, Chemical Ecology Laboratory, Brigham Young University, Provo, Utah 84602

F. S. Chew Department of Biology, Tufts University, Medford, Massachusetts 02155

Karen M. Clancy Department of Biological Sciences, Northern Arizona University, Flagstaff, Arizona 86011

Timothy P. Craig Department of Biological Sciences, Northern Arizona University, Flagstaff, Arizona 86011

Stanley H. Faeth Department of Zoology, Arizona State University, Tempe, Arizona 85287

Linda S. Fink Department of Zoology, University of Florida, Gainesville, Florida 32611

Fred Gould Department of Entomology, North Carolina State University, Raleigh, North Carolina 27650

Clive G. Jones Institute of Ecosystem Studies, The New York Botanical Garden, Mary Flagler Cary Arboretum, Millbrook, New York 12545

Richard L. Lindroth Department of Entomology, University of Wisconsin, Madison, Wisconsin 53706

Judith H. Myers The Ecology Group and Departments of Plant Science and Zoology, University of British Columbia, Vancouver, Canada V6T 1W5

Carolyn J. Nelson Pharmacy Department, University of Sydney, Sydney, Australia NSW 2006

Pekka Niemala Department of Biology, University of Turku, SF-20500 Turku, Finland

Peter W. Price Department of Biological Sciences, Northern Arizona University, Flagstaff, Arizona 86011

Richard A. Redak Department of Entomology, Colorado State University, Fort Collins, Colorado 80523

J. A. A. Renwick Boyce Thompson Institute, Ithaca, New York 14853

Christopher F. Sacchi Department of Biological Sciences, Northern Arizona University, Flagstaff, Arizona 86011

J. Mark Scriber Department of Entomology, Michigan State University, East Lansing, Michigan 48824

James N. Seiber Department of Environmental Toxicology, University of California, Davis, California 95616

Peter J. Silk Pheromone Research Group, New Brunswick Research and Productivity Council, Fredericton, New Brunswick, Canada E3B 5HI

Kevin C. Spencer Department of Medicinal Chemistry and Pharmacognosy, University of Illinois, Chicago, Illinois 60612

Mark Kwasny Stowe Department of Organismic and Evolutionary Biology, Harvard University, Boston, Massachusetts

Juha Tuomi Department of Biology, University of Turku, SF-20500 Turku, Finland

Gwendolyn L. Waring Department of Biological Sciences, Northern Arizona University, Flagstaff, Arizona 86011

Douglas W. Whitman USDA-ARS, Insect Biology and Population Management Research Laboratory, Tifton, Georgia 31793-0748

Arthur R. Zangerl Department of Entomology, University of Illinois, Urbana, Illinois 61801

Preface

The purpose of this book is to explore the degree to which chemicals are the currency of information exchange in coevolved systems and to reexamine present concepts of coevolution through interpretation of chemical parameters. Because the biosynthetic steps involved in the construction of chemical compounds can be determined, chemical as opposed to morphological characters have great potential quantitative utility in the determination of evolutionary rate, sequence, and direction. In systems of coradiating plants and herbivores, plant secondary chemicals are frequently biosynthesized in significant amounts and diversified in structure. If defensive in nature, these chemistries may be elicited as a result of ongoing coevolutionary processes, so it may be possible to predict system chemistries based on studies of the phylogenies of the element organisms of the system. On the other hand, these chemistries may in fact be causative parameters, and knowledge of defensive chemistry may allow us to infer the course of events in the development of a coevolved system.

In considering the problem from either perspective, we may wonder whether phylogenetically distinct taxa undergo similar production and diversification of defensive chemicals, and as a corollary, whether the various chemical classes differ intrinsically in their biological properties, and hence have potential as selective agents in coevolution. At higher trophic levels, predator–prey and social interactions can be mediated by chemical communication. The relationship between plant chemistry and edaphic factors, and its variation within individuals, populations, and species can be of critical importance in the determination of its coevolutionary role.

The papers collected here address the relationship between chemistry and coevolution at several levels and in so doing assess both the importance of chemistry in coevolved systems and the coevolutionary concept itself.

Recent volumes have analyzed the chemistry of plants: [*The Biochemistry of Plants, a Comprehensive Treatise,* E. E. Conn and P. K. Stumpf (eds.), Academic Press, 1983; *Encyclopedia of Plant Physiology,* E. A. Bell and B. V. Charlwood (eds.), Springer-Verlag, 1980], insects [*Comprehensive Insect Physiology, Biochemistry, and Pharmacology,* G. A. Kerkut and L. I. Gilbert (eds.), Pergamon Press, 1985; *Chemical Defenses of Arthropods,* M. S. Blum, Academic Press, 1981], and their interaction with one another [*Herbivores: Their Interaction with Plant Secondary Metabolites,* G. A. Rosenthal and D. H. Janzen (eds.), Academic Press, 1979; *Chemical Ecology of Insects,* W. J. Bell and R. T. Cardé (eds.), Sinauer Associates, 1984].

In addition, a recent systematic treatment of coevolutionary patterns (*Interaction and Coevolution,* J. N. Thompson, Wiley-Interscience, 1982) and a thorough treatment of the entire subject of coevolutionary theory [*Coevolution,* D. J. Futuyma and M. F. Slatkin (eds.), Sinauer Associates, 1983] exist, and many allusions are made to the possible influence of plant and animal chemicals in the systems analyzed.

This book is the first to evaluate processes of coevolution in terms of the chemistries of the organisms involved.

I am very grateful to Charles M. Chambers and the American Institute of Biological Sciences, Beverly Rathcke, the Ecological Section of the Botanical Society of America, the Southeastern Chapter of the Ecological Society of America, the Association for Tropical Biology, and the University of Florida, Gainesville, for their sponsorship of the original symposium "Chemical Mediation of Coevolution," on which this book is based. The symposium was held August 14–15, 1985, as part of the 36th annual AIBS meeting at the University of Florida. At that time, it was felt that a free exchange of ideas about the nature of coevolution could be useful in helping workers design tests of the role of chemistry in evolution. The surprising quantity of data already amassed by the speakers necessitated a reassessment of the work presented by all of us. The contents of this volume are the product of a considerable reorganization of and addition to the thought and knowledge contained in the original symposium. It is hoped that this collected work will provide an impetus for careful reconsideration of the possible roles played by chemistry in the establishment, maintenance, and fate of coevolutionary relationships.

A special acknowledgment is owed to the reviewers of the chapters, many of whom read several papers critically and all of whom have contributed immeasurably to the quality of the final product: S. C. H. Barrett, M. R. Berenbaum, E. Bernays, M. C. Birch, M. D. Bowers, L. B.

Brattsten, L. P. Brower, J. P. Bryant, R. T. Cardé, R. G. Cates, F. S. Chew, T. P. Craig, S. H. Faeth, D. J. Futuyma, F. Gould, J. D. Horner, J. E. Huheey, G. W. Ivie, C. G. Jones, T. T. Koslowski, D. E. Lincoln, R. L. Lindroth, M. M. Martin, J. C. Myers, J. R. Miller, S. T. Mitchell-Olds, K. F. Raffa, P. B. Reichardt, J. A. Renwick, J. C. Schultz, J. M. Scriber, F. R. Stermitz, M. K. Stowe, S. B. Vinson, and several anonymous reviewers.

I wish also to thank J. C. Junkus for valuable statistical evaluations and editorial comments. Finally, I wish to express particular thanks to my secretary, Marion Sitt, for her exceptional skill, patience, and fortitude in the face of unreasonable demands during the production of this volume.

The value of this collection lies in the efforts and contributions of the symposium participants. Thanks are due to each and every one of the authors for their formidable labors and exciting discourse. It has been a pleasure to learn from them.

Kevin C. Spencer
Chicago, Illinois

CHAPTER 1

Introduction: Chemistry and Coevolution

KEVIN C. SPENCER
Department of Medicinal Chemistry and Pharmacognosy
University of Illinois
Chicago, Illinois

I. Chemistry and the Development of Concepts of Coevolution

That insects eat plants is an observation of biology so ancient that to mention it here would be trite, except that it emphasizes our state of ignorance of the genesis of the phenomenon. Millennia have passed since historically cognizant and educable hominids have mentally recorded the obligate, omnipresent association between herbivores and their host plants, yet we are still unable both to describe with certainty the process by which these observed associations are established, or to predict with confidence what result will come of any such active interaction.

Eloquent descriptions of the details of plant and insect interactions were part and parcel of the birth of modern biology. These observations by older naturalists become essential components of both stimulus and evidence in the theses of Darwin and his contemporaries. Evolution was then named as the process, but a mechanistic description of the development of plant–insect interactions remained elusive even with the advent of the New Synthesis. Such interacting systems seemed to have been born of fortuitous involvement of disparate lineages of organisms, one with another, and usually were represented as one-sided specializations or mutualisms.

The realization of the necessity of formulating a generalized hypothesis to account for the frequent evolution of specialized herbivory awaited the presentation of a treatise (Brues, 1924) that held that related species of insects feed on related species of plants. This work brought together the many naturalistic observations of interacting systems with the nascent idea of genetic determinism to a synthetic conclusion: evolutionary relatedness in individual insect lineages may be reflected in genetically nonrelated but ecologically associated plant lineages.

Two generations later, Ehrlich and Raven (1964) described as "coevolution" the process by which plants and insects radiate and speciate in association with one another. The process so described entails the evolution of chemical defenses by plants directed against herbivores, which then frees the plant from herbivore pressure and permits the plant lineage to radiate and speciate. Interacting herbivores that evolve a counteradaptation or tolerance to this defense will be able to exploit the new competitor-free adaptive zone, with a resultant likelihood of radiation and speciation. This "reciprocal stepwise evolution" thus accounts for plant and insect diversification and observed patterns of specialization and association between insects and their hosts.

The importance of genetic reciprocity as a requirement for coevolution was emphasized by Janzen (1980). In this narrow definition, evolution of a trait in the individuals of one population in response to a trait of the individuals of a second population is followed by an evolutionary response in the second population to the change in the first. This is the coevolutionary scenario most amenable to detection and experimentation, particularly in measuring chemical changes, and it has been forcefully argued (Thompson, 1982) that this definition is the rational basis for the term coevolution as generally used to describe interactions between species pairs.

The pairwise reciprocal coevolutionary scenario of Janzen (1980), following Ehrlich and Raven (1964), predicts the development of a chemical "arms race" (Dawkins and Krebs, 1979), in which reciprocity between plant chemical defenses and insect counteradaptation takes place. Berenbaum (1979, 1981, 1983; Berenbaum and Feeny, 1981) tested this hypothesis and found that insect–plant specialization corresponded to host plant chemistry, and that this correspondence was maintained through lineages of insects and plants in a system comprising Apiaceae, *Papilio,* and secophorids and furanocoumarins. Insect–host plant affinities were seen to reflect single chemical changes through evolution of a detoxification system, consistent with a reciprocal arms-race scenario. Such reciprocity is also described in systems representing forms of interaction other than plants and their insect herbivores (see Thompson, 1982; and Craig *et al.*, this volume, for summaries). The examples used most frequently in support of this concept are perhaps the ant–acacia association (Janzen, 1966) and the rust and flax interaction (Flor, 1971). The first is a mutualism, the second is antagonistic, and these, as well as Berenbaum's chemical system, are interpreted (though not explicitly stated) as having arisen through "gene-for-gene" coevolutionary reciprocity.

While this definition may be best, the process so described may not be common (Futuyma and Slatkin, 1983c). Many authors feel that coevolution is a process that involves more than two species, and more than two foci for evolutionary change. Coevolution has thus been defined alternatively as a "diffuse" process (Fox, 1981).

Evidence for diffuse coevolutionary interactions between groups of plants and insects is presented by Gilbert and Raven (1975), Benson *et al.* (1976), Fox (1981), and Berenbaum (1983). This kind of interaction is concluded to be of more common and likely occurrence than paired species interactions by a consensus of recent authors (Futuyma and Slatkin, 1983c). Diffusion of interacting forces through a chemically diversified system is attributable partly to the age of the coevolutionary association, as documented for chrysomelid beetles feeding on Cucurbitaceae (Metcalf, 1979). Similarly, it has been concluded that the highly diversified patterns of specialization noted for *Heliconius* upon *Passiflora* involve plant chemistry (Spencer, 1987), not just ecological factors (Smiley, 1985), in a multispecies interaction in which selective pressures from many species determine the outcome of events (Benson *et al.*, 1976). Even in very large radiations of plants and insects where insect speciation events are apparently not a result of intense specialization upon plant resources, as in *Drosophila* upon Cactaceae, chemical factors are found to be important and possibly directly responsible for observed radiations (Barker and Starmer, 1982; Spencer, 1987).

Coevolution may thus be defined narrowly or broadly, but Futuyma and Slatkin (1983b) wisely warn us that coevolution defined too broadly becomes evolution and too narrowly becomes a biological improbability. Coevolution may differ from evolution in being a subset process defined only in terms of scale: limited species numbers, trophic levels, age, restriction in direction, rate of genetic change, and rate of change of evolutionary rate. Is it possible that chemistry somehow provides a basis for this difference?

The infinite capacity of plants to generate multitudinous chemotypes, and the corresponding ability of insects to recognize and exploit adaptively these segregates, are impetus enough to forward the hypothesis that aspects of plant chemistry can provide a sufficient mechanistic explanation for the generation of specialization and diversification in interacting systems of plants and insects. Indeed, Ehrlich and Raven state explicitly that secondary chemicals are the focus of coevolution between insects and plants, and much of the evidence presented in support of later concepts of coevolution is chemical.

II. Chemistry as a Coevolutionary Fulcrum

For perhaps 300 million years, insects have been eating plants (Wootton, 1981; Metcalf, 1979). The latter's persistence presents to us insistently the fact of their development of effective defenses against herbivory. The persistence of the insects is proof of their ability to generate counteradaptations. This is evolution. At a global scale, and at the taxonomic level of class and above, we can appreciate that this process is driven, directed, and responsive. However, the quantitative inestimability of selective forces and chaotic factors prevents us from confidently assuming that the overall process has been one of coevolution. We do not know the degree to which the genomic structures of herbivores have been constructed in direct response to the genomic structure of plants, and vice versa.

The primary goal of an herbivore is to select host plants that are optimal or at least adequate in fulfilling its nutritional requirements sufficiently that it may successfully reproduce (Lawton and Strong, 1981; Scriber, 1984). The precise identification of potential food sources is usually accomplished through plant chemistry and generally occurs with the assistance of historically determined associations of specific insect sensory systems with individual plant lineages (Ehrlich and Raven, 1964; Chew, 1975, 1977; Metcalf, 1979; Miller and Strickler, 1984). Primary metabolic chemical characteristics are continuously sensed and tracked by herbivores seeking nutrition. This may be independent of the degree of specialization and mode of sensory response shown toward secondary compounds, and we may expect arrays of primary compounds to vary selectively, as suggested for fruit sugars and frugivore/plant coevolution (Thompson and Willson, 1979).

Most classes of chemicals, and certainly proteins, possess enough inherent variability of structure to permit evolution thereupon. Chemical and biochemical changes are extremely specific and, once achieved, are highly constrained in alterability to conform to structure–function requirements. Speciation resulting from this impetus should yield derivative species that closely resemble the ancestral species in most respects. Hybridizability and species barriers should remain low, and interspecific competition very high except for differences in resource utilization due to chemical specialization. Pleiotropic effects of alterations of structure in enzymes of essential primary metabolism should be quite large, and those of the biosynthetically more distant secondary metabolism, smaller.

Plant chemistry may have itself evolved as a response to coevolutionary specialization by insects, simply because it is the only plant response that can create very major differences in the plant's appearance to the insect's sensory modalities with a single gene change (although most differences may be polygenic). Thus it is through chemistry that the plants become able to challenge threats or to move within a coevolutionary scenario with maximum rapidity. Development of a chemical response to coevolutionary events may therefore have occurred as an adaptation to maximize the rate of evolutionary change of the plant. While plant chemistry is not the only barrier, by any means, to herbivory, it may be the most critical in permitting rapid reciprocal speciation.

In a paired two-taxon coevolved system, selection might be expected toward the fixation of the mechanism that controls the rate of change: to a constant in each species of a pair in a mutualism, to a unilateral maximum in each in specialized herbivory. The detection of changes should also be efficient. Chemistry can change rapidly enough to have these effects.

In diffuse coevolution, the fill of species into the associative melange of interacting organisms may require differences in rates of evolution, in closeness of fit of cued response, and in distribution of character states across an adaptive norm. We may all too easily recognize this as a process of extinction and speciation. Extinction is not necessarily equivalent to speciation; in coevolution especially, extinction may have large transspecific effects. Biochemical changes can provide a means of rapid response to new selective pressures from several species. These changes can be accrued additively. While we may recognize speciation events along the lineage, we do not require extinction to explain them.

It can be imagined that in either a pairwise or a diffuse concept, coevolution as so far described diminishes diversity, whereas an examination at a larger scale shows that system diversity may be dramatically increased. If system diversity is proportional to system stability, we may find ourselves determining and predicting the stability of large ecosystems in terms of plant and animal coevolution through measures of chemical diversity. As a corollary to the logical statement that coevolution is diversity-dependent (Roughgarden, 1983), I would suggest that system diversity is utterly dependent on the form and process of coevolution.

Diversification of an ancestral species into several derivative species increases the frequency and persistence of the original genetic contribution over time. The more rapid and the more extensive the radiation of speciation from the ancestral genome, the more of that genome there will be living (expressed) in the environment. The more diversified the radiation, the greater the chances for persistence. Each derivative species may retain, without recourse to additional evolutionary change or expenditure, those biosynthetic pathways present in its progenitors. During speciation, the shared ancestral pathway may also be dropped, possibly with a single gene change, making the largest possible chemical change in that chemical series—obviation of the series.

Coevolution is a process that we may be able to observe and measure at least sometimes without resort to assumptions of selective regimes but that may usually require the description of selective events. These are not often experimentally measured. These selective regimes may be frequency- or density-dependent. If chemical, the selective agent may have effects based on frequencies of occurrence or absence as well as rate of change in individuals, population, species, and among many possibly unrelated species. These effects may constitute the deciding factors in determining whether an incipient coevolutionary interaction is stabilized, continuously evolves, or becomes extinct (Maynard-Smith, 1976).

As pointed out by Futuyma and Slatkin (1983b), the literature on coevolution concentrates on adaptations resulting from coevolution rather than on mechanistic interpretations of the process. Indeed, coevolution as clearly identifiable phenomenon is apparently a rare process and may require very special conditions in order to occur (Jermy, 1984; Fox, 1981; Thompson, 1982; Roughgarden, 1983). Might some aspect of the plant chemistry represent one such condition?

III. Chemical Study of Coevolution

Coevolution is a legitimate area of study as a "major point of contact between ecology and evolution" (Futuyma and Slatkin, 1983b). This is profoundly so as we select for study strong interactions, such as those between a common specialist insect and its favored foodplant, rather than rare contacts between uncommon plants and generalist browsers. Selective events and evolutionarily important processes might be expected to be more recognizable in such intense interactions. These intense specializations of insects upon their host plants are at a minimum often, and are possibly always, mediated by plant chemistry (Chambers, 1970; Feeny, 1976; Rosenthal and Janzen, 1979; Verschaffelt, 1910; Brues, 1924; Ehrlich and Raven, 1964; Futuyma, 1983). Chemical study is uniquely and wonderfully suited to the demonstration of reciprocal change in a coevolved plant–herbivore system. Chemical parameters can be completely quantitatively and qualitatively measured.

In lieu of historical data, any given interaction may be explained either by coevolution or by fortuitous ecological association of exapted forms (Janzen, 1980), and actual evidence for coevolution is equivocal (Strong *et al.*, 1984). Many interactions apparently explained by chemistry need much closer examination. As an example, Brown (1984) found that

Ithominiinae, which have long been known to be chemically defended by pyrrolizidine alkaloids, obtain these as adults, rather than from the larval foodplants that also contain pyrrolizidines. He also demonstrated their deterrency to predation by spiders, but the mechanism by which this defense could result from a coevolutionary relationship between the butterflies and their host plants versus adult forage plants cannot be easily inferred. Can we distinguish between exaptive and coevolutionary events through careful mapping of the developmental history of interacting biochemical pathways?

If historical interactions between herbivores and plants are important to subsequent evolution, and if indeed congruent phylogenies are the best evidence for coevolution (Mitter and Brooks, 1983; Gould, this volume), then to the degree that we are able to measure chemical changes over time or chemical differences between species within a lineage, we should be able to quantify coevolution through chemistry. In fact, we should be able to directly relate the complexity of the chemistry of a lineage to the age or intensity of coevolution. It may then become possible to distinguish satisfactorily coevolution (reciprocal gene changes) from cospeciation (reciprocal speciation events), two processes that are not necessarily similar (Futuyma, 1983).

It should also be possible to predict the end product of coradiation. The determination of the numbers of species and the amount of chemical diversification is the result of intraspecific competitive character displacement and resource partitioning. However, regardless of the age, strength, or type of interaction, these changes are responsive to other species' actions through competition and predation, all under the ultimate (edaphic) limitations of the environment. Resource limitation, in concert with the innate ability to evolve in terms of degree, direction, and rate of change, controls the strength of the coevolutionary interaction. This ability is also responsive to an evolutionary direction, and to coevolutionary events. Restriction of resource "slop" in the system can pay the price of restrictive change. Specialization should entail greater efficiency in coevolution, as coevolution represents a dedication to an evolutionary direction, and is a restrictive process. Measurement of chemical differences should relate to all other factors in a system.

Evoregulatory processes, evolutionary expenditure of energetic potential for control mechanisms for rate and direction of change, may explain why coevolved systems may not be optimally efficient in resource exploitation, or why diversification and speciation might replace optimization. These control processes are much more readily postulated upon consideration of restrictions on, and limited evolutionary choices available to, organisms that have invested in a particular biosynthetic pathway toward generation or tolerance of a chemical defense.

In using chemistry to probe coevolution, we will gain most from using a systems approach to bring ecological parameters into evolutionary consideration, phylogeny into character change assessments, biochemistry into interpretation of organismal interaction, and predation, competition, resource limitation, and edaphic factors to bear upon individual genetic change through progeny.

Several questions can be raised: (1) how and (2) when during the evolutionary course of development of a specialized interaction does plant chemistry manifest itself as a selective agent to either the plant or the insect; (3) is this necessary to the coevolutionary process; and (4) does this direct events in coevolution?

IV. Parameters and Probabilities in the Process of Coevolution

Coevolution is a directed, normative rate process, with an important component of variation. Each of the contributing ecological biotic and abiotic parameters is theoretically measurable over time. The *n*-dimensional hyperspace thus estimated is the ecological niche at a given time, but when summed over time it is the evolutionary process of the species reflected in adaptation and interaction.

The major difference between evolution and coevolution may be in terms of rate of change and in degree of constraint in direction. When codirectionality is imposed, the rate in each may not be the same, nor the magnitude of the response. Further, selective pressure to coevolve may be opposed by a continuing increase of a selective pressure to disengage from codependence in one or both parties. Lineages of both insects and plants may bounce in and out of specialization modes in cycles over evolutionary time in order to defeat the trap of codependence and to refresh their exhaustively dedicated genome.

What are the potentialities and probabilities for establishment of various coevolutionary scenarios? How does chemical variance in a lineage compare with covariance between interacting lineages? Most crop plant/major pest interactions represent coevolutionary interactions involving extreme specialization mediated by plant chemistry. Other pests that are generalist feeders may represent evolutionary responses to such interactions. Can we predict or control the outcome of pest/crop plant association? Are the plant–insect–chemical systems we see now what had existed, or have the original selective parameters leading to the establishment of the system changed? Are mammals very different from insects as herbivores in terms of their ability to form coevolutionary relationships? Has this difference evolved in response to selective pressure inherent in the coevolutionary process? Plant chemistry is discussed as a potentially important component of coevolution, but we have not addressed the biochemical responses of the insects. What are they? How do they relate to coevolution?

V. Our Direction

We shall certainly want to assess, in addressing the importance of chemistry in coevolution, the current definitions of coevolution. These are many and manifold, and the concept has been addressed thoroughly elsewhere (Jermy, 1976, 1984; Fox, 1981; Futuyma, 1983; Strong *et al.,* 1984; Thompson, 1982; Ehrlich and Raven, 1964; Futuyma and Slatkin, 1983a; Jones, 1981; Janzen, 1980; Schemske, 1983). We will additionally require that this assessment be derived from the data that have bearing upon the process, and the data we present here are chemical.

While the majority of the following chapters address plant–chemical–insect systems, as has the larger portion of coevolutionary literature to date, explorations are also made into mammalian systems, and into insect mimicry as that process derives ultimately from herbivory upon plants. The data thus presented will specifically address chemistry as a factor in the establishment and maintenance of coevolution, and test coevolutionary concepts for their pertinence to chemically mediated systems.

Where do hypotheses come from? As Morris Raphael Cohen (1954) observed, hypotheses come from people who think. The thinkers who have contributed their work to this volume have provided us with hypotheses derived from the careful observation of interactive processes, and with an intellectual stimulus to rethink our concept of coevolution.

References

Barker, J. S. F., and Starmer, W. T., eds. (1982). "Ecological Genetics and Evolution." Academic Press, New York.

Benson, W. W., Brown, K. S., and Gilbert, L. E. (1976). Coevolution of plants and herbivores: Passion flower butterflies. *Evolution* **29,** 659–680.

Berenbaum, M. R. (1979). Toxicity of furanocoumarin to armyworms: A case of biosynthetic escape from insect herbivores. *Science* **201,** 532–534.

Berenbaum, M. R. (1981). Patterns of furanocoumarin distribution and insect herbivory in the Umbelliferae: Plant chemistry and community structure. *Ecology* **62,** 1254–1266.

Berenbaum, M. R. (1983). Coumarins and caterpillars: A case for coevolution. *Evolution* **37,** 163–178.

Berenbaum, M. R., and Feeny, P. P. (1981). Toxicity of angular furanocoumarins to swallowtail butterflies: Escalation in a coevolutionary arms race? *Science* **212,** 927–929.

Brown, K. S., Jr. (1984). Adult-obtained pyrrolizidine alkaloids defend ithomiine butterfiles against a spider predator. *Nature (London)* **309,** 707–709.

Brues, C. T. (1924). The specificity of food plants in the evolution of phytophagous insects. *Am. Nat.* **58,** 127–144.

Chambers, K. L., ed. (1970). "Biochemical Coevolution." Oregon State Univ. Press, Corvallis, Oregon.

Chew, F. S. (1975). Coevolution of Pierid butterflies and their cruciferous foodplants. I. The relative quality of available resources. *Oecologia* **20,** 117–127.

Chew, F. S. (1977). Coevolution of Pierid butterflies and their cruciferous foodplants. II. The distribution of eggs on potential food-plants. *Evolution* **31,** 568–579.

Cohen, M. R. (1954). Scientific thought. *In* "American Thought: A Critical Sketch" (F. S. Cohen, ed.), Ch. 3. Collier Books, New York.

Dawkins, R., and Krebs, J. R. (1979). Arms races between and within species. *Proc. R. Soc. London Ser. B* **205,** 489–511.

Ehrlich, P. R., and Raven, P. H. (1964). Butterflies and plants: A study in coevolution. *Evolution* **18,** 586–608.

Feeny, P. P. (1976). Plant apparency and chemical defense. *Recent Adv. Phytochem.* **10,** 1–40.

Flor, H. H. (1971). Current status of the gene-for-gene concept. *Annu. Rev. Phytopathol.* **9,** 275–295.

Fox, L. R. (1981). Defense and dynamics in plant-herbivore systems. *Am. Zool.* **21,** 853–864.

Futuyma, D. J. (1983). Evolutionary interactions among herbivorous insects and plants. *In* "Coevolution" (D. J. Futuyma and M. Slatkin, eds.), Ch. 10, pp. 207–231. Sinauer, Sunderland, Massachusetts.

Futuyma, D. J., and Slatkin, M., eds. (1983a). "Coevolution." Sinauer, Sunderland, Massachusetts.

Futuyma, D. J., and Slatkin, M. (1983b). Introduction. *In* "Coevolution" (D. J. Futuyma and M. Slatkin, eds.), Ch. 1, pp. 1–13. Sinauer, Sunderland, Massachusetts.

Futuyma, D. J., and Slatkin, M. (1983c). The study of coevolution. *In* "Coevolution" (D. J. Futuyma and M. Slatkin, eds.), Epilogue. pp. 459–464. Sinauer, Sunderland, Massachusetts.

Gilbert, L. E., and Raven, P. H., eds. (1975). "Coevolution of Animals and Plants." Univ. of Texas Press, Austin, Texas.

Janzen, D. H. (1966). Coevolution of mutualism between ants and acacias in Central America. *Evolution* **20,** 249–275.

Janzen, D. H. (1980). When is it coevolution? *Evolution* **34,** 611–612.

Jermy, T. (1976). Insect-host-plant relationship: Co-evolution or sequential evolution? *In* "The Host-Plant in Relation to Insect Behavior and Reproduction" (T. Jermy, ed.), pp. 109–113. Plenum, New York.

Jermy, T. (1984). Evolution of insect/host plant relationships. *Am. Nat.* **124,** 609–630.

Jones, D. A. (1981). Cyanide and coevolution. *In* "Cyanide in Biology" (B. Vennesland, E. E. Conn, C. J. Knowles, J. Westley, and F. Wissing, eds.), pp. 509–516. Academic Press, New York.

Lawton, J. H., and Strong, D. R., Jr. (1981). Community patterns and competition in folivorous insects. *Am. Nat.* **118,** 317–338.

Maynard-Smith, J. (1976). A comment on the Red Queen. *Am. Nat.* **110,** 325–330.

Metcalf, R. L. (1979). Plants, chemicals, and insects: Some aspects of coevolution. *Entomol. Soc. Am. Bull.* **25,** 30–34.
Miller, J. R., and Strickler, K. L. (1984). Finding and accepting host plants. *In* "Chemical Ecology of Insects" (W. J. Bell and R. T. Cardé, eds.), Ch. 6, pp. 127–158. Sinauer, Sunderland, Massachusetts.
Mitter, C., and Brooks, D. R. (1983). Phylogenetic aspects of coevolution. *In* "Coevolution" (D. J. Futuyma and M. Slatkin, eds.), Ch. 4, pp. 65–98. Sinauer, Sunderland, Massachusetts.
Rosenthal, G. A., and Janzen, D. H., eds. (1979). "Herbivores: Their Interaction with Secondary Plant Metabolites." Academic Press, New York.
Roughgarden, J. (1983). The theory of coevolution. *In* "Coevolution" (D. J. Futuyma and M. Slatkin, eds.), Ch. 3, pp. 33–64. Sinauer, Sunderland, Massachusetts.
Schemske, D. W. (1983). Limits to specialization and coevolution in plant-animal mutualisms. *In* "Coevolution" (M. H. Nitecki, ed.), pp. 67–109. Univ. of Chicago Press, Chicago.
Scriber, J. M. (1984). Host-plant suitability. *In* "Chemical Ecology of Insects" (W. J. Bell and R. T. Cardé, eds.), Ch. 7, pp. 159–204. Sinauer, Sunderland, Massachusetts.
Smiley, J. T. (1985). Are chemical barriers necessary for evolution of butterfly-plant associations? *Oecologia* **65,** 580–583.
Spencer, K. C. (1987). Specificity of action of allelochemicals: Diversification of glycosides. *Am. Chem. Soc. Symp. Ser.* **330,** 275–288.
Strong, D. R., Jr., Lawton, J. H., and Southwood, T. R. E. (1984). "Insects on Plants." Harvard Univ. Press, Cambridge, Massachusetts.
Thompson, J. N. (1982). "Interaction and Coevolution." Wiley-Interscience, New York.
Thompson, J. N., and Willson, M. F. (1979). Evolution of temperate fruit/bird interactions: Phenological strategies. *Evolution* **33,** 973–982.
Verschaffelt, E. (1910). The cause determining the selection of food in some herbivorous insects. *Proc. Acad. Sci. Amsterdam* **13,** 536–542.
Wootton, R. J. (1981). Paleozoic insects. *Annu. Rev. Entomol.* **26,** 319–344.

CHAPTER 2

Genetics of Pairwise and Multispecies Plant–Herbivore Coevolution

FRED GOULD
Department of Entomology
North Carolina State University
Raleigh, North Carolina

I. Introduction

A. Terminology

The hypothesis that coevolution has been an important force in molding the phenotypic and ecological traits of the earth's biota (Ehrlich and Raven, 1964) is of obvious significance to evolutionary biologists and ecologists. Given that we accept evolution by natural selection as a general phenomenon, determining the major modes of natural selection is a logical next step. In this regard, Ehrlich and Raven's hypothesis is significant at the same level of inquiry as hypotheses about the importance of group selection, kin selection, and sexual selection. Designing experiments that rigorously test any of these very general hypotheses is difficult, but if we are to develop the concept of coevolution into a scientifically valid theory, we must strive to devise methods for testing its generality.

Precise definition of terms is important in this context, for, as pointed out by Futuyma and Slatkin (1983) and Van Valen (1983), if we define coevolution so broadly that it encompasses all adaptation to biotic aspects of the environment, it can without a doubt be shown to account for much of evolution, but then its power as a distinct predictive theory is diminished. In response to this problem, Janzen (1980) has "narrowly" defined coevolution as "an evolutionary change in a trait of the individuals in one population in response to a trait of the individuals of a second population, followed by an evolutionary response by the second population to the change in the first." When more than two species populations are involved, Janzen and others (Fox, 1981; Futuyma and Slatkin, 1983; May and Anderson, 1983) use the adjective "diffuse" to modify the noun "coevolution."

In attempting to understand coevolution more precisely, many adjectives have been used to describe specific coevolutionary processes. Adjectives such as intense, simultaneous, sequential, reciprocal, long-term, intermittent, and coupled are commonly used in discussions among workers in the field. For similar reasons, adverbs have become important in describing outcomes of coevolution, so, for example, we discuss "loosely" coevolved interactions and "highly" coevolved pairs of species. Most of these adjectives and adverbs can be useful for refined articulation of the general concept of coevolution. It may only be by such refinement that we will ever determine the validity and power of the general concept. It is currently less useful to ask the big question "how common is coevolution?" than to ask questions such as whether pairwise coevolution is as common as diffuse coevolution, or whether diffuse coevolution is possible.

Unfortunately, many of these adjectives and adverbs are vague and are used differently by various authors. As pointed out by Slatkin (1983), the precision of mathematical definitions of coevolution can serve us well in eliminating the vagaries of common language and in forcing us to think more critically. Roughgarden (1983) makes the important comment that only with the development of models "tailor-made" to mimic specific coevolutionary processes will we be able to develop and test coevolutionary predictions. Whether these models be verbal, graphical, or mathematical, they must be unambiguous if they are to be of value. Some models of this type have been developed to describe the coevolution of competitors (Levin, 1983; Roughgarden, 1983), but little work has been specifically aimed at plant/herbivore systems or plant/pollinator systems (Kiester *et al.,* 1984).

B. The Legacy of Ehrlich and Raven?

In their examination of butterfly/plant associations, Ehrlich and Raven (1964) found that closely related butterfly species often feed on closely related plant species. They used this as indirect evidence supporting a specific type of coevolution of plants and herbivores in which plants evolved physical and chemical defenses that were overcome by certain herbivores. They postulated that after a plant population evolved a novel and successful herbivore defense mechanism, a burst of ecological diversification and speciation occurred because plants with the novel defenses outcompeted other plant groups that suffered from more herbivory. Likewise, an herbivore population that breached this new defense could colonize the new plant group and speciate. Although not distinctly stated by Ehrlich and Raven (1964), it is often assumed that under the resumed selective pressure from herbivores, other novel defenses may have evolved leading to a branching pattern similar to that in Fig. 1a.

Since the publication of the Ehrlich and Raven (1964) paper, a number of scientists have used the apparent correlation between herbivore and host-plant taxonomy in certain systems as indirect evidence of coevolution (e.g., Benson *et al.,* 1975). The strength of such evidence has been widely questioned (Jermy, 1976, 1984; Gilbert, 1979; Janzen, 1980; Futuyma and Slatkin, 1983; Nitecki, 1983), for these phenotypic correlations could have arisen by evolutionary processes that did not even require historical association of the plant and herbivore taxa [e.g., sequential evolution (Jermy, 1976)]. Mitter and Brooks (1983) describe how cladistic techniques can help in testing whether or not plants and their herbivores evolved in parallel over a geological time period, but they and others (e.g., Futuyma, 1983) clearly state that such parallel evolution (or cladogenesis) cannot be used as unambiguous support of coevolution because

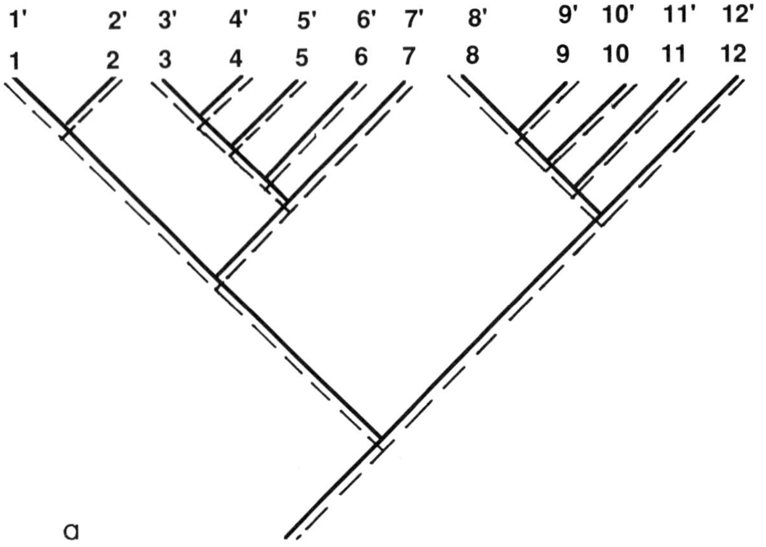

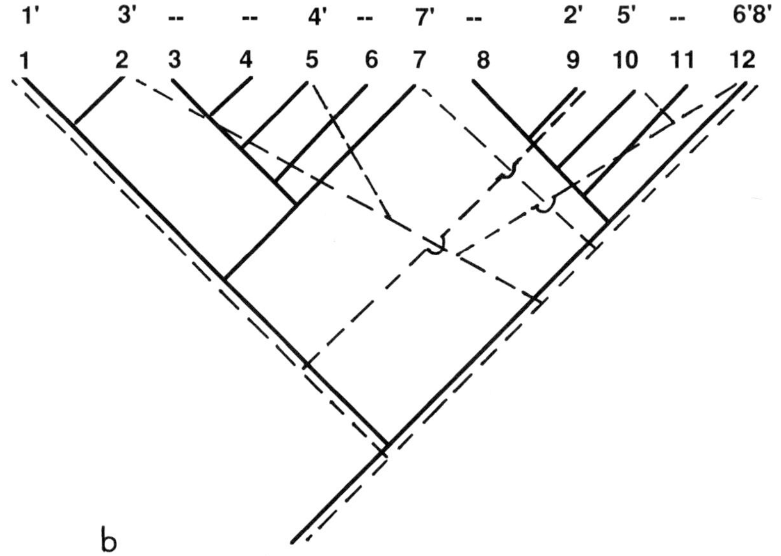

Fig. 1. Superimposed cladograms of herbivores (dotted lines) and plants (solid lines). (a) Perfectly congruent phylogenies of plants and herbivores. (b) An incongruent (or nonparallel) phylogeny of plants and herbivores. Numbers 1–12 represent currently existing plant taxa and their phylogenetic affinity (i.e., plants 1 and 2 are closely related). Numbers 1′–12′ represent existing herbivore taxa and their relatedness.

unilateral host-tracking adaptations by herbivores or vicariance events could produce similar patterns.

Perhaps as a legacy of Ehrlich and Raven's very important work, researchers have tended to focus their studies on explaining plant/herbivore species associations that appear to be coevolved (i.e., are ecologically and/or taxonomically intricate) instead of focusing on the process of coevolution itself. It is almost as if we assume that the only scientifically noteworthy outcomes of coevolution are coevolved species associations. In my opinion, this approach has caused us to ignore the fact that a complete lack of correlation between plant and herbivore phylogeny or chemotaxonomy in no way refutes the existence of, or intensity of, coevolution. Indeed, in some plant/herbivore interactions to be described in this chapter, there is good reason to expect that intense coevolutionary interactions (i.e., each species having a strong effect on the other species's fitness, and each species having the genetic variation and population structure to respond evolutionarily) will sometimes lead to incongruent phylogenies and destruction of highly coevolved interactions that the same process may have generated. When an herbivore is faced with a new, potent toxin by the plants it has been feeding on, the herbivore could respond by (1) becoming extinct, (2) adapting physiologically and remaining on the same host, or (3) it could evolve behaviors that allow it to choose another host taxon upon which it is now more fit than it is on the old, heavily defended host. I contend that plant/herbivore phylogenies as incongruent as that illustrated in Fig. 1b could result from combinations of these three responses over geological time, and that at least some of the herbivores in Fig. 1b may have been involved in intense "narrow-sense" (*sensu* Janzen, 1980) coevolution with their sequence of plant hosts.

We presently lack theory or empirical evidence to say whether or not the scenario in Fig. 1a is more likely than Fig. 1b when there is intense coevolution and extrinsic ecological factors do not interrupt the plant/herbivore interactions. The intuitive appeal of Fig. 1a may derive from its superficial neatness or from a belief that it requires less complex (and thus more likely) genetic changes on the part of an herbivore to adapt to one novel resistance mechanism than it does to adapt to a new host. Even if the pattern in Fig. 1a is the most likely result of intense coevolution, lack of this pattern does not indicate that coevolution was imperfect, unimportant, weak, intermittent, or nonexistent.

It has been widely recognized that evolution by individual selection can lead away from optimal adaptedness and may even lead to extinction (e.g., Lande, 1980). This outcome of evolution is as important to understanding the evolutionary process as is the evolutionary outcome of optimal adaptedness. Similarly, a "coevolved" species association is only one of a number of interesting outcomes of plant/herbivore coevolution (Futuyma, 1985a). For example, we can ask how common are extinctions resulting from the cost of defenses in a coevolved plant herbivore system.

Ehrlich and Raven's (1964) focus on a geological time span seems to have caused emphasis on long-term (more than 1000 years) coevolutionary arms races. While such unusual, long-term unidirectional coevolutionary interactions are certainly of interest, reciprocal, short-term, coevolutionary interactions occurring on a scale of 1–100 years may be as important to our understanding of present-day plant/herbivore community structure and dynamics. Coevolutionary interactions that involve only one cycle of reciprocal genetic interaction are a priori no less important to the structuring of communities than coevolutionary associations that have been built on 50 cycles of reciprocal genetic interactions. Thus, we must first take a broad view of diverse coevolutionary processes before determining which are most common and important.

I will discuss how we may use theoretical and empirical genetic techniques as tools for (1) more precisely describing the diverse coevolutionary interactions that could occur among plants and herbivores, and (2) determining parameters in the coevolutionary interactions between plants and herbivores that are most important in influencing short- and long-term evolutionary outcomes of these interactions (such as congruent or incongruent phylogenies). As will be emphasized in this chapter, verbal, graphical, and mathematical models are all important explanatory devices of theoretical genetics, but these explanatory devices are only useful if the biological thoughts and data used in developing them accurately reflect the real world. In the case of plant/herbivore coevolution, one important set of information involves how plants chemically defend themselves and how herbivores adapt to overcome these defenses. Although other factors such as predation and competition are also likely to be important (Bernays, 1988), they are not within the scope of this paper.

To date, most models of coevolution have been restricted to examining pairwise coevolution, even though most ecologists and evolutionary biologists believe that multispecies coevolution may be far more common (Futuyma and Slatkin, 1983). The difficulty of designing models of multiple species interactions as well as the complexity of the expected results have probably been impediments to the development of models of multispecies coevolution and collection of empirical information needed to test them. Since a major concern of this chapter is exploration of the diversity of potential coevolutionary interactions among plants and herbivores, I will begin by reviewing some of the models and empirical data that pertain mainly to pairwise coevolution of plants and herbivores, but I will then expand upon these pairwise approaches in discussing how we might be able to go about studying short-term multispecies coevolution.

II. Pairwise Coevolution

A. Basic Models

The most basic graphical model of unidirectional coevolution is presented in Fig. 2, with one herbivore and one plant species undergoing stepwise genetic changes through time. The graphical model presented in Fig. 1a is also basically a pairwise model of coevolution, since at any specific time each plant lineage is depicted as interacting with only one herbivore lineage. Both of these models are strictly concerned with simple population genetic processes and do not take into account the effects of population dynamic processes on coevolution. Levin and Udovic (1977) added the interaction of population dynamics with population genetics to a general model of pairwise coevolution and found that it generated six different types of biological interactions (Fig. 3) that do not

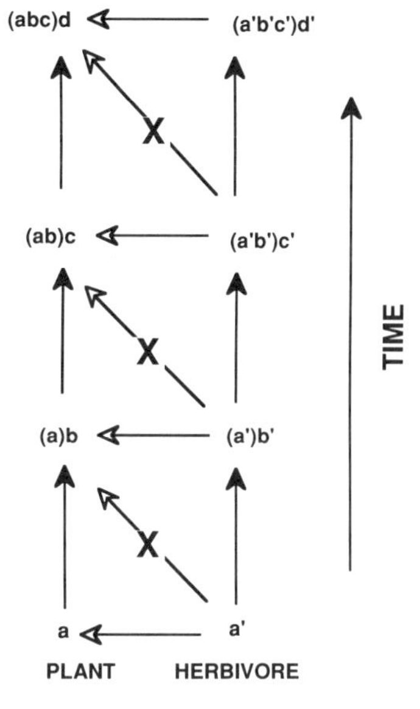

Fig. 2. A simplified graphical model of pairwise plant/herbivore coevolution. Vertical arrows indicate evolution through time. Letters a–d represent plant defenses; letters a′–d′ represent herbivore counteradaptations. Diagonal lines between herbivores and plants with X marks indicate lack of adaptation. Horizontal lines represent herbivores adapted to the specific plants. Thus, when the herbivore possesses counteradaptation c′, it can overcome defense c and feed on plants with that defense, but cannot feed on plants with other defenses. As coevolution proceeds, defenses and counteradaptations in parentheses may or may not be retained, depending on whether or not there is a cost involved in their maintenance.

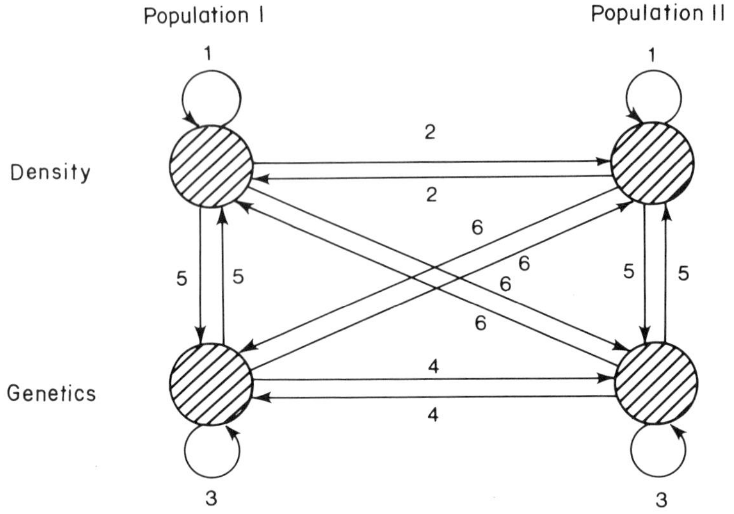

Fig. 3. Graphical model depicting six general types of interactions among densities and genetic systems of two interacting populations. The numbered arrows indicate the effect of one species's density on its own density (1), or its own genetics (5) (as in density-dependent selection); one species's density on a second species's density (2) or genetics; one species's genetics on a second species's genetics (4) or density (6); and one species's genetics on its own genetics (3) as in frequency-dependent selection. (From Levin and Udovic, 1977).

necessarily lead to unidirectional changes. The complexity of this general model was reasonable, given results of more specific, frequency-dependent models developed earlier (Mode, 1958, 1960, 1961; Person, 1966; Clarke, 1976).

Leonard and Czochor (1980), Kiyosawa (1982), and Levin (1983) summarize these frequency-dependent and age-structured models that describe the dynamics of pairwise coevolution. They generally consider the lower portion of Fig. 2, with plants in states A and B and herbivores in states A′ and B′. If there is no cost to producing the defenses in plant B, or the counteradaptation in herbivore B, all models predict eventual fixation of the herbivores with genes coding for state B′. The plant population will increase in frequency of plants with genes for state B until the herbivore population approaches fixation for state B′. Once even a small cost of plant defenses and herbivore adaptations is assumed, some of the models predict an oscillating frequency of plants A and B and herbivores A′ and B′ in the populations (e.g., general graphical representation in Fig. 4a). The oscillations occur because the selective advantage of a given genotype decreases as its frequency increases. One very important factor favoring maintenance of polymorphism is little lag time between gene

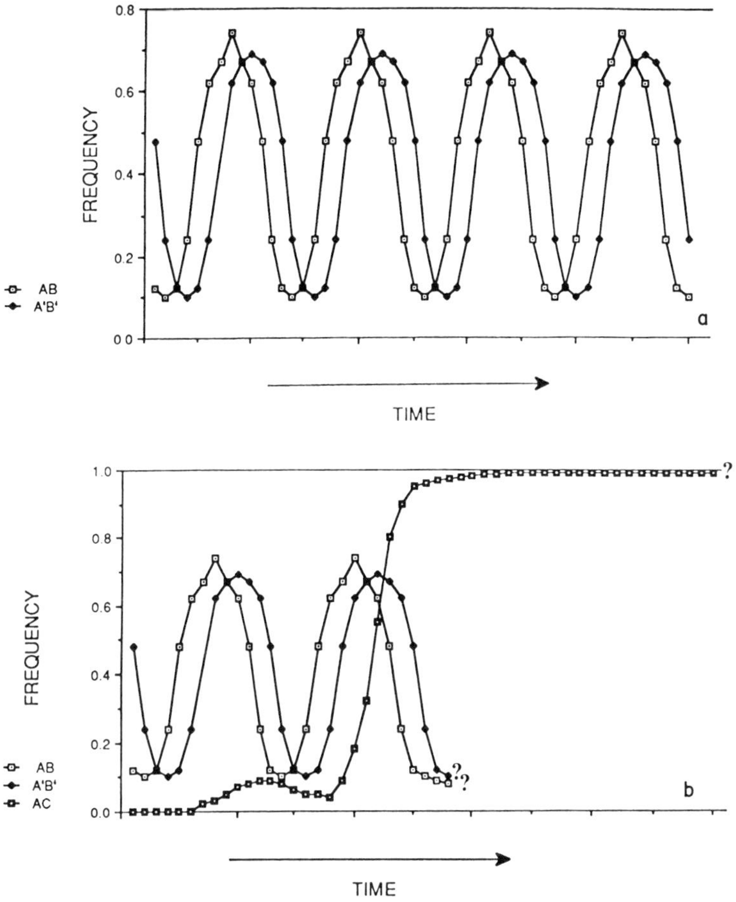

Fig. 4. Hypothetical changes in allelic frequencies of an asexual herbivore and plant that are interacting in a pairwise fashion. *A*, *B*, and *C* code for presence of distinct plant defenses; *a*, *b*, and *c* code for lack of defenses; *A′* and *B′* code for counteradaptation to defenses *A* and *B*, respectively. (a) Plant and herbivore populations are initially polymorphic for *b–B* and *b′–B′*, respectively, and frequencies of alleles in the population cycle over time as they track each other. (The frequency of $b = 1 - B$, the frequency of $b' = 1 - B'$). (b) The plant population is initially polymorphic for *b–B* and *c–C*, but the herbivore is only polymorphic for *b′–B′*. At first the frequencies of *B* and *B′* cycle, as in (a), since the frequency of *C* is very low, but since *C* is selected for whenever *B′* is at a high frequency, it comes to predominate in the plant population.

frequency changes and changes in fitness of the plant and herbivore genotypes. Once lag times are increased in the models, the oscillations in frequencies may lead to fixation of one or the other genotype. Indeed, a system that initially included some plants in state B and some herbivores in state B′ could evolve to a point where only plants in state A and herbivores in state A′ existed.

To date, mathematical models of pairwise plant/herbivore coevolution generally deal with two alleles in the plant and two alleles in the herbivore, and therefore fall short of describing the additional steps of coevolution that have been envisioned as leading to an evolutionary arms race (Whittaker and Feeny, 1971; Gilbert, 1971; Rathcke and Poole, 1975; Berenbaum and Feeny, 1981) with plants evolving at least first and second lines of defenses in response to herbivores that continue to amass new mechanisms for breaching these defenses. In its most basic textbook form (Fig. 2), after four cycles of stepwise coevolution, the plant species may be producing four defenses, while the herbivore has four distinct mechanisms for breaching these defenses. If old and partially to completely obsolete defensive and offensive mechanisms are maintained, the cost of these mechanisms might lead to ecological inefficiency [as suggested by the Red Queen hypotheses (Van Valen, 1973)]. In Fig. 2, I have put old defensive and offensive mechanisms in parentheses because evolutionary pressures could lead to the loss of obsolete armaments. This is thought to be the case with at least some ant-bearing *Acacia* spp. that do not produce the cyanide compounds produced by *Acacia* spp. without aggressive ant species (Rehr *et al.*, 1973). Whether coevolution is expected to lead to the development of a diverse arsenal or not will depend on many factors, including the physiological and ecological cost of each defense and the efficacy of each defense. These costs as well as effects of population dynamics will determine whether an "arms race" will occur at all, or if some kind of cyclic coevolutionary interaction will occur (see Gould, 1983; Thompson, 1986).

If the one-locus, two-allele mathematical models of plant/herbivore coevolution were expanded to multiple-locus models in which each genetic locus in the plant controlled a distinct defense and each locus in the herbivore a distinct counteradaptation, we might be able to develop a more detailed view of pairwise coevolution. Certainly mutations for a second or third defense are not held in abeyance until the first defense to evolve is totally overcome. When the interaction of alleles at one plant locus and one herbivore locus leads to oscillations in frequencies (Fig. 4a), an allele at a second locus in the plant or herbivore would be strongly selected for at certain points in the cycle and could change the nature of the coevolutionary interaction. Such multilocus models may seem fanciful, but since we know that plant and herbivore populations vary genetically in their defenses and counteradaptations, there is no reason to consider a one-locus model to be more reasonable. As shown graphically in Fig. 4b, a second locus in the plant could radically alter the cycle shown in Fig. 4a. These multilocus models have not been explored in depth (see Gould, 1986a,b), but it is doubtful that they would lead to the classic coevolutionary arms race depicted in Fig. 2.

B. Interaction of Behavior and Physiology

All of the models of coevolution described above assume closed genetic systems (i.e., no immigration or emigration of plants or herbivores from or to other populations) and no genetically mediated variation in host selection. Although microbial herbivores (generally called pathogens) do not show host selection capacity, most invertebrate and vertebrate herbivores have sophisticated host discrimination abilities (e.g., Ahmad, 1983), and for some species, individuals have been shown to differ genetically in their host preferences. A number of theoretical studies have examined some of the effects of herbivore host selection and migration between patches of different plant types on the rate and direction of change in herbivore genotype frequencies when plant genotype or species frequencies are held constant (Bush and Diehl, 1982; Gould, 1984a; Garcia-Dorado, 1986; Futuyma, 1986; Castillo-Chavez *et al.*, 1988; Rausher, 1988). These models can be viewed as assuming that herbivores are initially polymorphic at two loci, one for selecting between plant species 1 and 2, the other for physiological adaptation to plant species 2. Given that there is no competition (or other form of density-dependent mortality) among herbivores on the host plants, herbivores never remain polymorphic (i.e., > 5% of an allele). The panmictic herbivore population will either evolve behavioral avoidance of plant species 2 and not adapt to it physiologically, or will adapt physiologically to plant species 2 and not avoid it. An interesting result from these models is that fixation of either of the two two-locus genotypes can, in some cases, be importantly related to initial genotype frequencies as well as to the fitness differences between genotypes (see Wright, 1931). As shown in Fig. 5, there are some cases where a change in initial genotype frequency of less than 0.02 can radically alter the evolutionary trajectories of genes for physiological adaptation to a plant defense and genes for host-selection behavior. Given the genotype fitnesses used in this example, the two-locus genotype that becomes fixed in Fig. 5c, which avoids plant species 2, is about 70% as fit as the genotype fixed in Fig. 5a, which adapts physiologically. Thus the final host range and even population size of a small herbivore population invading a new plant community could, for some fitness values, depend on stochastic factors that determined its initial gene frequencies.

Rausher (1984a, 1985) examined the influence of frequency-dependent fitness on the evolution of herbivore preference for two plant species

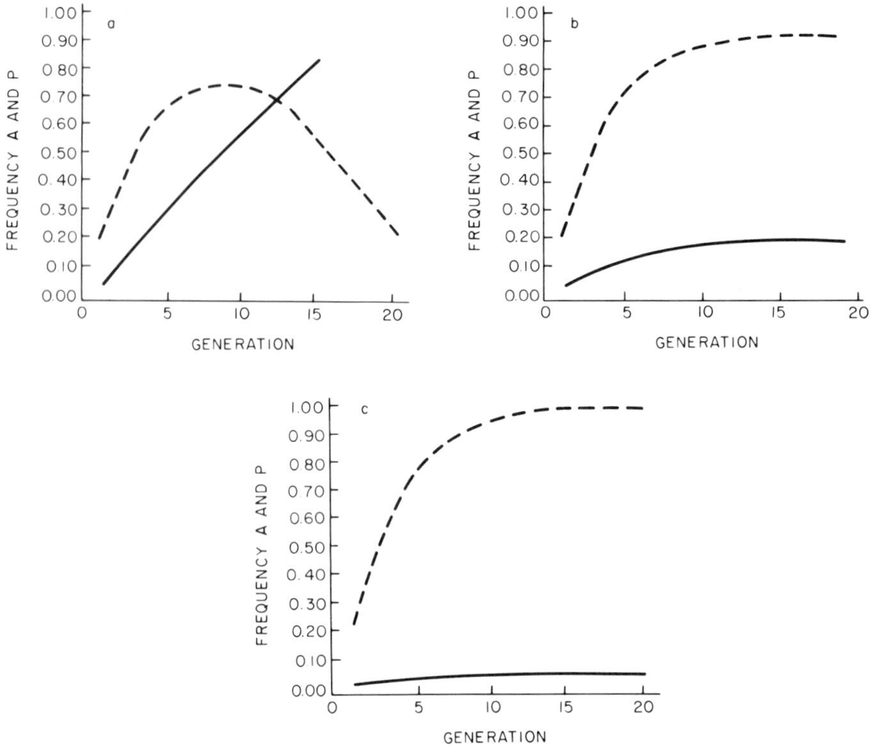

Fig. 5. Change in frequency of avoidance (*A*) and physiological adaptation (*P*) alleles over 20 generations assuming diploid sexual herbivore and additive inheritance. The *AA* individuals lay 10% of their eggs on plant species 2; *aa* individuals lay 90% of their eggs on plant species 2. Fitnesses of *PP* herbivore individuals on plant species 1 and 2 are 0.45 and 0.90, respectively. Fitnesses of *pp* individuals on species 1 and 2 are 0.55 and 0.10, respectively. Initial *A* allele frequency (---) is always 0.20. Initial *P* frequency (—) varies from 0.03 to 0.01 in a–c, respectively. (From Gould, 1984a).

when there was no genetic variation for physiological adaptation. Results of his models indicate that under certain circumstances, polymorphism for host preference may be maintained. Garcia-Dorado (1986) and Rausher (1988) have recently examined frequency-dependent cases with genetic variation for physiological and behavior characteristics. In many cases, the frequency dependence leads to polymorphism for both traits.

If, instead of allowing the herbivore to adapt to a nonevolving plant population, we allow both the plant and herbivore to evolve, we can begin to address the question of when we expect a reciprocal coevolutionary interaction between a plant and herbivore species to remain intact and when we expect it to end. Unlike Fig. 5, where the herbivore population is faced with a novel defense that is immediately distributed throughout the host-plant population, Fig. 6 presents a situation in which the novel defense is initially only found in 1% of the plants. The fitness of plants without the new defense is 20, 30, or 50% less than that of plants having the new defense (in Fig. 6a, b, and c, respectively) when faced solely with herbivores unadapted to the new defense. Change in frequency of alleles for avoiding the host plant is at first negative (the starting frequency shown is only for comparison with Fig. 5). Once selection on the plant population by the herbivore unadapted to the new defense causes an increase in the frequency of the new defense, frequencies of alleles for avoiding the host plant begin to increase. The relatively greater lag in increase of behavioral avoidance alleles compared to physiological counteradaptation to the defense leads to a higher probability of the coevolved species pair remaining in contact. It is interesting to note that in this very specific case the faster the plant defense spreads through the population (due to higher selection pressure by the herbivore), the less likely is the pairwise interaction to continue (compare Fig. 6a and 6c). The herbivore's new behavior in Fig. 6c causes it to switch host plant species as in Fig. 1b. A more general exploration of this model (F. Gould, unpublished data) indicates that "intensity" of coevolutionary interactions is not necessary for development of congruent phylogenies. In Fig. 6, I varied only a few important ecological variables. The alternate host(s) in this example (Fig. 6a–c) cause 45–55% mortality. When mortality on these hosts are set at higher values, there is a higher probability that the original pairwise interaction will remain intact. Broader exploration of models such as this, along with empirical estimation of selection pressures, is obviously needed before strong statements can be made about what factors would be most important in stabilizing or destabilizing a specific pairwise evolutionary interaction. On a more general level, these models bring up the question of whether coevolutionary interactions are more likely to lead to congruent phylogenies in herbivores with efficient host selection ability (e.g., Lepidoptera) or those lacking such abilities completely (e.g., microbes).

One important variable not considered in the above model is pleiotropy of genes affecting both physiology and behavior (but see Garcia-Dorado, 1986). As argued in Gould (1983), it is possible that such pleiotropy would lead to better host tracking and may be found more often in insects (e.g., some Orthoptera, Homoptera, and Coleoptera) in which juveniles and adults eat similar food, than in insects in which only one life stage is herbivorous (e.g., Lepidoptera). A related and equally important variable addressed by Gilbert (1979) is the mating structure of the herbivore population. Gilbert found that butterfly species that mated on or near their host plants were more likely to undergo radical host shifts than those in which the mating arena was independent of the host location.

C. Heritability of Plant Defenses and Herbivore Counteradaptations

Implicit in all of the models I have described is the assumption that variation in plant resistance and herbivore adaptation is heritable. Certainly there is evidence for the existence of heritable variation in these traits (e.g., Gould, 1983), but as pointed out in a number of this volume's chapters (i.e., Bryant; Craig *et al.*), there is a large environmental component to variation in plant defense that cannot respond to selection. Al-

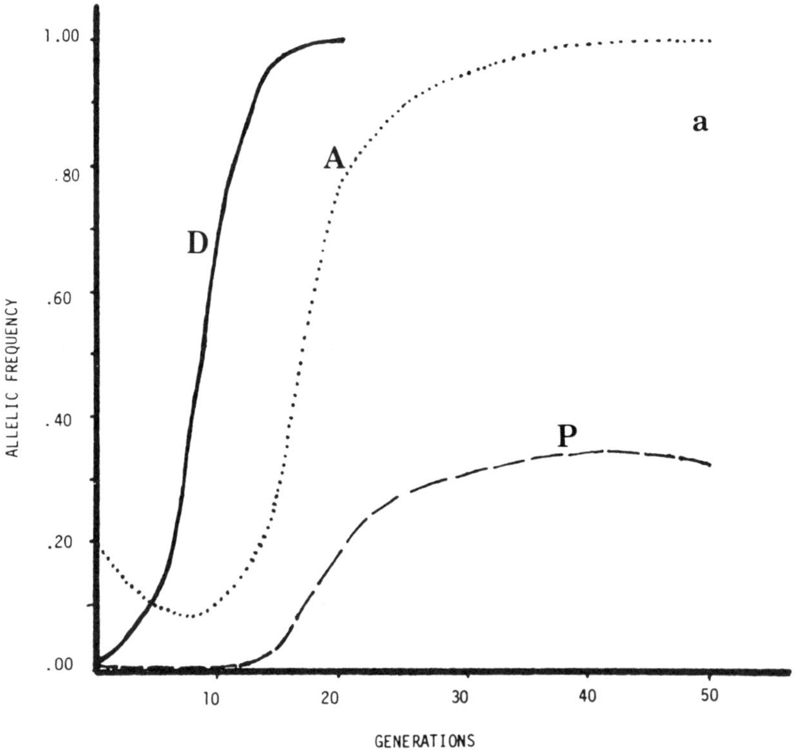

Fig. 6. Change in frequency of alleles for plant species 2 defense (*D*), alleles for herbivore avoidance of species 2 (*A*), and alleles for herbivore physiological adaptation to species 2 (*P*). Initial frequency of *A* is 0.20, *P* is 0.005, and *D* is 0.01. Fitnesses of *DD* and *dd* plants are equal in face of herbivore genotype *PP*; *dd* plant fitness is less than *DD* plant fitness in face of *pp* herbivore genotypes by 20, 30, and 40% in a, b, and c, respectively. Herbivores *PP* and *pp* have fitnesses of 0.90 and 0.10 on DD plants. Herbivores PP and pp have a fitness of 0.90 on dd plants. Additive inheritence of all traits is assumed. Other genetic and ecological assumptions are the same as those in Fig. 5. Simulation model is modified from Gould (1984a).

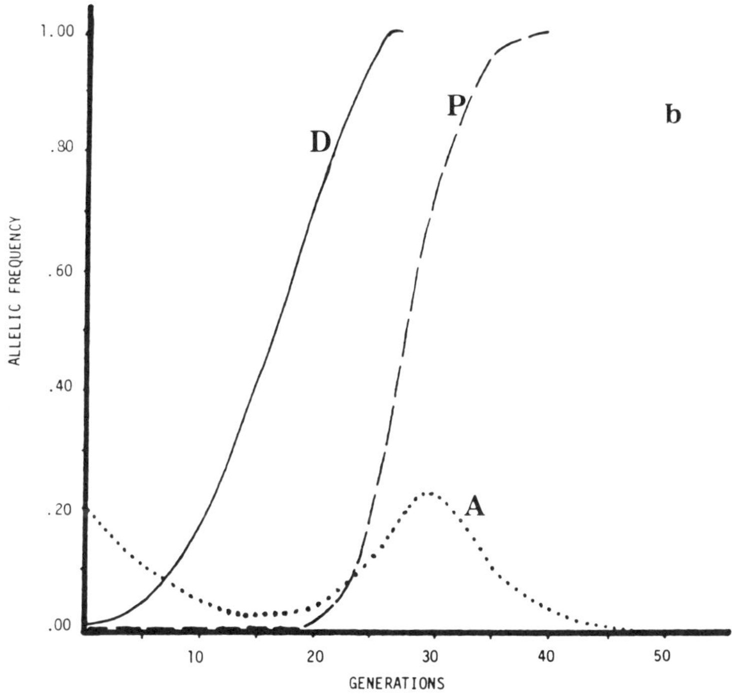

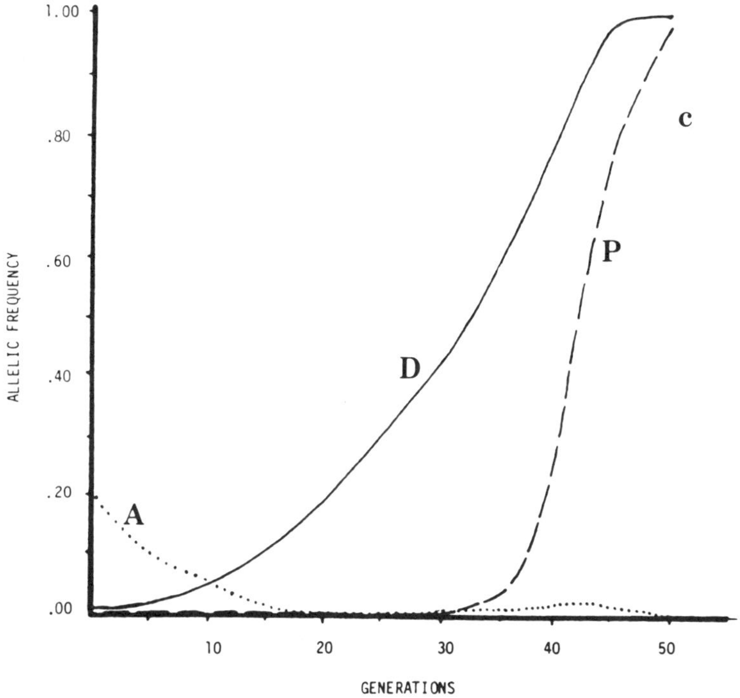

Fig. 6. (*continued*)

though not thoroughly documented, there is also an environmental component to herbivore adaptation.

It must be recognized that herbivores are not the only factor affecting plant fitness (see Craig *et al.*, this volume). If factors other than herbivory that affect plant fitness reduce the fitness of defended and undefended plants by an equal amount (on log scale), they should not affect the average rate of coevolution in large plant/herbivore populations. If other stresses decrease the relative difference in fitness of offspring from defended or undefended plants (or adapted versus unadapted herbivores), rates of microevolutionary change will tend to decrease. In some cases, abiotic stress can increase the relative difference in fitness of defended and undefended plants (e.g., Coley *et al.*, 1985) and thus favor increased rates of evolutionary change.

It is important to note that the best indicator of quantitative changes in a characteristic of a plant or animal is the additive genetic variance. If a trait has high heritability but low absolute additive genetic variance, selection may cause a rapid change but the magnitude of the change may be so small as to be biologically unimportant.

D. Relevant Data

There is evidence, in a few cases, that argues for long-term coevolutionary interactions leading to highly coevolved pairs of plants and herbivores (e.g., *Heliconius* butterflies, Benson *et al.*, 1975; fig wasps, Wiebes, 1979). While it is important to confirm these isolated cases of highly coevolved pairs of species, it may be equally important to find out more about genetic and environmental traits that would influence the short-term (1–100 years) dynamics and direction of reciprocal pairwise coevolutionary interactions. In the previous section, a number of these traits were discussed. A number of recent chapters and articles have reviewed empirical information on parameter values for these traits, and the reader is referred to them for detail (Futuyma and Peterson, 1985; Mitter and Futuyma, 1983; Gould, 1983; Morrow and Lamarche, 1979; Louda, 1984; Berenbaum *et al.*, 1986; Parker and Root, 1981; Rausher and Feeny, 1980).

We have some information on a few important parameters, such as fitness reductions due to herbivore and plant defenses, but for most parameters such as genetics of host-selection behavior we have little or no data. Perhaps not surprisingly, information on all the well-documented variables indicates a broad array of possible parameter values, which, when plugged into the existing models, can lead to divergent coevolutionary trajectories.

It is obvious from these data that there is a lot of genetic variation within and among populations of plants and herbivores for traits affecting their interactions. Consequently, whenever plants and herbivores affect each other's fitness, there is the potential for short-term coevolution. Since the data also indicate that plants and herbivores do affect each other's fitnesses at least sometimes, it is constructive to look for short-term coevolution and its effects on plant/herbivore interactions.

III. Multispecies Coevolution: Theory

A. What Is Meant by Diffuse?

As discussed in the introduction to this chapter, Janzen (1980) uses the adjective "diffuse" to modify the noun "coevolution" when referring to the coevolution of more than two species or more than two populations. Others have also used the term "diffuse" in this or a similar sense (e.g., Fox, 1981; May and Anderson, 1983; Futuyma, 1983). I think that once the term "diffuse" became embedded in the evolutionary ecology literature, the term itself became partially responsible for the focus of research emphasis on pairwise coevolution of plants and herbivores, even though it is not thought to be very common (Futuyma and Slatkin, 1983). "The Oxford English Dictionary" (1961) defines the adjective "diffuse" as "confused, distracted, perplexed; indistinct, vague, obscure, doubtful, uncertain." If these synonyms accurately reflect our psychological response to the word "diffuse," it is no wonder that we have shied away from studying diffuse coevolution. How could we ever hope to understand such an indistinct process? I intend to point out that although the process of multispecies coevolution of plants and herbivores may be complex, it need not be diffuse, and that we should reserve this term for cases of pairwise and multispecies coevolution that are indeed indecipherable.

B. Selection Intensity and Response to Selection

The most misleading implication of the word "diffuse" with regard to the process of multispecies coevolution may be that no single herbivore species within the herbivore complex will exert intense selection pressure on a single plant species (e.g., Fox, 1981). Indeed, the terms "diffuse" and "intense" seem antithetical. Counter to this, I would argue that the intensity of selection pressure of a single herbivore species on a plant may be greater when there is a complex of herbivores feeding on the plant. Certainly agricultural and forestry data indicate that plants can compensate for damage inflicted by one herbivore, while the damage due to interactions of several herbivores can be devastating (for reviews see Dickinson, 1979; Powell, 1979). As damage in terms of leaf area consumed increases, there may be a nonlinear response in the decrease of plant growth and reproduction. Even when herbivore damage is to reproductive tissue, the effect on reproductive output may be nonlinear (e.g., Fig. 7a). Thus, herbivore species 1 may decrease a plant's fitness from 1.0 to 0.9 when it attacks alone, while the same amount of damage by herbivore 1 could cause plant fitness to decrease from 0.7 to 0.3 if herbivore species 2 had caused an initial decrease from 1.0 to 0.7. The intensity of selection for plant adaptation (defense) to herbivore 1 in the latter case would be greater than it would be in the former (see Fig. 7b,c). More extreme cases of synergistic reductions in plant fitness are documented in the interaction of insect or nematode herbivores and microbial herbivores where the insects or nematodes predispose plants to microbial herbivore attack (Dickinson, 1979; Powell, 1979). In these cases, there would be strong selection for a gene or trait conferring resistance to either the microbial or invertebrate herbivore.

It is important to recognize that in herbivore communities strongly structured by competition, a specific defense that decreased a single her-

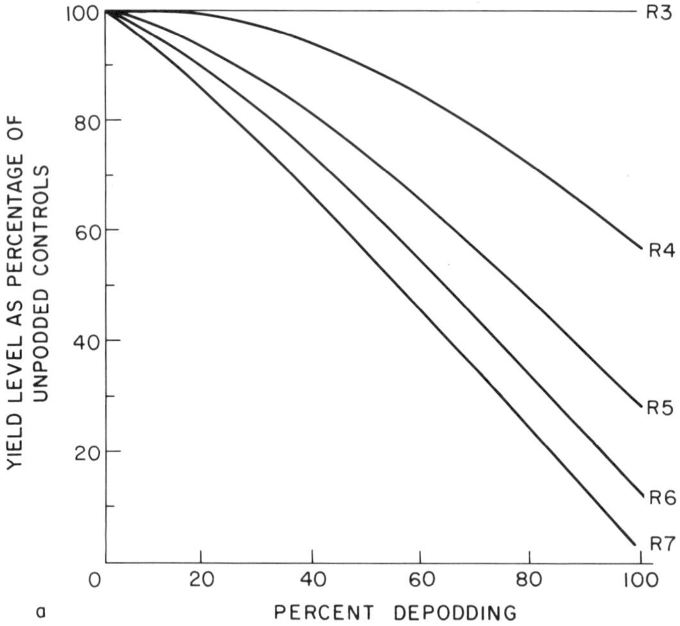

Fig. 7. (a) Relationship between pod damage and seed production in soybean. Solid and dashed lines are from two different studies. R2–R6 are codes for the phenological stage of the soybean plants when defoliated; R3 = pods beginning to form, R4 = pods formed but seed not developed, R5 = seeds partially developed, R6 = seeds developed. From Kogan and Turnipseed (1980). (b) Response of plant to selection by two herbivores (A and B) that attack separately. Dashed line is allelic frequency at the locus coding for defense against herbivore B, which causes a 30% decrease in plant fitness when it attacks by itself. Dotted line is allelic frequency at the locus that codes for adaptation to herbivore A, which causes 10% plant fitness decrease when it attacks alone. Simulation equations as in Gould (1986a). (c) Similar to (b) except that herbivores A and B attack together, causing a 70% reduction in plant fitness (see text for explanation).

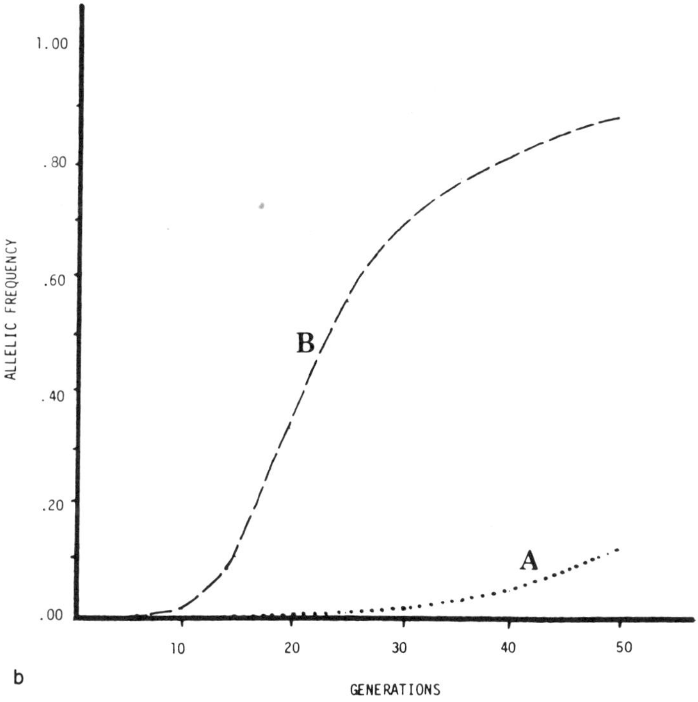

b

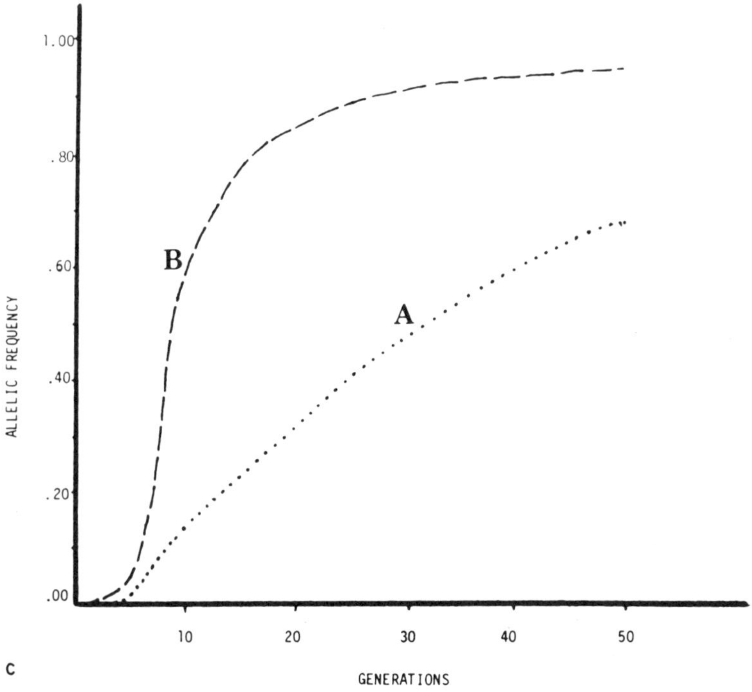

c

Fig. 7. (*continued*)

bivore species's density might not be selected for at all because other herbivore species would increase their density in response to decrease in the single herbivore's density. Support for such a general view of insect herbivore communities as being structured by density-dependent factors is not compelling (Lawton and Strong, 1981), but cases have been documented where interspecific competition is strong [e.g., *Drosophila* (Grimaldi and Jaenike, 1984), Lepidoptera and sawflies (Stamp, 1984; Karban, 1986)]. In such communities, more general plant defenses would be favored that reduce damage by the entire herbivore complex. Thus, we may expect different types of defenses to evolve depending on herbivore community interactions.

The existence of high intensity of selection (sensu Falconer, 1981) for adaptation of plants to specific herbivores in many herbivore community complexes does not seem improbable. Quite distinct from the question of selection intensity, however, is the question of response to selection. As long as the cost of a number of specific defenses is approximately additive, and a defense against one herbivore does not make a plant more susceptible to damage by the rest of the herbivore community, plant response to selection by single herbivores could be expected. As will be discussed below, available data pertinent to these assumptions are minimal.

The next question is whether most herbivores are too polyphagous to experience intense selection pressure from single plant species. If an herbivore fed on most of the plants in a diverse community, each plant would probably exert little selection pressure. But when we analyze the diets of many polyphagous herbivore species, their localized populations are not ecologically very polyphagous (Fox and Morrow, 1981). Therefore, localized selection pressure caused by single plant species could be strong. Unless the herbivores had long distance gene flow, local selection intensity would be an important variable in determining the probability and rate of at least short-term coevolution.

As in the case of plant response to herbivore selection pressure, two important factors may limit oligophagous and polyphagous herbivores from adapting to plant defenses: (1) a more than additive physiological cost of producing a number of mechanisms for counteradaptation, and (2) the likelihood that a mechanism that allows an herbivore to overcome one plant's defense makes it more susceptible to another plant's defense, thus causing the herbivore to incur an ecological cost of coping with a diversity of defenses.

C. Patterns of Herbivore and Plant Fitness in Multispecies Interactions

If we look at an oligophagous or polyphagous herbivore when feeding on various host plants in a single community, we generally find significant variation in host suitability vis à vis herbivore fitness. The pattern of variation in host utilization varies from one herbivore to the next (e.g., Futuyma and Gould, 1979). Four of the many possible patterns of variation in host suitability are depicted in Fig. 8. The interesting question with both ecological and evolutionary implications that arises is, why do these patterns take on a specific form in each case? Are they simply patterns that would be expected based on a nonevolutionary association, or are they structured by evolutionary and/or coevolutionary processes?

The same approach can be taken in asking about the resistance of a plant to members of the herbivore community that feed on it. Figure 9 presents possible patterns of a plant population's reduced fitness in face of the herbivores that damage it. Interactions between pairs and groups of herbivores are not considered here. As with the pattern of a single herbivore population's fitness on a variety of hosts, we can ask whether the plant's pattern of fitness reduction is purely ecological or is in part the result of evolutionary and/or coevolutionary processes.

Answering these questions will certainly not be simple (if there are answers at all). Using the patterns of host use and defenses depicted in Figs. 8 and 9 as a conceptual guide, at least two general interacting approaches for answering these questions are possible. One involves determining what changes in patterns of herbivore host use and plant defense could possibly result from multispecies evolutionary interactions. The second involves the development of falsifiable hypotheses about expected patterns of host use that would result from ecological versus coevolved interactions. Graphical and mathematical models of multispecies evolutionary interactions could be useful in developing these hypotheses.

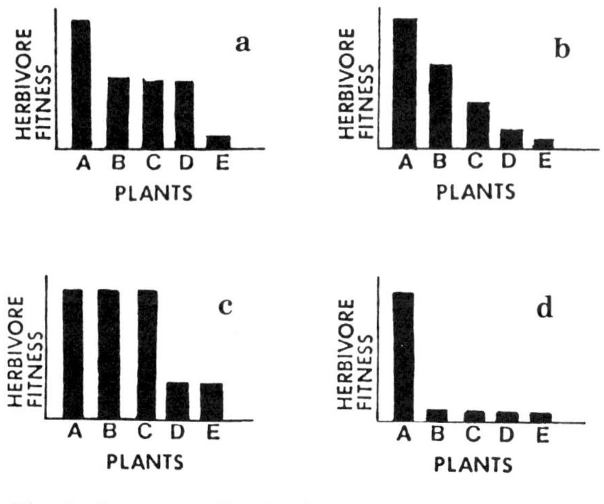

Fig. 8. Some possible herbivore fitness profiles on a set of plant species A, B, C, D, E found in the herbivore's habitat.

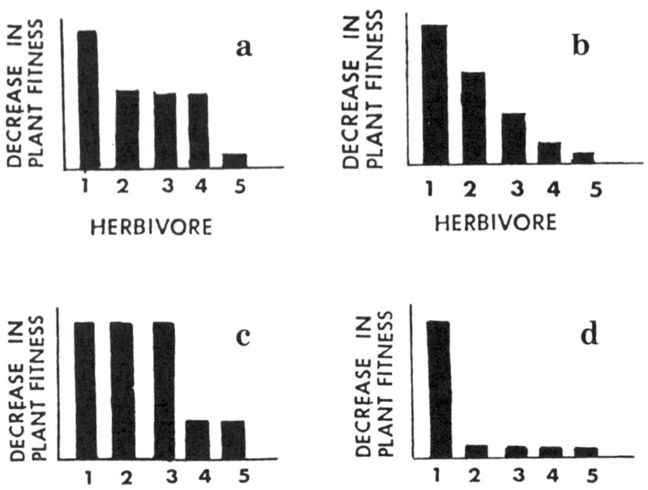

Fig. 9. Some possible profiles of fitness-reducing damage to single plant species caused by a complex of herbivores 1–5.

D. Models of Multispecies Evolutionary Dynamics

As in the modeling of pairwise coevolution, it is very important in considering multispecies evolutionary processes to know whether there is a significant physiological cost to the defenses produced by the plants and the counteradaptations produced by the herbivores. If there is a physiological cost, we also need to know whether the cost of multiple defenses and counteradaptations is more than, equal to, or less than the sum of their individual costs.

In multiple species interactions there is an added dimension to the question of cost: that is, the question of ecological cost. We need to know whether high or low fitness of an herbivore on one plant species is independent of fitness on other plant species. Similarly, is plant resistance to damage by one herbivore species independent of its resistance to other damaging herbivore species? The ecological slogan, "a jack of all trades is a master of none," would imply a lack of independence. If there is a lack of independence, we must ask how strong the dependence is and what form the dependence takes on. Does a large increase in an herbivore's fitness on one plant cause small decreases in its fitness on all of its other hosts? Does it cause decreased fitness on some hosts, but increased fitness on others? Figure 10 presents some plausible consequences of selection for adaptation by an herbivore to one host plant. In Fig. 10a, the herbivore's adaptation is independent; in Figs. 10b and 10c, the herbivore becomes more of a specialist and generalist, respectively.

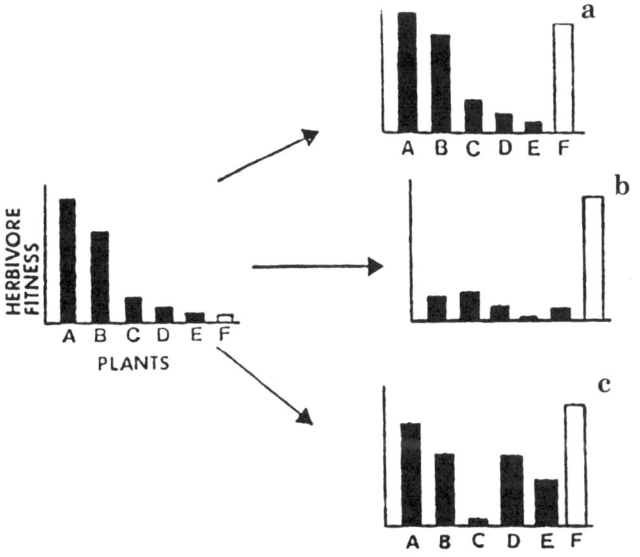

Fig. 10. Some possible effects of selection for physiological adaptation by an herbivore to one marginal host (F) on that herbivores' adaptation to its other hosts. (a) Adaptation to host (F) is independent of other host range components; (b) there is a strong negative effect of adaptation to host (F) on the herbivore's fitness on other hosts; (c) shows positive and negative effects.

In addition to physiological interactions, we must also consider behavioral adaptations of herbivores in plant communities. If behaviors for accepting one host are not independent of behaviors for accepting other hosts, genes that cause avoidance of a toxic host plant may also cause avoidance of a good host (Chew and Rodman, 1979; Levins and MacArthur, 1969). In such cases, the mean fitness of herbivores with genes for avoidance may be higher or lower than that of the rest of the population. Our predictions about how herbivore host ranges evolve would certainly be improved by a better understanding of all of these components of herbivores' abilities to utilize hosts.

From the plant perspective we may ask, how independent or interactive (genetically and physiologically) are defenses against various herbivores and pathogens? If a plant evolves a defense (or defenses) against one herbivore, does it become on average more or less defended against the rest of its herbivore community? Figure 11 presents some plausible scenarios. Here, I separate the component of net fitness decrease that is due to herbivore damage from the component due to cost (C) of producing constitutive and inducible defenses. Interactions among herbivores that influence plant fitness cannot be described by this model.

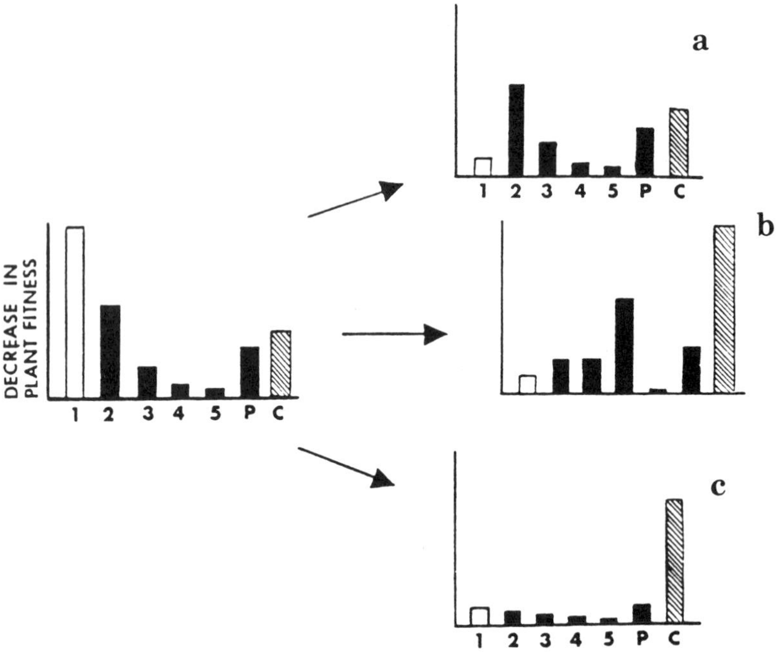

Fig. 11. Some possible effects of a plant's new defense against one herbivore (1) on the damage to that plant caused by other herbivores (2–5) and pathogens (P). Also, the average physiological cost of defense against herbivore 1 to the plant's fitness (C) is depicted. (a) The new defense only affects the interaction of the plant with herbivore 1, and does not involve an additional physiological cost to the plant. (b) The defense causes less damage by some herbivores and more by others; it entails physiological cost. (c) The defense lowers damage by all herbivores and the pathogen; there is a physiological cost to this defense. (There are many other possibilities not depicted here.)

The graphical models in Figs. 10 and 11 could, respectively, help us visualize the evolution of herbivore host ranges in the face of genetically fixed and ecologically stable plant communities, and the evolution of plant defenses in the face of genetically fixed, ecologically stable herbivore communities. We could predict, perhaps naively, that evolution would lead to host ranges or plant defenses that optimized overall fitness of plants and herbivores within the limits of genetically imposed constraints.

If we had at least a general understanding about the form taken by genetic constraints on plant and herbivore adaptations, we might be able to predict which patterns are evolutionarily more or less likely outcomes. If we could do this, we might be able to a priori predict certain qualities to be expected of the herbivores and plants in certain communities that were more specific and testable than the ones we presently have (e.g., Rhoades and Cates, 1976; Feeny, 1976; Rhoades, 1985). Roughgarden (1983) and Rummel and Roughgarden (1983) provide predictions of expected structures of an *Anolis* lizard community if the structure were the result of purely ecological interactions, or also involved short-term evolution and/or coevolution. Although not fully satisfying, this type of approach could be used to make predictions about differences in plant resistance patterns that a plant species would exhibit in a community that lacked one specific herbivore. Similarly, we could predict the type of evolutionary changes in host ranges that an herbivore community would undergo if a specific plant were removed (e.g., chestnut blight) or were naturally absent (see Gould, 1979). Rathcke (1985) has conducted an interesting experiment on host choice of slugs that is relevant to this type of approach.

IV. Short-Term Multispecies Coevolution: Relevant Data

All the data described or alluded to in the section on pairwise coevolution are relevant to studying short-term multispecies coevolution. In addition to these data, we need information on the extent to which adaptations to use one host species affect usage of other hosts, and how a plant defense against one herbivore species affects the plant's resistance to damage by other herbivore species. Studies of ecological interactions between herbivore species that affect plant fitness would be important in establishing how addition or deletion of a specific herbivore from an herbivore community would affect selection pressure on plants.

A. Correlated Effects of Herbivore Adaptation to Host Plants

1. Physiological Adaptations

According to the "jack of all trades is a master of none" adage, we should find a trade-off between adaptation to one host and fitness on a second host. A number of studies have searched for such a trade-off with minimal success. Gould (1979) found that a mite strain (*Tetranychus urticae* Koch.) that had adapted to resistance factor(s) in cucumber plants was slightly less adapted to lima bean than the control strain from which it was derived. This slight trade-off was more than counterbalanced by an increase in the cucumber-adapted strain's survival on potato and tobacco.

Additionally, adaptation to cucumber seemed to be independent of fitness on another unrelated host, plantain. The genetic factor(s) involved in adaptation to cucumber was shown to have a large additive component to its inheritance.

A number of detailed field studies have revealed that in general, when two or more populations (or subpopulations) of an insect herbivore are collected from different hosts, the offspring are relatively most fit on the host from which their parents were collected (see Bowers, this volume). Tavormina (1982) found that subpopulations of *Liriomyza brassicae* Riley from a specific host [either *Barbarea vulgaris* R. Br., *Thlaspi arvense* L., or *Brassica nigra* (L). Koch] developed slightly faster on that host than did subpopulations collected from the other hosts. Tabashnik (1983) tested a *Colias philodice eriphyle* Edw. population from alfalfa fields and another from a native habitat (which included the hosts *Vicia americana* Mohl. and *Lathyrus leucanthus* Rydb.). He found that larvae from the two populations generally grew best on plants found in the habitats from which they were collected. Rausher (1982) compared two populations of *Euphydryas editha* Bdv. from populations that fed mostly on *Pedicularis densiflora* Benth. or *Collinsia tinctoria* Hartw. ex Benth. He found that the herbivore grew relatively better on its native host than the novel host, but cautioned that this type of result did not necessarily indicate a trade-off in adaptation to the various hosts. Indeed, all the patterns described in these studies of field populations could be generated by independent adaptation to the local hosts with no ecological costs of adaptation at all.

Rausher (1984a,b) and Via (1984a,b) did detailed quantitative genetic analyses in search of more definitive evidence of ecological trade-offs within single herbivore populations. Rausher found positive but no negative correlations in the performance of tortoise beetles on two morning-glory species. Via (1984b) found that pupal weights of full-sib families of leaf miners on tomato were positively correlated with pupal weights on pea, but that development time on the two crops was not correlated. Thus, these two detailed quantitative genetic analyses (Rausher, 1984a; Via, 1984a,b) of within-population variation detected no trade-offs (negative genetic correlations) between larval performance on different host species.

Taking a genetically less sophisticated, but nonetheless rigorous, approach, Weber (1985) tested over 1000 clones of the polyphagous aphid *Myzus persicae* Sulz. for clone-specific rates of population increase on sugar beet and potato. Although clones varied in their rate of increase by a factor of 5× and 8× on sugar beet and potato, respectively, Weber found absolutely no correlation between clone growth rates on these two taxonomically distant hosts. This would indicate independent inheritance of adaptation to sugar beet and tomato as in Fig. 10a. Futuyma and Philippe (1987) found a similar lack of support for ecological trade-off between fitness of "clones" of the fall cankerworm on two oak species, chestnut, and red maple. Additionally, Hare and Kennedy (1986) recently found that a population of Colorado potato beetles that did well developing on horsenettle developed as well on tomato as a population that developed poorly on horsenettle.

Service's (1984) study of intrinsic rate of increase of clones of the aphid *Uroleucon rudbeckiae* Fitch on plant clones of *Rudbeckia laciniata* L. is relevant to adaptation at the intraspecific host level. He found an aphid clone by plant clone interaction effect in his analysis of variance, but one aphid clone was equal to or superior to the other three aphid clones on all host-plant clones. Again, evidence for a trade-off in host-specific adaptations is equivocal.

The only study of arthropods revealing a within-herbivore population trade-off is that of Gould (1979), and since that trade-off appeared to be small and involved experiments with only one mite population, its generality is certainly questionable. The literature on microbial herbivores (pathogens) reveals extreme costs to adaptation (i.e., virulence) in some cases and no measurable costs in other cases (Parlevliet, 1983; Leonard, 1986). This has led to some debate in the plant pathology literature regarding the agricultural and evolutionary importance of such costs, but Leonard (1986) has argued convincingly for the evolutionary importance of even small costs of adaptation in these organisms.

Why haven't we found more significant trade-offs in host-specific adaptations of invertebrate herbivores? Maybe they are very rare, or generally too small to detect. If this is the case, we can deemphasize the hypothesis of physiological trade-offs in predicting patterns of host range evolution and/or coevolution. Assuming the absence of trade-offs would certainly be useful in simplifying models of host range evolution and/or coevolution. But are our data sufficient to justify abandonment of the ecological trade-off hypothesis (e.g., Levins, 1969; Krebs, 1972; Roughgarden, 1979) that has gained the status of a paradigm in the general ecology and genetics literature? How strong are our apparently anomalous results regarding the trade-offs in herbivore host utilization ability?

Because most of these studies involve laboratory testing of field populations of herbivores (or populations that were not given sufficient time to adapt to lab conditions), Rose's (1984) and Service and Rose's (1985) general criticism of genetic correlation studies conducted on life-history traits is applicable to some of the data on herbivore host use. Some of the positive correlations found in the various studies of wild populations tested in the lab could, in part, be due to variation among families in adaptedness to novel, lab environmental conditions and to the greenhouse-grown host plants, which can be of a very different quality than field-grown plants. Taking this perspective to the extreme, it can be argued that a positive genetic correlation significantly less than +1.0 is the result of an ecological trade-off. For cases where no positive or negative correlation is found and the power of the statistical test is strong, such an interpretation would, however, not be convincing. For example, in the study of Weber (1985) where over 1000 clones were tested, the chance that lab-rearing artifacts balanced against the biologically important trade-offs to yield an insignificant correlation of $r = 0.05$ is unlikely. Since Weber's study is an exception in terms of statistical power, other ways of circumventing the problems with artifacts in genetic correlation studies are needed (see Bell, 1986; Reznick *et al.,* 1986; Futuyma, 1988a,b).

One useful approach toward alleviating these problems would be determining genetic correlations of herbivores living under field conditions. Of course even this approach has limitations, since an experiment conducted in July measures genotype adaptation to climate as well as host plant, so such studies must be carefully designed.

We have taken the reductionistic approach of trying to determine whether adaptation by larvae of a polyphagous lepidopteran species (*Heliothis virescens* Fabr.) to a single fitness-reducing plant compound (quercetin) would affect their fitness when faced with other plant compounds. In our initial study (Gould, unpublished data), two populations of *H. virescens* were each split in two. One subpopulation from each population was selected for rapid growth on artificial diets containing quercetin. The other two subpopulations were maintained on artificial diets lacking this phenolic compound. Both selected subpopulations adapted to the quercetin-containing diets, but the inheritance of adaptation to the quercetin diet was mostly additive in one case and mostly dominant in the other case, implicating different genetic mechanisms in the two cases. When one adapted strain and a recurrently selected line of the other strain that was backcrossed to the control subpopulation of the first population were tested on artificial diets containing a variety of secondary plant compounds, it was obvious that the subpopulations that had adapted to quercetin were crossadapted to some other plant compounds (Fig. 12). In a couple of cases, negatively correlated responses were found. More interesting was the fact that the two adapted strains differed in their patterns of cross-adaptation and negative cross-adaptation (Fig. 12). Analysis of selected hybrid crosses substantiated the genetic basis of these genetic interactions.

What conclusions can be drawn from these data? (1) There are positive and negative correlations in adaptation to the various chemicals. (2) The similarity of the chemicals at the level of class of compound (e.g., phenolic, alkaloid) does not clearly predict the probability or type of genetic interaction. (3) The genetic interactions found within one population are not necessarily generalizable to other populations.

To elaborate on conclusion (1), negatively correlated responses are less common than positively correlated responses; thus, physiological trade-offs may not be common. Conclusion (2) is consistent with the results of Gould (1979) where mites that probably adapted to a triterpenoid in cucumber were cross-adapted to potato and tobacco, which are in general defended against herbivores by alkaloids and phenolics (Tingey, 1984; Isman and Duffey, 1982). These results indicate that adaptation to an array of chemically dissimilar compounds may not require independent physiological mechanisms. Conclusion (3) is somewhat troublesome in that it indicates that within a single herbivore species, the set of genetic correlations between adaptations to various plant defenses may differ from population to population. This means that if we want to predict host-range patterns of herbivore species based on mechanisms of adaptation, it is not sufficient to analyze the correlations among plant-specific fitnesses in only one herbivore species population. How many populations need to be measured and how one should use the population specific matrices to build an evolutionary and/or coevolutionary model remain unsolved. The preliminary work presented in Fig. 12 indicates that some genetic correlations (e.g., quercetin–rutin, quercetin–nicotine) may be stable, but only two populations were examined. Experiments with more herbivore populations are being carried out, but where do we stop? Is there any way of developing a set of reliable guidelines for predicting the extent and pattern of variation to be expected in correlations between adaptive traits of an herbivore species? A more detailed biochemical understanding of herbivore adaptations would be useful. Studies in this area are progressing (e.g., Brattsten, 1983; Yu, 1983), but are as yet too preliminary to answer this question (see Dowd *et al.*, 1983; Gould, 1984b; Rose, 1985). A recent

	COMPOUND						
	QUERC	RUT	GRA	NIC	TRI	GOS	DIG
SELECTED 1	+	+	0	+	0	0	-
SELECTED 2	+	+	+	+	0	-	-

Fig. 12. Correlated responses to selection for rapid growth on artificial diets containing quercetin. Two strains of *Heliothis virescens* were selected for adaptation to quercetin containing diets. After selection, strain 2 was backcrossed for six generations to the control stock of selected strain 1. Growth of control 1, selected 1, and selected 2 were compared on diets containing quercetin (querc), rutin (rut), gramine (gra), nicotine (nic), tridecanone (tri), gossypol (gos), and digitoxin (dig). (+) The selected strain was larger than the control strain after 10 days; (−) the selected stock was smaller; (0) no significant difference was found.

study by Neal (1987) has come up with the interesting result that increased mixed-function oxidase activity does not lower growth efficiency in *Heliothis zea* Boddie. If this result is generalizable, it may explain why we do not find strong ecological trade-offs.

2. Behavioral Adaptations

Thus far I have discussed physiological adaptation, but genetic correlations among host preference traits are at least equally important in understanding multispecies coevolution. Are there alleles specific for attraction to one host that have no effect on attraction to another host? What is the structure of the attraction/repulsion genetic correlation matrix, and is it stable across populations within a species? Answers to these questions are slow in coming. We have data indicating variation in host preference within and among herbivore populations (e.g., Tabashnik *et al.,* 1981; Wasserman and Futuyma, 1981; Singer, 1983; Stanton and Cook, 1984; Prokopy *et al.,* 1984), and in some cases, a genetic basis for this variation has been established (e.g., Carson and Ohta, 1981; Jaenike, 1985). In the work of Carson and Ohta (1981) genetic differences (probably at one-locus) between populations of *Drosophila silvestris* are responsible for a change from monophagy to polyphagy, so host preference does not seem to be inherited as a one-gene/one-host relationship. Genes for preferring one suitable new host could therefore also cause acceptance of an unsuitable host.

In a recent detailed study, Wasserman (1986) analyzed oviposition behavior of 22 strains of the southern cowpea weevil (*Callosobruchus maculatus* Fabr.) on six host seeds. He, like Carson and Ohta (1981), found that most of the genetic variance in behavior involved differences in how selective the strains were, not in their rank ordering of preference. Thus, some strains seemed more polyphagous than others. In Wasserman's study, one exceptional strain differed from the others in the way it ranked host-seed acceptability.

Beyond these studies, little is known regarding genetics of herbivore host choice. However, one lab study of a single gene mutant of *Drosophila melanogaster* L. is instructive. Kikuchi (1973) chemically induced and then selected a mutant strain of *D. melanogaster* that was attracted to a volatile compound [4-(*O*-hydroxyphenyl)-3-buten-2-one] that was deterrent to the parent strain. Kikuchi then tested the response of the two strains to 64 odors. In 12 cases the mutant was attracted to the novel odor, while the parent strain was not. In all of the other cases, the two strains were similar in their behavioral response. This supports the theme that host-range evolution does not proceed by independent acquisition or loss of single hosts. Kikuchi (1973) proposed that there may have been some relationship between the chemical structure of the 12 specific compounds, indicating some predictiveness to the cross-attraction.

Arnold's (1981a,b) work on genetics of garter snake host preference offers an example of how one might develop a matrix of genetic correlations related to host preference. He found both positive and negative correlations in preferences for prey odors, but, as with herbivore physiological adaptations, there was a predominance of positive correlations. He argues that there is stability of the genetic correlation matrix across garter snake populations. It would be valuable to assess the stability of such a host-preference correlation matrix for an herbivore.

3. Interactions between Preference and Suitability

There are some interesting data showing a correlation between an herbivore genotype's oviposition host preference and the host's suitability to that herbivore genotype (Service, 1984; Via, 1986; also see Gould, 1983). In all of the species examined, adult and larval stages fed on the same hosts. Such correlations may be more common in cases where both adults and juveniles are herbivorous than in cases where only juveniles are herbivorous, but this hypothesis has not been tested (Gould, 1983).

B. Variability in the Genetics of Plant Defenses

Berenbaum and Zangerl (this volume) have reviewed the topic of genetics of plant defenses, so a general review will not be presented here. Instead, I will focus on some data regarding genetics of phytochemical synthesis that are important in developing models of multispecies coevolution.

Models of multispecies coevolution require knowledge of the form taken by sets of genetic correlations between an herbivore's adaptations to plants (e.g., plants A–F, Fig. 10), and a plant's defenses against herbivores (e.g., herbivores 1–5, Fig. 11). As indicated in the previous section, work aimed at understanding the matrix of genetic correlations of herbivore adaptations had led to some puzzling conclusions regarding ecological trade-offs. Moreover, the work has shown that within an herbivore species, populations may possess differently shaped genetic correlation matrixes.

If we now consider the plant component of coevolution, similar questions may be asked about the stability of sets of genetic correlations within a plant's profile of defense against various herbivores. There is some indirect but extensive data in the phytochemistry literature that suggests population-to-population variation in these genetic correlations. In some respects, these data on plants are actually more informative than the data on herbivores, if we accept the generality that secondary chemicals and combinations of secondary chemicals contribute significantly to plant defenses (see Berenbaum and Zangerl, this volume).

As pointed out by Berenbaum and Zangerl, phytochemists have long been interested in patterns of variation in the profile of secondary chemicals within plant species. Studies aimed at understanding how these profiles of chemicals are generated have taken both a biochemical (e.g., Croteau, 1984) and a genetic approach (Hanover, 1966a,b; Lokki *et al.*, 1973; Baradat *et al.*, 1975; Vernet *et al.*, 1977a,b; Lincoln and Murray, 1978; Berenbaum *et al.*, 1986). Although many of these studies have a bearing on the question of stability of genetic correlations between quantities of various compounds in a plant, I will focus on terpenoids of the mint family (Labiatae) as a relevant example.

Because of their economic importance, the monoterpenes of the mint family have been studied extensively at the inter- and intraspecific level in a number of genera (e.g., *Mentha, Thymus,* and *Satureja*). In *Satureja douglasii* Briq., for example, a single plant may contain more than 10 distinct monoterpenes (e.g., Rice *et al.*, 1978), but most plants have one or two predominent monoterpenes. The predominant monoterpenes of plants within and among populations may vary (Rhoades *et al.*, 1976; Lincoln and Langenheim, 1981). Plants that share a similar profile of monoterpene concentrations are referred to as chemotypes (see Tetenyi, 1970).

There is evidence that variability in the monoterpene composition of mints influences herbivory. Rice *et al.* (1978) found that the chemotype of *S. douglasii,* characterized by high concentrations of pulegone, was less preferred by a native herbivore (*Ariolimax dolichophallus* Mead) than three other chemotypes, and that purified pulegone was more deterrent to the herbivore than the major monoterpene components of the other chemotypes. Similarly, Gouyon *et al.* (1983) found variation in herbivory related to monoterpene composition of *Thymus vulgaris* L.

The various monoterpenes and combinations of monoterpenes in plants of the mint family may affect individual herbivores differently. If so, understanding the structure of the matrix of genetic correlations among monoterpene concentrations of plants within a single population is important in understanding trade-offs in the defense profiles of these plants. For example, if the concentrations of pulegone and carvone in plants were negatively correlated and pulegone was avoided by herbivore A and carvone avoided by herbivore B, a single plant could not be optimally defended against both herbivores. If pulegone and carvone concentrations were independent or positively correlated, a single plant could be defended against both herbivores (see Berenbaum *et al.*, 1986). If there was a negative correlation in one population and a positive correlation in another, we would expect different evolutionary trajectories in the two populations in response to the same herbivore community.

Extensive genetic analyses of terpene concentrations and biosynthetic pathways in a number of species (*T. vulgaris,* Gouyon and Vernet, 1980;

Mentha citrata Ehrh. Lincoln *et al.*, 1986; Lincoln and Murray, 1978; *S. douglasii,* Lincoln and Langenheim, 1981; *Chrysanthemum vulgare* (Lam.) Gaterau, Lokki *et al.*, 1973; *Pinus monticola* Dougl. ex D. Don, Hanover, 1966a,b; *P. pinaster* Saland., Baradat *et al.*, 1975) are extremely useful in this regard. In these studies, specific enzymatic conversions commonly appear to be under control of single genes with additive to completely dominant expression of alleles (this does not rule out some polygenic control). Due to the generally sequential nature of biosynthetic pathways, an allele of a gene acting early in the biosynthetic pathway can have epistatic effects on genes acting on conversions further along the pathway (see Vernet, 1977, for an extreme example).

The work of David Lincoln and his colleagues on the genetics of monoterpenes in *Mentha* is the most extensive and conclusive effort to date in this field and will be used here as an example. Lincoln *et al.* (1986, and references within) have examined the inheritance and interactions between nine genes in the monoterpene biosynthetic pathway of *Mentha* using within- and between-species genetic crosses followed by quantitative chemical analyses. Figure 13, which summarizes their work, depicts the hypothesized genetic control over the biosynthetic pathway based on genetic and radiotracer/enzyme studies (Croteau, 1984). In Fig. 13, each large capital symbol refers to the dominant allele of a specific gene. For example, allele *C* controls conversion of limonene to carvone (probably by coding for a specific enzyme). Most alleles are specific for one type of conversion, but the *R* allele is involved in a number of conversions.

How does all this relate to correlations between concentrations of various terpenes? Consider two populations, one polymorphic at locus *I* (having *I* and *i* alleles), and the second lacking *I* (only having *i*), but polymorphic at locus P^s (P^s and p^s alleles). In the population polymorphic at the *I* locus, there would be a negative relationship between production of the linalool/linalyl acetate complex and a large number of other monoterpenes, including isopinocamphone, limonene, carvone, dihydrocarvone, piperitenone, pulegone, menthone, and isomenthone. In the second population, there would be a negative correlation between the concentrations of pulegone and menthol [given the presence of *R* alleles, plants with P^s produce 78% menthol and less than 1% pulegone; plants with p^s produce less than 1% menthol and 79% pulegone (Lincoln and Murray, 1978)]. In this second population, there would be no negative correlation between linalool/linalyl acetate and the 3-oxygenated compounds.

I have presented a very simplified view of this complex biosynthetic pathway. If we consider the possibilities for polymorphism at other loci, it becomes apparent that a number of possible correlation matrixes could be produced. Additionally, the quantity of all of the aforementioned compounds can be affected by factors (environmental and genetic) that influ-

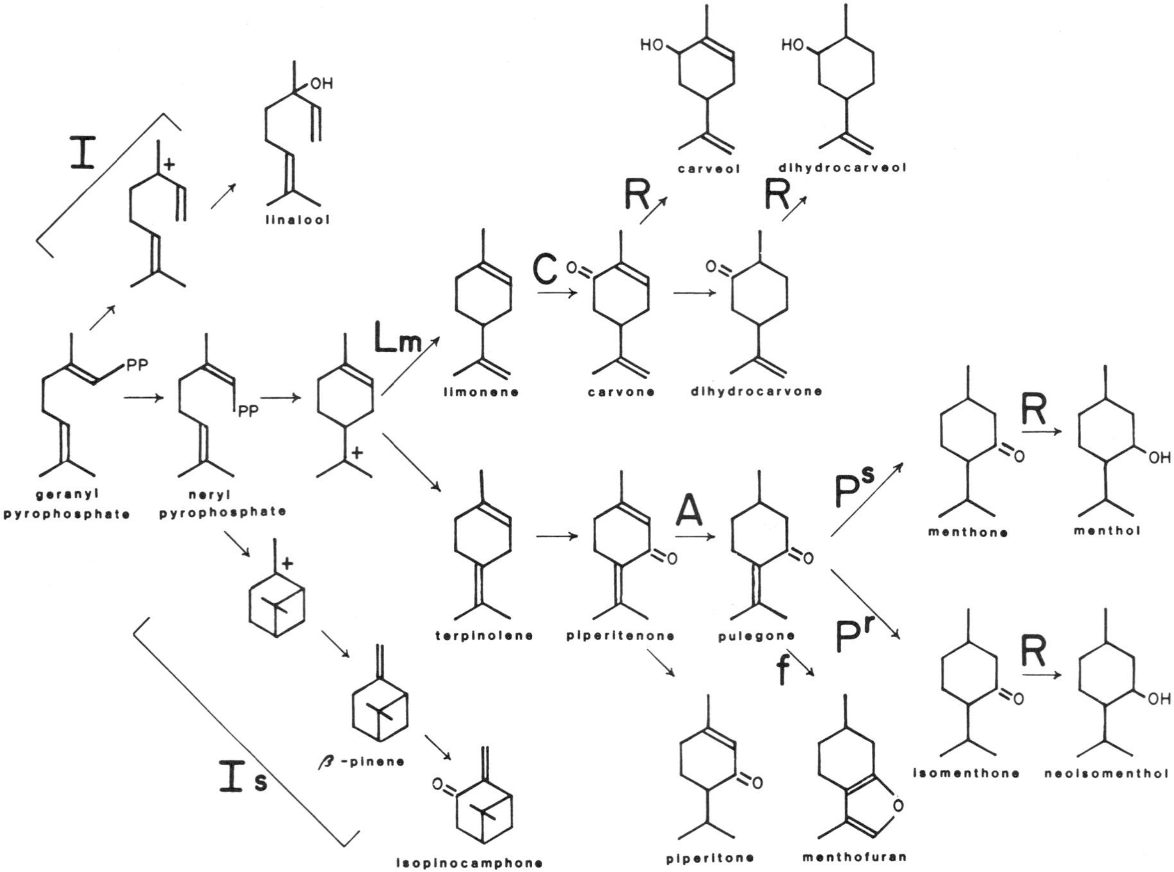

I
linalool
geranyl pyrophosphate
neryl pyrophosphate
PP
Lm
limonene
C
carvone
R
carveol
dihydrocarvone
dihydrocarveol
terpinolene
piperitenone
A
pulegone
P^s
P^r
f
menthone
menthol
isomenthone
neoisomenthol
piperitone
menthofuran
β-pinene
isopinocamphone
Is

ence the abundance of precursor compounds. Even though the genetic analysis presented here did not involve natural plant populations, analogous pathways and correlation matrixes could be expected in natural populations. Indeed, work on genetics of secondary chemistry in natural populations of other mint species supports the prediction of population to population variation in genetic correlations between secondary compounds that can affect herbivory [*S. douglasii,* Rhoades *et al.* (1976), Rice *et al.* (1978); *T. vulgaris,* Vernet (1977), Vernet *et al.* (1977a,b), Gouyon and Vernet (1980), Gouyon *et al.* (1983)].

We may conclude from these studies that genetic correlations between the level of defense of individual plants to an array of herbivores may be population-specific. That is to say the amounts of damage caused by herbivore 1 and herbivore 2 to individual plants in population A may be negatively correlated, while in population B the genetic correlation may be positive or totally lacking. The further away two compounds are in a biosynthetic pathway, the lower the stability of their genetic correlation is apt to be (also see Berenbaum *et al.,* 1986). These results indicate that studies of genetic correlations in herbivory patterns from a number of plant populations will be needed to determine the stability of a specific genetic correlation in defense against multiple herbivores.

These phytochemical data also offer helpful insights into measuring physiological costs of producing defensive chemicals. In some plant populations, the physiological cost of producing a defensive chemical will appear to be high if the biosynthetic step for which the population is polymorphic is an energy-requiring step. In another population, quantitative genetic analyses could reveal no cost (or even a benefit) to producing the same defensive chemical if the biosynthetic step for which it is polymorphic does not require energy or saves energy.

V. Conclusions

One goal motivating studies of plant herbivore interactions is determining the importance of specific types of evolutionary and coevolutionary processes in structuring plant–herbivore interactions. We are far from

---

Fig. 13. Genetic control of the monoterpene biosynthetic pathway in mint. *I, Is, Lm, C, R, A,* P^s, P^r, and *f* are alleles that control the biosynthetic conversions indicated by the arrows, probably by coding for production of specific enzymes. Genes controlling the production of geranyl pyrophosphate, the precursor in this pathway, have not been studied, and polymorphism of these genes could cause positive correlations among concentrations of many of the compounds.

achieving this goal, and in some respects the more work we do in this area, the further away we seem to be from attaining this goal.

Naively optimistic hypotheses predicting generality of long-term, narrow-sense coevolution between plants and herbivores are being abandoned, as are vague hypotheses about "general coevolution." Replacing them are hypotheses about specific types of coevolutionary processes that should have distinct short-term (and possibly long-term) effects on the structure and dynamics of plant–herbivore interactions. These more specific hypotheses are theoretically testable, but the tests will require data on many ecological and genetic parameters of the systems being examined.

Dissecting the contemporary ecological and genetic workings of a few tractable plant–herbivore systems with at least two herbivores and hosts would be useful. Practical development of such studies would be difficult and they would certainly require long-term commitments with no assurance of obtaining clear-cut results. To the extent that understanding if and how the processes of multispecies coevolution influence plant/herbivore interactions is important to us, such detailed studies would be worth the effort they require.

Acknowledgments

It is difficult to accurately recall all of the people who contributed to this left-handed view of coevolution. Among them are E. Bernays, D. Futuyma, G. Kennedy, D. Landis, K. Leonard, C. Mitter, C. Nalepa, M. Rausher, A. Sheck, E. Simms, K. Suiter, and a very thorough anonymous reviewer. This work was partially funded by National Science Foundation grant BSR-8214786A02 and U.S. Department of Agriculture CGRP grant 85-CRCR-1-1687.

References

Ahmad, S. (1983). "Herbivorous Insects. Host-Seeking Behavior and Mechanisms." Academic Press, New York.

Arnold, S. J. (1981a). Behavioral variation in natural populations. I. Phenotypic, genetic and environmental correlations between chemoreceptive responses to prey in the garter snake, *Thamnophis elegans*. *Evolution* **35,** 489–509.

Arnold, S. J. (1981b). Behavioral variation in natural populations. II. The inheritance of a feeding response in crosses between geographic races of the garter snake, *Thamnophis elegans*. *Evolution* **35,** 510–515.

Baradat, Ph., Bernard-Dagan, C., Pauly, G., and Zimmermann-Fillon, C. (1975). Les terpenes du pin maritime: aspects biologiques et genetiques. *Ann. Sci. For.* **32,** 29–54.

Bell, G. (1986). Reply to Reznick *et al. Evolution* **40,** 1344–1346.

Benson, W. W., Brown, K. S., and Gilbert, L. E. (1975). Coevolution of plants and herbivores: Passion flower butterflies. *Evolution* **29,** 659–680.

Berenbaum, M., and Feeny, P. (1981). Toxicity of angular furanocoumarins to swallowtail butterflies: Escalation in a coevolutionary arms race? *Science* **212,** 927–929.

Berenbaum, M. R., Zangerl, A. R., and Nitao, J. K. (1986). Constraints on chemical coevolution: Wild parsnips and the parsnip webworm. *Evolution* **40,** 1215–1228.

Bernays, E. A. and Graham, M. (1988). On the evolution of host specificity in phytophagous arthropods. *Ecology,* in press.

Brattsten, L. B. (1983). Cytochrome P-450 involvement in the interactions between plant terpenes and insect herbivores. *Am. Chem. Soc. Symp. Ser.* **208,** 173–195.

Bush, G. L., and Diehl, S. R. (1982). Host shifts, genetic models of sympatric speciation and the origin of parasitic insect species. *In* (J. H. Visser and A. K. Minks, eds.). "Insect-Plant Relationships" Cent. Agric. Publ. Doc., Wageningen.

Carson, H. L., and Ohta, A. T. (1981). Origin of the genetic basis of colonizing ability. *In* "Evolution Today" (G. G. E. Scudder and J. L. Reveal, eds.), pp. 365–370. Carnegie Mellon Univ., Pittsburg.

Castillo-Chavez, C., Levin, S. A., and Gould, F. (1988). Physiological and behavioral adaptation to varying environments: A mathematical model. *Evolution,* in press.

Chew, F. S., and Rodman, J. E. (1979). Plant resources for chemical defense. *In* "Herbivores, Their Interaction with Secondary Plant Metabolites" (G. A. Rosenthal and D. H. Janzen, eds.), pp. 271–307. Academic Press, New York.

Clarke, B. C. (1976). The ecological genetics of host-parasite relationships. *In* "Genetic Aspects of Host-Parasite Relationships" (A. E. R. Taylor and R. Muller, eds.), pp. 87–103. Blackwell, Oxford.

Coley, P. D., Bryant, J. P., and Chapin, F. S. (1985). Resource availability and plant antiherbivore defense. *Science* **230,** 895–899.

Croteau, R. (1984). *In* "Isopentenoides in Plants" (W. D. Nes, G. Fuller, and L. Tsai, eds.), pp. 31. Dekker, New York.

Dickinson, C. H. (1979). External synergisms among organisms inducing disease. *In* "Plant Disease. An Advanced Treatise" (J. G. Horsfall and E. B. Cowling, eds.), pp. 97–111. Academic Press, New York.

Dowd, P. F., Smith, C. M., and Sparks, T. C. (1983). Detoxification of plant toxins by insects. *Insect Biochem.* **13,** 453–468.

Ehrlich, P. R., and Raven, P. H. (1964). Butterflies and plants: A study in coevolution. *Evolution* **18,** 586–608.

Falconer, D. S. (1981). "Introduction to Quantitative Genetics," 2nd Ed. Longman, New York.

Feeny, P. P. (1976). Plant apparency and chemical defense. *Recent Adv. Phytochem.* **10,** 1–40.

Fox, L. R. (1981). Defense and dynamics in plant-herbivore systems. *Am. Zool.* **21,** 853–864.

Fox, L. R., and Morrow, P. A. (1981). Specialization: Species property or local phenomenon? *Science* **211,** 887–893.

Futuyma, D. J. (1983). Evolutionary interactions among herbivorous insects and plants. *In* "Coevolution" (D. J. Futuyma and M. Slatkin, eds.), pp. 207–231. Sinauer, Sunderland, Massachusetts.

Futuyma, D. J. (1985). Evolution and coevolution in communities. *In* "Patterns and Processes in the History of Life." (D. Raup and D. Jablonski, eds.), pp. 369–381. Springer-Verlag, Berlin.

Futuyma, D. J. (1986). The role of behavior in host-associated divergence in herbivorous insects. *In* "Evolutionary Genetics of Invertebrate Behavior." (M. D. Huettel, ed.), pp. 295–302. Plenum, New York.

Futuyma, D. J., and Gould, F. (1979). Associations of plants and insects in a deciduous forest. *Ecol. Monogr.* **49,** 33–60.

Futuyma, D. J., and Peterson, S. C. (1985). Genetic variation in the use of resources by insects. *Annu. Rev. Entomol.* **30,** 217–238.

Futuyma, D. J., and Philippe, T. H. (1987). Genetic variation and covariation in responses to host plants by *Alsophila pometaria* (Lepidoptera: Geometridae). *Evolution* **41,** 269–279.

Futuyma, D. J., and Slatkin, M. (1983). "Coevolution." Sinauer, Sunderland, Massachusetts.

Futuyma, D. J. (1988). The role of behavior in host-associated divergence in herbivorous insects. *In* "Evolutionary Genetics of Arthropod Behavior" (M. Huettel, ed.). Plenum, New York.

Gardia-Dorado, A. (1986). The effect of niche preference on polymorphism protection in a heterogeneous environment. *Evolution* **40,** 936–945.

Gilbert, L. E. (1971). Butterfly-plant coevolution: Has *Passiflora adenopoda* won the selectional race with Heliconiine butterflies? *Science* **172,** 585–586.

Gilbert, L. E. (1979). Development of theory in the analysis of insect-plant interactions. *In* "Analysis of Ecological Systems" (D. Horn, R. Mitchell, and G. Stairs, eds.), pp. 117–154. Ohio State Univ. Press, Columbus.

Gould, F. (1979). Rapid host range evolution in a population of the phytophagous mite *Tetranychus urticae* Koch. *Evolution* **33,** 791–802.

Gould, F. (1983). Genetics of plant-herbivore systems: Interactions between applied and basic study. *In* "Variable Plants and Herbivores in Natural and Managed Systems" (R. Denno and B. McClure, eds.), pp. 599–653. Academic Press, New York.

Gould, F. (1984a). Role of behavior on the evolution of insect adaptation to insecticides and resistant host plants. *Bull. Entomol. Soc. Am.* **30,** 34–41.

Gould, F. (1984b). Mixed function oxidases and herbivore polyphagy: The devil's advocate position. *Ecol. Entomol.* **9,** 29–34.

Gould, F. (1986a). Simulation models for predicting durability of insect resistant germ plasm: A deterministic diploid, two-locus model. *Environ. Entomol.* **15,** 1–10.

Gould, F. (1986b). Simulation models for predicting durability of insect-resistant germ plasm: Hessian fly (Diptera: Cecidomyiidae)-resistant winter wheat. *Environ. Entomol.* **15,** 11–23.

Gouyon, P. H., and Vernet, Ph. (1980). Etude de la variabilité genétique dans une population de *Thymus vulgaris* L. *Acta Ecol.* **1,** 165–178.

Gouyon, P. H., Fort, Ph., and Caraux, G. (1983). Selection of seedlings of *Thymus vulgaris* by grazing slugs. *J. Ecol.* **71,** 299–306.

Grimaldi, D., and Jaenike, J. (1984). Competition in natural populations of mycophagous *Drosophila*. *Ecology* **65,** 1113–1120.

Hanover, J. W. (1966a). Genetics of terpenes 1. Gene control of monoterpene levels in *Pinus monticola* Dougl. *Heredity* **21,** 73–84.

Hanover, J. W. (1966b). Environmental variation in the monoterpenes of *Pinus monticola* Dougl. *Phytochemistry* **5,** 713–717.

Hare, J. D., and Kennedy, G. G. (1986). Genetic variation in plant-insect associations: Survival of *Leptinotarsa decemlineata* populations on *Solanum carolinense*. *Evolution* **40,** 1031–1043.

Isman, M. B., and Duffey, S. S. (1982). Toxicity of tomato phenolic-compounds to the tomato fruitworm, *Heliothis zea*. *Entomol. Exp. Appl.* **31,** 370–376.

Jaenike, J. (1985). Genetic and environmental determinants of food preference in *Drosophila tripunctata*. *Evolution* **39,** 362–369.

Janzen, D. H. (1980). When is it coevolution? *Evolution* **34,** 611–612.

Jermy, T. (1976). Insect-host-plant relationship—Co-evolution or sequential evolution? *Symp. Biol. Hung.* **16,** 109–113.

Jermy, T. (1984). Evolution of insect/host plant relationships. *Am. Nat.* **124,** 609–630.

Karban, R. (1986). Interspecific competition between folivorous insects on *Erigeron glaucus*. *Ecology* **67,** 1063–1072.

Kiester, A. R., Lande, R., and Schemske, D. W. (1984). Models of coevolution and speciation in plants and their pollinators. *Am. Nat.* **124,** 220–243.

Kikuchi, T. (1973). Specificity and molecular features of an insect attractant in a *Drosophila* mutant. *Nature (London)* **243,** 36–38.

Kiyosawa, S. (1982). Genetics and epidemiological modeling of breakdown of plant disease resistance. *Annu. Rev. Phytopathol.* **20,** 93–117.

Kogan, M., and Turnipseed, S. G. (1980). Soybean growth and assessment of damage by arthropods. *In* "Sampling Methods in Soybean Entomology" (M. Kogan and D. C. Herzog, eds.), pp. 3–29. Springer-Verlag, New York.

Krebs, C. J. (1972). "Ecology." Harper & Row, New York.

Lande, R. (1980). Sexual dimorphism, sexual selection and adaptation in polygenic characters. *Evolution* **34,** 292–305.

Lawton, J. H., and Strong, D. R., Jr. (1981). Community patterns and competition in folivorous insects. *Am. Nat.* **118,** 317–338.

Leonard, K. J. (1986). The host population as a selective factor. *In* "Populations of Plant Pathogens: Their Dynamics and Genetics" (M. S. Wolfe and C. E. Caten, eds.), pp. 163–179. Blackwell, Oxford.

Leonard, K. J., and Czochor, R. J. (1980). Theory of genetic interactions among population of plants and their pathogens. *Annu. Rev. Phytopathol.* **18,** 237–258.

Levin, S. A. (1983). Some approaches to the modelling of coevolutionary interactions. *In* "Coevolution" (M. H. Nitecki, ed.), pp. 21–65. Univ. of Chicago Press, Chicago.

Levin, S. A., and Udovic, J. D. (1977). A mathematical model of coevolving populations. *Am. Nat.* **111,** 657–675.

Levins, R. (1969). "Evolution in Changing Environments." Princeton Univ. Press, Princeton, New Jersey.

Levins, R., and MacArthur, R. (1969). An hypothesis is to explain incidence of monophagy. *Ecology* **50,** 910.

Lincoln, D. E., and Langenheim, J. H. (1981). A genetic approach to monoterpenoid compositional variation in *Satureja douglasii*. *Biochem. Syst. Ecol.* **9,** 153–160.

Lincoln, D. E., and Murray, M. J. (1978). Monogenic basis for reduction of (+)-pulegone to (−)-methone in *Mentha* oil biogenesis. *Phytochemistry* **17,** 1727–1730.

Lincoln, D. E., Murray, M. J., and Lawrence, B. M. (1986). Chemical composition and genetic basis for the isopinocamphone chemotype of *Mentha citrata* hybrids. *Phytochemistry* **25,** 1857–1863.

Lokki, J., Sorsa, M., Forsen, K., and Schantz, M. V. (1973). Genetics of monoterpenes in *Chrysanthemum vulgare*. I. Genetic control and inheritance of some of the most common chemotypes. *Heredity* **74,** 225–232.

Louda, S. M. (1984). Herbivore effect on stature, fruiting, and leaf dynamics of a native crucifer. *Ecology* **65,** 1379–1386.

May, R. M., and Anderson, R. M. (1983). Parasite-host evolution. *In* "Coevolution" (D. J. Futuyma and M. Slatkin, eds.), pp. 186–206. Sinauer, Sunderland, Massachusetts.

Mitter, C., and Brooks, D. R. (1983). Phylogenetic aspects of coevolution. *In* "Coevolution" (D. J. Futuyma and M. Slatkin, eds.), pp. 65–98. Sinauer, Sunderland, Massachusetts.

Mitter, C., and Futuyma, D. J. (1983). An evolutionary-genetic view of host plant utilization by insects. *In* "Herbivores and Their Variable Hosts in Natural and Managed Systems" (R. F. Denno and M. S. McClure, eds.), pp. 427–459. Academic Press, New York.

Mode, C. J. (1958). A mathematical model for the coevolution of obligate parasites and their hosts. *Evolution* **12,** 158–165.

Mode, C. J. (1960). A model of a host-pathogen system with particular reference to the rusts of cereals. *In* "Biometrical Genetics, pp. 84–96. Pergamon, New York.

Mode, C. J. (1961). A generalized model of a host-pathogen system. *Biometrics* **17,** 386–404.

Morrow, P. A., and Lamarche, V. C. (1979). Tree ring evidence for chronic insect suppression of productivity in subalpine *Eucalyptus*. *Science* **201,** 1244–1246.

Neal, J. (1987). The metabolic cost of induced mixed function oxidase activity in *Heliothis zea*. *Entomol. Exp. Appl.* **43,** 175–179.

Nitecki, M. H. (1983). "Coevolution." Univ. of Chicago Press, Chicago.

Parker, M. A. (1985). Local population differentiation for compatibility in an annual legume and its host-specific fungal pathogen. *Evolution* **39,** 713–723.

Parker, M. A. and Root, R. B. (1981). Insect herbivores limit habitat distribution of a nature composite, *Machaeranthera canescens*. *Ecology* **62,** 1390–1392.

Parlevliet, J. E. (1983). Models explaining the specificity and durability of host resistance derived from the observations on the barley-*Puccinia hordei* system. *In* "Durable Resistance in Crops" (F. Lamberti, J. M. Waller, and N. A. van der Graaf, eds.), pp. 57–80. Plenum, New York.

Person, C. (1966). Genetic polymorphism in parasitic systems. *Nature* (*London*) **212,** 266–267.

Powell, N. T. (1979). Internal synergisms among organisms inducing disease. *In*

"Plant Disease" (J. G. Horsfall and E. B. Cowling, eds.), pp. 113–133. Academic Press, New York.
Prokopy, R. J. McDonald, P. T., and Wong, T. T. Y. (1984). Interpopulation variation among *Ceratitis capitata* flies in host acceptance patterns. *Entomol. Exp. Appl.* **35,** 65–69.
Rathcke, B. (1985). Slugs as generalist herbivores: Tests of three hypotheses on plant choices. *Ecology* **66,** 828–836.
Rathcke, B. J., and Poole, R. W. (1975). Coevolutionary race continues: Butterfly larval adaptation to plant trichomes. *Science* **187,** 175–176.
Rausher, M. D. (1982). Population differentiation in *Euphydryas editha* butterflies: Larval adaptation to different hosts. *Evolution* **36,** 581–590.
Rausher, M. D. (1984a). Tradeoffs in performance on different hosts: Evidence from within- and between-site variation in the beetle *Deloyala guttata*. *Evolution* **38,** 582–595.
Rausher, M. D. (1984b). The evolution of habitat preference in subdivided populations. *Evolution* **38,** 595–608.
Rausher, M. D. (1985). Variability for host preference in insect populations: Mechanistic and evolutionary models. *J. Insect Physiol.* **31,** 873–889.
Rausher, M. D. (1988). The evolution of habitat preference. III. The evolution of avoidance and adaptation. *In* "Evolution of Insect Pests: The Patterns of Variation" (K. C. Kim, ed.). Wiley, in press.
Rausher, M. D., and Feeny, P. (1980). Herbivory, plant-density, and plant reproductive success—The effect of *Battus philenor* on *Aristolochia reticulata*. *Ecology* **61,** 905–917.
Rehr, S. S., Feeny, P. P., and Janzen, D. H. (1973). Chemical defense in Central American non-ant-acacias. *J. Anim. Ecol.* **42,** 405–416.
Reznick, D. N., Perry, E., and Travis, J. (1986). Measuring the cost of reproduction: A comment on papers by Bell. *Evolution* **40,** 1338–1344.
Rhoades, D. F. (1985). Offensive-defensive interactions between herbivores and plants: Their relevance in herbivore population dynamics and ecological theory. *Am. Nat.* **125,** 205–238.
Rhoades, D. F., and Cates, R. G. (1976). Toward a general theory of plant antiherbivore chemistry. *Rec. Adv. Phytochem.* **10,** 168–213.
Rhoades, D. F., Lincoln, S. W., and Langenheim, J. H. (1976). Preliminary studies of monoterpenoid variability in *Satureja douglasii*. *Biochem. Syst. Ecol.* **4,** 5–12.
Rice, R. L., Lincoln, D. E., and Langenheim, J. H. (1978). Palatability of monoterpenoid compositional types of *Satureja douglasii* to a generalist molluscan herbivore, *Ariolimax dolichophallus*. *Biochem. Syst. Ecol.* **6,** 45–53.
Rose, H. A. (1985). The relationship between feeding specialization and host plants to aldrin epoxidase activities of midgut homogenates in larval lepidoptera. *Ecol. Entomol.* **10,** 455–467.
Rose, M. R. (1984). Genetic covariation in *Drosophila* life history: Untangling the data. *Am. Nat.* **123,** 565–569.
Roughgarden, J. (1979). "Theory of Population Genetics and Evolutionary Ecology: An Introduction." Macmillan, New York.
Roughgarden, J. (1983). The theory of coevolution. *In* "Coevolution" (D. J.

Futuyma and M. Slatkin, eds.), pp. 33–64. Sinauer, Sunderland, Massachusetts.
Rummel, J. D., and Roughgarden, J. (1983). Some differences between invasion-structured and coevolution-structured competitive communities: A preliminary theoretical analysis. *Oikos* **41,** 477–486.
Service, P. (1984). Genotypic interactions in an aphid-host plant relationship: *Uroleucon rudbeckiae* and *Rudbeckia laciniata. Oecologia (Berlin)* **61,** 271–276.
Service, P. M., and Rose, M. R. (1985). Genetic covariation among life-history components: The effect of novel environments. *Evolution* **39,** 943–945.
Singer, M. (1983). Determinants of multiple host use by a phytophagous insect population. *Evolution* **37,** 389–403.
Slatkin, M. (1983). Models of coevolution: Their use and abuse. *In* "Coevolution" (M. H. Nitecki, ed.), pp. 339–370. Univ. of Chicago Press, Chicago.
Stamp, M. E. (1984). Effect of defoliation by checkerspot caterpillars (*Euphydryas phaeton*) and sawfly larvae (*Macrophya nigra* and *Tenthredo grandis*) on their host plants (*Chelone* spp.). *Oecologia* **63,** 275–280.
Stanton, M. L., and Cook, R. E. (1984). Sources of intraspecific variation in the hostplant-seeking behavior of *Colias* butterflies. *Oecologia* **60,** 365–370.
Tabashnik, B. E. (1983). Host range evolution: The shift from native legume hosts to alfalfa by the butterfly, *Colias philodice eriphyle. Evolution* **37,** 150–162.
Tabashnik, B. E., Wheelock, H., Rainbolt, J. D., and Watt, W. B. (1981). Individual variation in oviposition preference in the butterfly, *Colias eurytheme. Oecologia (Berlin)* **50,** 225–230.
Tavormina, S. J. (1982). Sympatric genetic divergence in the leaf-mining insect *Liriomyza brassica* (Diptera: Agromyzidae). *Evolution* **36,** 523–534.
Tetenyi, P. (1970). "Infraspecific Chemical Taxa of Medicinal Plants." Chemical Publ., New York.
Thompson, J. N. (1986). Patterns in coevolution. *In* "Coevolution and Systematics" (A. R. Stone and D. L. Hawksworth, eds.), pp. 119–143. Clarendon, Oxford.
Tingey, W. M. (1984). Glycoalkaloids as pest resistance factors. *Am. Potato J.* **61,** 157–167.
Van Valen, L. M. (1973). A new evolutionary law. *Evol. Theory* **1,** 1–30.
Van Valen, L. M. (1983). How pervasive is coevolution? *In* "Coevolution" (M. H. Nitecki, ed.), pp. 1–19. Univ. of Chicago Press, Chicago.
Vernet, Ph. (1977). Genétique—les variations de composition de l'essence de *Thymus vulgaris* L.: Mode de transmission héréditaire de trois terpènes. *C.R. Acad. Sci. Paris* **284,** 1289–1292.
Vernet, Ph., Guillerm, J. L., and Gouyon, P. H. (1977a). Le polymorphisme chimique de *Thumus vulgaris* L. (Labiee) I. Repartition des formes chimiques en relation avec certains facteurs écologiques. *Ecol. Plant.* **12,** 159–179.
Vernet, Ph., Guillerm, J. L., and Gouyon, P. H. (1977b). Le polymorphisme chimique de *Thymus vulgaris* L. (Labiee) II. Carte à l'échelle 1/25000 des formes chimiques dans la région de Saint-Martin-de-Londres (Herault-France). *Ecol. Plant* **12,** 181–194.
Via, S. (1984a). The quantitative genetics of polyphagy in an insect herbivore, I.

Genotype-environment interaction in larval performance on different host plant species. *Evolution* **38,** 881–895.
Via, S. (1984b). The quantitative genetics of polyphagy in an insect herbivore, II. Genetic correlations in larval performance within and among host plants. *Evolution* **38,** 896–905.
Via, S. (1986). Genetic covariance between oviposition preference and larval performance in an insect herbivore. *Evolution* **40,** 778–785.
Wasserman, S. S. (1986). Genetic variation in adaptation to foodplants among populations of the southern cowpea weevil, *Callosobruchus maculatus:* Evolution of oviposition preference. *Entomol. Exp. Appl.* **42,** 201–212.
Wasserman, S. S., and Futuyma, D. J. (1981). Evolution of host plant utilization in laboratory populations of the southern cowpea weevil, *Callosobruchus maculatus* Fabricius (Coleoptera: Bruchidae). *Evolution* **35,** 605–617.
Weber, G. (1985). Genetic variability in host plant adaptation of the green peach aphid, *Myzus persicae. Entomol. Exp. Appl.* **38,** 49–56.
Whittaker, R. H., and Feeny, P. P. (1971). Allelochemicals: Chemical interactions between species. *Science* **171,** 757–770.
Wiebes, J. J. (1979). Co-evolution of figs and their insect pollinators. *Annu. Rev. Ecol. Syst.* **10,** 1–12.
Wright, S. (1931). Evolution in Mendelian populations. *Genetics* **16,** 97–159.
Yu, S. J. (1983). Induction of detoxifying enzymes by allelochemicals and host plants in the fall armyworm. *Pest. Biochem. Physiol.* **19,** 330–336.

CHAPTER 3

Forces Preventing Coevolution in the Three-Trophic-Level System: Willow, A Gall-Forming Herbivore, and Parasitoid

TIMOTHY P. CRAIG, PETER W. PRICE, KAREN M. CLANCY, GWENDOLYN L. WARING, AND CHRISTOPHER F. SACCHI
Department of Biological Sciences
Northern Arizona University
Flagstaff, Arizona

I. Introduction

Ever since Ehrlich and Raven (1964) emphasized the term coevolution it has been frequently invoked to explain the evolution of traits in plant–herbivore interactions, but its existence has only rarely been demonstrated. Recent reviews by Jermy (1976, 1984), Fox (1981), Futuyma (1983), and Strong *et al.* (1984) have concluded that tight coevolution between plants and herbivores may be relatively rare. The selective pressures that result in coevolution have usually been unidentified, or untested. Futuyma and Slatkin (1983) note that coevolutionary literature concentrates on adaptations that may have resulted from coevolution rather than the processes that produce coevolved traits. Ecologists need to understand processes facilitating and preventing coevolution to make predictions about its occurrence and frequency. General hypotheses need to be generated specifying conditions necessary for coevolution and tested in specific systems to advance our understanding of coevolution. Studies of forces facilitating and preventing coevolution are needed to understand why evidence for such an attractive idea is infrequently found in the field.

Coevolution can only be detected if we define acceptable evidence for coevolution. Detecting the presence or absence of coevolved traits is not easy. With enough imagination most data could be interpreted in either a coevolutionary or non-coevolutionary way. We therefore rely on the principle of parsimony: hypotheses should be tested on the simplest responses by one member of an interaction to the other. When expected coevolutionary responses are not found we can often learn more by understanding the forces preventing coevolution than by constructing increasingly complex coevolutionary scenarios. Demonstrating one organism's optimization to some degree in exploiting an interaction with another is not sufficient evidence for coevolution: most organisms have general strategies that can be applied with some success to a specific

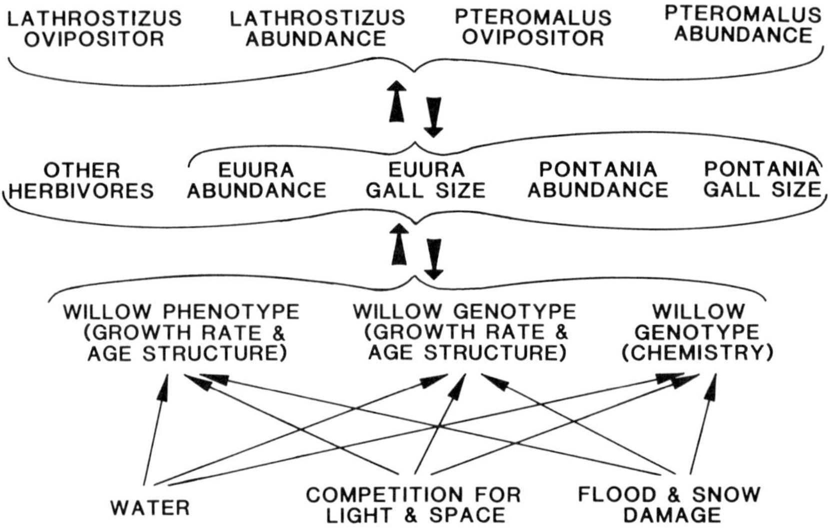

Fig. 1. Possible coevolutionary interactions in the willow community. Thick arrows between trophic levels indicate that interactions are possible between any pair of population traits. The category "other herbivores" includes a free-feeding sawfly, a leaf-folding sawfly, cicadellids, grasshoppers, tortricids, other lepidopterans, small rodents, beavers, and additional uncommon herbivores. *Pontania* sp. is a leaf-galling sawfly that occurs at high densities in some years. *Lathrostizus euurae* is an ichneumonid parasitoid that attacks both *Euura* and *Pontania*.

interaction. Conclusive evidence is demonstration of mutual adaptations of two organisms.

We examined forces facilitating and preventing coevolution in the system consisting of the arroyo willow *Salix lasiolepis* Bentham (Salicaceae), the shoot-galling sawfly *Euura lasiolepis* Smith (Tenthredinidae), and the parasitoid *Pteromalus* sp. We found no evidence of coevolution in this system. We use the definition proposed by Janzen (1980) for coevolution: "an evolutionary change in a trait of the individuals of one population in response to a trait of the individuals of a second population, followed by an evolutionary response by the second population to the change in the first."

Many possibilities exist for coevolved traits in any system; we have studied the most likely possibilities in the willow community. Like most ecological communities, the one centered on the arroyo willow is very complex (Fig. 1). Several species are involved, and each species has

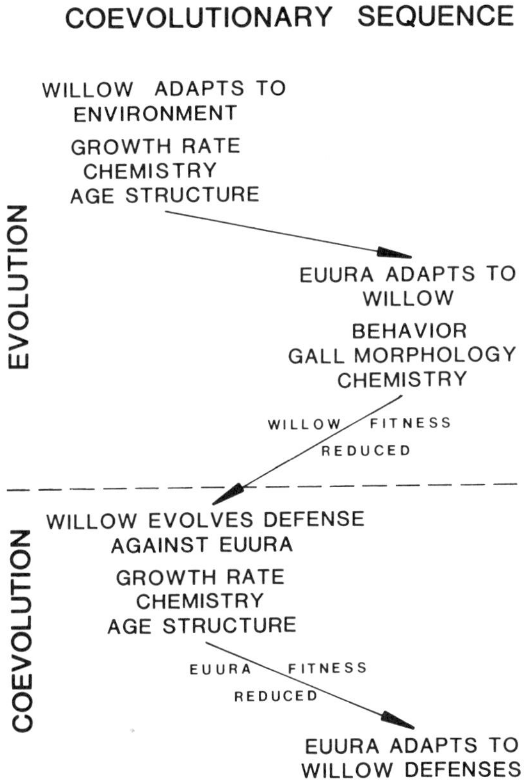

Fig. 2. Potential coevolutionary pathway in the willow system, showing the minimal requirements to demonstrate the presence of coevolved traits.

numerous traits that could coevolve with numerous traits of other species. Each species can also interact with many environmental factors.

Coevolution requires at least three evolutionary steps (Fig. 2) in plant–herbivore interactions. Plants adapt to their environment and the herbivores adapt to their host plant in every plant–herbivore system. The crucial evidence for coevolution in plant–herbivore systems is specific plant defenses against the herbivore, and specific adaptations to these defenses by the herbivore. Evidence of coevolution would be specific genetic willow defenses against *Euura*, and specific genetic adaptations to these defenses by *Euura*. The most likely willow defenses would be (1) willow toxins specifically active against *Euura* and (2) a willow growth form with shorter shoots, which specifically defends against *Euura*. Craig *et al.* (1986) have shown *Euura* form galls on the longest shoots. The willow might evolve from its shrubby growth form with long shoots, to a tree-like growth form with growth divided into a larger number of short shoots. The most likely counteradaptations by the *Euura* would be (1) sensitivity to and adaptations to specific toxins and (2) the ability to use shorter shoots. A more potent gall stimulus might be evolved so that *Euura* could induce galls on short shoots.

II. Natural History

Salix lasiolepis is adapted to a high-disturbance environment. Arroyo willows are found predominantly in steep canyons, between 1370 and 2290 m in Arizona. They are subjected to high seasonal floods from spring run-off. A characteristic willow excavated at the Museum of Northern Arizona near Flagstaff, Arizona, illustrates adaptations to floods (Fig. 3). A large portion of the willow is underground in rootstocks and underground stems, which are protected from floods that frequently damage aboveground stems. The willow has a shrublike growth form, with branches reaching a maximum age of 12–14 years. Craig *et al.* (1986) termed the number of years between the rootstock and the shoot, the shoot "branch age." Lower mean "branch age" of a clone was highly correlated with long shoot lengths. When the distal portion of a branch is damaged, indefinitely dormant buds sprout from near the base, lowering the mean "branch age" of the clone and producing young fast-growing shoots. Rapid regrowth is an advantage in competing for space and light following a flood.

The life history of *Euura lasiolepis* was described in Price and Craig (1984). The sawfly emerges from the gall in the spring and attacks young fast-growing shoots. A gall is formed on the shoot, and overwintering occurs in the gall, with emergence the following spring leaving an exit

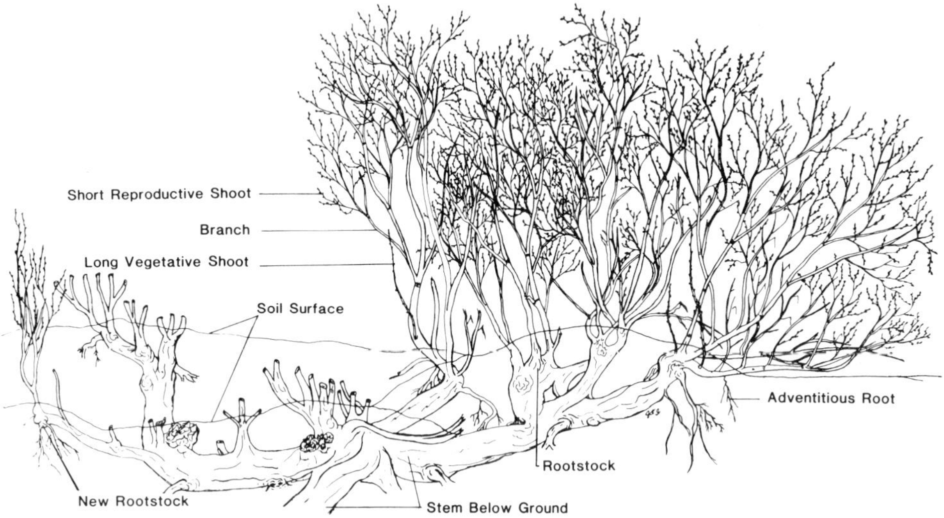

Fig. 3. Arroyo willow, *Salix lasiolepis,* excavated on Schultz Creek on the property of the Museum of Northern Arizona. Note that the term shoot refers to current year's growth, and branch refers to accumulated growth on one stem over several years.

hole. The gall can stunt or kill the shoot either by disrupting vascular tissue or by allowing pathogens into plant tissues.

Euura is parasitized by *Pteromalus,* an ectoparasitoid that oviposits through the gall wall onto larvae in July and August. Preliminary information on *Lathrostizus,* another parasitoid (cf. Fig. 1), shows no indication of coevolution with the two lower trophic levels and will not be discussed further.

III. Results

A. Factors Favoring Coevolution

1. Willows and Sawflies Have the Potential for Strong Mutual Reduction in Individual Fitness

Individual willows can have fitness significantly reduced by the sawfly. Sawfly attack can control willow growth, making it susceptible to further attack, termed "resource regulation" (Craig *et al.,* 1986), which occurs in the following way. Sawflies form more galls on rapidly growing shoots, which are more frequent on plants with young "branch ages." Galls frequently stunt or kill shoots, leading to sprouting of indefinitely dormant buds near the base of the willow, which lowers the mean branch age of the willow. A heavily attacked clone is kept in a juvenile condition, with fast-growing shoots providing good resources for the sawfly; we term this the juvenilization cycle. Sexual reproduction is reduced by galling in the long term by "resource regulation" and in the short term by reduction in reproductive bud number on galled shoots (Craig *et al.*, 1986; Sacchi *et al.*, 1988). *Euura* decreases willow fitness by reducing both vegetative and sexual reproduction.

Individual sawflies can have their fitness significantly reduced by willow characteristics. *Euura* are dependent on the willow during their entire life cycle. Price and Craig (1984) found that about 40% of a sawfly cohort died due to interactions with the willow, and survival of cohorts ranged from 0.5% to 80% in different years on different clones (Price and Clancy, 1986). However, we will demonstrate that willow characters relating to sawfly survival relate primarily to willow phenotype, not genotype.

2. The Willow and Sawfly Have a Highly Specific Interaction; Each is the Most Important Species the Other Interacts With

The willow and the sawfly have a highly specific interaction. *Euura lasiolepis* occurs only on *Salix lasiolepis,* and *Euura* has the largest impact of any common herbivore observed on the willow. We have found no evidence of other common insect herbivores on the willow that cause extensive death of shoots and branches. Mammalian herbivory has been observed, but it is extremely rare and patchy.

The potential for high reciprocal impact and high specificity are frequently assumed to be sufficient to produce coevolution, but we have not detected coevolved traits in the willow system.

B. Factors Preventing Coevolution

1. Coevolutionary Forces are Weak Relative to Other Selective Forces Acting on Willow and Sawfly

Noncoevolutionary forces compromise selection for coevolved *Euura* and *Salix* traits. Disturbance and *Euura* exert opposing selective forces on willow shoot length. High disturbance favors fast-growing shoots because rapidly growing shade-intolerant plants like the willow must quickly regrow following above-ground destruction to compete for light and space (Bryant *et al.*, 1983). A potential willow defense against *Euura* would be evolution of shorter shoots, since rapidly growing shoots are highly vulnerable to sawfly damage (Craig *et al.*, 1986). We measured the relative impact of disturbance and *Euura* attack, by counting branch mortality on willow clones from each source. Two sites were sampled, one known to have a high sawfly density (Schultz Creek) and one with a low sawfly density (Oak Creek). Disturbance caused significantly more branch mor-

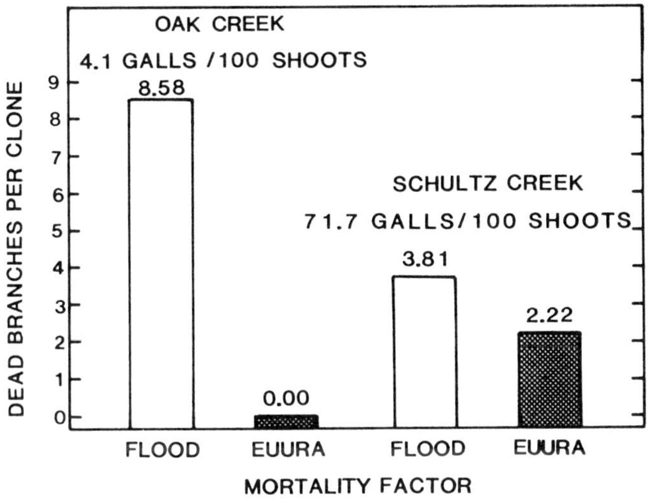

Fig. 4. Branch mortality from floods and snow (floods), and from *Euura lasiolepis*. Disturbance mortality was significantly higher than *Euura* mortality at both Oak Creek (n = 50 clones), and at Schultz Creek (paired t-test, t = 2.84, n = 50 clones, $p < 0.01$).

tality at both sites (Fig. 4). Sawflies killed branches at Schultz Creek, but branch death from physical factors was still the more important selective force. Because rapid shoot growth is adaptive in response to floods and snow breakage, shorter shoots cannot evolve as a defense against *Euura*. The sawfly population does not have a sufficient impact to counter selection on growth form by physical factors. The autecology of the plant is more important than its synecology with insects.

2. Physical Factors Produce Spatial and Temporal Variation in the Interaction between the Sawfly and the Willow that Is Not Genetically Determined

Water availability strongly influences *Euura* population density by determining resource availability and sawfly survival. Drought is typical in northern Arizona during sawfly oviposition and development of early instars (May–June), so water for willow shoot growth and sawfly survival depends on winter precipitation. Death of eggs and early instars is the largest source of mortality in the sawfly life cycle (Price and Craig, 1984). Price and Clancy (1986) show that years with low winter precipitation have high rates of sawfly egg death. They found willow resources for *Euura* decline in drought years when lack of water reduces shoot growth. Experiments on potted willows demonstrated that high water treatments had higher rates of oviposition and survival.

Precipitation patterns over several years determine sawfly population densities at most sites. High gall densities were maintained at sites with permanent water sources, while along intermittent streams gall densities were high only 1 out of 5 years studied (Fig. 5). Water availability at oviposition time influences gall densities in that year by determining shoot growth and in following years through impact on *Euura* survival. A severe drought in the winter of 1980/1981 (28 cm winter precipitation, October

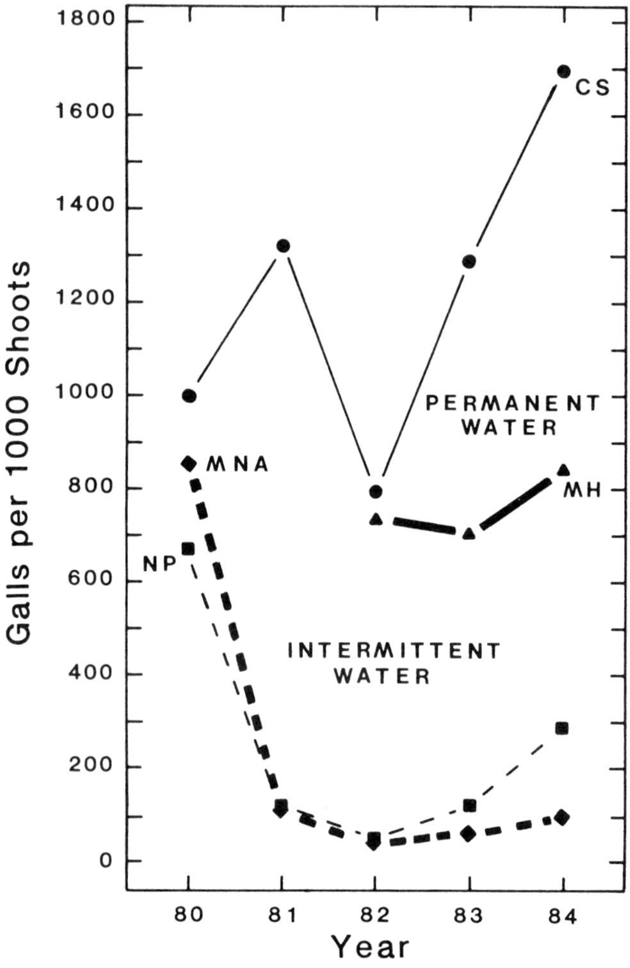

Fig. 5. Mean gall density, 1980–1984, at four sites on the Schultz Creek drainage. Two sites had constant water: CS = Coyote Spring (clones CS 1–2) and MH = McMillan House Spring (MH 1–10). Two sites had intermittent water on Schultz Creek: NP = Northland Press (NP 4,5,7,8,9) and MNA = Museum of Northern Arizona (MNA 1–7).

through May) led to a sharp decline in gall densities on intermittent stream sites in 1981. Gall densities were slow to recover during 2 years with wet winters. A second dry winter in 1983/1984 (18 cm winter precipitation) again drastically reduced sawfly survival, which will again depress populations.

Strong selection on the willow for defenses against *Euura* occurs rarely and unpredictably in sites with intermittent water. A series of years with high winter precipitation may be necessary for gall densities to reach the level of the 1980 generation. Using data from Price and Clancy (1986) we calculate that winter precipitation must be at least 35 cm to prevent population decline in dry sites. The frequency of sawfly generations likely to reach the 1980 level was calculated using records from the Flagstaff Weather Bureau. Three years with winter precipitation above 35 cm preceded the large 1980 generation, while 2 years with winter precipitation above 35 cm produced only a slight population increase in 1984. We conclude that 3 years of above 35 cm precipitation are necessary for high gall densities in dry sites. Three consecutive years of above 35 cm precipitation occurred only five times in 36 years (1947–1984), and did not occur at all during a 20-year period (1947–1967). Strong selection by sawflies is very intermittent, and it does not cause death of genets, only ramets. Weakened genet reproduction output may be important, but propagation by seed is also extremely unpredictable (Sacchi *et al.*, 1988), further weakening *Euura*'s selective impact.

Strong selection on the willow for defenses against *Euura* also operates only in restricted areas in most years. Willow clones were divided into three categories based on *Euura* impact (Fig. 6): mean gall densities above 1 gall per shoot are necessary to induce extensive shoot death and pro-

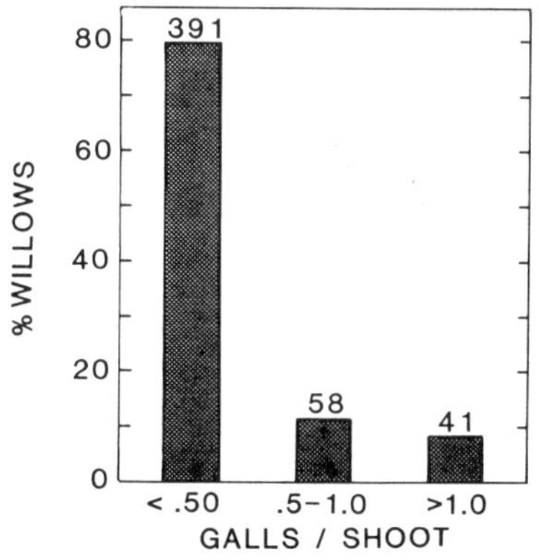

Fig. 6. Frequency of 1984 gall densities on 490 clones in the Schultz Creek drainage between Schultz Pass and Highway 180, near Flagstaff, Arizona. Numbers above the bars are the number of willows in each class.

duce "resource regulation" (Craig *et al.*, 1986), gall densities between 0.5 and 1.0 galls per shoot influence sexual reproduction (Sacchi *et al.*, 1988), and mean gall densities below 0.5 galls per shoot have little detectable impact. Gall densities that have a large impact on willow growth and reproduction are relatively rare. High-gall-density willows are patchily distributed (Fig. 7). The large majority of willows in our study area are in sites with intermittent water, and high gall densities were found only in areas that retain water through the critical May–June period.

Experiments showed that willow phenotype is much more important than the genotype in determining susceptibility to *Euura* attack. Gall densities on plants grown in pots under identical conditions showed no significant relationship to densities on wild clones (Fig. 8). Experimental plants were started from cuttings, which gave them a young "branch age" structure; they were fertilized frequently and watered daily (high water/high fertilizer treatment). Under these conditions all plants had vigorous juvenile growth and high gall densities, regardless of genotype. Waring and Price (unpublished data) looked at the relationship between gall density on the experimental plants and numerous variables including clone, treatment (water and fertilizer level), phenols, proteins, and shoot growth. Multivariate regression analysis showed shoot length was the only significant and independent variable. Shoot growth explained 67% to 91% of the variation in gall density in different treatments (Fig. 9).

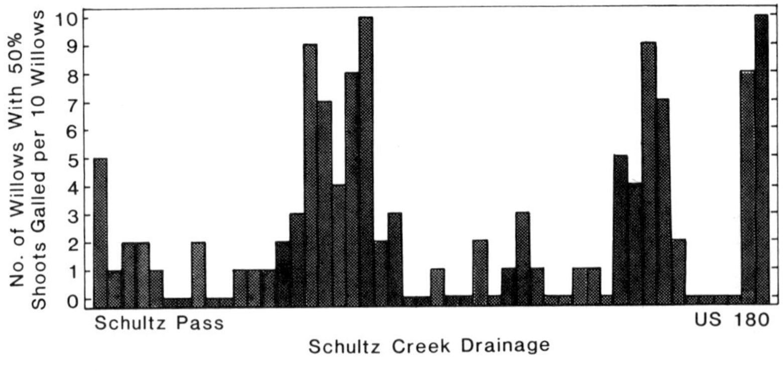

Fig. 7. Distribution of willow clones with high gall densities (greater than 50% of shoots galled) in the Schultz Creek Drainage between Schultz Pass and Highway 180, near Flagstaff Arizona. Bars represent the number of clones with high gall densities out of each group of 10 clones sampled sequentially down the drainage. Note that in only 10 groups of 10 plants out of 48 such groupings did gall densities reach 0.5 galls/shoot on average. The distribution was tested against a Poisson distribution with a chi-square goodness of fit, and was nonrandom (χ^2 = 33.87, df = 5). A variance to mean ratio of 1.286 indicated the distribution was clumped.

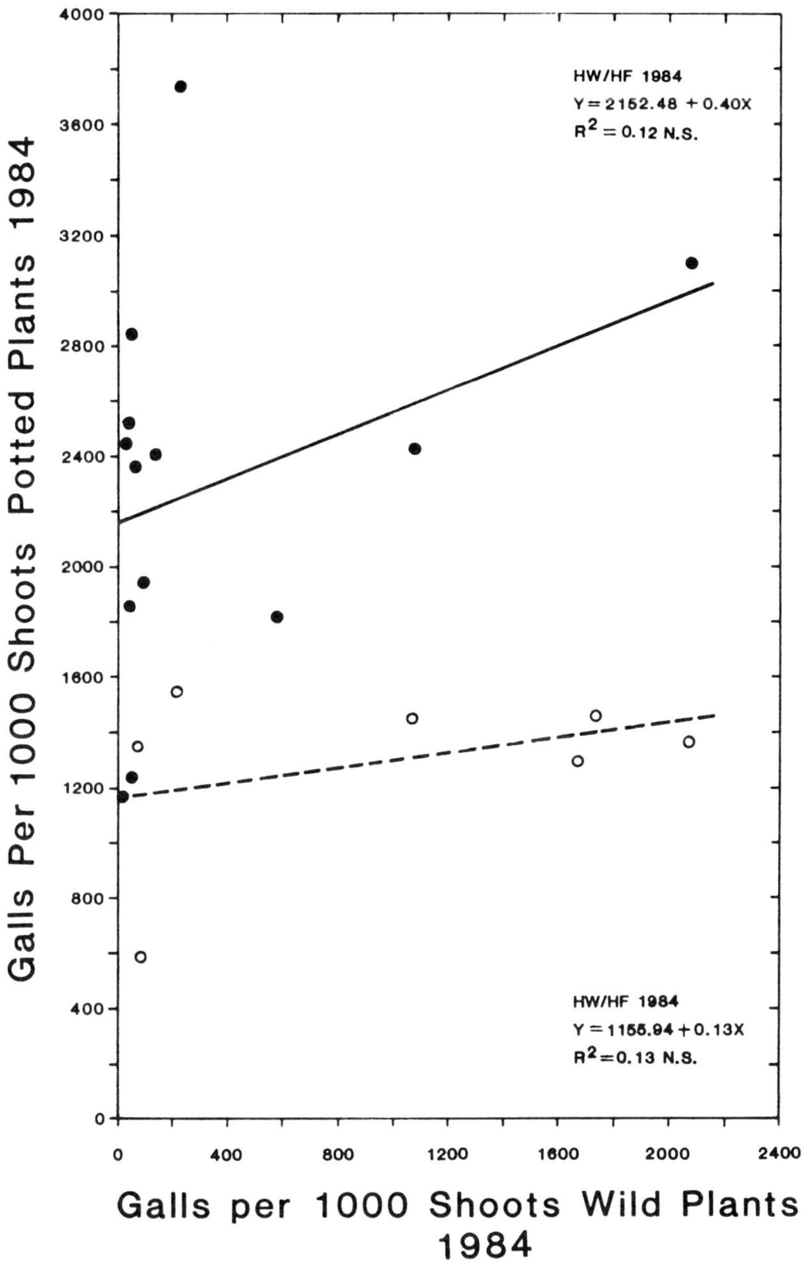

Fig. 8. Regression of gall densities on potted willows given high water (HW) and high fertilizer (HF) on gall densities of source clones in the field. Solid circles are for 3-year-old plants representing 13 clones in pots with 4 plants per pot; open circles are for 2-year-old plants representing 7 clones in pots with 1 plant per pot. All pots were placed on platforms in the canopy of high-gall-density clones (CS 1–2).

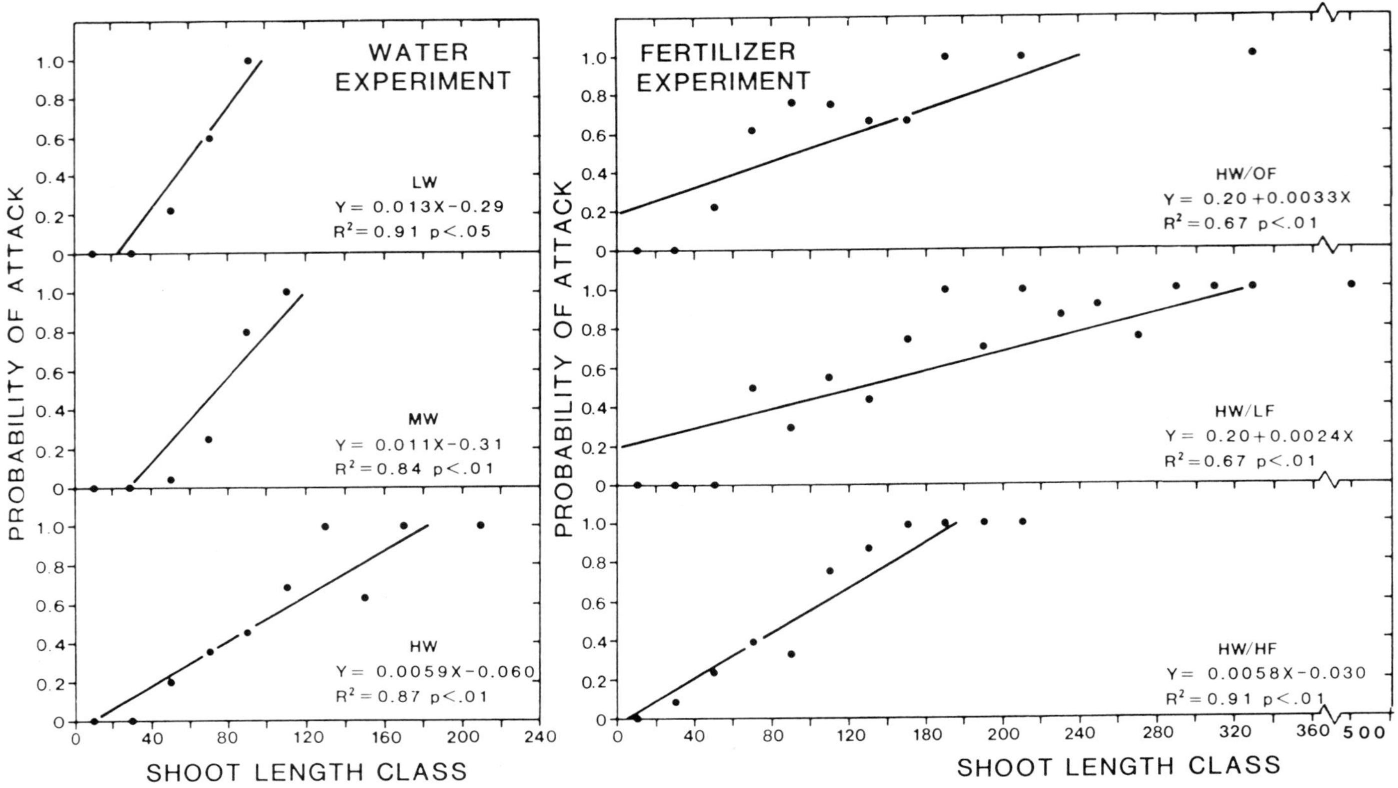

Fig. 9. Regressions of probability of attack on shoot length classes in water and fertilizer experiments on potted plants. HW = high water, watered once a day. MW = medium water, watered once every 7 days. LW = low water, watered once every 14 days. No fertilizer was applied in the water experiment. OF = zero fertilizer, not fertilized. LF = low fertilizer. HF = high fertilizer. Plants were fertilized with Peters Professional All Purpose fertilizer (PNK 20–20–20).

Euura are not choosing oviposition sites primarily on the basis of genetically determined willow characters. In the field galls are formed most frequently on fast-growing shoots, which are found on clones with low mean branch ages in sites with high water availability. Any vigorously growing shoot is susceptible to damaging levels of attack. Genetic variation for shoot length and therefore galling susceptibility does exist; however, the genetic component is very small. For example, all the experimental clones (Fig. 7) had severe sawfly damage, and would probably undergo "resource regulation." While a very small proportion of clones seem to have genetically based resistance to attack, all clones can become phenotypically resistant if shoot growth is reduced.

3. The Sawfly Modifies the Willow Resources, Making the Willow Genotype Irrelevant to Sawfly Survival

Euura galls contain better resources than unmodified plant tissue. Phenols and proteins were measured in galled tissue and in ungalled tissue (Waring and Price, unpublished data). Proteins, which are probably important in sawfly nutrition, were higher in galled tissue than in ungalled tissue (Table I). Phenol glycosides are a logical group to examine for chemical defenses because they are the most widespread group of secondary compounds in the Salicaceae, and they function as both attractants and deterrents to herbivores (Rowell-Rahier, 1984; Tahvanainen *et al.*, 1985). Phenols were lower in galled tissue than in ungalled tissue (Table I). The sawfly exists in an improved environment regardless of the willow genotype for phenols and proteins. We term this "resource manipulation," which we define as active improvement of an herbivore's environment initiated by the herbivore. Host-plant manipulation by gallers has been discussed by Weis and Kapelinski (1984) and Weis and Abrahamson (1986). Our studies, to date, indicate that plant chemistry is unimportant in the *Euura–Salix* interaction in terms of plant defense.

Sawflies have evolved the ability to recognize the willow and to choose suitable oviposition sites probably on the basis of plant chemistry. When

Table I. Comparison of Phenols and Proteins in Galled and Ungalled Tissues

	Ungalled Tissue, Mean (± SE)	Galled Tissue, Mean (± SE)	Probability
Total phenol[a]	0.70 (± 0.02)	0.14 (± 0.02)	0.0000
	$n = 26$	$n = 23$	df(1, 47)
Total protein[b]	11.72 (± 0.67)	13.57 (± 0.44)	0.0430
	$n = 26$	$n = 23$	df(1, 47)

[a] Mean total phenol concentration in optical density units, (OD_{725})(absorbance/mg).
[b] Mean percent total protein/mg.

given a choice between rapidly growing shoots and more slowly growing shoots, sawflies oviposit in the rapidly growing shoots even when deprived of potential physical cues in a laboratory study (Craig, unpublished). Preliminary results lead us to hypothesize that sawflies may use phenol glycoside concentrations as cues in making these decisions.

Sawflies may be responding in two ways to phenol glycosides: using them as oviposition cues, and then modifying them to avoid their defensive effects. Many plants contain chemicals that are defenses against generalists but are feeding or oviposition stimulants for specialists (Futuyma, 1983). Phenol glycosides are found widely in willows (Rowell-Rahier, 1984) and are not a specific defense against *E. lasiolepis*. The sawfly, a specialist, may have adapted to phenol glycosides which were evolved as defenses against generalists. "Resource manipulation" and the use of oviposition cues represent an evolutionary response by *Euura*, not coevolution between the willow and sawfly. Coevolution could be demonstrated only if the sawfly were responding to phenol glycosides that were specific willow defenses evolved against *Euura*. Alternatively, it could be argued that there has not been sufficient time since the sawfly radiated onto the willow for the willow to respond. However, both willows and sawflies are ancient groups, willows dating from at least the Eocene (Cronquist, 1981) and sawflies from the Jurassic (Strong *et al.*,

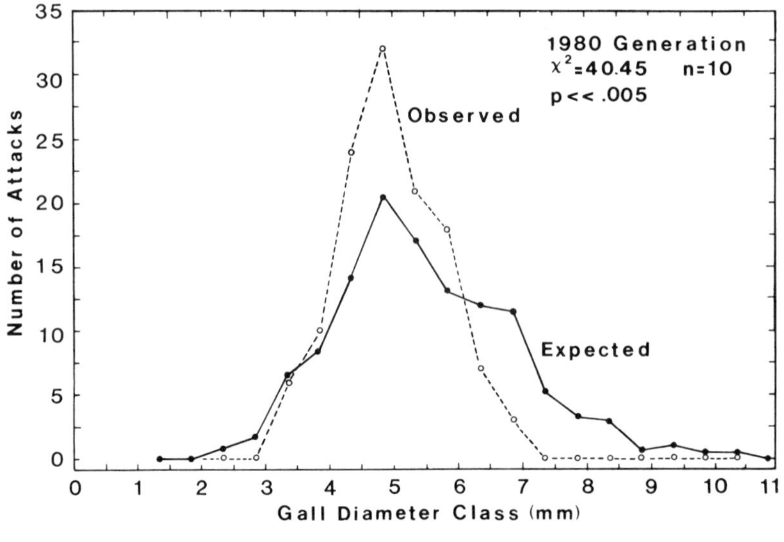

Fig. 10. Number of attacks by *Pteromalus* on *Euura* galls. Dashed line is the distribution of parasitoid attack. Solid line is the expected distribution of attacks if attack is random. Sample consisted of subsamples on clones MNA 1–4 and NP 8–9, 800 galls total, from which 121 *Pteromalus* emerged. Note that attacks are much lower than expected on the larger size class galls.

1984). Although we have no specific information on the age of the interaction we have studied, it seems highly probable that there has been sufficient time for coevolution.

The system certainly has the potential for coevolution: the first two steps of the coevolutionary sequence have been accomplished (Fig. 2). The willow could respond by evolving novel defensive compounds that could not be altered through "resource manipulation." Since we do not understand the mechanism behind "resource manipulation," we cannot speculate on what pathways are open to the plant. The willow could also manipulate chemical cues used by the sawfly to choose oviposition sites. Craig *et al.* (1986) and Price and Clancy (1986) have shown that sawflies choose high-quality sites for oviposition; if the willow were able to alter the ability of the sawfly to recognize high-quality sites, it could lead to increased sawfly mortality and avoid losing expensive rapidly growing shoots.

The sawfly has adapted to the plant genotype in at least two ways: it probably selects oviposition sites based on a chemical cue, and it has adapted to modify plant defenses and nutrition. We have no evidence that these adaptations have brought reciprocal changes from the willow.

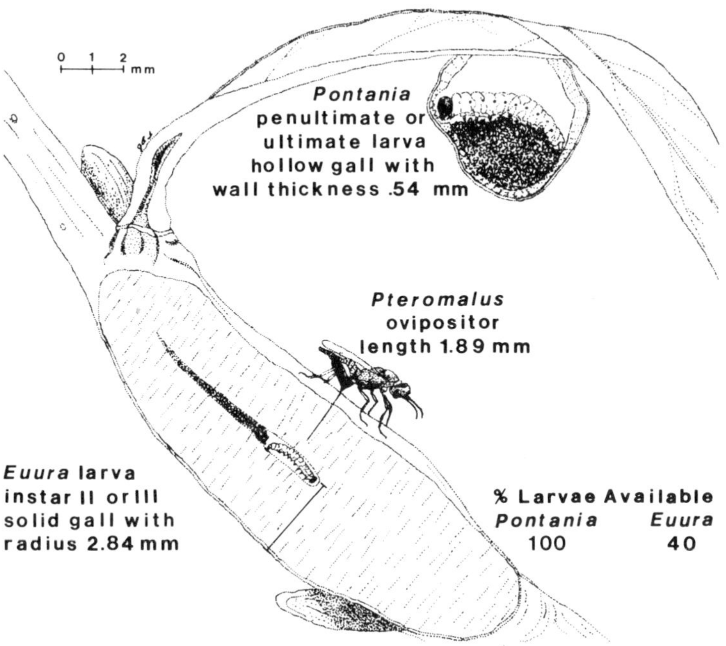

Fig. 11. Comparison of gall diameters and *Pteromalus* ovipositor length. Estimated percent of larvae of each species available to *Pteromalus* is based on ovipositor length, and distance to larva centered in mid gall.

Table II. Comparison of *Pteromalus* Densities on *Euura lasiolepis* and *Pontania* sp. on Clones MNA 1–7 and NP 1–9

	1981		1982		1983	
	D[a]	P	D	P	D	P
On *Euura*	4.00	0.05	3.00	0.06	8.50	0.09
On *Pontania*	46.20	0.05	64.43	0.04	24.08	0.07

[a] D, number per thousand willow shoots. Densities were calculated by multiplying the proportion of galls parasitized (P) by *Pteromalus* by gall density per 1000 shoots.

4. The Parasitoid *Pteromalus* Has Stronger Relationships with Other Community Members and Has Not Coevolved with *Euura*

Pteromalus has not evolved to efficiently exploit *Euura*. Large *Euura* galls are not parasitized at the rate expected if attack were random (Fig. 10). The parasitoid attacks both *Euura* galls and the leaf galls of the closely related sawfly *Pontania* (near *pacifica;* see Clancy *et al.*, 1986). The leaf gall is hollow and has a thin wall; thus, all larvae can be easily reached by the *Pteromalus* ovipositor (Fig. 11). The stem gall is solid and the parasitoid ovipositor cannot reach most *Euura* larvae.

We hypothesize that *Pteromalus* has evolved to exploit the more prevalent leaf galler. *Pontania* is attacked more frequently than *Euura* and *Pontania* is more abundant than *Euura* (Table II). The parasitoid opportunistically attacks small-diameter stem galls. If the parasitoid and *Euura* had coevolved, a slightly longer *Pteromalus* ovipositor would be expected, making all *Euura* available for attack. Ovipositor evolution to efficiently attack the stem gall might decrease its fitness on the more common host. An evolutionary change in ovipositor length alone may be impossible. The overall body size might change also, so that adaptation to hosts with thicker galls would mean larger body size. Larger body size necessarily requires larger hosts, which would be found in larger galls; therefore this avenue for adaptive response may have severe evolutionary constraints. The expected coevolutionary response of *Euura* would be evolution of a slightly larger gall diameter protecting all larvae from the parasitoid. Low levels of selection by *Pteromalus* are probably insufficient to balance other selective forces and physiological constraints that influence this character.

IV. Discussion

We have identified specific forces preventing coevolution in the willow–sawfly–parasitoid system. We believe that these are specific examples of forces widely operating in plant–herbivore interactions. Several reviews have suggested coevolution is possible only under very restrictive conditions (Fox, 1981; Thompson, 1982; Strong *et al.,* 1984). We have attempted to synthesize information from the literature to propose a list of the minimum requirements for coevolution. We propose that five qualities must be present in interactions for coevolution to occur.

1. *Strong interaction.* Each species must have the potential to significantly affect the fitness of individuals of the other species, so that coevolved adaptations can be detected as different from the background of other adaptations, and as divergent from phylogenetically closely related species (Ehrlich and Raven, 1964; Fox, 1981; Thompson, 1982; Schemske, 1983; Strong *et al.,* 1984). To coevolve, each species must exert strong selection on the other, or selection for coevolved traits will be swamped by the myriad of other selective forces.

   In the *Salix–Euura* interaction each species can have a large impact on the other's fitness, but this rarely happens. The willow adapts to other, stronger, impacts; the sawfly has relatively low impact. Most willow traits that affect sawfly survival are related to environmentally induced or age-related plant phenotype. The sawfly reduces its interaction with the willow genotype through "resource manipulation." There is little selection on the genotype of one species by the other.
2. *High specificity.* Each species must tightly interact with the other species, so that mutually induced adaptations are not compromised by other selective pressures, making coevolved traits weakly developed and difficult to detect (Fox, 1981; Futuyma, 1983; Feinsinger, 1983; Gilbert, 1983; Schemske, 1983; Strong *et al.,* 1984). Specifically, coevolved traits are unlikely to result if each species in an interaction is evolving in response to numerous other species. In a plant–herbivore interaction coevolution is most likely if the herbivore feeds primarily on one plant, and herbivory on the plant is primarily due to that herbivore.

   Euura is the most important herbivore, but not the most important selective force on the willow. Environmental factors are stronger selective forces. The willow phenotype, not genotype, is a strong selective force on *Euura.*
3. *Widespread.* The interaction must involve a large proportion of each population, because if the interaction involves only a small segment of each population, gene flow will swamp the evolving trait and stabilize the trait at the population norm (Feinsinger, 1983; Strong *et al.,* 1984). Gene flow from noninteractive areas would counteract selection for coevolution, even if there were no cost for a coevolving character.

   Intense sawfly–willow interaction is not widespread; the conditions favorable for sawfly survival are patchy. Willows undergoing "resource regulation" should be selected to evolve defenses. Traits for defense against *Euura,* advantageous in locations with high gall densities, would confer no advantage where most willows grow. If defense against *Euura* is costly, sawfly resistance would lower willow fitness in most environments.
4. *Predictable.* The interaction must be predictable in frequency and intensity to coevolve, for only then is selection likely to be strong enough to yield

detectable traits, and requirements 1 and 2 will be fulfilled (Schemske, 1983; Strong *et al.,* 1984). If selection for coevolution is unpredictable, other selective pressures and genetic drift will produce non-coevolved traits in the intervening spaces and times.

The *Salix–Euura* interaction is highly unpredictable in time, making coevolution of traits unlikely. Willow susceptibility is determined by stochastic environmental factors such as precipitation. The majority of willows are under inconsistent selection for defenses against *Euura*. Rarely, strong selection may operate on the majority of willows for defenses against *Euura* for several years. An unpredictable number of years may follow with very low selection for sawfly resistance, during which other forces act on traits such as shoot length.

5. *Dominant selective force.* Selection for coevolution must not be compromised by other stronger selective forces or coevolutionary traits will remain inconspicuous, or absent, and the requirements of quality 2 will not be met. Low consistent levels of selection can result in coevolved traits. But many traits have numerous selective forces acting on them, and strong selective pressures opposing coevolution from outside the interaction will prevent coevolution (Strong *et al.,* 1984).

   Uncompromised selective forces for coevolution have not been found in the willow system. Disturbance selects for rapid shoot growth directly opposing sawfly selection for shorter shoots. *Euura* galls do not exert an uncompromised selection on *Pteromalus* ovipositor length. *Pteromalus* has evolved with an ovipositor adapted to efficiently exploit the more prevalent leaf galler.

The willow–sawfly–parasitoid system has not coevolved because its interactions lack qualities needed to coevolve. The relative importance of each quality cannot be quantified without experiments designed to isolate the quality. We predict that the absence of even one interaction quality would prevent coevolution. Several forces interact to prevent coevolution in the willow system.

We propose that coevolution is rare because most interactions lack the minimum qualities necessary for species to coevolve. Plant–herbivore interactions are typically complex and contain many forces preventing coevolution. Rarely is a system so simple that it lacks such forces and has the qualities to coevolve.

We have tried to assess whether our list of qualities necessary for coevolution predicts its existence in a limited number of other intensely studied systems. Making comparisons is difficult because we have not set quantitative limits on each of the qualities and because the research we reviewed was not designed to answer the questions we raise.

Obligate mutualisms such as ants and acacias (Janzen, 1966), figs and fig wasps (Janzen, 1979; Weibes, 1979), and yucca moth and yucca (Powell and Mackie, 1966; Aker and Udovic, 1981) have some of the most widely accepted examples of coevolved traits. Such mutualisms create intense selection for coevolution because obligate relations are highly specific, have a high mutual impact on fitness, are widespread and predictable, and because selection to maintain the interaction will outweigh any force opposing coevolution.

In contrast to the obligate mutualisms, other mutualisms, such as most plant–pollinator interactions, lack evidence of coevolution (Feinsinger, 1983). Feinsinger argues that all three systems he reviews in detail, orchids–euglossines, bumblebees–flowers, and tropical hummingbirds–flowers, are not tightly coevolved. Many of the examples lack most or all of the qualities we propose. Most pollinator–flower interactions lack specificity; pollinators use many different flowers, and most flowers have many different pollinators. Because of this low specificity, the reciprocal impact on fitness of each interacting species is low. Many plants' pollinator complexes vary over their range and through time. Likewise, most pollinators shift their flower species use over time and space, resulting in interactions that are not widespread or predictable. Low specificity means that many different selective forces act on each interacting species and selection for coevolution is compromised.

Gene-for-gene interactions between hosts and parasites documented in flax and flax rust reviewed by Flor (1971) and in hessian fly and wheat (Hatchett and Gallun, 1970) provide strong evidence for coevolution. So far, gene-for-gene relationships have been found only in agricultural systems (Day, 1974; Barrett, 1983) that are artificially simplified, increasing the intensity of all the coevolutionary forces we list. Crop plants frequently experience "resistance breakdown" (Barrett, 1983), when a pathogen produces a new mutation for virulence indicating a new coevolutionary step has occurred. Agronomists can stop pest outbreaks by introducing new genes for resistance into the plant. The gene-for-gene relationship demonstrates that these associations are extremely specific. A high mutual impact on fitness is indicated by the widespread reduction in yield of crops during outbreaks, and the great reduction in parasite density when a new resistant variety is produced. Epidemics occur when genetically uniform crops are grown over wide areas for several years, making the interaction of host and parasite widespread and predictable. Agricultural systems eliminate many of the other selective forces that could oppose selection for coevolution found in natural systems.

As we pointed out in the introduction, evidence for plant–herbivore coevolution in natural systems is not conclusive. These systems are generally much more complex, involving many more organisms and selective forces than the obligate mutualisms and gene-for-gene interactions. It is more difficult to gather the necessary ecological information to evaluate coevolutionary forces in these systems.

Some studies have documented relationships between traits such as chemistry and herbivory but have not closely examined ecological factors. For example, Berenbaum's studies of the association of insect herbivores and host plants in the Umbelliferae (1979, 1981, 1983, this volume) have shown that plant chemistry corresponds to insect specialization in a way that closely fits the Erhlich and Raven coevolutionary scenario (Strong *et al.*, 1984). She has documented highly specific interactions between plants and herbivores and a mutual impact on fitness. It would be interesting to know how widespread and predictable these interactions are and what other important selective forces act on this system.

Butterfly–host plant interactions are particularly interesting to examine because the coevolutionary scenario was first outlined for these systems (Ehrlich and Raven, 1964) and because extensive data are available. Butterflies of the genus *Euphydryas* have been proposed as coevolving with their host plants, although specific coevolutionary sequences have not been proposed. In general this group does not have the qualities we propose for coevolved systems. There is not a strong indication of a mutual reduction in fitness. A complex set of ecological factors determines oviposition choice and larval survival, not host-plant traits alone (Singer, 1971, 1972, 1983; Ehrlich *et al.*, 1975; Williams, 1983). In the cases studied so far, larvae frequently do as well on host as nonhost plants and there is no evidence that plant choice has evolved in response to plant nutrition and biochemical differences (Holdren and Ehrlich, 1982; Williams *et al.*, 1983; Bowers, 1986). Butterfly host-plant preference differs between sites; a preferred host plant at one site may not be preferred at another site (Singer, 1971, 1983; Bowers, 1986). The interactions are therefore not widespread. Populations of some species fluctuate widely and even go locally extinct (Ehrlich *et al.*, 1975, 1980), making the frequency and intensity of the interaction unpredictable. Relatively little attention has been paid to the impact on plant fitness and on other selective forces in the system. There is little indication of highly specific chemical coevolutionary interactions (Holden and Ehrlich, 1982; Bowers, 1986), and the results suggest a very diffuse coevolution (as defined by Fox, 1981), rather than the more restrictive definition of coevolution we are using in this chapter.

In contrast to the *Euphydryas* studies, a sequence of coevolutionary diversification has been proposed for *Heliconius* butterfly herbivory on *Passiflora* vines (Benson *et al.*, 1976; Benson, 1978), and specific coevolved characters of butterflies and host plants have been identified (Gilbert, 1971, 1975; Williams and Gilbert, 1981). There is evidence that these associations are highly specific (Gilbert, 1975; Benson *et al.*, 1976; Benson, 1978; Smiley, 1985), that there is a potential for mutual impact on fitness (Gilbert, 1971, 1975; Williams and Gilbert, 1981; Smiley, 1985), and that the interactions are widespread and predictable (Gilbert, 1975; Benson, 1978; Smiley, 1985). Selective forces that might oppose coevolution have received limited attention. It is interesting to note again that host-plant specialization is not solely chemically based, and other ecological factors are important in determining oviposition preference (Smiley, 1978, 1985). Quantification of all forces promoting coevolution in this system would be valuable.

The qualities we propose as necessary for coevolution are only a first step in understanding coevolutionary processes. Qualities preventing or leading to coevolution need to be examined in many systems, and the list of qualities enlarged or narrowed. We believe that much can be learned about the mechanism of coevolution by studying ecological forces in both kinds of systems where evidence for coevolution is present and absent. Studies should quantify the forces preventing or leading to coevolution, and answers to the following questions would clarify coevolutionary processes. How many species can interact and coevolve? What mutual change in fitness is necessary for species to coevolve? What proportion of each population must interact for them to coevolve? How frequent and predictable must interactions be for species to coevolve? And, how important are opposing selective forces in preventing coevolution?

If it is discovered that ecological forces commonly prevent coevolution of specific defenses, then ecologists will have to reexamine the view of plant–herbivore interactions developed over the last 20 years. Forces preventing coevolution may facilitate evolution in other directions. "Resource regulation" and "resource manipulation" mechanisms by which *Euura lasiolepis* increases exploitation of the willow are only possible in the absence of coevolved willow defenses. Factors that prevent *Euura* populations from exerting strong selective forces on the willow allow selection of traits in the sawfly that increase their impact on individual willows. Attention paid to the possibility of coevolved plant defenses may have obscured understanding of important herbivore strategies for exploiting plants in the absence of defenses, or when defenses are severely curtailed.

Acknowledgments

Financial support was provided through grants from the National Science Foundation (DEB-8021754 and BSR-8314594), and a Sigma Xi grant to Timothy P. Craig. Joanne K. Itami assisted in data collection, prepared most of the figures, and commented on the manuscript. We thank the Museum of Northern Arizona and its director Phillip Thompson for allowing us to set up study sites on their property. Comments from two anonymous reviewers, Christopher C. Smith, and Kevin Spencer improved the paper.

References

Aker, C. L., and Udovic, D. (1981). Oviposition and pollination behavior of the yucca moth, *Tegeticula maculata* (Lepidoptera: Prodoxidae), and its relation to the reproductive biology of *Yucca whipplei* (Agavaceae). *Oecologia* **49,** 96–101.

Barret, J. A. (1983). Plant-fungus symbiosis. *In* "Coevolution" (D. J. Futuyma and M. Slatkin, eds.), pp. 1–13. Sinauer, Sunderland, Massachusetts.

Benson, W. W. (1978). Resource partitioning in passion vine butterflies. *Evolution* **32,** 493–518.

Benson, W. W., Brown, K. S., Jr., and Gilbert, L. E. (1976). Coevolution of plants and herbivores: Passion flower butterflies. *Evolution* **29,** 659–680.

Berenbaum, M. (1979). Toxicity of furanocoumarin to armyworms: A case of biosynthetic escape from insect herbivores. *Science* **201,** 532–534.

Berenbaum M. (1981). Patterns of furanocoumarin distribution and insect herbivory in the umbelliferae: Plant chemistry and community structure. *Ecology* **62,** 1254–1266.

Berenbaum M. (1983). Coumarins and caterpillars: A case for coevolution. *Evolution* **37,** 163–169.

Bowers, M. D. (1986). Population differences in larval host plant use in the checkerspot butterfly, *Euphydryas chalcedona*. *Entomol. Exp. Appl.* **40,** 61–69.

Bryant, J. P., Chapin, F. S., and Kline, D. P. (1983). Carbon/nutrient balance of plants in relation to vertebrate herbivory. *Oikos* **40,** 357–368.

Clancy, K. M., Price, P. W., and Craig, T. P. (1986). Life history and natural enemies of an undescribed sawfly near *Pontania pacifica* (Hymenoptera: Tenthredinidae) that forms leaf galls on arroyo willow, *Salix lasiolepis*. *Ann. Entomol. Soc. Am.* **79,** 884–892.

Craig, T. P., Price, P. W., and Itami, J. K. (1986). Resource regulation by a stem-galling sawfly on the arroyo willow. *Ecology* **67,** 419–425.

Cronquist, A. (1981). "An Integrated System of Classification of Flowering Plants." Columbia Univ. Press, New York.

Day, P. R. (1974). "Genetics of Host-Parasite Interactions." Freeman, San Francisco.

Ehrlich, P. R., and Raven, P. H. (1964). Butterflies and plants: A study in coevolution. *Evolution* **18,** 586–608.

Ehrlich, P. R., White, R. R., Singer, M. C., McKechnie, S. W., and Gilbert, L. E. (1975). Checkerspot butterflies: A historical perspective. *Science* **188,** 221–228.

Ehrlich, P. R., Murphy, D. D., Singer, M. C., Sherwood, C. B., White, R. R., and Brown, I. L. (1980). Extinction, reduction, stability, and increase: The responses of the checkerspot butterfly (*Euphydryas*) to the California drought. *Oecologia* **46,** 101–105.

Feinsinger, P. (1983). Coevolution and pollination. *In* "Coevolution" (D. J. Futuyma and M. Slatkin, eds.), pp. 282–310. Sinauer, Sunderland, Massachusetts.

Flor, H. H. (1971). Current status of the gene-for-gene concept. *Phytopathology* **9,** 275–296.

Fox, L. R. (1981). Defense and dynamics in plant-herbivore systems. *Am. Zool.* **21,** 853–864.

Futuyma, D. J. (1983). Evolutionary interactions among herbivorous insects and plants. *In* "Coevolution" (D. J. Futuyma and M. Slatkin, eds.), pp. 207–231. Sinauer, Sunderland, Massachusetts.

Futuyma, D. J., and Slatkin, M. (1983). Introduction. *In* "Coevolution" (D. J. Futuyma and M. Slatkin, eds.), pp. 1–13. Sinauer, Sunderland, Massachusetts.

Gilbert, L. E. (1971). Butterfly-plant coevolution: Has *Passsiflora adenopoda* won the selectional race with Heliconiine butterflies? *Science* **172,** 585–586.

Gilbert, L. E. (1975). Ecological consequences of a coevolved mutualism between butterflies and plants. *In* "Coevolution of Animals and Plants" (L. E. Gilbert and P. H. Ravens, eds.), pp. 210–240. Univ. of Texas Press, Austin.

Gilbert, L. E. (1983). Coevolution and mimicry. *In* "Coevolution" (D. J. Futuyma and M. Slatkin, eds.), pp. 263–281. Sinauer, Sunderland, Massachusetts.

Hatchett, J. H., and Gallun, R. L. (1970). Genetics of the ability of the Hessian fly, *Mayetiola destructor,* to survive on wheats having different genes for resistance. *Ann. Entomol. Soc. Am.* **63,** 1400–1407.

Holdren, C. E., and Ehrlich, P. R. (1982). Ecological determinants of food plant choice in the checkerspot butterfly *Euphydryas editha* in Colorado. *Oecologia* **52,** 417–423.

Janzen, D. H. (1966). Coevolution of mutualism between ants and acacias in Central America. *Evolution* **20,** 249–275.

Janzen, D. H. (1979). How to be a fig. *Annu. Rev. Ecol. Syst.* **10,** 13–51.

Janzen, D. H. (1980). When is it coevolution? *Evolution* **34,** 611–612.

Jermy, T. (1976). Insect-host-plant relationship: Co-evolution or sequential evolution? *Symp. Biol. Hung.* **16,** 109–113.

Jermy, T. (1984). Evolution of insect/host plant relationships. *Am. Nat.* **124,** 609–630.

Powell, J. A., and Mackie, R. A. (1966). Biological interrelationships of moths and *Yucca whipplei. U. Calif. Publ. Entomol.* **42,** 1–46.

Price, P. W., and Clancy, K. M. (1986). Multiple effects of precipitation on *Salix lasiolepis* and populations of the stem-galling sawfly, *Euura lasiolepis. Ecol. Res.* **1,** 1–14.

Price, P. W., and Craig, T. P. (1984). Life history, phenology, and survivorship of a stem-galling sawfly, *Euura lasiolepis* (Hymenoptera: Tenthredinidae), on the arroyo willow, *Salix lasiolepis,* in northern Arizona. *Ann. Entomol. Soc. Am.* **77,** 712–719.

Rowell-Rahier, M. (1984). The presence or absence of phenolglycosides in *Salix* (Salicaceae) leaves and the level of dietary specialization of some of their herbivorous insects. *Oecologia* **62,** 26–30.

Sacchi, C. F., Price, P. W., Craig, T. P., and Itami, J. K. (1988). Impact of the shoot-galler *Euura lasiolepsis* on sexual reproduction in the arroyo willow *Salix lasiolepsis. Ecology,* in press.

Schemske, D. W. (1983). Limits to specialization and coevolution in plant-animal mutualisms. *In* "Coevolution" (M. H. Nitecki, ed.), pp. 67–109. Univ. of Chicago Press, Chicago.

Singer, M. C. (1971). Evolution of food-plant preferences in the butterfly *Euphydryas editha. Evolution* **25,** 383–389.

Singer, M. C. (1972). Complex components of habitat suitability within a butterfly colony. *Science* **176,** 75–77.

Singer, M. C. (1983). Determinants of multiple host use by a phytophagous insect population. *Evolution* **37,** 389–403.

Smiley, J. T. (1978). Plant chemistry and the evolution of host specificity: New evidence from *Heliconius* and *Passiflora*. *Science* **201,** 745–747.

Smiley, J. T. (1985). Are chemical barriers necessary for evolution of butterfly-plant associations? *Oecologia* **65,** 580–583.

Strong, D. R., Lawton, J. H., and Southwood, T. R. E. (1984). "Insects on Plants." Harvard Univ. Press, Cambridge, Massachusetts.

Tahvanainen, J., Julkunen-Titto, R., and Kettunen, J. (1985). Phenolic glycosides govern the food selection pattern of willow feeding leaf beetles. *Oecologia* **67,** 52–56.

Thompson, J. N. (1982). "Interaction and Coevolution." Wiley, New York.

Weibes, J. T. (1979). Co-evolution of figs and their insect pollinators. *Annu. Rev. Ecol. Syst.* **10,** 1–12.

Weis, A. E., and Abrahamson, W. G. (1986). Evolution of host-plant manipulation by gall makers: Ecological and genetic factors in the *Solidago-Eurosta* system. *Am. Nat.* **127,** 681–695.

Weis, A. E., and Kapelinski, A. (1984). Manipulation of host plant development by the gall-midge *Rhabdophaga strobiloides*. *Ecol. Entomol.* **9,** 457–465.

Williams, K. S. (1983). The coevolution of *Euphydryas chalcedona* butterflies and their larval host plants III. Oviposition behavior and host plant quality. *Oecologia* **56,** 336–340.

Williams, K. S., and Gilbert, L. E. (1981). Insects as selective agents on plant vegetative morphology: Egg mimicry reduces egg laying by butterflies. *Science* **212,** 467–468.

Williams, K. S., Lincoln, D. E., and Lincoln, P. R. (1983). The coevolution of *Euphydryas chalcedona* butterflies and their larval host plants II. Maternal and host plant effects on larval growth, development and food-use efficiency. *Oecologia* **56,** 330–335.

CHAPTER 4

Searching for Defensive Chemistry in the Cruciferae, or, Do Glucosinolates Always Control Interactions of Cruciferae with Their Potential Herbivores and Symbionts? No!

F. S. CHEW
Department of Biology
Tufts University
Medford, Massachusetts

I. Introduction

A. Why the Interest in Glucosinolates?

The Cruciferae (= Brassicaceae) are defined by several shared attributes, including the glucosinolate–myrosinase system found in all members that have been examined so far (Ettlinger and Kjaer, 1968; Kjaer, 1973; Hedge, 1976; Rodman, 1981). The glucosinolates have figured prominently in our studies of this and related families in the Capparales for several reasons. First, the coincidence of taxonomic distribution of glucosinolates and patterns of insect and fungal specificity toward crucifers has made this family and these compounds the focus of much research on these interactions (reviewed by Feeny, 1977; VanEtten and Tookey, 1979). Second, the glucosinolates are important characters for examining biosynthetic pathways and relationships among genera that contain specific compounds (reviewed by Rodman, 1981). Third, the glucosinolates contribute to the importance of some genera, especially *Brassica,* as sources of condiments and vegetables (Tsunoda *et al.*, 1980), and create problems for mammalian use of *Brassica* seed oils and seed meals (Vaughan *et al.*, 1976; VanEtten and Tookey, 1979; Fenwick *et al.*, 1983).

The taxonomic distribution and structural variation of glucosinolates and their hydrolysis products make them conspicuous as candidates for chemical mediation of interactions between crucifers and other organisms. In this chapter, I discuss evidence for the role of glucosinolates in the host-plant specificity of crucifer interactions with herbivores and pathogens. In discussing specificity I include both the failure of some organisms to utilize any Cruciferae as potential hosts as well as the differential utilization of individual species within the Cruciferae by some organisms (cf. Singer, 1986). First, I briefly review evidence that glucosinolates are determinants of specificity in some of these interactions, that is, that the suitability of individual host-plant species as food sources for specific heterotroph species is determined by host possession of specific glucosinolates. I also review evidence to the contrary, that is, that glucosinolates, their hydrolysis products, and other secondary compounds whose taxonomic distributions within the Cruciferae coincide with intrafamilial host specificity patterns are not necessarily determinants of specificity in those cases. Second, for two crucifer–heterotroph systems, I present evidence that determinants of specificity are probably not glucosinolates or related hydrolysis products. I discuss evidence that other, as yet unidentified, primary or secondary compounds may be responsible for specificity in these cases. Third, I explore implications of these data for chemical defense of the Cruciferae against adapted insects, and the probability that these interactions are the result of chemically mediated coevolution between crucifers and adapted insects.

B. Glucosinolates in the Cruciferae

The glucosinolates comprise some 80 naturally occurring thioglucosides with side chains of varying structure (reviewed by Larsen, 1981; Rodman, 1981). These compounds are stored in glucosidic form in plant tissues, separately from the thioglucosidases (= myrosinases) that co-occur in all plants that contain glucosinolates, as far as is known (Rodman, 1981). The precise sites of glucosinolate and thioglucosidase storage in plant tissues remain uncertain, but recent interpretations support the idea that thioglucosidases occur associated with the cytosol and cytosolic membranes while glucosinolates are stored in vacuoles (Luthy and Matile, 1984).

The glucosinolates can be ordered biosynthetically by side-chain structure (reviewed by Ettlinger and Kjaer, 1968; Larsen, 1981; Rodman, 1981). Most glucosinolates belong to a series generated by one of three mechanisms: (1) hydroxylation; (2) chain lengthening; (3) oxidation of sulfur in the side chain. These series represent stepwise analogues separated

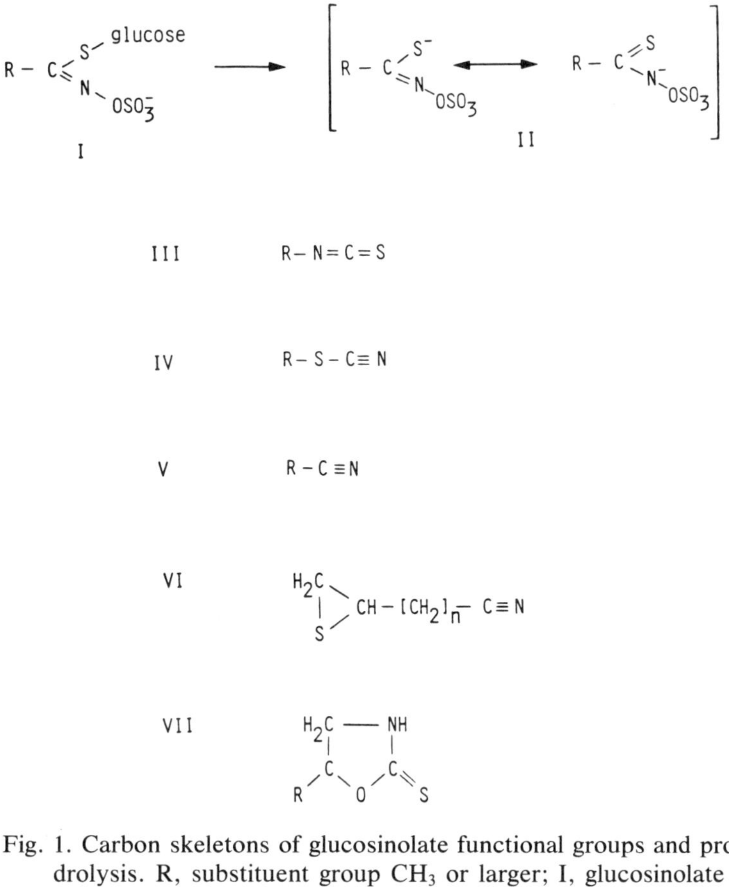

Fig. 1. Carbon skeletons of glucosinolate functional groups and products of hydrolysis. R, substituent group CH_3 or larger; I, glucosinolate (mustard oil glucoside); II, unstable intermediate resulting from enzymatic hydrolysis of *S*-glucose bond; III, isothiocyanate (mustard oil); IV, thiocyanate; V, nitrile; VI, cyanoepithioalkane; VII, oxazolidine-2-thione (goitrin).

by one or a few enzymatic reactions. Glucosinolate profiles of individual plants or crucifer populations usually comprise a few to a dozen compounds that are part of such series (e.g., glucosinolate profiles published by Al-Shebaz, 1973; Rodman, 1974, 1980; Rodman and Chew, 1980; Rodman *et al.*, 1981; Feeny and Rosenberry, 1982; Kruckeberg *et al.*, 1982; Louda and Rodman, 1983). Glucosinolate profiles (types and proportions of individual constituents) are taxon-specific, despite extensive intrapopulation variability of some taxa (e.g., Heaney and Fenwick, 1980; reviewed by Rodman, 1981). At generic and tribal levels, the fragmentary data available suggest that morphologically related clusters of genera contain characteristic arrays of glucosinolates that are biosynthetically related (Rodman, 1981). However, presence or absence of individual compounds is rarely diagnostic, with a few exceptions, such as rhamnosides in Resedaceae and methylglucosinolate in the Capparaceae (reviewed by Kjaer, 1973; Rodman, 1981).

Glucosinolate hydrolysis occurs when the *S*-glucose bond of the parent glucosinolate (Fig. 1, I) is enzymatically cleaved by myrosinase (thioglucoside glucohydrolase, E.C. 3.2.3.1) to yield glucose and an unstable aglycone (Fig. 1, II) (reviewed by Ettlinger and Kjaer, 1968). The aglycone rearranges to yield sulfate and to form a number of products that vary in functional group as well as side-chain substituents. These include isothiocyanates (mustard oils) (Fig. 1, III), thiocyanates (Fig. 1, IV), nitriles (Fig. 1, V), cyanoepithioalkanes (Fig. 1, VI), and oxazolidine-2-thiones (formed spontaneously from 2- and 3-hydroxy-substituted isothiocyanates, Fig. 1, VII) (reviewed by Ettlinger and Kjaer, 1968; Cole, 1976; Larsen, 1981). Accessory enzymes and variation in myrosinase isozymes have roles in generating diverse hydrolysis products that are only beginning to be understood (e.g., Bjorkman, 1976; Larsen, 1981; Spencer, 1987).

II. Glucosinolates as Determinants of Host-Plant Specificity

A. Evidence in Favor

The glucosinolates and their hydrolysis products were identified early as determinants of the host-plant specificity of crucivorous insects (reviewed for many heterotrophs by Ettlinger and Kjaer, 1968; for insects by Schoonhoven, 1972; for butterflies by Feeny *et al.*, 1983; Chew and Robbins, 1984). Early work demonstrated the stimulatory effects of these compounds on feeding behavior of larvae of *Pieris brassicae* L. and *P. rapae* L. (Verschaffelt, 1911). Positive responses to glucosinolates were observed in egg-laying behavior of *P. brassicae* adults (David and Gardiner, 1962). Because ovipositing *Pieris* butterflies choose cruciferous host-plants with nearly unerring precision (Verschaffelt, 1911; Remington, 1952; Dethier, 1959), and early instar *Pieris* larvae have limited mobility (Remington, 1952; Chew, 1980; Cappuccino and Kareiva, 1985), the idea that ovipositing females and feeding larvae should use these compounds to locate and identify host plants was attractive. In addition, these compounds have been implicated in the attack of crucifers by some pathogenic fungi (Walker *et al.,* 1937). This evidence implied that glucosinolate mediation of interactions between crucifers and heterotrophs seeking hosts might be a general phenomenon. The idea that specificity of crucifer–heterotroph interactions is mediated by glucosinolates and their hydrolysis products (e.g., Fraenkel, 1959; Ehrlich and Raven, 1964) provided the impetus for examining the responses of heterotrophs to these compounds.

Negative effects on heterotrophs include glucosinolates acting as barriers to successful herbivory or pathogen attack. Glucosinolates are toxic and unpalatable to some nonadapted insects (e.g., umbellifer specialist lepidopteran *Papilio polyxenes* Fabr.) (Erickson and Feeny, 1974; Blau *et al.,* 1978) and a generalist lepidopteran *Spodoptera eridania* Cramer, but only at artificially high concentrations (Blau *et al.,* 1978). Polyphagous aphids that are not pests of crucifers are deterred (Nault and Styer, 1972). Concentrations of specific glucosinolates are negatively correlated with feeding by adapted chrysomelid beetles and a psyllid homopteran that normally feed on crucifers (Louda and Rodman, 1983). Among glucosinolate hydrolysis products, isothiocyanates are reported to be bacteriostatic (reviewed by Virtanen, 1965; Ettlinger and Kjaer, 1968). They are reported to be toxic to insects (Lichtenstein *et al.,* 1962, 1964, reviewed by Feeny, 1977). Isothiocyanates are also reported to be fungistatic (Walker *et al.,* 1937; Hooker *et al.,* 1943). Downy mildew [*Perenospora parasitica* (Pers.) ex Fr.], an economically important crucifer pathogen, is inhibited by high concentrations of allylisothiocyanate (Greenhalgh and Mitchell, 1976). Holley and Jones (1985) report fungitoxic effects of in situ isothiocyanate formation in crucifer seed to *Menatospora* sp. Other reports, however, suggest that quantitative variation in specific glucosinolates or hydrolysis products in crucifer cultivars or their wild relatives is not systematically correlated with resistance to other pathogenic fungi [*Plasmodiophora brassicae* Woron. (Chong *et al.,* 1985); *Leptosphaeria maculans* (Desm.) Ces. et de Not. (Mithen *et al.,* 1984)]. Another class of hydrolysis products, thiocyanates, are reportedly unattractive to crucivorous flea beetles (P. Feeny, 1977; and personal communication) and toxic to some mammals (reviewed by VanEtten and Tookey, 1979).

Positive effects of glucosinolates include their acting as attractants or stimulants to adapted specialist herbivores (reviewed by Schoonhoven, 1972). Examples include glucosinolates stimulating feeding by larvae of

Pieris brassicae (David and Gardiner, 1966; Schoonhoven, 1967). Examples also include positive responses of adult *P. brassicae* (David and Gardiner, 1962; Ma and Schoonhoven, 1973), the aphids *Brevicoryne brassicae* (L.) (van Emden, 1972; Nault and Styer, 1972; Petterson, 1973) and *Myzus persicae* (Sulz.) (Nault and Styer, 1972; cf. van Emden, 1972, 1978), cabbage rootfly (Traynier, 1967), and flea beetles (*Phyllotreta* spp.) (Hicks, 1972). Allylisothiocyanate is attractive or stimulatory to germinating spores of the crucifer pathogen clubroot, *Plasmodiophora brassicae* Woron. (Hooker *et al.,* 1945), larvae of *Pieris rapae* L. (Hovanitz and Chang, 1962), and adults of crucivorous flea beetles (Feeny *et al.,* 1970; Hicks, 1972; Burgess and Wiens, 1980) and an anthomyid fly (Nair and McEwen, 1976). Rodman and Chew (1980) provided evidence that glucosinolate profiles are closely correlated with the behavior of egg-laying pierid butterflies toward eight crucifer species in a natural community. *Pieris occidentalis* Reakirt and *P. napi macdunnoughii* Remington preferentially laid eggs on the two crucifer species whose profiles contain allyl- and 3-butenylglucosinolate. One of the two species (a naturalized weedy species) fails to support larval development to pupation (Chew, 1975). Other crucifers in this community contained other glucosinolates. These data support the idea that ovipositing butterflies confuse the naturalized species with the native one on the basis of glucosinolate profiles. However, there is no direct experimental evidence for the involvement of these or other glucosinolates in butterfly preference among these crucifers.

B. Contradictory Observations

Recent observations suggest that observed effects of glucosinolates or their derivatives on one or a few species have been generalized to other species to a greater degree than evidence now supports. First, glucosinolates do not necessarily represent a barrier to nonadapted specialist herbivores. Allylglucosinolate acts as a weak larval feeding stimulant for nonadapted specialist tobacco hornworms (*Manduca sexta* (L.): Sphingidae) (J. L. Frazier, personal communication; F. S. Chew, unpublished observations). This effect may be due to the glucose moiety, as has been suggested for the weak stimulation of *M. sexta* larvae by the glycoalkaloid tomatine (G. De Boer and F. Hanson, personal communication). Newly hatched larvae from laboratory-maintained stocks develop normally to adulthood on semisynthetic diets containing 0.001 *M* allylglucosinolate. Pupal weights of *M. sexta* reared on 0.0001, 0.001, and 0.01 *M* sinigrin did not differ significantly from weights of individuals reared on sinigrin-free semisynthetic diets (one-way ANOVA $F[3,70] = 2.30$, $p > 0.05$). However, larvae reared on higher concentrations of sinigrin fed for a longer period before pupating (one-way ANOVA $F[3,72] = 3.49$, $p < 0.01$).

This result is similar to the longer larval development times observed for *M. sexta* reared on *Brassica napus* L. (De Boer and Hanson, 1984). Larvae appear to egest or excrete most if not all of the allylglucosinolate with their frass (F. S. Chew and W. Mechaber, unpublished observations). Possibly the ability of the *M. sexta* gut to egest and excrete most of the poisonous alkaloid nicotine found in its host plant, tobacco, contributes to its ability to utilize food containing allyglucosinolate [Self *et al.* (1964); see also Nijhout (1975) on other unique aspects of the *Manduca* gut; see Isman and Duffey (1983) and Gould (1984) for other examples of egestion or excretion of glucosides and other xenobiotics; see Turunen (1985) for review]. The egg-laying specificity of wild *M. sexta* for solanaceous plants (e.g., *Nicotiana tabacum* L., on which *M. sexta* is a pest) precludes most encounters of these larvae with plants that contain glucosinolates. Further, because laboratory-reared *M. sexta* accept a broader range of food than wild larvae, the behavior and growth of wild larvae toward glucosinolate-containing food needs to be examined. However, the barrier to utilization of these plants by lab-reared larvae is evidently not physiological intolerance or behavioral rejection of allylglucosinolate.

Second, glucosinolates may stimulate oviposition or feeding behavior without being responsible for herbivore discrimination among plant species (see Jermy, 1966). Renwick (Renwick and Radke, 1983; Renwick, this volume; and personal communication) has shown that painting a substrate with allylglucosinolate alone stimulates oviposition in *Pieris rapae* but does not promote butterfly discrimination among different potential egg-laying sites. Further, laboratory-reared *P. rapae* laid fewer eggs on substrates painted with allylglucosinolate alone (even at artificially high concentrations) than on substrates painted with a water-soluble extract of cabbage. Although egg-laying butterflies are stimulated by allylglucosinolate, other compounds appear to be involved. Ma and Schoonhoven (1973) demonstrated that individual glucosinolates elicit varying electrophysiological activity patterns from tarsal chemoreceptors in the butterfly *P. brassicae*. Butterflies often scratch the surface of potential host plants with tarsal claws before laying eggs (reviewed by Chew and Robbins, 1984). However, these electrophysiological responses may represent stimulatory effects rather than discriminatory responses that contribute to specificity (Schoonhoven, 1977). Given the reported attractiveness of allylisothiocyanates to *P. rapae* larvae (Hovanitz and Chang, 1962), these results suggest that different compounds stimulate larval feeding behavior and adult egg-laying behavior of *P. rapae*.

Third, the evidence that some species in the Cruciferae escape attack by adapted specialist insect herbivores (e.g., pierid butterflies) because they contain additional classes of compounds is equivocal and incomplete. Usher and Feeny (1983) found that mature laboratory-reared *P. rapae* larvae ate and grew well on semisynthetic diets containing compounds whose taxonomic distribution among crucifer species (Rodman *et al.*, 1982; Usher and Feeny, 1983) coincides with genera refused by pierid larvae (Chew, 1975) and crucivorous flea beetles (Feeny *et al.*, 1970). The diets contained variously the cardenolide k-strophanthin, cucurbitacins E and I, and the alkaloid lunarine. The cardenolide and cucurbitacins are avoided (though not refused) by several crucivorous flea beetles (Nielsen, 1978; Nielsen *et al.*, 1977), a result that lends weak support to the hypothesis that these compounds are responsible for the observed feeding specificity. The result Usher and Feeny (1983) observed suggests the possibility that compounds with different side chains may have different biological activity, but also suggests that the responsible compounds may not be the conspicuous ones whose taxonomic distribution coincides with herbivore specificity. This work also leaves open the possibility that younger larval instars of *P. rapae* or more stenophagous pierid species are sensitive to these compounds. Usher and Feeny chose the most euryophagous leaf feeder among pierid butterflies (Chew, 1980; Courtney and Chew, 1987), and the most euryophagous instar (Chew, 1975, 1980). Their work demonstrates that the pure compounds do not mimic the effects of the intact plants on *P. rapae*.

This review, while not exhaustive, suggests that glucosinolates (and their hydrolysis products) coincide with heterotroph specificity among angiosperm families and within the Cruciferae but are responsible for it in only some cases. These compounds protect Cruciferae from attack by many nonadapted heterotrophs. But for other nonadapted heterotrophs, lack of specific attractant compounds (found in normal hostplants) rather than presence of glucosinolates may be responsible for apparent protection (see Thorsteinson, 1960; Jermy, 1966, 1984; Dethier, 1980). For many adapted heterotrophs, glucosinolates and their derivatives are attractive and stimulate host-seeking or attacking behaviors. Evidence for their involvement in heterotroph specificity within the Cruciferae exists but is equivocal, and leaves many cases of infrafamilial specificity unresolved.

III. Two Cases of Negative Evidence

Among the Cruciferae, the heaviest damage results from attack by adapted heterotrophs (Root, 1973; Williams, 1981). Some taxa, however, are avoided by adapted heterotrophs (Slansky and Feeny, 1977; Usher and Feeny, 1983, and references therein). What mediates the defense or escape of these crucifer species? What accounts for this intrafamilial specificity? In the following sections, I present evidence for two cases in which members of the Cruciferae now avoid interaction with potential herbivores adapted to using Cruciferae, and potential fungal symbionts.

A. Crucifer Incompatibility with Potential Fungal Symbionts

Crucifers, and their capparalean relatives, represent one of a few angiosperm orders that are not symbiotic with mycorrhizal fungi (reviewed by Gerdemann, 1968; Malloch *et al.,* 1980). This symbiotic condition in angiosperms is ancient, widespread (Malloch *et al.,* 1980), and generally mutualistic, involving plant–fungal exchange of mineral nutrients, especially phosphate, and photosynthates (reviewed by Hayman, 1983). Where angiosperms are not associated with vesicular–arbuscular mycorrhizal (VAM) fungi, symbiosis with other fungi (more specialized mycorrhizal fungi of more recent origin) occurs (Malloch *et al.,* 1980). The co-occurrence of two family-wide traits in the Cruciferae—presence of glucosinolates and lack of infection by VAM fungi—naturally suggested the hypothesis that glucosinolates prevent the establishment of symbiosis (e.g., Hayman *et al.,* 1975; Medve, 1983). The reputations some crucifers have as harmful companion plants (reviewed by Rice, 1984) also suggest the possibility that these allelopathic effects are mediated by glucosinolates or their hydrolysis products, which may act as a specific diffusable barrier to the establishment of the plant–fungal symbiosis so common among other plants.

An experimental examination of this hypothesis was made by Glenn and her colleagues in this laboratory (Glenn, 1983; Glenn *et al.,* 1985, and in press). Cultivars of the genus *Brassica,* containing varying levels of glucosinolates in root tissue (7–524 ppm fresh weight), were grown in agar with representatives of the VAM fungal genera *Glomus* and *Gigaspora,* in the presence or absence of normal VAM fungal hosts. We monitored progressive growth of the VAM fungi from germinating spores *in situ* in sterile agar culture for 2–6 weeks (Glenn *et al.,* 1985). Hyphal growth rates and branching frequency of VAM fungi were measured near host (tobacco or tomato) and nonhost (brassica) roots. Liquid media used to grow sterile cultures of crucifers contain glucosinolates (Elliott and Stowe, 1971); during 6 weeks of sterile culture, liquid medium in which roots were grown accumulated more than six times the total amount of glucosinolates contained in root tissue. Similar proportions of the glucosinolate types observed in root tissues were excreted [*sic*] into the liquid medium (Elliott and Stowe, 1971).

VAM fungi exhibited reduced growth in the presence of brassica alone, compared to growth with host plants alone, regardless of glucosinolate level in the roots. We did not observe cell-wall penetration of brassicas growing in agar (Glenn *et al.,* 1985), but fungal hyphae sometimes encircled or grew parallel to the root axis. Reports of occasional penetration of crucifer cell walls and formation of single arbuscles (thought to be the site of mineral nutrient–photosynthate exchange) in field samples (Bevege and Bowen, 1975; Hirrel *et al.*, 1978; Medve, 1983) and in sterile culture (Tommerup, 1984) are consistent with the idea of limited VAM fungal growth near plants in the Cruciferae.

When grown in agar with both brassica and host plant together, however, no differences were observed in VAM fungal growth compared with growth in the presence of normal hosts alone (Table I) (Glenn, 1983). The presence of the brassica cultivars inhibited neither establishment of VAM symbiosis nor the proliferation of already-established symbioses in normal host plants.

These results suggest that the root exudates of *Brassica* cultivars lack some diffusable compound(s) that stimulates VAM fungal growth. Rather than producing exudates containing diffusable inhibitors of VAM fungal growth, the brassica roots fail to exude or secrete substances that are necessary for normal VAM fungal growth. We suggest (Glenn, 1983; Glenn *et al.*, 1985) that because brassica tolerate very low soil phosphate levels, their root cell membranes are less "leaky" than those of other plants. "Leaky" cell membranes that permit exudation of various carbohydrates and related compounds are one effect of phosphate deficiency on normal hosts (Graham *et al.*, 1981; Schwab *et al.*, 1983), and low soil phosphate levels promote VAM fungal infection of normal hosts (reviewed by Hayman, 1983). We also suggest that specialized crucifer capabilities for phosphate absorption (Nye, 1981; Grinsted *et al.*, 1982) and ecological constraints on crucifer phenology may make VAM fungal symbiosis disadvantageous for crucifers (Glenn, 1983; Glenn *et al.*, 1985, and in press).

Finding that glucosinolates do not play the hypothesized role in this plant–fungal interaction does not exclude them as a cause of nonmycorrhizal status in this group, but implicating them or their hydrolysis products in nonmycorrhizal status requires further evidence. Further informa-

Table I. Effect of Presence or Absence of Host (Tomato) Companion Roots on the Density of *Glomus mosseae* (Nicol. & Gerd.) Germ-Tube Branches Near Nonhost (Brassica) Roots

Plants	Intersections	Branches/mm[a]
Brassica and tomato	168	2.7 ± 0.01
Brassica alone	262	0.8 ± 0.02
Tomato alone	40	3.0 ± 0.35

[a] Within a 2-mm radius surrounding the intersection of a brassica root and a germ tube, the number of branch points per millimeter of germ tube was counted. The mean number of such branch points per millimeter of germ tube ±95% confidence interval (CI) is tabulated.

tion is needed on how glucosinolate hydrolysis products are formed *in situ* (cf. Holley and Jones, 1985). Despite considerable uncertainty about the precise anatomy of glucosinolate and thioglucosidase storage in crucifer tissues, we know that these substrates and the enzymes that hydrolyze them are localized in the cytoplasm. Parenchyma cell vacuoles are the probable storage sites for glucosinolates in root tissue (Grob and Matile, 1979). Thioglucosidases have been associated with various membrane components of cells, e.g., endoplasmic reticulum and cell membrane (Pihakaski and Iversen, 1976). The proposal that thioglucosidases are cytosolic enzymes with a marked affinity for membranes is an attractive hypothesis that reconciles much contradictory evidence about the localization of these enzymes (Luthy and Matile, 1984; Chew, 1988). Some evidence exists for endogenous production of volatile isothiocyanates by intact plants, at very low levels (reviewed by Rodman, 1981), but their occurrence in media surrounding roots is uninvestigated.

The anatomy of symbiosis between plant and VAM fungus and the localization of the glucosinolate–thioglucosidase system in the cytoplasm makes it unlikely that *in situ* glucosinolate hydrolysis is responsible for nonmycorrhizal status of crucifers (cf. Holley and Jones, 1985). In symbioses with normal hostplants, VAM fungal arbuscules are confined to the apoplast (the compartment inside the cell wall but outside the cell membrane) of root cortical cells (Bonfante-Fasolo, 1982). Glucosinolates might be hydrolyzed and isothiocyanates or other aglycone products might be released *in situ* upon damage to the cell membrane. Such damage might occur during the occasional VAM fungal infections of Cruciferae. However, given the rarity of VAM fungal penetration of crucifer cell walls, contact between VAM fungal hyphae and the crucifer cell membrane is likely to be a rare occurrence (Glenn, 1983). The glucosinolates and their hydrolysis products are unlikely to be responsible for the nonmycorrhizal status of the Cruciferae.

B. Crucifer Escape from Specialized Insect Herbivores

Wild crucifers escape from specialist herbivores by being ephemeral and relatively hard to find (Feeny, 1975; Rhoades and Cates, 1976). Low concentrations of compounds having potent effects (e.g., glucosinolates) would deter many potential herbivores (Feeny, 1975; Rhoades and Cates, 1976), but many adapted herbivores tolerate or use these compounds to find their host plants. Some crucifers escape from herbivory by chemical deterrence or toxicity to adapted herbivores, even though the compounds responsible for such effects have not yet been identified (Usher and Feeny, 1983). Interestingly, at least some of these chemically defended crucifers contain lower levels of glucosinolates than most wild crucifers [*Erysimum asperum* (Nutt.) DC. (Rodman and Chew, 1980); *Capsella bursa-pastoris* (L.) Medic. (J. A. A. Renwick, personal communication; F. S. Chew, unpublished observations) and may be chemically less apparent to adapted herbivores. Other crucifer species escape because they are phenologically unavailable (Feeny, 1977). In some crucifers, plant parts that support larval development when tender become too tough for young larvae [e.g., if ground up, tough siliques of *Hesperis matronalis* L. support growth of *Anthocharis cardamines* L.; however, females continue to lay eggs on this visually conspicuous plant! (Courtney, 1981); cf. *Lupinus amplus* Greene and its herbivorous butterfly *Glaucopsyche lygdamus* Doubleday (Breedlove and Ehrlich, 1968)]. In others, egg-laying by adapted herbivores whose larvae are cannibalistic is deterred by plant morphological mimics of eggs [e.g., *Streptanthus glandulosus* Hook. leaf-tip callosities successfully mimic eggs of *Pieris sisymbrii* Bdv. (Shapiro, 1981a,b); cf. Williams and Gilbert, 1981)].

What are the evolutionary prospects for crucifers and other angiosperms that are attacked by adapted specialist herbivores? We might suppose, given the results reviewed by Feeny (1975, 1977) and Rhoades and Cates (1976), that nonapparent crucifers would become fugitive species, condemned by their palatability to hide from herbivores that tolerate glucosinolates. To explore this question, we have recently examined interactions of Capparales-feeding pierid butterflies and their capparalean host plants in the Atlas Mountain region of Morocco (Courtney and Chew, 1987). This region is species-rich in both pierid butterflies and members of the Cruciferae and Resedaceae (Capparales). We examined two aspects of interactions between adapted herbivores and their potential host plants. First, we examined possible relationships between larval performance (feeding acceptance and survival) and plant classification within the Capparales. We reared larvae of eight pierid butterflies on appropriate plant parts of nearly 30 capparalean species, including representatives of the Capparaceae, the Resedaceae, and the six major tribes of the Cruciferae. Second, we examined possible relationships between patterns of host-plant utilization by pierid butterflies and other crucivorous herbivores in the field, and the density and numerical stability of crucifer populations from year to year. We documented crucifer population densities and utilization by pierid butterflies for 3 years (1981–1983) and damage patterns to crucifer leaves in 1981.

Rearing larvae on a wide range of capparalean species revealed few differences between the responses of pierid species (Table II). Most butterfly species refused to eat the same several crucifer species and generally did well on the same several plants. Plant species could therefore be categorized as "good" or "bad" for pierid larvae. Flower-feeding and leaf-feeding pierids responded similarly to the array of capparalean species. The differences among pierid butterflies reflect the stenophagy or euryophagy of individual pierid species rather than partitioning of poten-

Table II. Performance of Pierid Larvae on Moroccan Capparales

		Pierid Species Alive/Tested			
Species	Tribe/Family	Reject	Feed	2nd	4th
Allysum minus	Alysseae	6/6			
A. minutum	Alysseae	3/3			
A. atlanticum	Alysseae	3/3			
Cheiranthus cheiri	Hesperidae	3/3			
Matthiola parviflora	Hesperidae	4/4			
Reseda alba	Resedaceae	1/1			
R. attenuata	Resedaceae	1/2			1/2
R. phyteuma	Resedaceae			1/1	
R. lutea	Resedaceae	2/5	1/5	1/5	1/5
Descurainia sophia	Sisymbrieae	1/4	3/4		
Capsella bursa-pastoris	Lepidieae	2/6	3/6	1/6	
Capparis spinosa	Capparaceae	1/6	1/6		3/6[a]
Trachystoma balli	Brassicaceae		2/4	1/4	1/4
Crambe hispanica	Brassicaceae		1/5	2/5	2/5
Moricandia arvensis	Sisymbrieae			2/4	2/4
Hirschfeldia incana	Brassicaceae		1/6	3/6	2/6
Isatis tinctoria	Sisymbrieae			3/7	4/7
Cardaria draba	Lepidieae				2/2
Neslia paniculata	Sisymbrieae				2/2
Brassica rapa	Brassicaceae			2/5	3/5
Sisymbrium irio	Sisymbrieae	1/5[a]		2/5	2/5
S. officinale	Sisymbrieae			2/5	3/5
Diplotaxis tenuisiliqua	Sisymbrieae			3/6	3/6
Nasturtium officinale	Arabideae			2/6	4/6
Sinapis arvensis	Brassicaceae	1/7[a]		2/7	4/7
Biscutella didyma	Lepidieae			2/7	5/7
Erucastrum laevigatum	Lepidieae	1/7[a]		1/7	5/7
S. erysimoides	Sisymbrieae				6/6

Notes: Species tested are *Anthocharis belia, Colotis evagore, Euchloe ausonides, E. belemia, Pieris brassicae, Pontia daplidice*, and *Zegris eupheme*. Nomenclature of butterfly species follows Higgins and Riley (1980); that of crucifer species follows Tutin et al. (1964).

[a] Categories that include responses of *Colotis evagore*.

tial host species among pierid species. Similar results were observed for troidine swallowtail butterflies and their *Aristolochia* host-plants by Brown *et al.* (1981), and for *Itaballia* spp. and their capparaceous host plants by Jordan (1981). The exception to thc general pattern of pierid responses is *Colotis evagore* Lucas, a representative of a large genus that radiated in the Ethiopian biogeographic province [53 described species in Africa south of the Sahara desert (Talbot, 1939; D'Abrera, 1980)]. Recorded host plants of *Colotis* are various genera in the Capparaceae (Van Son, 1949). Because the phenologies of *C. evagore* and crucifer species in Morocco have limited overlap, in most years, Cruciferae are no longer

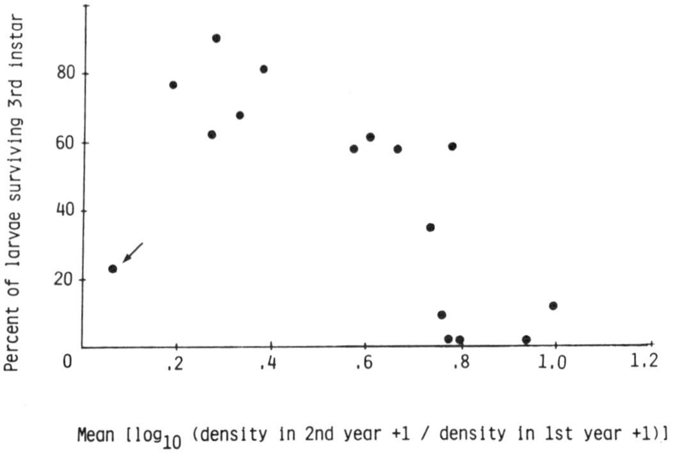

Fig. 2. Survival of flower-feeding pierid larvae on capparalean species as a function of year-to-year stability in plant density. In the laboratory, larvae that survive the third instar nearly always survive to pupation. The arrow indicates *Crambe hispanica,* whose populations are numerically stable, but whose flowers (on which larvae feed) are very small, visually inconspicuous, and ephemeral. All plants and larvae are native to Morocco and were reared in Morocco; $r = -0.773$ ($p < 0.001$, $N = 15$), excluding the point for *C. hispanica;* including the point, $r = -0.559$ ($p < 0.05$, $N = 16$).

available as potential host plants when ovipositing *C. evagore* fly, and the butterflies must rely on perennial Capparaceae.

Association of some cruciferous tribes with good or poor larval performance occurs (Table II). Three congeners of the tribe Alysseae, two in the Hesperidae, and four members of the family Resedaceae are successfully utilized only by the most euryophagous butterflies (*Pontia daplidice* L. and *Pieris rapae,* for flower- and leaf-feeding guilds, respectively). Members of other tribes, especially Brassicaceae and Sisymbrieae, are generally associated with good performance of pierid larvae.

However, examination of these data in relation to year-to-year changes in population densities of crucifers suggests another relationship between larval performance and characteristics of crucifer species (Fig. 2). These data suggest that plant species on which larvae do very poorly are those that have the most highly variable densities from year to year. Another interpretation of Table II, together with Fig. 2, is that members of some crucifer tribes, which possess feeding deterrents to pierid larvae, also have populations whose densities vary enormously from year to year. Crucifers whose populations have large-amplitude variations in density from year to year apparently do escape herbivore attack (Table III for flower-feeding pierids, Fig. 3 for other invertebrate herbivores). Further,

Table III. Rates of Host Infestation by Five Flower-Feeding Pierid Butterflies

Butterfly	Host Numerical Stability (mean)[a]	Infestation Rate (Number Eggs/Number Plants)
Euchloe belemia	1.07	0.10
	0.87	0.095
	0.74	0.15
	0.63	0.12
	0.45	0.49
Pontia daplidice	0.83	0.098
	0.45	0.075
	0.41	0.12
E. ausonia	0.81	0.06
	0.72	0.085
	0.45	0.10
	0.41	0.13
	0.40	0.79
	0.35	0.39
	0.32	0.53
	0.27	0.62
Anthocharis belia	0.45	0.10
	0.42	0.39
	0.36	0.23
Zegris eupheme	0.28	1.05

[a] Host mean numerical stability = mean [$\log_{10}$(density in 2nd year + 1/density in 1st year + 1)].

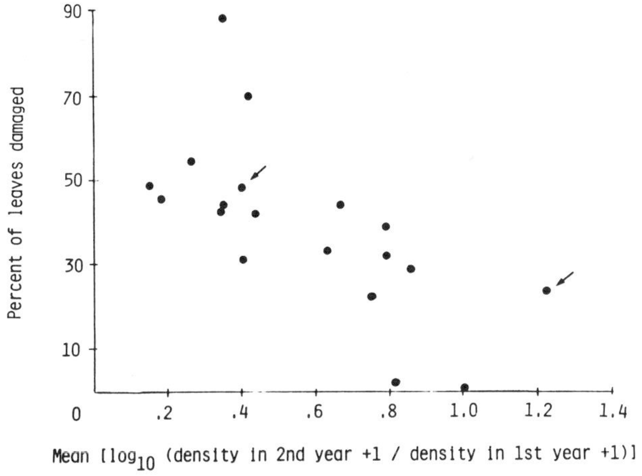

Fig. 3. Proportion of leaves of capparalean species damaged by invertebrate herbivores as a function of year-to-year stability in plant density. The two arrows indicate points for two North American species, *Barbarea orthoceras* Ledeb. and *Dentaria diphylla* Michx., and their invertebrate herbivores. All other points represent Moroccan species observed in Morocco; $r = -0.663$ ($p < 0.01$, $N = 19$).

Table IV. Effect of Host-Plant Abundance and Numerical Stability on Utilization of Host Populations by Flower-Feeding Butterflies

	Numerical Stability	Abundance		
		High	Medium	Low
A. *Anthocharis belia*	High	6/14	23/35	0/1
	Low	0/1	1/4	2/7
	Low	4/17	9/12	0/0
B. *Pontia daplidice*	High	1/9	4/29	0/0
	Low	0/0	0/1	6/9
	Low	2/12	5/17	0/0

Notes: Abundance = $\log_{10}$(density in 1st year + 1); numerical stability = $\log_{10}$(density in 2nd year + 1/density in 1st year + 1). The fraction of populations utilized in each density–stability category is tabulated.

butterfly species differ in the range of crucifer densities they utilize (Table IV) (Courtney and Chew, 1987): note that the most euryophagous flower-feeding butterfly species (*Pontia daplidice,* Table II) utilizes populations with much wider ranges of densities and year-to-year variation than its more stenophagous flower-feeding relative *Anthocharis belia* (Table IV).

IV. Evolution of Chemical Defense in the Cruciferae

Why should crucifers with strongly fluctuating population dynamics support poor larval performance in rearing experiments? To put the question another way, why should a plant have two types of defense against specialized herbivores—ecological and intrinsic—either of which would be effective alone? To explore the central issue presented by this question, we must ask what evidence there is to support the hypothesis that each defense is an exaption that presently functions as a defense, but that evolved under other circumstances. Is plant–herbivore coevolution necessary to account for the observed characteristics of these plants? The salient features of these plants are those related to ecological and intrinsic defenses: (1) rapid growth, rapid-cycling life histories, displaced phenologies and population dynamics characterized by large-amplitude fluctuations—the effect of these traits on herbivores depends on the presence of other crucifers that cannot exhibit these traits; (2) attributes of the aboveground parts that deter or poison specialized herbivores and attributes of below-ground parts that do not support successful symbiosis with vesicular–arbuscular mycorrhizal fungi—these traits affect herbivores and fungi regardless of whether other crucifers exhibit similar traits. For each of these attributes, I consider what abiotic factors might have been selective agents and then consider the possible role(s) of herbivores and other heterotrophic organisms.

A. Evolution of Attributes That Function as Ecological Defenses

Mediterranean-type climates are the norm for the geographic centers of crucifer diversity (Hedge, 1976). The relatively recent origin of the Cruciferae during Eocene times (Muller, 1981) in the Palearctic (Hedge, 1976) makes it likely that adaptation to seasonal drought was an important selective factor during crucifer evolution. For crucifers, this highly seasonal and often abbreviated availability of resources for plant growth favors rapid-cycling life histories. Selection for rapid-cycling life histories, with the accompanying attributes of rapid growth, small plant size, and displaced phenologies, can be thus accounted for by consideration of abiotic factors related to Mediterranean climates. This is particularly plausible when we observe that among crucifers, the species with the smallest sizes, most rapid-cycling histories, and most displaced phenologies are those that occur in open, less mesic habitats.

Rapid-cycling life histories, displaced phenologies, and the associated traits make some crucifers, especially those that occur in open habitats, less apparent to herbivores and provide a mechanism for ecological escape due to reduced apparency (e.g., Shapiro, 1981a). Reduced apparency also occurs in some of these rapid-cycling annual crucifers in the form of between-year variation in plant density (Section III,B). How might we account for variation in another trait that now functions as an ecological defense in some species—viz., the considerable variation among crucifer species in the stability of plant density (in a given locality) from one year to the next?

Three lines of argument suggest different mechanisms by which variation in population dynamics among crucifer species might be accounted for. First, the highly seasonal rains of the Mediterranean winter growing season, leading to a highly seasonal input of moisture and dissolved minerals, would favor two types of growth patterns (Chapin, 1980): (1) a pattern that minimizes time to complete a seed-to-seed life cycle—this pattern would be favored in years when sparse or early rainfall and early seasonal drought limited plant access to moisture and mineral resources for growth; (2) a pattern that maximizes fecundity (seed production)—this pattern would be favored in years when prolonged or ample rainfall extended plant access to resources for growth. Given the spatially patchy and likely partially isolated population structure of these crucifer species, selection favoring these patterns might shift from year to year (Wright, 1931; Levins, 1968; Wilson, 1983). In some years, the onset of seasonal drought would truncate the growing season and would favor individuals that contribute something (however little) to the seed bank by minimizing life cycle time. Other, more slowly growing individuals in the population, no matter how potentially fecund, would be unable to contribute to the seed bank in such a year if they had not formed seed by the onset of seasonal drought. (A continuum of these traits would be expected to occur among individuals in a population, but the most rapid-cycling individuals could not also be the most fecund.) This would be especially true in open, more xeric habitats where seasonality is more extreme. In each partially isolated group, we would expect that in some years, the proportion of seeds contributed to the seed bank by time-minimizers would rise in comparison to those contributed by fecundity-maximizers. In other years, particularly wet ones, however, fecundity-maximizers in a population would contribute proportionately more seed. If dormancy and germination requirements of related seeds are similar to those of their maternal parents, we might expect that in some years the multigenerational seed bank would yield many of the progeny of the fecundity-maximizers, as well as many progeny of time-minimizers, thus generating a large, dense local population, while in other years the plants emerging from the seed bank would be only the progeny of time-minimizers.

While we have no data on seed size or dormancy polymorphism in these taxa, somatic seed polymorphs that differ in size (or other morphological character) and dormancy requirements are especially common among short-lived species of the Compositae, Chenopodiaceae, Gramineae, and Cruciferae (Harper, 1977). Differences in seed size alone would affect germination because size affects how crucifer seeds respond to microtopographical variation (White, 1968, cited by Harper, 1977). Somatic seed size polymorphism might thus generate germination polymorphism. Harper suggests that because the proportion of morphs may be altered directly by environmental conditions under which the maternal plant develops, somatic polymorphism permits a greater degree of sensitivity to environmental influences than does genetic polymorphism (Harper, 1977).

Second, if it is true that seeds that enter the seed bank tend to germinate under conditions similar to those of their maternal parents, insect herbivory on developing fruits and seeds may reinforce fluctuation amplitude. If pierid larvae feeding on developing fruits and seeds affect the allocation of maternal resources to the remaining seeds, crucifer reproduction may be affected by altering fecundity of individuals and the time period over which seeds enter the seed bank in a given year. Although the specific effects of herbivores depend on the timing and site of herbivory (for review see Watson and Casper, 1984; for crucifers, see Free and Williams, 1978), it is reasonable to expect such an effect some of the time. These effects would be generated by the pattern of pierid larval feeding on developing crucifer fruits and seeds: pierid larvae tend to eat the youngest seeds first, at least partly because they are unable to feed successfully on tough, more mature fruit tissue. In linear fruits (siliques), larvae eat developing young fruits directionally from distal end to proximal point of attachment, so that the risk of herbivory for seeds at the proximal end is delayed and they may become tough enough to resist herbivory [e.g., *Hesperis matronalis* L., (Courtney, 1982); cf. lupines (Breedlove and Ehrlich, 1968)]. In short growing seasons, fecundity will be reduced below the level expected in the absence of the herbivores, and the time over which seeds enter the seed bank may be shortened by herbivory. In wet extended growing seasons, fecundity might be increased above the levels expected in the absence of herbivory. This increase would occur despite the depredations of herbivory because the "clipping" of terminal apices by the insect herbivores would enhance formation of axillary flowering apices that would eventually produce mature seed during an extended growing season.

Third, formation of large monocultures of crucifers may be an adaptation for more effective insect pollination in plants with phenologies that are displaced in relation to insect phenologies (Auerbach and Shmida, 1985). These authors observed that some crucifers in open (desert) habitats in Israel grow in very large monoculture patches—very probably a different aspect of the population fluctuation phenomenon observed by us (Courtney and Chew, 1987; and unpublished data). In particular, Auerbach and Shmida (1985) suggest both that the monoculture "patchiness" observed in some crucifer species may have evolved in response to competition for insect pollinators, and that the displaced phenologies of these same crucifer species have the effect of minimizing encounter with (insect) herbivores. The connection between selection for more effective cross-pollination and selection for escape from herbivory (ecological or constitutive) may be an important one if insect pollination promotes outcrossing and more open recombination systems (Levin, 1975). Unfortunately for our present consideration, data do not yet exist to permit comparison of the pollination requirements of these crucifer species with others that do not form large monocultures (M. J. Auerbach, personal communication).

While it is possible that interaction with insects or other biotic agents may have affected evolution of these life history traits, crucifer adaptation to the seasonal Mediterranean climate can satisfactorily account for the evolution of plants with rapid-cycling traits. The question of what accounts for variation in the population dynamics attributes of these plants is less clear. Data on the heritability of dormancy and germination requirements of these crucifer species would enable us to evaluate the likelihood that the first two hypotheses presented above correctly account for the mechanisms by which interspecific differences in population dynamics of crucifers are maintained. Data on pollination requirements of the desert taxa that form large monocultures would permit evaluation of the third.

B. Intrinsic Barriers to Symbiosis and Herbivory

What factors have contributed to the evolution of intrinsic barriers to symbiosis and herbivory in these crucifers? I consider specifically (a) the incompatibility of crucifers with VAM fungi and (b) the attributes that are responsible for poor larval growth performance on some crucifer species. As with the consideration of ecological attributes that now have defensive functions, I consider what evidence exists to support the hypothesis that these intrinsic attributes are exaptions that now serve as barriers to interaction.

In the case of crucifer incompatibility with VAM fungi, selection for rapid-cycling life history traits opposed evolution of compatibility for symbiosis. This interpretation argues that evolution of rapid-cycling traits also favored evolution of barriers to VAM fungal symbiosis because the symbiosis was not mutually beneficial (Glenn, 1983; Glenn *et al.*, 1985).

Two observations provide support for the view that the benefits of VAM symbiosis, advantageous for normal hosts, are superfluous and possibly disadvantageous for crucifers. First, crucifers may derive little benefit from the enhanced phosphate nutrition provided by symbiosis with VAM fungi. Selection for rapid growth would favor selection for rapid mineral acquisition (reviewed by Chapin, 1980). In the case of *Brassica* (and perhaps other crucifers), these plants tolerate low phosphate fertility because their roots have specialized mechanisms for depleting soil phosphate to lower levels than other flowering plants (Nye, 1981; Grinsted *et al.*, 1982). Second, VAM fungal symbiosis may not benefit rapid-cycling crucifers because they would set seed before the beneficial effects of symbiosis [often delayed for several weeks until symbiosis is well established (Bethlenfalvey *et al.*, 1982)] are realized. It is more difficult to determine whether VAM fungi would impose a photosynthate drain large enough to detrimentally affect the crucifers. If crucifer roots supplied mycorrhizal symbionts with the same levels of primary nutrients provided by normal host plants, the drain might have an effect on a small, rapid-cycling plant. Marschner (1986) reviews studies showing that the proportion of organic carbon released into the rhizosphere (presumably via the root cell apoplast, where it would be available to symbiotic VAM fungi) varies widely, from less than 1% to more than 30% of the total dry matter production of a young plant. Although crucifers have among the highest photosynthetic rates measured per plant (Mooney, 1983), it is not known how photosynthetic rate and compensatory photosynthetic capability are related, making impossible an estimate of such an effect.

This interpretation thus suggests that crucifer failure to support symbiosis with VAM fungi is a consequence of selection for rapid growth and adaptation by crucifers to highly seasonal climates. This apparent failure of *Brassica* roots to provide sufficient primary nutrients to support the fungal symbiont is not the only barrier to establishment of symbiosis (for further discussion of properties of crucifers in relation to compatibility with fungi, see Heath, 1977; Tommerup, 1984), but it is clearly a major constitutive barrier. Although VAM fungi sometimes establish single arbuscles in crucifers, these incipient mycorrhizae never become fully developed (Tommerup, 1984). It would be instructive to examine the compatibility of VAM fungi with roots of longer-lived relatives of the crucifers, such as some of the woody Capparaceae, whose life histories and habitats suggest that mycorrhizal symbiosis might be advantageous.

In the case of constitutive attributes that are responsible for poor larval growth performance, we must consider two questions concerning the acquisition of a constitutive trait that acts as a barrier to a specialized herbivorous insect. First, how did a novel constitutive attribute arise and spread? Second, how did this novel constitutive trait acquire or retain its effect as a barrier to specialized herbivorous insect exploitation of the plants that contained it? As with the attributes that function as ecological defenses, we must ask, are these traits exaptations that now function as constitutive defense against specialized herbivorous insects?

Acquisition and spread of a novel metabolic product in a crucifer population may have occurred in response to a biotic factor, such as a pathogen, or an abiotic one, such as mineral nutrition. In the former case, the novel attribute can be regarded as the product of crucifer evolution in response to a selective agent that currently exerts selection, such as the specialized herbivorous insects, or to one that no longer exerts influence (cf. Janzen, 1968, 1973; Rhoades, 1985). The maintenance of the trait then depends on its being neutral or advantageous, and its effect on specialized herbivorous insects that attempt colonization of the plant is fortuitous. Either way, by whatever specific mechanism the novel attribute arose (see Levy and Levin, 1971; McKey, 1980; Chew and Rodman, 1979, for discussion of mechanisms), its spread and maintenance involved interaction with herbivores or pathogens. In the latter case, by contrast, plant response to mineral nutrition may result in novel metabolic attributes in response to soil mineral nutrition (McKey *et al.,* 1978; Coley *et al.,* 1985). In this case, the effects of the novel attributes on specialized herbivorous insects are fortuitous. Further, these attributes are associated not with specific crucifer taxa, but with specific habitats and soil types, and might involve primary as well as secondary compounds [see Scriber and Slansky (1981) for effects of plant attributes such as water content on lepidopterous larvae].

In the case of perennial crucifers that are refused by feeding larvae, the compounds that act as constitutive defenses against specialized insect herbivores are extractable from plants grown in cultivation (Nielsen, 1978; Renwick and Radke, 1983; Renwick, this volume). The species that are rapid-cycling annuals have not yet been so tested, however, and it remains to be seen whether the association is with these specific crucifer taxa, or with conditions under which they are grown. Recent work on the effects of mineral environments on trees and some perennials shows that phenolic compounds may differ dramatically among individuals in a taxon that grows in different mineral environments (McKey *et al.*, 1978; Coley *et al.*, 1985; see Gershenzon, 1984, for review of other classes of compounds and the effects of other environmental factors). This interpretation opens the possibility that the constitutive defenses possessed by these rapid-cycling crucifers are a conditional product of their mineral environment rather than a defense produced as part of their biosynthetic repertoires, independent of abiotic environmental variables.

C. Components of Successful Crucifer Escape from Herbivory

Why have specialized herbivorous insects failed to adapt successfully, in terms of larval performance, to the constitutive defenses in these rapid-cycling crucifers with large-amplitude population fluctuations? Why have these crucifer taxa escaped herbivory from specialized insects when a large proportion of crucifer taxa has not? What prevents these specialized insects from successfully adapting to these as-yet-unidentified defenses, in contrast to the complex adaptations of many specialist insects to chemical defenses of their hosts? In cases of both perennials and rapid-cycling annuals in this group of crucifers, certain attributes reduce the chance of plant–insect encounter in these taxa in comparison with normally used crucifer hosts. This reduced opportunity for encounter slows natural selection for overcoming this chemical defense by preventing selection from being consistent and persistent, thus leaving the spread of any genes for successful (larval) adaptation to chance (Chew, 1977; Futuyma, 1983).

Two factors in particular contribute to reduced encounter rate. First, in the perennial species examined so far, the levels of glucosinolates, which are attractive to specialized insect herbivores, are reduced compared to those of other crucifers (Rodman and Chew, 1980; J. A. A. Renwick, personal communication). This is also apparently the case for at least one rapid-cycling annual species, *Capsella bursa-pastoris* (J. A. A. Renwick, personal communication; Chew, unpublished data). Because glucosinolates and their hydrolysis products are attractants and behavioral stimulants for some specialized herbivorous insects, lower levels of such compounds may reduce the attractiveness of the plant to the insect, thus further reducing encounter rate between plant and insect. Such quantitative reduction in attractant compounds, accompanied by the presence of new classes of chemical constituents, might eventually result in substitution of one class of compounds by another as constitutive defense (Gardner, 1977).

Second, for the rapid-cycling annuals, morphological inconspicuousness (e.g., small size, divided leaves) contributes to reduced encounter rates that are counteracted by the opportunistic behavior of only the most euryophagous pierid butterfly species (e.g., *Pontia daplidice* L.). The behavioral tendency of the more stenophagous butterflies to oviposit only in crucifer stands with certain "threshold" densities (Section III,B) further reduces the encounter rate between these herbivores and crucifers.

The failure of these specialized insects to adapt to these particular crucifers as potential host plants and the attributes that reduce encounter rates between the two suggest that these particular rapid-cycling and perennial crucifers have now successfully and permanently escaped pierid butterfly herbivory. This escape apparently depends as much on mechanisms for reducing encounter rate between insect and plant as on the possession of intrinsic barriers against feeding by these insects (cf. Feeny, 1977). The reduction of glucosinolate levels by both perennial and rapid-cycling taxa may be a response to abiotic factors (e.g., mineral resources) or a result of interaction with biotic agents. In either case, the uncoupling of possible coevolution between these insect species and some crucifer species but not others will enable us to isolate plant determinants of specificity by comparing responses of different insect species. It would be of critical interest to know, for example, whether the intrinsic barriers to pierid butterfly colonization of these crucifers are taxon-specific and of taxon-wide distribution [an expectation predicted by coevolutionary hypotheses involving "reciprocal radiation" from ancestral taxa (Mitter and Brooks, 1983)]. An alternative possibility is that the barriers possessed by each crucifer taxon will differ from each other even though all may exert similar effects on susceptible species of pierid butterflies. This possibility might be expected if constitutive barriers in the rapid-cycling plants arose after successful ecological escape due to greatly reduced encounter rates.

The mechanisms by which these crucifers exert their detrimental effects on specialized herbivores such as pierid butterfly larvae are not known. In many cases, the larvae starve in preference to feeding on the plant, suggesting either that they do not recognize it as a potential host plant or that the plant contains some constituent that acts as a powerful feeding deterrent. Work to isolate and identify plant constituents with both positive and negative effects on adult and larval behavior, based on an approach of comparing responses of different pierid species to specific fractions isolated from plants, is now in progress (see Renwick, this volume). In addition, work on the responses of individual pierid species to such constituents will be required, since it is now apparent that biochemically mediated specificity in the interactions of plants and herbivorous insects undoubtedly involves the specific insect systems to a degree not previously suspected (Spencer, 1987).

D. Concluding Remarks

I have argued here that specificity of interactions between crucifers and some of their potential herbivores and symbionts is mediated by factors other than glucosinolates and their hydrolysis products. Further, I have argued that these as-yet-unidentified attributes may have evolved in response to abiotic factors, and that the specificity observed in these examples may be a fortuitous byproduct. But the question of what agents are responsible for specificity in these examples remains. If not glucosinolates, related classes of compounds (e.g., Nielsen, 1978), or other conspicuous taxon-specific constituents, then what? By accurately isolating and identifying the plant constituent determinants of heterotroph specificity, dissecting them from the ecological "defenses" of plants, and documenting the responses of potential herbivores (e.g., Singer, 1984), symbionts, or pathogens to these constituents, we may provide the evidence that will lead to a better understanding of how specificity evolves.

Acknowledgments

I am grateful to A. Renwick, F. Hanson, G. De Boer, and J. Frazier for sharing unpublished results; to J. Glyphis and F. Gould for reference to relevant literature; to J. Rodman, D. Bowers, and P. Feeny for stimulating discussion; to G. Puttick, R. Robbins, P. Feeny, W. Mechaber, and four reviewers for improving the manuscript; to G. Ellmore for help with the figures; and to R. Bryan, for abiding faith. I remain responsible for any errors of fact or interpretation. Research described in this paper was partially supported by the National Geographic Society (2305-81 to F. S. Chew), U.S. Dept. of Agriculture (SEA-CGRO 7900435 to F. S. Chew; University of Wisconsin Cooperative Agreement 58-519-B-0-904 to P. H. Williams) and Tufts University (Dickens–Olmsted Fund to M. G. Glenn). This chapter is lovingly dedicated to the memory of Lee Richmond Girton Snyder.

References

Al-Shebaz, I. A. (1973). The biosystematics of the genus *Thelypodium* (Cruciferae). *Contrib. Gray Herb. Harvard Univ.* **204,** 3–148.

Auerbach, M. J., and Shmida, A. (1985). Israeli crucifers: Patch size, phenology, and defensive chemistry. *Bull. Ecol. Soc. Am.* **66,** 137.

Bethlenfalvey, G. J., Brown, M. S., and Pacovsky, R. S. (1982). Parasitic and mutualistic associations between a mycorrhizal fungus and soybean: Development of the hostplant. *Phytopathology* **72,** 889–893.

Bevege, G. I., and Bowen, G. D. (1975). *Endogone* strains and hostplant differences in development of vesicular–arbuscular mycorrhizas. *In* "Endomycorrhizas" (F. E. Sanders, B. Mosse, and P. B. Tinker, eds.), pp. 77–86. Academic Press, New York.

Bjorkman, R. (1976). Properties and function of plant myrosinases. *In* "The Biology and Chemistry of the Cruciferae" (J. G. Vaughan, A. J. MacLeod, and B. M. G. Jones, eds.), pp. 191–205. Academic Press, New York.

Blau, P. A., Feeny, P., and Contardo, L. (1978). Allylglucosinolate and herbivorous caterpillars: A contrast in toxicity and tolerance. *Science* **200,** 1296–1298.

Bonfante-Fasolo, P. (1982). Cell wall architecture and mycorrhizal association as revealed by cryoultramicrotomy. *Protoplasma* **111,** 113–120.

Breedlove, D. E., and Ehrlich, P. R. (1968). Plant-herbivore coevolution: Lupines and lycaenids. *Science* **162,** 671–672.

Brown, K. S., Damman, A. J., and Feeny, P. (1981). Troidine swallowtails (Lepidoptera: Papilionidae) in southeastern Brazil: Natural history and foodplant relationships. *J. Res. Lepid.* **19,** 199–226.

Burgess, L., and Wiens, J. E. (1980). Dispensing allyl isothiocyanate as an attractant for trapping crucifer-feeding flea beetles. *Can. Entomol.* **112,** 93–97.

Cappuccino, N., and Kareiva, P. (1985). Coping with a capricious environment: A population study of a rare pierid butterfly. *Ecology* **66,** 152–161.

Chapin, F. S. (1980). The mineral nutrition of wild plants. *Annu. Rev. Ecol. Syst.* **11,** 233–260.

Chew, F. S. (1975). Coevolution of pierid butterflies and their cruciferous foodplants. I. The relative quality of available resources. *Oecologia* **20,** 117–127.

Chew, F. S. (1977). Coevolution . . . II. The distribution of eggs on potential foodplants. *Evolution* **31,** 568–579.

Chew, F. S. (1980). Foodplant preferences of *Pieris* caterpillars (Lepidoptera). *Oecologia* **46,** 347–353.

Chew, F. S. (1988). Biological effects of glucosinolates. *In* "Biologically Active Natural Products for Potential Use in Agriculture" (H. G. Cutler, ed.), American Chemical Society, in press.

Chew, F. S., and Robbins, R. K. (1984). Egg-laying in butterflies. *Symp. R. Entomol. Soc. London* **11,** 65–79.

Chew, F. S., and Rodman, J. E. (1979). Plant resources for chemical defense. *In* "Herbivores: Their Interaction with Secondary Plant Metabolites" (G. A. Rosenthal and D. H. Janzen, eds.), pp. 271–307. Academic Press, New York.

Chong, C., Chiang, M. S., and Crete, R. (1985). Studies on glucosinolates in clubroot resistant selection and susceptible commercial cultivars of cabbage. *Euphytica* **34,** 65–73.

Cole, R. A. (1976). Isothiocyanates, nitriles, and thiocyanates as products of autolysis of glucosinolates in Cruciferae. *Phytochemistry* **15,** 759–762.

Coley, P. D., Bryant, J. P., and Chapin, F. S. (1985). Resource availability and plant antiherbivore defense. *Science* **230,** 895–899.

Courtney, S. P. (1982). Coevolution of pierid butterflies and their cruciferous foodplants. IV. Crucifer apparency and *Anthocharis cardamines* oviposition. *Oecologia* **52,** 258–265.

Courtney, S. P., and Chew, F. S. (1987). Coexistence and hostplant use by a large community of pierid butterflies: Habitat is templet. *Oecologia* **71,** 210–220.

D'Abrera, B. (1980). "Butterflies of the Afrotropical Region." Lansdowne, East Milbourne, Australia.

David, W. A. L., and Gardiner, B. O. C. (1962). Oviposition and the hatching of eggs of *Pieris brassicae* in a laboratory culture. *Bull. Entomol. Res.* **53,** 91–109.

David, W. A. L., and Gardiner, B. O. C. (1966). Mustard oil glycosides as feeding stimulants for *Pieris brassicae* larvae in a semisynthetic diet. *Entomol. Exp. Appl.* **9,** 247–255.

De Boer, G., and Hanson, F. E. (1984). Food plant selection and induction of feeding preference among host and non-host plants in larvae of the tobacco hornworm *Manduca sexta*. *Entomol. Exp. Appl.* **35,** 177–193.

Dethier, V. G. (1959). Egg-laying habits of Lepidoptera in relation to available food. *Can. Entomol.* **91,** 554–561.

Dethier, V. G. (1980). Evolution of receptor sensitivity to secondary plant substances with special reference to deterrents. *Am. Nat.* **115,** 45–66.

Ehrlich, P. R., and Raven, P. H. (1964). Butterflies and plants: A study in coevolution. *Evolution* **18,** 586–608.

Elliott, M. C., and Stowe, B. B. (1971). Distribution and variation of indole glucosinolates in woad (*Isatis tinctoria* L.). *Plant Physiol.* **48,** 498–503.

Erickson, J. M., and Feeny, P. (1974). Sinigrin: A chemical barrier to the black swallowtail butterfly, *Papilio polyxenes*. *Ecology* **55,** 103–111.

Ettlinger, M. G., and Kjaer, A. (1968). Sulfur compounds in plants. *Recent Adv. Phytochem.* **1,** 59–144.

Feeny, P. (1975). Biochemical coevolution between plants and their insect herbivores. *In* "Coevolution of Plants and Animals" (L. E. Gilbert and P. H. Raven, eds.), pp. 3–19. Univ. of Texas Press, Austin.

Feeny, P. (1977). Defensive ecology of the Cruciferae. *Ann. Missouri Bot. Gard.* **64,** 221–234.

Feeny, P., and Rosenberry, L. (1982). Seasonal variation in the glucosinolate content of *Brassica nigra* and three *Dentaria* species in central New York state. *Biochem. Syst. Ecol.* **10,** 23–32.

Feeny, P., Paauwe, K. L., and Demong, N. J. (1970). Flea beetles and mustard oils: Hostplant specificity of *Phyllotreta cruciferae* and *P. striolata* adults (Coleoptera: Chrysomelidae). *Ann. Entomol. Soc. Am.* **63,** 832–841.

Feeny, P., Rosenberry, L., and Carter, M. (1983). Chemical aspects of oviposition behavior in butterflies. *In* "Herbivorous Insects: Host-seeking Behavior and Mechanisms" (S. Ahmad, ed.), pp. 27–76. Academic Press, New York.

Fenwick, G. R., Heaney, R. K., and Mullin, W. J. (1983). Glucosinolates and their breakdown products in food and food plants. *CRC Crit. Rev. Food. Sci. Nutr.* **18,** 123–201.

Fraenkel, G. S. (1959). The raison d'être of secondary plant substances. *Science* **129,** 1466–1470.

Free, J. B., and Williams, I. H. (1978). A survey of the damage caused by crops of oil-seed rape (*Brassica napus* L.) by insect pests in south-central England and its effect on seed yield. *J. Agric. Sci.* (*Cambridge*) **90,** 417–424.

Futuyma, D. J. (1983). Evolutionary interactions among herbivorous insects and plants. *In* "Coevolution" (D. J. Futuyma and M. Slatkin, eds.), pp. 207–231. Sinauer, Sunderland, Massachusetts.

Gardner, R. Y. (1977). Systematic distribution and ecological function of the secondary metabolites of the Rosidae–Asteridae. *Biochem. Ecol. Syst.* **5,** 29–35.

Gerdemann, J. W. (1968). Vesicular–arbuscular mycorrhizar and plant growth. *Annu. Rev. Phytopathol.* **6,** 397–418.

Gershenzon, J. (1984). Changes in the levels of plant secondary metabolites under water and nutrient stress. *Recent Adv. Phytochem.* **18,** 273–320.

Glenn, M. G. (1983). Hyphal growth patterns of symbiotic fungi (vesicular–arbuscular mycorrhizae) near host and non-host (Cruciferae: *Brassica*) roots. Ph.D. dissertation, Tufts University, Medford, Massachusetts.

Glenn, M. G., Chew, F. S., and Williams, P. H. (1985). Hyphal penetration of *Brassica* (Cruciferae) roots by a vesicular–arbuscular mycorrhizal fungus. *New Phytol.* **99,** 463–472.

Glenn, M. G., Chew, F. S., and Williams, P. H. (1988). Influence of glucosinolate content of *Brassica* (Cruciferae) roots on growth of vesicular–arbuscular mycorrhizal fungi. *New Phytol.*, in press.

Gould, F. (1984). Mixed-function oxidases and herbivore polyphagy—The devil's advocate position. *Ecol. Entomol.* **9,** 29–34.

Graham, J. H., Leonard, R. T., and Menge, J. A. (1981). Membrane-mediated decrease in root exudation responsible for phospohorus inhibition of vesicular–arbuscular mycorrhizae formation. *Plant Physiol.* **68,** 548–552.

Greenhalgh, J. R., and Mitchell, N. D. (1976). The involvement of flavour volatiles in the resistance to downy mildew of wild and cultivated forms of *Brassica oleracea. New Phytol.* **77,** 391–398.

Grinsted, M. J., Hedley, M. J., White, R. E., and Nye, P. H. (1982). Plant-induced changes in the rhizosphere of rape (*Brassica napus* var. Emerald) seedlings. I. pH Change and the increase in P concentration in the soil solution. *New Phytol.* **91,** 199–229.

Grob, K., and Matile, P. (1979). Vacuolar location of glucosinolates in horseradish root cells. *Plant Sci. Lett.* **14,** 327–335.

Harper, J. L. (1977). "Population Biology of Plants." Academic Press, London.

Hayman, D. S. (1983). The physiology of vesicular–arbuscular endomycorrhizal symbiosis. *Can. J. Bot.* **61,** 944–963.

Hayman, D. S., Johnson, A. M., and Ruddlesdin, I. (1975). The influence of phosphate and crop species on *Endogone* spores and vesicular–arbuscular mycorrhizas under field conditions. *Plant Soil* **43,** 489–495.

Heaney, R. K., and Fenwick, G. R. (1980). The glucosinolate content of *Brassica* vegetables. A chemotaxonomic approach to cultivar identification. *J. Sci. Food Agric.* **31,** 794–801.

Heath, M. C. (1977). A comparative study of non-host interaction with rust fungi. *Physiol. Plant Pathol.* **10,** 73–88.

Hedge, I. C. (1976). A systematic and geographical survey of the Old World Cruciferae. *In* "The Biology and Chemistry of the Cruciferae" (J. G. Vaughn, A. J. MacLeod, and B. M. G. Jones, eds.), pp. 1–45. Academic Press, London.

Hicks, K. P. (1972). Host plant and habitat selection by crucifer feeding flea beetles (Coleoptera: Chrysomelidae). Ph.D. dissertation, Cornell University, Ithaca, New York.

Higgins, L. G., and Riley, N. D. (1980). "A Field Guide to the Butterflies of Britain and Europe," 4th Rev. Ed. Collins, London.

Hirrel, M. C., Mehravaran, H., and Gerdemann, J. W. (1978). Vesicular–arbuscular mycorrhizae in the Chenopodiaceae and Cruciferae: Do they occur? *Can. J. Bot.* **56,** 2813–2817.

Holley, R. A., and Jones, J. D. (1985). The role of myrosinase in the development of toxicity toward *Nematospora* in mustard seed. *Can. J. Bot.* **63,** 521–526.

Hooker, W., Walker, J. C., and Smith, F. G. (1943). Toxicity of beta-isothiocyanate to certain fungi. *Am. J. Bot.* **30,** 632–637.

Hooker, W., Walker, J. C., and Link, K. P. (1945). Effects of two mustard oils on *Plasmodiophora brassicae* and their relation to resistance to clubroot. *J. Agric. Sci.* **70,** 63–78.

Hovanitz, W. and Chang, V. C. S. (1962). Three factors affecting larval choice of foodplant. *J. Res. Lepid.* **1,** 21–42.

Isman, M. B., and Duffey, S. S. (1983). Pharmacokinetics of chlorogenic acid and rutin in larvae of *Heliothis zea. J. Insect Physiol.* **29,** 295–300.

Janzen, D. H. (1968). Hostplants as islands in evolutionary and contemporary time. *Am. Nat.* **107,** 768–790.

Janzen, D. H. (1973). Hostplants as islands. II. Competition in evolutionary and contemporary time. *Am. Nat.* **107,** 786–790.

Jermy, T. (1966). Feeding inhibitors and food preference in chewing phytophagous insects. *Entomol. Exp. Appl.* **9,** 1–12.

Jermy, T. (1984). Evolution of insect/hostplant relationships. *Am. Nat.* **124,** 609–630.

Jordan, C. T. (1981). Population biology and hostplant ecology of caper-feeding pierid butterflies in northeastern Mexico. Ph.D. dissertation, Univ. of Texas, Austin.

Kjaer, A. (1973). The natural distribution of glucosinolates: A uniform group of sulfur-containing glucosides. *In* "Chemistry in Botanical Classification" (G. Bendz and J. Santesson, eds.), pp. 229–234. Academic Press, New York.

Kruckeberg, A. R., Rodman, J. E., and Worthington, R. D. (1982). Natural hybridization between *Streptanthus arizonicus* and *S. carinatus* (Cruciferae). *Syst. Bot.* **7,** 291–299.

Larsen, P. O. (1981). Glucosinolates. *In* "The Biochemistry of Plants" (E. E. Conn, ed.), Vol. 7, pp. 501–525. Academic Press, New York.

Levin, D. A. (1975). Pest pressure and recombination systems in plants. *Am. Nat.* **109,** 437–451.

Levins, R. (1968). "Evolution in Changing Environments." Princeton Univ. Press, Princeton, New Jersey.

Levy, M., and Levin, D. A. (1971). Origin of novel flavonoids in *Phlox* allotetraploids. *Proc. Natl. Acad. Sci. U.S.A.* **68,** 1627–1630.

Lichtenstein, E. P., Strong, F. M., and Morgan, D. G. (1962). Identification of 2-phenylethylisothiocyanate as an insecticide occurring naturally in the edible part of turnips. *J. Agric. Food Chem.* **10,** 30–33.

Lichtenstein, E. P., Morgan, D. G., and Mueller, C. H. (1964). Naturally occurring insecticides in cruciferous crops. *J. Agric. Food Chem.* **12,** 128–161.

Louda, S. M., and Rodman, J. E. (1983). Ecological patterns in the glucosinolate content of a native mustard, *Cardamine cordifolia,* in the Rocky Mountains. *J. Chem. Ecol.* **9,** 397–422.

Luthy, B., and Matile, P. (1984). The mustard oil bomb: rectified analysis of the subcellular organization of the myrosinase system. *Biochem. Physiol. Pflanz.* **179,** 5–12.

Ma, W. C., and Schoonhoven, L. M. (1973). Tarsal chemosensory hairs of the large white butterfly *Pieris brassicae* and their possible role in oviposition behavior. *Entomol. Exp. Appl.* **16,** 343–357.

McKey, D. (1980). Origins of novel alkaloid types—A mechanism for rapid phenotypic evolution of plant secondary compounds. *Am. Nat.* **115,** 754–759.

McKey, D., Watean, P. G., Mbi, C. N., Gartlan, J. S., and Struhsaker, T. T. (1978). Phenolic content of vegetation in two African rain forests: Ecological implications. *Science* **202,** 61–64.

Malloch, D. W., Pirozynski, K. A., and Raven, P. H. (1980). Ecological and evolutionary significance of mycorrhizal symbioses in vascular plants (a review). *Proc. Natl. Acad. Sci. U.S.A.* **77,** 2113–2118.

Marschner, H. (1986). "Mineral Nutrition of Higher Plants." Academic Press, New York.

Medve, R. J. (1983). The mycorrhizal status of the Cruciferae. *Am. Midl. Nat.* **109,** 406–408.

Mithen, R. F., Lewis, B. G., Heaney, R. K., and Fenwick, G. R. (1984). Preliminary investigations of the relationship between leaf glucosinolates and susceptibility to *Leptosphaeria maculans* in wild and cultivated brassicas. *Eucarpia Cruciferae Newslett.* **9,** 82–83.

Mitter, C., and Brooks, D. R. (1983). Phylogenetic aspects of coevolution. *In* "Coevolution" (D. J. Futuyma and M. Slatkin, eds.), pp. 65–98. Sinauer, Sunderland, Massachusetts.

Mooney, H. A. (1983). Carbon-gaining capacity and allocation patterns of Mediterranean-climate plants. *In* "Mediterranean-type Ecosystems: The Role of Nutrients" (F. J. Kruger, D. T. Mitchell, and J. U. M. Jarvis, eds.), pp. 103–119. Springer-Verlag, Berlin.

Muller, J. (1981). Fossil pollen records of extant angiosperms. *Bot. Rev.* **47,** 1–142.

Nair, K. S. S., and McEwen, F. L. (1976). Host selection by the adult cabbage maggot, *Hylemya brassicae* (Dipera: Anthomyidae): effect of glucosinolates and common nutrients on oviposition. *Can. Entomol.* **108,** 1021–1030.

Nault, L. R., and Styer, W. E. (1972). Effects of sinigrin on host selection by aphids. *Entomol. Exp. Appl.* **15,** 423–427.

Nielsen, J. K. (1978). Host plant discrimination within Cruciferae: Feeding responses of four leaf beetles (Coleoptera: Chrysomelidae) to glucosides, cucurbitacins, and cardenolides. *Entomol. Exp. Appl.* **24,** 41–54.

Nielsen, J. K., Larsen, L. M., and Sorenson, H. (1977). Curcurbitacin E and I in *Iberis amara:* Feeding inhibitors for *Phyllotreta nemorum. Phytochemistry* **16,** 1519–1522.

Nijhout, H. F. (1975). Excretory role of the midgut in larvae of the tobacco hornworm *Manduca sexta* (L.). *J. Exp. Biol.* **62,** 221–230.

Nye, P. H. (1981). Changes in pH across the rhizosphere induced by roots. *Plant Soil* **61,** 7–26.

Petterson, J. (1973). Olfactory reactions of *Brevicoryne brassicae* (L.) (Homoptera: Aphidae). *Swed. J. Agric. Res.* **3,** 95–103.

Pihakaski, K., and Iversen, T.-H. (1976). Myrosinase in Brassicaceae. I. Localization of myrosinase in cell fractions of roots of *Sinapis alba* L. *J. Exp. Bot.* **27,** 242–258.

Remington, C. L. (1952). The biology of nearctic Lepidoptera I. Foodplants and life histories of Colorado Papilionoidea. *Psyche* **59,** 61–70.

Renwick, J. A. A., and Radke, C. D. (1983). Chemical recognition of host plants for oviposition by the cabbage butterfly, *Pieris rapae* (Lepidoptera: Pieridae). *Environ. Entomol.* **12,** 446–450.

Rhoades, D. F. (1985). Offensive-defensive interactions between herbivores and plants: Their relevance in herbivore population dynamics and ecological theory. *Am. Nat.* **125,** 205–238.

Rhoades, D. F., and Cates, R. G. (1976). Toward a general theory of plant antiherbivore chemistry. *Recent Adv. Phytochem.* **10,** 168–213.

Rice, E. L. (1984). "Allelopathy," 2nd Ed. Academic Press, New York.

Rodman, J. E. (1974). Systematics and evolution of the genus *Cakile* (Cruciferae). *Contrib. Gray Herb. Harvard Univ.* **205,** 3–146.

Rodman, J. E. (1980). Population variation and hybridization in sea-rockets (*Cakile,* Cruciferae): Seed glucosinolate characters. *Am. J. Bot.* **67,** 1145–1159.

Rodman, J. E. (1981). Divergence, convergence, and parallelism in phytochemical characters: The glucosinolate–myrosinase system. *In* "Phytochemistry and Angiosperm Phylogeny" (D. A. Young and D. A. Seigler, eds.), pp. 43–79. Praeger, New York.

Rodman, J. E., and Chew, F. S. (1980). Phytochemical correlates of herbivory in a community of native and naturalized Cruciferae. *Biochem. Syst. Ecol.* **8,** 43–50.

Rodman, J. E., Kruckeberg, A. R., and Al-Shebaz, I. A. (1981). Chemotaxonomic diversity and complexity in seed glucosinolates of *Caulanthus* and *Streptanthus* (Cruciferae). *Syst. Bot.* **6,** 197–222.

Rodman, J. E., Brower, L., and Frey, J. (1982). Cardenolides in North American *Erysimum* (Cruciferae), a preliminary chemotaxonomic report. *Taxon* **3,** 507–516.

Root, R. B. (1973). Organization of a plant-arthropod association in simple and diverse habitats: The fauna of collards (*Brassica oleracea*). *Ecol. Monogr.* **43,** 95–124.

Schoonhoven, L. M. (1967). Chemoreception of mustard oil glucosides in some lepidopterous larvae. *Entomol. Exp. Appl.* **12,** 555–564.

Schoonhoven, L. M. (1972). Secondary plant substances and insects. *Recent Adv. Phytochem.* **5,** 197–224.

Schoonhoven, L. M. (1977). Individuality of insect feeding behavior. *Proc. Sci. K. Akad. Wet. Amsterdam. Ser. C* **70,** 556–568.

Schwab, S. M., Menge, J. A., and Leonard, R. T. (1983). Quantitative and qualitative effects of phosphorus on extracts of sudangrass roots in relation to vesicular–arbuscular mycorrhizal formation. *Plant Physiol.* **73,** 761–765.

Scriber, J. M., and Slansky, F. (1981). The nutritional ecology of immature insects. *Annu. Rev. Entomol.* **26,** 183–211.

Self, L. S., Guthrie, F. E., and Hodgson, E. (1964). Adaptation of tobacco hornworms to the ingestion of nicotine. *J. Insect Physiol.* **10,** 907–914.

Shapiro, A. M. (1981a). The pierid red-egg syndrome. *Am. Nat.* **117,** 276–294.
Shapiro, A. M. (1981b). Egg-mimics of *Streptanthus* (Cruciferae) deter oviposition by *Pieris sisymbrii. Oecologia* **48,** 142–143.
Singer, M. C. (1984). Butterfly-hostplant relationships: Host quality, adult choice and larval success. *Symp. R. Entomol. Soc. London* **11,** 81–88.
Singer, M. C. (1986). The definition and measurement of oviposition preference in plant-feeding insects. *In* "Methods for Studying Mechanistic Interactions between Insects and Plants" (J. Miller and T. A. Miller, eds.), pp. 65–94. Springer-Verlag, Berlin.
Slansky, F., and Feeny, P. (1977). Stabilization of the rate of nitrogen accumulation by larvae of the cabbage butterfly on wild and cultivated food plants. *Ecol. Monogr.* **47,** 209–228.
Spencer, K. C. (1987). Specificity of action of allelochemicals: Diversification of glycosides. *Am. Chem. Soc. Symp. Ser.* **330,** 275–288.
Talbot, G. (1939). Revisional notes on the genus *Colotis* Hubn. (Lepidoptera: Pieridae). *Trans. R. Entomol. Soc. London* **88,** 173–246.
Thorsteinson, A. J. (1960). Host selection in phytophagous insects. *Annu. Rev. Entomol.* **5,** 193–218.
Tommerup, I. C. (1984). Development of infection by a vesicular–arbuscular mycorrhizal fungus in *Brassica napus* and *Trifolium subterraneum. New Phytol.* **98,** 497–502.
Traynier, R. M. M. (1967). Effect of host plant odor on the behavior of the adult cabbage rootfly *Eroischia brassicae. Entomol. Exp. Appl.* **10,** 321–328.
Tsunoda, S., Hinata, K., and Gomez-Campo, C. (1980). "*Brassica* Crops and Wild Allies." Japan Sci. Press, Tokyo.
Turunen, S. (1985). Absorption. *In* "Comprehensive Insect Physiology, Biochemistry and Pharmacology" (G. A. Kerkut and L. I. Gilbert, eds.), Vol. 4, pp. 241–277. Pergamon, Oxford.
Tutin, T. G., Heywood, V. H., Bruges, N. A., Valentine, D. H., Walters, S. M., and Webb, D. A., eds. (1964). "Flora Europaea," Vol. 1. Cambridge Univ. Press, Cambridge, Massachusetts.
Usher, B. F., and Feeny, P. (1983). Atypical secondary compounds in the family Cruciferae: Tests for toxicity to *Pieris rapae,* an adapted crucifer-feeding insect. *Entomol. Exp. Appl.* **34,** 257–265.
van Emden, H. F. (1972). Aphids as phytochemists. *In* "Phytochemical Ecology" (J. B. Harborne, ed.), pp. 25–43. Academic Press, New York.
van Emden, J. F. (1978). Insects and secondary plant substances—an alternative viewpoint with special reference to aphids. *In* "Biochemical Aspects of Plant and Animal Coevolution" (J. B. Harborne, ed.), pp. 309–323. Academic Press, New York.
VanEtten, C. H., and Tookey, H. L. (1979). Chemistry and biological effects of glucosinolates. *In* "Herbivores: Their Interactions with Secondary Plant Metabolites" (G. A. Rosenthal and D. H. Janzen, eds.), pp. 471–500. Academic Press, New York.
Van Son, G. (1949). The butterflies of Southern Africa. Papilionidae and Pieridae. *Transvaal Mus. Mem.* **1,** (3).
Vaughan, J. G., MacLeod, A. J., and Jones, B. M. G. (1976). "The Biology and Chemistry of the Cruciferae." Academic Press, London.

Verschaffelt, E. (1911). The cause determining the selection of food in some herbivorous insects. *Proc. K. Ned. Akad. Wet.* **13,** 536–542.
Virtanen, A. I. (1965). Studies on organic sulphur compounds and other labile substances in plants. A review. *Phytochemistry* **4,** 207–228.
Walker, J. C., Morell, S., and Foster, H. H. (1937). Toxicity of mustard oils and related sulfur compounds to certain fungi. *Am. J. Bot.* **24,** 536–541.
Watson, M. A., and Casper, B. B. (1984). Morphogenetic constraints on patterns of carbon distribution in plants. *Annu. Rev. Ecol. Syst.* **15,** 233–258.
White, J. (1968). Studies on the behaviour of plant populations in model systems. M.Sc. thesis, Univ. of Wales. Cited by Harper, J. L. (1977). "Population Biology of Plants," p. 119. Academic Press, New York.
Williams, K. S., and Gilbert, L. E. (1981). Insects as selective agents on plant vegetative morphology: Egg mimicry reduces egg-laying by butterflies. *Science* **212,** 467–469.
Williams, P. H. (1981). "Screening Crucifers for Multiple Disease Resistance." Univ. of Wisconsin Press, Madison.
Wilson, D. S. (1983). The group selection controversy: History and current status. *Annu. Rev. Ecol. Syst.* **14,** 159–187.
Wright, S. (1931). Evolution in Mendelian populations. *Genetics* **16,** 97–159.

CHAPTER 5

Stalemates in the Coevolutionary Arms Race: Syntheses, Synergisms, and Sundry Other Sins

MAY R. BERENBAUM AND ARTHUR R. ZANGERL
Department of Entomology
University of Illinois
Urbana, Illinois

SWINGING PENDULUMS have long been an occupational hazard in the field of ecology, and they are especially hazardous in the study of insect plant interactions. The pendulum is currently in motion, for example, around whether herbivorous insects do or do not compete for food (e.g., Lawton and Strong, 1981), whether pugnacious insect bodyguards do or do not enhance fitness of the plants they protect (Bentley, 1977; O'Dowd and Catchpole, 1983; Temple, 1983), and, of most immediate concern in the context of this volume, whether herbivorous insects and their host plants do or do not coevolve. One motive force behind this particular controversy is that the term "coevolution" means different things to different people. The "coevolutionary arms race" concept, since it was first proposed by Ehrlich and Raven (1964) and analogized by Whittaker and Feeny (1971), has been promiscuously applied not only to all *levels* of interaction, ranging from the family to the individual, but also to all *types* of interaction, ranging from predator–prey to herbivore–plant to competitors to symbionts to communities (Futuyma and Slatkin, 1983). In this chapter we discuss the obstacles to recognizing coevolution in nature—in particular, to distinguishing between observational limitations and real genetic constraints in both theory and practice.

I. Constraints on Experimental Detection of Coevolution

Janzen's (1980) reinterpretation of coevolution as reciprocal genetic changes in interacting populations provided a testable hypothesis for identifying a coevolutionary interaction, according to which much in the literature previously described as "coevolution" could not be considered as such. One vocal critic of the concept of herbivore–plant coevolution, Tibor Jermy (1984), maintains that the crucial element of any coevolutionary scenario, that herbivorous insects act as selective agents on their host plants, has yet to be demonstrated incontrovertibly. Intuitively, the concept that insects can differentially reduce plant fitness or even cause differential mortality among genotypes is far from unreasonable. Its intuitive appeal makes all the more puzzling the fact that demonstrations of the phenomenon are few and far between; there are, however, complicating factors that make such demonstrations difficult. First, herbivorous insects are not the only selective agents with which plants must contend; in any given season, response to herbivory, while present, may well be masked by responses to other unidentified selective agents such as abiotic factors (Craig *et al.*, this volume), allelopathic interactions, or competition for nutrients or pollinators. In addition, in our ignorance, we may well be measuring the wrong parameters in order to estimate selection response; our a priori guesses as to how a plant *should* respond may well be partially or even totally divorced from reality. Apparent stalemates in any "arms race," then, may result both from our inability to predict or detect reciprocal responses and from genuine genetic constraints on responses.

II. Constraints on Selection Response in a Coevolutionary Interaction

Plant and animal breeders, for whom selection is artificial and therefore a foregone conclusion, are familiar with the fact that selection response does not always follow expectations even in a carefully controlled experiment. The selection response, the difference in mean phenotype between the offspring of selected parents and the entire parental generation, can be viewed mathematically as,

$$R = h^2S$$

where h^2 is the heritability of a trait and S is the selection differential (Falconer, 1981).

Heritability in the narrow sense is the proportion of the total phenotypic variance attributable to additive genetic variance, or variance contributed by genes affecting the phenotype in a plus-or-minus fashion (Falconer, 1981). Additive gene action is distinct and uncorrelated with dominant and epistatic gene action; the net effect of additive gene action on the phenotype is predictable, whereas the effect of recessive and epistatic gene action is not (Falconer, 1981). Heritabilities do not necessarily reflect the relative contribution of each gene to the trait; not all genes at all loci affect the phenotype equally. Allelic substitutions at "major genes" (Hohenboken, 1985a) have a larger effect on phenotype; favorable alleles at those loci are more likely to reach fixation early in any selection scheme because their effect relative to environmental background will be greater. Thus, a failure to find variable resistance traits does not necessarily rule out a past coevolutionary relationship; a lack of variation could in fact easily result *from* a past coevolutionary relationship if resistance genes are "major genes." An investigator examining an interaction between a plant and an insect may find no heritable variation for a trait that would logically be considered important in mediating the interaction. Such a finding could reflect reduced variability due to prior coevolution, or reduced variability independent of the interaction; discriminating between these two conclusions would require knowledge of the past history of the relationship.

Several other genetic factors mediate selection response (Hohenboken, 1985a). Directional dominance, for example, is one such factor. If a trait is positively influenced primarily by dominant genes, response to selection for an increase in that trait (assuming intermediate frequencies at the start of selection) will be slower than if the trait is influenced by recessive alleles, since heterozygotes, which are phenotypically equivalent to the dominant homozygotes, will persist in the population. Directional gene frequency resulting from a population's previous selection history also mediates selection response (Hohenboken, 1985a). Response to selection will be slow for an allele present in high frequency at the beginning of a selection episode, but selection against that allele would produce rapid change in gene frequency. In addition, heterozygote advantage can influence selection response. If the heterozygote is consistently favored by selection in a given direction, then a population can reach an equilibrium in which genetic variation exists but there is no response to selection. These three factors—directional dominance, directional gene frequency, and heterozygote advantage—can all cause deviations from expected rates of response to selection.

One additional complicating factor that can obscure the detection of or prevent a response to selection is that certain phenotypic traits are component traits, that is, traits determined by two or more contributing characters (Hohenboken, 1985b). In these traits, the phenotype measured is either biologically or mathematically the combined product of several traits. The concentration of any given plant chemical is likely to be a component trait, reflecting the combined action of biosynthetic enzymes and the combined effects of selection on all intermediates in the biosynthetic pathway. Plant secondary chemicals are the products of multiple enzyme systems. These systems often represent groups of functionally related yet distinct enzymes that are compartmentalized, ostensibly to enhance catalytic efficiency, allow separation of competing pathways, protect unstable intermediates, and to increase the potential for metabolic "fine tuning" (Stafford, 1981). The biosynthesis of flavonoids, coumarins, cyanogenic glycosides, certain alkaloids, and terpenoids is thought to involve such multienzyme complexes (Stafford, 1981). Selection on any component trait—such as the concentration of a particular chemical—may not bring about the same genetic changes as selection on any of the several characters contributing to that component trait.

While the response to selection depends in part on the quantity of genetic variation (estimated by heritability), the selection differential, the deviation of the mean phenotype of selected individuals from the population mean (Falconer, 1981), also influences selection response. The selection differential S can be expressed as

$$S = i\sigma_p$$

where σ_p is the phenotypic standard deviation of the trait, and i is the intensity of selection or the "standardized selection differential," the number of phenotypic standard deviations by which the selected individuals differ from the population mean (Falconer, 1981). The selection differential is subject to environmental influences that can bring about a reduced selection response. Population size, which is at least partially environmentally determined, affects the selection intensity in that increasing the population size increases the probability that phenotypically extreme individuals will occur in a population. In addition, environmental factors can influence generation time, which in turn influences the selection response over a given time period. Although response to selection can be expressed in terms of a single generation ($R = h^2S$), the response in a given time interval necessarily depends on t, the generation time in years (so R per year $= h^2S/t$) (Hohenboken, 1985a). Experimental studies may not be long enough to detect the selection response, depending on the complicating influence of environmental variables such as temperature or rainfall.

The selection differential is also influenced by the number of individuals contributing to the next generation. While selection progresses rapidly when only a small proportion of the population is contributing, the process reaches a point of diminishing returns. In small populations, due to genetic drift, selection response is actually decreased because alleles with small but beneficial effects on a trait can be lost and alleles with unfavorable effects are more likely to be fixed. Since very little is known about most breeding systems in wild plant populations (Mitchell-Olds and Rutledge, 1986), expected selection response in a plant–insect interaction is difficult to predict.

Finally, S, the selection differential, is subject to the vagaries of natural selection. A change in a trait responding to one selective agent may well enhance fitness with respect to that selective agent but it may at the same time decrease fitness with respect to another selective agent. For example, an increase in the concentration of a plant chemical may reduce plant vulnerability to attack by generalist herbivores but it at the same time may increase the likelihood of detection and attack by specialist herbivores which use the chemical as a kairomone (e.g., Blau *et al.*, 1978).

Heritability is one constraint that can limit not only the response to selection per se but also the detection of a response to selection. Environmental variance affects any trait with a heritability less than unity, and, as any ecologist knows, environments can change from one day to the next. Thus, heritabilities are unique to the environment in which they are measured, and the actual response to selection is more likely to match the expected response in situations in which parental and progeny generations share a similar environment. Moreover, as selection proceeds, gene frequencies change and alleles become fixed, reducing genetic variation and increasing environmental influence in subsequent generations on a given phenotypic trait. A change in the amount of additive genetic variance changes the heritability of a trait, so heritability estimates apply only to the generation in which they are measured.

Independent contributions of each of the three factors affecting selection response, h^2, S, and t, cannot be easily isolated. While an animal or plant breeder may have some success sorting them out in a laboratory, it is likely that in nature changes in one factor can effectively cancel out changes in another of the three factors. Even under artificial selection with carefully controlled conditions, expectations can fall far short of reality; replicated selection experiments under controlled conditions often produce wildly different results (e.g., Clayton and Robertson, 1957). Genetic changes taking place may simply be undetectable, or selection response may fail to take place due to conflicting factors such as heterozygote advantage, epistasis, linkages among component traits, or opposing forms of selection.

III. Genetics of Resistance in Wild Parsnip to Parsnip Webworms

It is not altogether surprising, given the operational and genetic obstacles to detection, that examples of selection response to herbivory are rare. In fairness, however, it must be said that the number of studies examining *genetic* changes in herbivore–plant interactions are even rarer. Those studies that do exist focus primarily on the genetics of the insect populations (Via, 1984; Rausher, 1984; Gould, 1983, and this volume). We have been studying one particular insect–plant interaction for several years expressly to determine whether insects can indeed effect genetic changes in their host plants. Along the way, we have found evidence that practically every sort of obstacle to detecting selection responses exists in the plant population, including genetic correlations among component traits, opposing forces of natural selection, and environmental effects on phenotypic traits. All these notwithstanding, we did find evidence that insects can indeed act as a selective force on plant chemistry and that there were significant selection differentials in a plant–insect system within an ecological time frame consistent with a coevolutionary scenario (Berenbaum *et al.*, 1986).

Pastinaca sativa L. is a weed introduced from Europe that has become widely established along roadsides and in waste places throughout eastern North America. Throughout its range it is attacked by one principal herbivore, *Depressaria pastinacella* (Duponchel), the parsnip webworm, also believed to be introduced from Europe. *D. pastinacella* feeds on flowers and developing fruits of the plant and can reduce seed production of individual plants to zero (Berenbaum *et al.*, 1986). However, some plants in every population successfully set abundant seed in the presence of the herbivore.

Like many species in the Umbelliferae, *P. sativa* contains furanocoumarins, secondary chemicals with light-associated toxic and deterrent properties to a number of insects (Berenbaum, 1978; Kagan and Chan, 1983; Yajima *et al.*, 1977). Since within-population variation in furanocoumarin content is known to exist in *P. sativa* (Berenbaum *et al.*, 1986) the possibility exists that such chemical variation may underlie the observed variation in resistance to *D. pastinacella*. In order to determine the genetic basis, if any, for resistance to the herbivore, we conducted a series of experiments on families of parsnip seeds collected from the primary umbels of different individuals. Primary umbels set the majority of seeds (Hendrix, 1984) and are consumed by the parsnip webworms.

Since parsnip, as a temporally dioecious umbel, obligately outcrosses in the primary umbel (Cruden and Hermann-Barker, 1977; Berenbaum *et al.*, 1986), seeds on the same umbel are likely to be half-sibs, having the same maternal parent but different paternal parents, although some proportion of full sibs may be present. We employed half-sib estimates of narrow-sense heritability, realizing that these are maximum estimates of additive genetic variance (Berenbaum *et al.*, 1986). Ten seeds from each of 20 families were planted in a completely randomized design in an old-field habitat adjacent to natural stands of parsnip. Ten additional seeds from each of the same 20 families were planted in a greenhouse; these plants were used to compare half-sib and parent offspring estimates of heritability. Plants in the old-field garden were exposed to natural levels of herbivory and plants in the greenhouse were kept free of insects.

We examined numerous plant characteristics and used stepwise multiple regression to identify traits that could best account for significant amounts of variation in seed production in the presence of the herbivore. These traits included phenological traits such as flowering date, physical characteristics (plant size and weight), and chemical characteristics, including the absolute (μg or μg/mg dry weight) and relative amounts (percentage of total) of five furanocoumarins (Fig. 1) in leaves and seeds. Dependent variables in the regression, reflecting herbivore resistance, were the mean number of seeds in the primary umbel per family and per individual, and the proportion of damaged umbellets per family and per individual (Berenbaum *et al.*, 1986).

We also estimated the selection differential for resistance traits once they were identified. Since the covariance between relative fitness and a character is equivalent to the selection differential for that character (Lande and Arnold, 1983), we estimated selection differentials in both the field and greenhouse populations as the covariance between relative fitness and the resistance characters. Moreover, to estimate the selection gradient, we calculated the partial regression coefficients of relative fit-

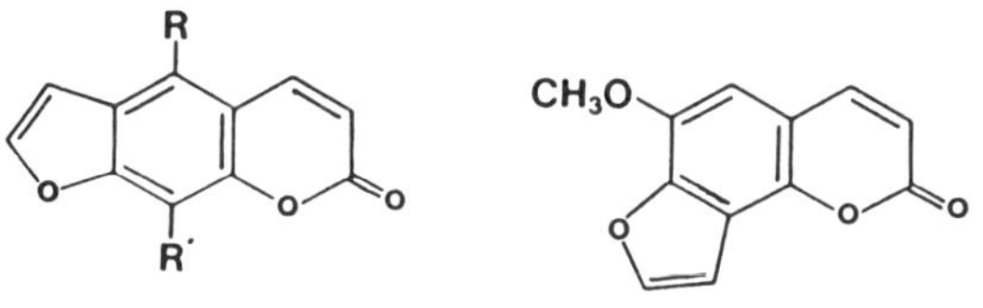

V

I R=OCH_3, R′=H
II R=R′=OCH_3
III R=H, R′=OCH_3
IV R=H, R′=O–

Fig. 1. Furanocoumarins in the wild parsnip.

Table I. Stepwise Multiple Regression to Identify Significant Resistance Components in *Pastinaca sativa*

Independent Variable	Cumulative Variation Explained (%)	Regression Coefficient	F(df)	p
Percent bergapten in seeds	36	+0.98	10.5(4, 15)	<0.01
Sphondin in seeds	50	+0.51		
Percent sphondin in leaves	60	−0.37		
Bergapten in leaves	74	−0.37		

Notes: Dependent variable is number of seeds produced. From Berenbaum *et al.* (1986).

ness on each character; the slope of the selection gradient is equivalent to the magnitude of the regression coefficent and serves as an estimate of the selection intensity (Lande and Arnold, 1983; Berenbaum *et al.*, 1986). We used Lande and Arnold's technique after establishing that furanocoumarins in seeds are not induced by insect feeding (Berenbaum *et al.*, 1986; but see next section).

We found by stepwise multiple regression that four furanocoumarin characters accounted for about three-fourths of the variation in resistance to webworm damage (Table I). All four factors had a heritability significantly different from zero—that is, a significant amount of the phenotypic variance in the trait was directly attributable to additive genetic variance (Table II). In the field, significant selection differentials were found for at least one of these factors (the proportion of the furanocoumarin bergapten in the seeds), suggesting that selection by webworms changed the distribution of the character either directly, or indirectly via selection on a correlated character. Significant selection gradients for this character as well as the other three resistance factors (Table III) imply that the herbivore acted as a selective agent directly on each of those traits.

So genetic variation for resistance to *D. pastinacella* exists in a wild parsnip population. This finding is remarkable given the number of compelling reasons that such genetic variation should neither have existed nor (if it existed) have been detected (Mitchell-Olds and Rutledge, 1986). For example, the fact that susceptible genotypes persist in a population expe-

Table II. Heritability (h^2) of Resistance Components in *Pastinaca sativa*

Character	h^2	SE
Percent bergapten in seeds	0.94	0.308
Sphondin in seeds	0.62	0.292
Percent sphondin in leaves	1.09	0.342
Bergapten in leaves	0.53	0.225

Source: From Berenbaum *et al.* (1986).

Table III. Directional Selection Differentials and Selection Gradients for Resistance Characters in Field Parsnips Exposed to Herbivory by Parsnip Webworms

Character	Selection Differential[a]	Selection Gradient
Flowering date	−0.1891[b]	−0.148
Percent bergapten in seeds	0.0107[b]	0.883[c]
Bergapten in leaves	−0.0027	−0.354[b]
Percent sphondin in leaves	−0.0086	−0.351[b]
Sphondin in seeds	0.0106	0.500[c]

Source: From Berenbaum *et al.* (1986).
[a] Significance level is for the product-moment correlation of relative fitness with the character.
[b] Significant at the 0.05 level.
[c] Significant at the 0.01 level.

riencing herbivory suggests that there is some disadvantage to possessing these resistance traits. The greenhouse family members can be examined to give some indication of these disadvantages, keeping in mind that the greenhouse environment differs from the field environment in many ways other than the absence of herbivory (Berenbaum *et al.*, 1986). One indication is the negative selection differential in greenhouse plants for the proportion of bergapten in seeds (Table IV). This negative selection differential means that plants high in bergapten have lower fitness. This trade-off has a genetic basis, as indicated by genetic correlation. Genetic correlations, the correlations between breeding values (see Falconer, 1981), influence the effect of selection on two traits in the same individual. A high negative genetic correlation, for example, indicates that selection for an increase in one trait results in a simultaneous decrease in the other trait. In the greenhouse, in the absence of *D. pastinacella*, several of the

Table IV. Directional Selection Differentials and Selection Gradients for Resistance Characters in Greenhouse Parsnip Free of Herbivores

Character	Selection Differential[a]	Selection Gradient
Percent bergapten in seeds	−0.00142[b]	−0.463
Bergapten in leaves	−0.00082	−0.044
Percent sphondin in leaves	−0.00091	−0.180
Sphondin in seeds	0.03104	0.158

Source: Berenbaum *et al.* (1986).
[a] Significance level is for the product-moment correlation of relative fitness with the character.
[b] Significant at the 0.05 level.

Table V. Significant Genetic Correlations between Seed or Leaf Furanocoumarins and Secondary Ray Number (Seed Potential) in Greenhouse Parsnips

	r	SE
Bergapten in leaves × number 2° rays	−0.632	0.309
Xanthotoxin in leaves × number 2° rays	−0.790	0.340
Sphondin in leaves × number 2° rays	−0.688	0.267
Percent bergapten in seeds × number 2° rays	−1.528	0.689
Percent isopimpinellin in seeds × number 2° rays	−0.888	0.408

Source: From Berenbaum *et al.* (1986).

major resistance factors are negatively genetically correlated with the production of secondary rays; since each secondary ray bears two seeds, the number of secondary rays is an indication of potential seed set or fitness (Table V). Thus, an increase in resistance, such as an increase in the proportion of bergapten in the seeds, results in a decrease in the potential seed set. Resistant individuals then are potentially less competitive in the absence of herbivory. Opposing selection pressures can act to stabilize the distribution of a trait in the population and mediate the response to selection.

IV. Syntheses as Constraints

The trade-off between fitness in the sense of yield (in our case, seed number) and resistance to insect herbivory hardly counts as a novel observation—plant breeding literature abounds with examples of resistant cultivars that are for one reason or another less than agronomically desirable (Rhoades, 1979). But the implications of a fitness cost of chemical responses to herbivory have not been widely recognized. Trade-offs between alternative biosynthetic pathways or between individual constituents are known to exist (Chew and Rodman, 1979) but the effects of such tradeoffs on susceptibility to insect herbivory are subject to speculation. In the wild parsnip, for example, the biosynthetic pathways leading to furanocoumarin production have for the most part been elucidated (Brown, 1970, 1979; Steck and Brown, 1970; Brown and Steck, 1973; Brown *et al.*, 1970; Thompson *et al.*, 1978) (Fig. 2). In both the greenhouse and the field, there are negative genetic correlations between the proportions of furanocoumarins along branch points in a pathway (Table VI)— angular versus linear, 5- versus 8-substituted, and 8-methoxylated versus 8-*O*-prenylated. In contrast, there are positive genetic correlations between biosynthetic precursors and products—for instance, 5- versus 5,8-substituted furanocoumarins.

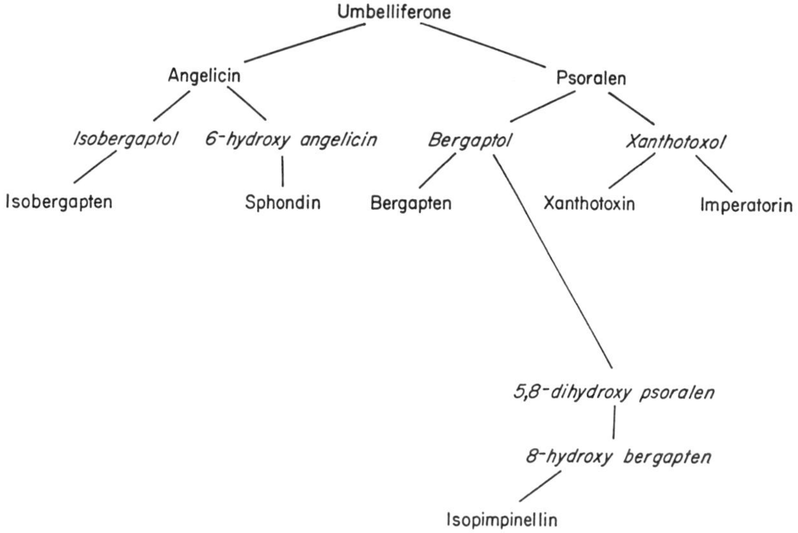

Fig. 2. Biosynthetic pathway leading to furanocoumarins of *Pastinaca sativa* (Brown, 1970, 1979; Steck and Brown, 1970; Brown and Steck, 1973; Brown *et al.*, 1970; Thompson *et al.*, 1978).

These trade-offs suggest that selection cannot easily act to increase one component of a furanocoumarin mixture without altering the concentration of an associated compound in the mixture (robbing Peter to pay Paul, as it were). There appears to be substrate limitation involved in biosynthetic response to selection. This limitation is consistent with the observa-

Table VI. Significant Genetic Correlations among Furanocoumarins along Biosynthetic Branch Points in *Pastinaca sativa*

Structural Type[a]	Compounds[b]	Source of Compounds	Genetic Correlation
Angular vs. linear	SPH vs. BER	Greenhouse leaves	−0.93 (0.14)[c]
	SPH vs. XAN	Greenhouse seeds	−0.39 (0.17)
8-MeO vs. 8-OR	XAN vs. IMP	Field leaves	−0.83 (0.25)
	XAN vs. IMP	Greenhouse leaves	−1.01 (0.20)
5-MeO vs. 5,8-MeO	BER vs. ISO	Greenhouse seeds	0.88 (0.15)
5-MeO vs. 8-MeO	BER vs. XAN	Greenhouse leaves	0.87 (0.27)
	BER vs. XAN	Greenhouse seeds	−0.69 (0.23)

[a] MeO, Methoxylated; OR, *O*-prenylated.
[b] IMP, Imperatorin; BER, bergapten; ISO, isopimpinellin; XAN, xanthotoxin; SPH, sphondin.
[c] SE in parentheses.

Table VII. Effects of Primary Umbel Defloration on Flower Furanocoumarins in *Pastinaca sativa* (Two-Way ANOVA)

Characters	Control	Deflorated
Primary umbel		
Percent bergapten	11.0 ± 0.5	10.4 ± 1.5
Percent xanthotoxin	44.3 ± 3.7	38.2 ± 5.2
Percent sphondin	1.8 ± 0.5	2.4 ± 0.8
Secondary umbel		
Amount of isopimpinellin	11.2 ± 2.7	16.0 ± 1.1[a]
Amount of bergapten	9.8 ± 1.4	11.5 ± 1.0
Percent xanthotoxin	37.0 ± 1.9	39.4 ± 1.6
Percent sphondin	2.1 ± 0.8	2.1 ± 0.6

[a] Significant at $p < 0.05$ (t-test).

tion that all correlations between absolute amounts of furanocoumarins are positive. If there is an increase in a constituent at one point of the pathway, there are corresponding increases everywhere subsequent to that constituent in the pathway (Peter and Paul take their windfall profits and invest jointly, as it were). When more substrate is available, it is channeled proportionately into all parts of the pathway.

Additional evidence of biosynthetic limitations to chemical response to herbivory comes from defloration studies (Nitao, 1987). Removing florets from the primary umbel of parsnip results not in an overall increase in the production of furanocoumarins [viz., cucurbitacins in cucurbits (Carroll and Hoffman, 1980; Tallamy, 1985) or phenolics in many trees (Schultz and Baldwin, 1983)] but rather in only a slight change in the composition of the furanocoumarins present, including an increase in the proportion of isopimpinellin in the remaining flowers, not one of the resistance factors identified earlier by stepwise regression (Table VII). If selection is not acting to increase the overall concentration of biosynthetic precursors, an increase in a resistance factor can occur only at the expense of other constituents in a mixture.

V. Synergisms as Constraints

In addition to limitations on chemical responses due to biosynthetic constraints, toxicological efficacy is limited due to the milieu in which the chemical is produced. Almost certainly there is no plant alive that contains only a single secondary chemical—wild parsnip, for example, contains secondary chemicals from at least seven distinct biosynthetic pathways (Berenbaum, 1985). The toxicity of one class of chemicals may depend to a great extent upon the presence or absence of a co-occurring chemical. In other words, synergistic and antagonistic interactions among secondary chemicals present an additional constraint both on the action and detection of chemical coevolution. For example, another parsnip herbivore, the black swallowtail *Papilio polyxenes asterius* Stoll, metabolizes furanocoumarins via mixed-function oxidases (MFOs), membrane-bound enzymes that perform a variety of oxidative reactions (Bull *et al.*, 1986). In addition to furanocoumarins, wild parsnip produces a class of phenylpropene derivatives possessing a methylenedioxyphenyl moiety (MDP). Some chemicals with this structure are capable of competitively inhibiting MFO activity. According to Bull *et al.* (1986), swallowtail MFOs are sensitive to MDP inhibition. The MDP-containing synthetic synergist piperonyl butoxide administered *in vitro* reduces detoxication ability. Myristicin, an MDP compound that comprises a significant proportion of wild parsnip essential oil (Stahl and Kubeczka, 1979), co-occurs with furanocoumarins in leaves and seeds of wild parsnip; the possibility therefore exists that furanocoumarin toxicity is influenced by the amount of myristicin present. In the laboratory, this effect can be demonstrated with the corn earworm *Heliothis zea* (Boddie) (Berenbaum and Neal, 1985), in which as little as 0.1% myristicin in the diet results in a fivefold decrease in LD_{50}, the dose at which 50% of the population dies. With a tenfold variation within a parsnip population in furanocoumarin content and a thousandfold variation in myristicin content (Neal, 1988), and with all possible combinations thereof, the selective force acting on the herbivore would logically be the combined toxicological effect. Any lack of correlation between one single chemical constituent and a herbivore may simply reflect unaccounted-for synergisms with other chemicals. Selection response may be detectable only as a concerted response among several biosynthetically unrelated traits. If unmeasured chemicals act as significant synergists, the Lande and Arnold (1983) regression approach to measuring selection gradients may not be valid since the outcome of regression analysis depends largely on the choice of characters entered into the regression. This caveat should be acknowledged in studies of selection response.

MFOs are not the only detoxication system in insects prone to inhibition and synergism. In addition to MFOs, many insects possess active glutathione *S*-transferase (GST) enzymes; these largely soluble enzymes effect a number of metabolic conversions associated with detoxication (Matsumura, 1976). Many extremely widespread plant constituents—oxygenated aromatic flavonoids such as myricetin and quercitin—can inhibit GST activity (Yu, 1984), raising the possibility that they synergize toxicants metabolized by GST. Xanthotoxin is a powerful inducer of GST activity and may be at least partially metabolized by that route; the fact that *D. pastinacella* sensitivity to xanthotoxin is unaffected by myristicin

(Nitao, 1987) suggests that such an alternate metabolic route may be present.

Co-occurring substances may not only synergize toxicants, they may also reduce their efficacy. Yu (1984) and others have demonstrated that certain plant constituents, such as coumarins or terpenes, can induce enzyme activity to the point that toxicity is greatly reduced, ostensibly due to more efficient detoxication. The long and short of this discussion is that variation in the concentrations of a particular class of chemical compounds may appear to have little or no correlation with herbivore distribution or abundance as measured phytochemically, but when covariation in synergist or antagonist chemicals is considered, they may account for a great deal more variation than when considered in isolation. Rarely is simultaneous variation in several classes of chemicals measured in this context; for that matter, simultaneous variation even within a class of biosynthetically related compounds is rarely measured. In the case of furanocoumarins, we have found that this within-class variation is an essential component of resistance—no single furanocomarin can account for resistance, nor can the total amount of furanocoumarins in any plant part account satisfactorily for resistance (Berenbaum *et al.*, 1986). Toxicologically, a mixture of furanocoumarins is quite different from an equimolar amount of a single furanocoumarin (Berenbaum, 1985), perhaps due to some kind of analog synergism (McKey, 1979).

In all likelihood, resistance to herbivory in *P. sativa* depends on qualitative and quantitative variation in a great many chemicals. The rarely stated but widely held conviction that coevolution is a gene-for-gene cycle (whereby one gene conceptually equals one chemical) may be an artifact from artificial selection studies, in which plant breeders can select most efficiently for mono- or oligogenic ("major gene") resistance traits. As Gould (1983) stated, polygenic resistance is at least theoretically more likely to arise and persist in a natural situation.

VI. Sundry Other Sins

Finally, there are sundry other sins—ecological or environmental limitations on genetically based chemical responses to herbivory. Again, this is a phenomenon quite familiar to plant breeders. Resistant cultivars often lose their resistance under environmental conditions different from those used to select for resistance. These conditions can act both on the plant and on the herbivore. Temperature is one factor (Tingey and Singh, 1980) that can affect performance of the insect, alter physiological processes in the plant that indirectly affect insect performance through host suitability, or change the toxicological efficacy of plant chemicals [e.g., negative temperature dependence of pyrethrins (Matsumura, 1976)].

Table VIII. Effect of Light and Nutrient Variation on Mean Total Furanocoumarin Concentration in *Pastinaca sativa* Leaves

Light Level	Nutrient Levels		
	Low	Intermediate	High
Low	0.567[a]	0.398	0.437
Intermediate	0.494	0.795	0.576
High	0.491	0.495	0.481

Notes: Analysis of variance for table data: Nutrients, $p = 0.330$; light, $p = <0.001$; light × nutrients, $p = <0.001$. From Zangerl and Berenbaum (1987).
[a] Concentration in μg/mg.

In virtually all cases, heritabilities of furanocoumarin characters were less than 1 (Berenbaum *et al.*, 1986), allowing for environmental modification of these traits to varying degrees. Light intensity and light quality are two such environmental modifiers. Furanocoumarin toxicity as well as quantity are influenced by ultraviolet light (Berenbaum, 1978). Not only does *P. sativa* increase production of furanocoumarins in response to

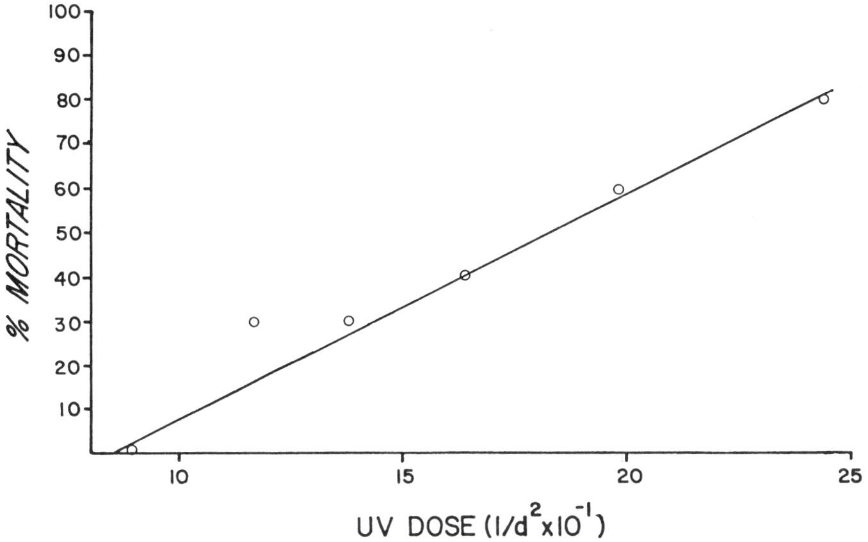

Fig. 3. Toxicity of xanthotoxin to *Heliothis zea*, the corn earworm, at different doses of UV radiation. First-instar larvae were placed inside Pyrex tubes with artificial diets containing 0.1% (by fresh weight) xanthotoxin. Ten tubes were placed at each of five distances from a fluorescent sunlamp emitting UVA radiation peaking at 320 nm. The regression of mortality on UVA dose was significant ($p < 0.001$). UVA dose accounts for 96% of the variation in mortality.

ultraviolet (UV), but the efficacy of the furanocoumarins, due to their mode of action, is enhanced as well (Zangerl and Berenbaum, 1987) (Table VIII and Fig. 3). The location of an individual in a population with respect to incident UV levels could well affect its susceptibility to herbivore damage. Soil nutrients can also affect resistance (Tingey and Singh, 1980, and references therein). Again, soil nutrient quality affects the extent and nature of the furanocoumarin response to light in *P. sativa*. Parsnips grown in the greenhouse under a restricted nutrient regime fail to demonstrate an induction response to increasing light levels (Table VIII) (Zangerl and Berenbaum, 1987). Abiotic factors in the microhabitat then can greatly influence the expression of resistance and the detection of a selection response.

VII. Conclusions

Even though environmental factors influence resistance, resistance may nonetheless be genetically based and subject to selection by the herbivore. There is a plethora of environmental selective forces acting on any given plant trait, and it is not at all surprising that many of these forces, aside from insect herbivory, affect the expression of these traits. It is entirely possible that different agents of plant stress select for particular combinations of chemicals; changes in these *combinations*, rather than individual chemicals, may be specific to the stress agent. One need not argue that insects are the sole selective force maintaining furanocoumarin variation in parsnips at all. Furanocoumarins are broadly biocidal and have demonstrable effects on such ecologically relevant selective forces as bacteria, fungi, nematodes, insects, mammals, birds, and even inter- and intraspecific plant competitors (Murray et al. 1982). Different insects are known to respond differently to furanocoumarins (Berenbaum, 1978; 1981; Berenbaum *et al.*, 1986), and changes in chemical phenotype will affect each species differently.

Berry (1985) stated that the relative fitness of an individual varies with the environment and that "mutualism, parasitism, predation [and] competition are all part of the environment." Particular chemical fitness traits in a plant may be adaptive only under a particular set of environmental conditions. It is now widely recognized in plant breeding that assessment of genotypes for disease resistance should be made in more than one environment (Arnold and Brown, 1968). In all probability, resistance to a specific insect herbivore in any plant species is continuously assessed in many different populations in changing environments. An investigator interested in evaluating plant resistance to herbivores must therefore couch his or her findings in terms of the particular ecological conditions under which it was measured and recognize that they may not pertain in all populations of the plant. After all, according to Janzen (1980), coevolution is a populational phenomenon.

The pendulum of popular sentiment on the subject of coevolution is still swinging, which is as it should be. It is clear, however, from our experience that a priori expectations based on instantaneous and idealized responses to even a well-defined selective agent are almost destined to be disappointing. While sweeping generalizations and grand paradigms may be stimulating, they are a little premature at this point; the study of coevolution requires a few more years of painstaking grunt work to document the genetics of resistance to herbivory, the nature of the responses to selection, and the effects of plant resistance on the genetic structure of the insect population. If the arms race analogy holds to its logical extreme, it seems likely that any stalemates (in terms of both research and selection response) are only temporary at best.

Acknowledgments

For comments on the manuscript we thank J. K. Nitao, J. J. Neal, E. Heininger, S. Sandberg, S. Passoa, and K. Spitze. This research was supported by National Science Foundation grants BSR 82–14713 and BSR 860–6015. We also thank editor Kevin Spencer for his incredible patience and forbearance.

References

Arnold, M. H., and Brown, S. J. (1968). Variation in the host-parasite relationship of a crop disease. *J. Agric. Sci.* **71,** 19–36.

Bentley, B. (1977). Extrafloral nectaries and protection by pugnacious body guards. *Annu. Rev. Ecol. Syst.* **8,** 407–478.

Berenbaum, M. (1978). Toxicity of a furanocoumarin to armyworms: A case of biosynthetic escape from insect herbivores. *Science* **201,** 532–534.

Berenbaum, M. (1981). Effects of linear furanocoumarins on an adapted specialist insect (*Papilio polyxenes*) *Ecol. Entomol.* **6,** 345–351.

Berenbaum, M. R. (1985). Brementown revisited: Allelochemical interactions in plants. *Recent Adv. Phytochem.* **19,** 139–169.

Berenbaum, M., and Neal, J. (1985). Synergism between myristicin and xanthotoxin, a naturally co-occuring plant toxicant. *J. Chem. Ecol.* **11,** 1349–1358.

Berenbaum, M. R., Zangerl, A. R., and Nitao, J. K. (1986). Constraints on chemical coevolution: Wild parsnips and the parsnip webworm. *Evolution* **40,** 1215–1228.

Berry, R. J. (1985). The process of pattern: Genetical possibilities and constraints in coevolution. *Oikos* **44,** 222–228.

Blau, P. A., Feeny, P., Contardo, L., and Robson, D. S. (1978). Allylglucosinolate and herbivorous caterpillars: A contrast in toxicity and tolerance. *Science* **200,** 1296–1298.

Brown, S. A. (1970). Biosynthesis of furanocoumarins in parsnips. *Phytochemistry* **9,** 2471–2479.

Brown, S. A. (1979). Biochemistry of the coumarins. *Recent Adv. Phytochem.* **12,** 249–286.

Brown, S. A., and Steck, W. (1973). 7-Demethylation and osthenol as intermediates in furanocoumarin biosynthesis. *Phytochemistry* **12,** 1315–1324.

Brown, S. A., El-Dakhakny, M., and Steck, W. (1970). Biosynthesis of linear furanocoumarins. *Can. J. Biochem.* **48,** 863–871.

Bull, D. L., Ivie, G. W., Beier, R. C., and Prior, N. W. (1986). *In vitro* metabolism of a linear furanocoumarin (8-methoxypsoralen, xanthotoxin) by mixed function oxidases of larvae of black swallowtail butterfly and fall armyworm. *J. Chem. Ecol.* **12,** 855–892.

Carroll, C. R., and Hoffman, C. A. (1980). Chemical feeding deterrent mobilized in response to insect herbivory and counteradaptation by *Epilachna tredecimnotata. Science* **209,** 414–416.

Chew, F. S., and Rodman, J. E. (1979). Plant resources for chemical defense. *In* "Herbivores: Their Interaction with Secondary Plant Metabolites" (G. Rosenthal and D. H. Janzen, eds.), pp. 271–307. Academic Press, New York.

Clayton, G. A., and Robertson, A. (1957). An experimental check on quantitative genetical theory. II. The long term effects of selection. *J. Genet.* **55,** 152–170.

Cruden, R. W., and Hermann-Barker, S. M. (1977). Temporal dioecism: An alternative to dioecism? *Evolution* **31,** 863–866.

Ehrlich, P. R., and Raven, R. H. (1964). Butterflies and plants: A study in coevolution. *Evolution* **18,** 586–608.

Falconer, D. S. (1981). "Introduction to Quantitative Genetics," 2nd Ed. Longman, New York.

Futuyma, D. J., and Slatkin, M., eds. (1983). "Coevolution." Sinauer, Sunderland, Massachusetts.

Gould, F. (1983). Genetics of plant-herbivore systems: Interactions between applied and basic study. Ch. 17. *In* "Variable Plants and Herbivores in Natural and Managed Systems" (R. F. Denno and M. S. McClure, eds.), Ch. 17, pp. 599–653. Academic Press, New York.

Hendrix, S. D. (1984). Compensatory reproduction in a biennial herb following insect defloration. *Oecologia* **42,** 107–118.

Hohenboken, W. D. (1985a). Prediction and measurement of response to selection. *In* "General and Quantitative Genetics" (A. B. Chapman, ed.), Ch. 8, pp. 167–186. Elsevier, New York.

Hohenboken, W. D. (1985b). Quantitative, threshold and component characteristics. *In* "General and Quantitative Genetics" (A. B. Chapman, ed.), Ch. 4, pp. 61–76. Elsevier, New York.

Janzen, D. H. (1980). When is it coevolution? *Evolution* **34,** 611–612.

Jermy, T. (1984). Evolution of insect/host relationships. *Am. Nat.* **124,** 609–630.

Kagan, J. H., and Chan, G. (1983). The photoovicidal activity of plant components towards *Drosophila melanogaster. Experientia* **39,** 402–403.

Lande, R., and Arnold, S. J. (1983). The measurement of selection on correlated characters. *Evolution* **37,** 1210–1226.
Lawton, J. H., and Strong, D. R. (1981). Community patterns and competition in folivorous insects. *Am. Nat.* **118,** 317–338.
McKey, D. (1979). The distribution of secondary compounds within plants. *In* "Herbivores: Their Interaction with Secondary Plant Metabolites" (G. Rosenthal and D. Janzen, eds.), pp. 56–133. Academic Press, New York.
Matsumura, F. (1976). "Toxicology of Insecticides." Plenum, New York.
Mitchell-Olds, T., and Rutledge, J. J. (1986). Quantitative genetics in natural plant populations: A review of theory. *Am. Nat.* **127,** 379–402.
Murray, R. D. H., Mendez, J., and Brown, S. A. (1982). "The Natural Coumarins." Wiley, Chichester.
Neal, J. (1988). In preparation.
Nitao, J. (1987). Adaptations to furanocoumarins of *Pastinaca sativa* by *Depressaria pastinacella* (Lepidoptera: Oecophoridae). Doctoral dissertation, University of Illinois at Urbana-Champaign.
O'Dowd, D. J., and Catchpole, E. (1983). Ants and extrafloral nectaries—No evidence for plant protection in *Helichrysum* spp.—ant interactions. *Oecologia* **59,** 191–200.
Rausher, M. D. (1984). Tradeoffs in performance on different hosts: Evidence from within- and between-site variation in the beetle *Deloyala guttata*. *Evolution* **38,** 582–595.
Rhoades, D. F. (1979). Evolution of plant chemical defense against herbivores. *In* "Herbivores, Their Interaction with Secondary Plant Metabolites" (G. A. Rosenthal and D. H. Janzen, eds.), pp. 3–54. Academic Press, New York.
Schultz, J. C., and Baldwin, I. (1983). Oak leaf quality declines in response to defoliation by gypsy moth larvae. *Science* **217,** 149–151.
Stafford, H. A. (1981). Compartmentation in natural product biosynthesis by multienzyme complexes. *In* "The Biochemistry of Plants" (E. E. Conn, ed.), Vol. 7, pp. 117–137. Academic Press, New York.
Stahl, E., and Kubeczka, K. H. (1979). Über aetherische Öle der Apiaceae (Umbelliferae). VI Untersuchungen zum Vorkommen von Chemotypen bei *Pastinaca sativa* L. *Planta Med.* **37,** 49–56.
Steck, W., and Brown, S. A. (1970). Biosynthesis of angular furanocoumarins. *Can. J. Biochem.* **48,** 872–880.
Tallamy, D. W. (1985). Squash beetle feeding behavior: An adaptation against induced cucurbit defenses. *Ecology* **66,** 1574–1579.
Temple, A. (1983). Bracken fern (*Pteridium aquilinum*) and nectar-feeding ants: A nonmutualistic interaction. *Ecology* **64,** 1411–1422.
Thompson, H. J., Sharma, S. K., and Brown, S. A. (1978). *O*-Methyltransferases of furanocoumarin biosynthesis. *Arch. Biochem. Biophys.* **188,** 272–287.
Tingey, W., and Singh, J. J. (1980). Environmental factors influencing the magnitude and expression of resistance. *In* "Breeding Plants Resistant to Insects" (F. G. Maxwell and P. R. Jennings, eds.), pp. 87–114. Wiley, New York.
Via, S. (1984). The quantitative genetics of polyphagy in an insect herbivore. II. Genetic correlations in larval performance within and among host plants. *Evolution* **38,** 896–905.

Whittaker, R. H., and Feeny, P. P. (1971). Allelochemics: Chemical interactions between species. *Science* **171,** 757–770.

Yajima, T., Kato, N., and Munakata, K. (1977). Isolation of insect anti-feeding principles in *Orixa japonica*. Thunb. *Agric. Biol. Chem.* **41,** 1263–1268.

Yu, S. J. (1984). Interactions of allelochemicals with detoxication enzymes of insecticide-susceptible and resistant fall armyworms. Pest *Biochem. Physiol.* **22,** 60–68.

Zangerl, A. R., and Berenbaum, M. R. (1987). Furanocoumarins in wild parsnip: Effects of photosynthetically active radiation, ultraviolet light, and nutrients. *Ecology* **68,** 516–520.

CHAPTER 6

Chemistry and Coevolution: Iridoid Glycosides, Plants, and Herbivorous Insects

M. DEANE BOWERS
Museum of Comparative Zoology, Harvard University
Cambridge, Massachusetts

I. Introduction

Much of the criticism of the importance of coevolutionary processes in the evolution of the relationships of insects and their host plants, particularly with regard to the role of plant allelochemistry, has considered only long-term adaptive radiation schema involving on the order of 10^5 or 10^6 years (Janzen, 1980; Mitter and Brooks, 1983; Jermy, 1984). Such long-term coevolutionary interactions (Ehrlich and Raven, 1964) may be obscured by background noise in the system, by conflicting selective pressures, and by changes in the herbivore community over time (Janzen, 1980). Despite these considerations, Berenbaum (1983) developed a very convincing case for such long-term coevolution as a major force in shaping the herbivore community on furanocoumarin-containing plants.

Janzen (1980) defines coevolution as reciprocal evolutionary changes between interacting populations. Such coevolutionary changes, I suggest, do not require the intervention of millennia to be an important component of the interactions of insects and their host plants but, on the contrary, can occur over very short periods of time, on the order of decades or even years. Work on a variety of plant–herbivore systems has provided data that support this idea (Louda and Rodman, 1983b; Bowers and Puttick, 1988; refs. in Gould, 1983; and Mitter and Futuyma, 1983). For example, herbivorous insects may show substantial variation in host plant use among different populations (Hsiao, 1978; Fox and Morrow, 1981; Rausher, 1982; Scriber, 1983; Bowers, 1986). Their short generation time allows insects the potential to evolve quickly in response to changes in the host plant, and to changes in plant allelochemicals. In addition, plant populations or individuals with different allelochemical profiles may be differentially susceptible to herbivore attack (Breedlove and Ehrlich, 1968, 1972; Dolinger *et al.*, 1973; Rice *et al.*, 1978; Lincoln and Langenheim, 1979; Louda and Rodman, 1983a,b). These characteristics suggest that short-term reciprocal evolutionary responses of insect and host-plant populations may be important in the dynamics of these interactions and may be amenable to study and experimental manipulation.

Recent work on the interactions of several species of herbivorous insects and plants containing a group of plant allelochemicals, the iridoid glycosides, provides a system with potential for examining such short-term coevolution. In this chapter I will first introduce the iridoid glycosides and describe ways in which these compounds are important in the interactions of insects and their host-plants and predators. Second, I will discuss evidence that plant allelochemistry acts as a selective agent on insect populations and that insect herbivores act as selective agents on plant populations by acting specifically on plant allelochemistry.

II. The Iridoid Glycosides

The iridoids are a group of cyclopentanoid monoterpene-derived compounds, which usually, but not always, occur as O-linked glucosides (Fig. 1, I). About 300 iridoids (El-Naggar and Beal, 1980) have been described from about 50 families of plants (Table I). The term iridoid is derived from iridodial (Fig. 1, II), iridomyrmecin, and other similar compounds that

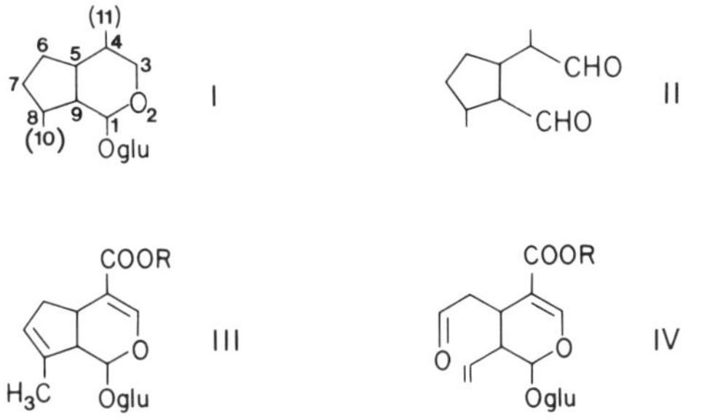

Fig. 1. I, Numbering system of iridoid glycosides; II, iridodial; III, general structure of a carbocyclic iridoid glycoside; IV, general structure of a seco-iridoid.

serve as defense compounds of ants in the genus *Iridomyrmex* (Cavill *et al.*, 1956). These compounds are volatile monoterpenes, while the iridoids are nonvolatile, generally polar compounds.

The iridoids as a group can be divided into two classes, the carbocyclic iridoids (Fig. 1, III) which have a ring containing only carbon, and the seco-iridoids (Fig. 1, IV), which lack such a ring. Jensen *et al.* (1975) further divide the iridoids into several groups based on their chemical structure and biosynthesis. In addition, seco-iridoids and close derivatives can form the more complex indole alkaloids by combining with an amino acid or amino acid derivative, usually tryptophan or tryptamine (Cordell, 1974). Recent experiments suggest that there are two major biosynthetic pathways for iridoid production in plants (Damtoft *et al.*, 1983). The first [Group 1 of Damtoft *et al.* (1983)] produces iridoids derived from deoxyloganic acid, including the seco-iridoids and indole alkaloids. The second [Group 2 of Damtoft *et al.* (1983)] produces iridoids derived from 8-*epi*-deoxyloganic acid.

In recent years the iridoids have come into prominence for several reasons: (1) they have proven to be useful characters for indicating phylogenetic relationships among plant families and orders (Jensen *et al.*, 1975; Hegnauer and Kooiman, 1978; Dahlgren *et al.*, 1981; Gershenzon and Mabry, 1983); (2) new iridoids are continually being isolated and identified (e.g., Bianco *et al.*, 1985; Jensen and Nielsen, 1985; Marco, 1985; Gardner *et al.*, 1987); (3) iridoids have been shown to be viable precursors for prostaglandin synthesis (Berkowitz *et al.*, 1979; Tixidre *et al.*, 1982); most relevant here, (4) they are mediators of the interactions of insects and plants (Bowers, 1980, 1981, 1983b, 1984; Bowers and Puttick, 1986; Sha'aban *et al.*, 1980; Bernays and DeLuca, 1981; Stephenson, 1981; 1982; Stermitz *et al.*, 1986a); and (5) they are important in the interaction of insects with their predators (Bowers, 1980, 1981, 1983a; Bowers *et al.*, 1985).

Table I. Orders and Families of Plants Containing Iridoid Glycosides

Order, Family	Order, Family
Hamamelidales	Dipsacales
Hamamelidaceae	Caprifoliaceae
Buxales	Viburnaceae
Daphniphyllaceae	Valerianaceae
Rutales	Dipsaceae
Meliaceae	Calyceraceae
Polygalales	Loasales
Malpighiaceae	Loasaceae
Fouquieriales	Goodeniales
Fouquieriaceae	Goodeniaceae
Ericales	Oleales
Actinidiaceae	Oleaceae
Ericaceae	Gentianales
Monotropaceae	Loganiaceae
Pyrolaceae	Rubiaceae
Epacridaceae	Menyanthaceae
Roridulaceae	Gentianaceae
Eucommiales	Apocynaceae
Eucommiaceae	Scrophulariales
Sarraceniales	Bignoniaceae
Sarraceniaceae	Myoporaceae
Cornales	Buddlejaceae
Garryaceae	Scrophulariaceae
Alangiaceae	Globulariaceae
Nyssaceae	Selaginaceae
Cornaceae	Stilbaceae
Aucubaceae	Retziaceae
Davidiaceae	Plantaginaceae
Toricelliaceae	Lentibulariaceae
Symplocaceae	Pedaliaceae
Icacinaceae	Martyniaceae
Escalloniaceae	Acanthaceae
Montiniaceae	Hippuridales
Stylidiaceae	Hippuridaceae
Hydrangeaceae	Lamiales
Sambucaceae	Verbenaceae
Adoxaceae	Callitrichaceae
	Lamiaceae

Source: Taxonomy follows Dahlgren (1980) as described in Gershenzon and Mabry (1983).

Iridoid glycosides or iridoid-containing plants have been used medicinally as purgatives, antibiotics, diuretics, sedatives, and, perhaps most intriguing, as an anti-aphrodisiac known as monk's pepper (Table II). Of most relevance here, plants containing iridoid glycosides have been used as natural insecticides (Hegnauer, 1966) (Table II).

Table II. Pharmacological and Biological Activity of Iridoids

Activity	Reference
Bitter tonic	Hegnauer (1966)
Wound healing	Hegnauer (1966)
Antibiotic agent	Rombouts and Links (1956)
Diuretic	Hegnauer (1966)
Hypotensive agent	Petkov and Manolov (1972)
Sedative	Inouye *et al.* (1974b)
Laxative	Inouye *et al.* (1974a)
Purgative	Inouye *et al.* (1974a)
Anti-aphrodisiac	Hansel and Winde (1959)
Insecticide	Hegnauer (1966)
Insect defensive compound	Pavan and Finzi (1974), Pavan (1975), Bowers and Puttick (1986)

Note: More complete discussions of these may be found in Hegnauer (1966), Bobbitt and Segebarth (1969), and Sticher (1977).

One notable characteristic of iridoids is their bitterness, which is evident from their taste to humans and their use as "bitter tonics" (Thomas, 1961; Hegnauer, 1966; Bobbitt and Segebarth, 1969), although not all iridoids have this quality. Particularly notable in this regard are some of the iridoids from gentians, *Gentiana* (Gentianaceae) species and the Indian gentian, *Swertia chirata* Buch.-Ham., which are some of the bitterest substances known (Wagner and Vasirian, 1974; Sticher, 1977). The bitter quality of the iridoids has been linked to their chemical structure (Kubota and Kubo, 1969). This characteristic of the iridoids may be of critical importance in their use as defensive compounds by insects feeding on plants containing them (Bowers and Puttick, 1986).

The iridoid glycosides are of particular interest to botanists because of their potential value in elucidating phylogenetic relationships among certain groups of angiosperms (Jensen *et al.*, 1975; Dahlgren *et al.*, 1981; Gershenzon and Mabry, 1983). Most notably, Dahlgren (Jensen *et al.*, 1975; Dahlgren, 1977; 1980; Dahlgren *et al.*, 1981) arranged the major orders of the Asteridae [*sensu* Cronquist, (1968)] into two groups, the iridoid and noniridoid groups. He has suggested that the division between those families with and without iridoids constitutes a major split in the Asteridae. In addition, Dahlgren has used the presence of iridoids, coupled with other morphological characters, in families outside the Asteridae to suggest that these families are more closely related to the Asteridae, for example, the Loasaceae (Dahlgren *et al.*, 1981) and the Cornales (Gershenzen and Mabry, 1983). Gershenzon and Mabry (1983) provide a comparison of the various classifications of the angiosperms, with particular attention to chemical features, notably the iridoids. They also very importantly point out that iridoids occur in some unrelated families such as the Daphniphyllaceae, Hamamelidaceae, Meliaceae, and Malpighiaceae, suggesting that iridoids may have arisen several times in the evolution of the angiosperms (Gershenzon and Mabry, 1983, p. 27). Kaplan and Gottlieb (1982) have also used iridoids as systematic markers for the classification of the dicotyledons.

III. Iridoid Glycosides, Plants, and Insects

Research on the importance of iridoid glycosides in the interaction of insects and plants is a recent development. However, in one of the early demonstrations of the importance of plant allelochemicals in insect feeding behavior, Nayar and Fraenkel (1963) suggested that iridoid glycosides were responsible for the restriction of larvae of the Catalpa sphinx, *Ceratomia catalpae* (Boisduval) (Lepidoptera: Sphingidae), to feeding on species of *Catalpa* (Bignoniaceae), although at that time they were unable to identify the particular iridoids involved. In addition, Hegnauer (1973, p. 352) suggested that iridoids might be important in the host-plant specificity of the butterfly *Euphydryas editha* (Boisduval) (Nymphalidae) based on the findings of Singer (1971). Subsequent research has shown that, indeed, iridoid glycosides are important in a variety of ways in the interactions of plants containing those compounds, insects feeding on those plants, and the predators of those insects.

A. Iridoid Glycosides and Host-Plant Specialization

The host-plant records of a variety of insect species suggest that they are specialists on iridoid glycoside-containing plants. For example, the host-plant records for butterflies in the genera *Euphydryas* (Bowers, 1981) and *Anartia* (Table III) (Nymphalidae) and for flea beetles in the genus

Table III. Larval Host-Plant Families of *Anartia* Species

Butterfly Species	Host-Plant Families
Anartia amathea (L.)	Acanthaceae Labiatae
Anartia chrysopelea (Hübner)	Verbenaceae
Anartia lytrea (Godart)	unknown
Anartia fatima (Godart)	Acanthaceae
Anartia jatrophe (L.)	Acanthaceae Labiatae Verbenaceae Scrophulariaceae

Notes: From Silberglied *et al.* (1979). The four plant families used by these butterflies contain species that have been shown to contain iridoid glycosides (Jensen *et al.*, 1975).

Dibolia (Chrysomelidae) (Table IV) are restricted to iridoid-containing genera or families. In addition, the buckeye butterfly, *Junonia coenia* (Hübner) (Nymphalidae), specializes on plant species in four families that contain iridoid glycosides: Scrophulariaceae, Plantaginaceae, Acanthaceae, and Verbenaceae (Bowers, 1984). The specialization of these insects on iridoid glycoside-containing plants suggested that these compounds were important larval feeding stimulants and attractants. Indeed, feeding experiments using artificial diets showed that larvae of *Euphydryas chalcedona* (Doubleday & Hewitson) and *J. coenia* grew and survived best on diets with iridoid glycosides and preferred those diets in choice tests (Bowers, 1983a, 1984).

The implication of these results on growth and feeding preference of *J. coenia* and *E. chalcedona* larvae, coupled with host-plant records for these insects, was that ovipositing females also used these compounds as cues for appropriate host plants. However, work with other systems had suggested that female oviposition cues may be more complex than simply plant allelochemicals (Feeny *et al.*, 1983). Factors such as leaf shape (Rausher, 1978), host-plant distribution (Singer, 1971), and time since last oviposition (Singer, 1983) can all affect a female's oviposition behavior. Despite these confounding factors, Pereyra and Bowers (1988) found that *J. coenia* females use pure iridoid glycosides as oviposition cues. In both choice and nonchoice tests using agar discs as oviposition substrates, females laid virtually no eggs on control discs without iridoid glycosides and laid dozens of eggs on discs with dried ground leaf material or pure iridoid glycosides typical of the host plant *Plantago lanceolata* L. (Planta-

Table IV. Host Plants of Leaf Beetles in the Genus *Dibolia* (Chrysomelidae)

Species	Host Plant
D. borealis Chev.	*Plantago major* L. (Plantaginaceae), *P. lanceolata* L.
D. ovata LeConte	unknown
D. championi Jacoby	unknown
D. sinuata Horn	unknown
D. lybonoti Horn	unknown
D. catherinia Mignot	*Penstemon* sp. (Scrophulariaceae)
D. reyheria Mignot	*Penstemon* sp.
D. kansana Parry	unknown
D. penstemonis Parry	*Penstemon* sp.
D. californica Parry	unknown
D. obscura Parry	unknown
D. chelones Parry	*Chelone glabra* L. (Scrophulariaceae)
D. melampyri Parry	*Melampyrum americanum* (Beauverd) (Scrophulariaceae)

Source: From Parry (1974).

ginaceae). In addition, females were able to distinguish discs with the highest concentration of iridoid glycoside and preferred to oviposit on those.

B. Iridoid Glycosides and Generalist Insects

Although serving as attractants to some insect species specializing on iridoid glycoside-containing plants, iridoid glycosides are deterrent to at least some generalist insects. For example, the iridoid glycoside ipolamiide was found to inhibit feeding in three species of generalist insects, *Spodoptera littoralis* (Lepidoptera: Noctuidae), and the acridid grasshoppers *Schistocerca gregaria* (Forskål) and *Locusta migratoria* L. (Orthoptera: Acrididae) (Bernays and DeLuca, 1981). In addition, two other iridoid glycosides, catalposide and specioside, were found to inhibit feeding in the gypsy moth, *Lymantria dispar* L. (Lymantriidae) (El-Naggar and Doskotch, 1980; Bowers and Puttick, unpublished data). Three other iridoid glycosides, aucubin, catalpol, and loganin, significantly decreased the growth rate of larvae of the gypsy moth and the southern armyworm, *Spodoptera eridania* (Cramer) (Bowers and Puttick, 1988; Puttick and Bowers, 1988). A fourth iridoid glycoside, asperuloside, also significantly reduced larval growth in the gypsy moth, but not in *S. eridania* (Bowers and Puttick, 1988; Puttick and Bowers, 1988). Thus, although most iridoid glycosides tested appear to reduce feeding and/or growth of generalist insects, different species of generalist insects may respond differently to specific iridoid glycosides.

A different but analogous system was studied by Stephenson (1981, 1982) who found that iridoid glycosides in the nectar of *Catalpa speciosa* Warder (Bignoniaceae) deterred nectar thieves. Specifically, he found that incorporation of catalpol (Fig. 2, IX) and catalposide (Fig. 2, VI), two iridoid glycosides found in *C. speciosa*, into sucrose solutions in amounts similar to those reported for the leaves of *Catalpa* species inhibited feeding by nectar thieves [ants and the skipper, *Poanes hobomok* (Harris) (Lepidoptera: Hesperiidae)]. Legitimate pollinators such as bumblebees and carpenter bees were not deterred from feeding on either *C. speciosa* nectar or the sucrose solution containing the iridoid glycosides. The iridoid glycosides also appeared to be toxic to the nectar thieves—those species exhibited abnormal behavior, such as an inability to fly, falling off the substrate, inability to orient correctly, and general intoxication (Stephenson, 1982). It is interesting that the nectar-robbing ant species Stephenson used were those that visit the extrafloral nectaries of *C. speciosa*. His results suggest that the nectar composition in the extrafloral nectaries is different from that of the flowers and does not contain iridoid glycosides or contains them in amounts too low to be deterrent to the ants (Stephenson, 1982).

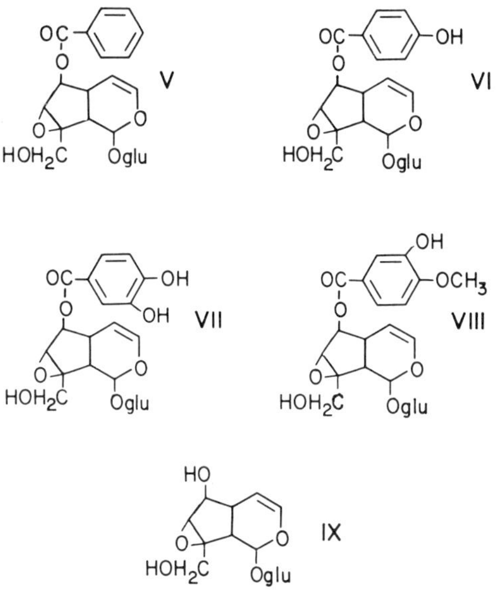

Fig. 2. V, Veronicoside; VI, catalposide; VII, verproside; VIII, 6-isovanillylcatalpol; IX, catalpol.

C. Sequestration of Iridoid Glycosides and Unpalatability

Laboratory feeding experiments have shown that butterflies of the genus *Euphydryas* are generally unpalatable to birds (Bowers, 1980, 1981; M. D. Bowers and S. Farley, unpublished), although substantial bird predation has been documented in one population of *E. chalcedona* (Bowers *et al.*, 1985). *Euphydryas phaeton* (Drury) the eastern representative of the genus, is not only unpalatable (i.e., tastes bad and is rejected on the basis of this bad taste), but is also toxic, causing vomiting in bluejays that have ingested the butterfly (Bowers, 1980) [see Brower (1984) for a discussion of the distinction between unpalatability and toxicity]. However, the unpalatability of the butterfly depends on the host plant fed upon by the larva: butterflies that fed as larvae on the primary host plant, *Chelone glabra* L. (Scrophulariaceae), are unpalatable and emetic, while those which fed as larvae on an alternate host plant, *Plantago lanceolata*, are relatively palatable (Bowers, 1980).

Euphydryas butterflies sequester iridoid glycosides from the host plant during larval feeding and retain these compounds through to the adult stage (Bowers and Puttick, 1986; Stermitz *et al.*, 1986a) (Table V). Sequestration of iridoid glycosides appears to be responsible for the unpalatability of these butterflies (Bowers and Puttick, 1986). One iridoid glycoside, catalpol (Fig. 2, IX) seems to be particularly important in the

Table V. Occurrence of Iridoid Glycosides in Plants and Insect Herbivores Feeding on those Plants Containing Iridoid Glycosides

Insect	Host Plant	Iridoids[a]	Sample	Iridoids	Reference[b]
Euphydryas phaeton (Drury) (Nymphalidae)	*Chelone glabra* L. (Scrophulariaceae)	auc & cat	Egg	cat	1
			Larva	cat	
			Pupa	cat	
			Adult	cat	
			Meconium	cat	
			Frass	0	
	Aureolaria flava (L.) (Scrophulariaceae)	auc	Larva	auc	1
			Pupa	auc	
			Adult	auc	
			Meconium	auc	
			Frass	auc	
	Plantago lanceolata L. (Plantaginaceae)	auc & cat	Eggs	cat	1
			Larva	cat & auc	
			Pupa	cat & auc	
			Adult	cat & auc	
			Meconium	cat & auc	
			Frass	0	
Euphydryas chalcedona (Doubleday & Hewitson)	*Scrophularia californica* (Chambers & Schlecht) (Scrophulariaceae)	auc	Adult	auc	1
Euphydryas anicia (Doubleday & Hewitson)	*Besseya plantaginea* James (Rydb.) (Scrophulariaceae)	auc, cat, & others	Adult	auc & cat	2
	Castilleja integra Gray (Scrophulariaceae)	cat, mac, & others	Adult	cat & mac	2
Poladryas arachne (Edwards) (Nymphalidae)	*Penstemon virgatus* Gray (Scrophulariaceae)	cat & others	Adult	cat	4

Platyptilia pica (Wlsm.) (Pterophoridae)	*Castilleja sulphurea* Rydb.	cat, auc, & 5 others	Larva	cat[c]	3
			Adult	0	
			Frass	+[d]	
Meris alticola Hulst (Geometridae)	*Besseya plantaginea*	auc, cat, & others	Pupa	cat	5
			Cocoon silk	cat	
			Adult	0	
			Frass	0	
Junonia coenia (Hübner) (Nymphalidae)	*Plantago lanceolata*	auc & cat	Larva	auc & cat	1
			Pupa	auc & cat	
			Adult	0	
			Meconium	auc & cat	
			Frass	auc & cat	
Ceratomia catalpae (Boisduval) (Sphingidae)	*Catalpa bignonioides* Walt. (Bignoniaceae)	cat, cat'side, & others	Larva	cat	1
			Pupa	cat	
			Adult	0	
			Meconium	cat, cat'side	
			Frass	cat, cat'side & others	
Cotesia congregata[e] (Say) (Braconidae)			Adult	0	1
Endothenia hebesana (Walker) (Tortricidae)	*Chelone glabra*	auc & cat	Pupa	0	1
Tenthredo grandis (Norton) (Tenthredinidae)	*Chelone glabra*	auc & cat	Larva	cat	1
			Frass	cat	
Dibolia chelones Parry (Chrysomelidae)	*Chelone glabra*	auc & cat	Adult	auc & cat	

[a] Cat, catalpol; auc, aucubin; mac, macfadienoside; cat'side, catalposide.
[b] References: 1, Bowers and Puttick (1986); Bowers, Puttick, and Grunow (unpublished); 2, Stermitz *et al.* (1986a); 3, Stermitz *et al.* (1986b); 4, Stermitz *et al.* (unpublished); 5, Gardner (1987).
[c] A trace of catalpol was found in one larval sample and was probably due to food present in the gut.
[d] Iridoids were reported in the frass (+ = positive), but compounds were not identified.
[e] *Cotesia congregata* is a parasitoid of *C. catalpae*.

unpalatability of *Euphydryas*. Chemical analyses of *E. phaeton* and the primary and an alternate host plant support this idea. *Chelone glabra* (primary host plant) and *P. lanceolata* (alternate host plant) contain the same two major iridoid glycosides, aucubin and catalpol (Bowers and Puttick, 1986, and refs. therein); however, the relative amounts of these two compounds differ substantially in the two plant species. HPLC analysis showed that *C. glabra* contains about six times more catalpol than aucubin, while *P. lanceolata* contains more aucubin than catalpol (Bowers and Puttick, unpublished data). *Euphydryas phaeton* butterflies fed on these two plants reflect the relative concentrations of these two iridoid glycosides in the hostplants: *E. phaeton* butterflies fed on *C. glabra* contain more catalpol than aucubin, while those fed on *P. lanceolata* contain about equal amounts of the two compounds. Thus the emetic quality of *E. phaeton* fed on *C. glabra* appears to be due to the much higher catalpol content of that plant species and the resulting higher catalpol content of butterflies reared on that plant. In addition, *E. phaeton* selectively sequesters catalpol from *P. lanceolata*: although there is variation from one plant sample to another and from one butterfly to another in the total and relative amounts of catalpol and aucubin, the mean ratio of aucubin to catalpol in *P. lanceolata* is 19 : 1, while that in the butterflies is about 1 : 1 (M. D. Bowers and G. M. Puttick, 1986; unpublished data).

Stermitz *et al.* (1986a) found that another species of *Euphydryas, E. anicia* (Doubleday & Hewitson), also contained a much higher proportion of catalpol than was found in the host plant, *Besseya plantaginea* (James) (Rydb.) (Scrophulariaceae). *Besseya plantaginea* contains six iridoid glycosides: catalpol, 8-epiloganic acid, mussaenoside, verproside, 6-isovanillylcatalpol, and veronicoside. The latter three compounds are all esters of catalpol: veronicoside is 6-benzoylcatalpol (Fig. 2, V), and verproside is 6-dihydroxybenzoylcatalpol (Fig. 2, VII). Thus these four iridoid glycosides have the same basic structure, differing only in the substituent group on C-6 (Fig. 2). *Euphydryas anicia* may be selectively sequestering catalpol, but it also appears to be metabolizing the catalpol esters to catalpol. Stermitz *et al.* (1986a) did not find veronicoside, 6-isovanillylcatalpol, or verproside in the butterflies, although these compounds comprised 78% of the total iridoid glycoside content of the plant, suggesting that conversion is indeed occurring. Gardner corroborated this suggestion by feeding larvae of *E. anicia* an artificial diet containing an ester of catalpol, 6-isovanillylcatalpol (Gardner, 1987; Gardner and Stermitz, 1988). High-performance liquid chromatography analyses showed that larvae metabolized the 6-isovanillylcatalpol into catalpol, which was sequestered, and isovanillic acid, which was excreted (Gardner and Stermitz, 1988). This sequestration of catalpol may be linked to the apparent importance of catalpol to the unpalatability of these butterflies or it may be a function of the chemical structure and consequent metabolism of these esters.

A European species of *Euphydryas*, *E. cynthia* Hb., fed on *P. lanceolata* and *P. alpina* L., has recently been shown to sequester iridoid glycosides (Franke *et al.*, 1987). Although adults were not analyzed, a combined sample of eight larvae and eight pupae was found to contain aucubin, catalpol, and 6-*O*-glucopyranosyl aucubin (Franke *et al.*, 1987).

Other insect species that feed on iridoid glycoside-containing plants also sequester these compounds (Table V). Two species of sawfly feed as larvae on *Chelone glabra: Tenthredo grandis* (Norton) and *Macrophya nigra* (Norton) (Tenthredinidae) (Stamp, 1984). The larvae are superficially difficult to distinguish, and both are black and white. They rest on the host plant, are extremely conspicuous, and appear to be Müllerian mimics (personal observations). The adults of these two species look quite different, but both are black. Feeding tests with bluejays did not distinguish between the larvae of these two species, but showed that they were unpalatable and emetic (Bowers, 1980). Thin-layer chromatography (TLC) has shown that these larvae contain the iridoid glycoside catalpol (Table V). The adults have not been tested.

Flea beetles in the genus *Dibolia* (Chrysomelidae: Alticinae) specialize on plants containing iridoid glycosides (Table IV). These beetles feed on the same plant species as larvae and adults. The larvae are leaf miners, while the adults are leaf skeletonizers (Parry, 1974), feeding on leaf surfaces and appearing extremely conspicuous. Adults of the species that feeds on *C. glabra*, first identified as *D. borealis* Chev., but probably *D. chelones* Parry (Parry, 1974), were assayed for the presence of iridoid glycosides by TLC and found to contain aucubin and catalpol, the two primary iridoid glycosides known from *C. glabra* (Table V). This is particularly interesting in that the flea beetles (Alticinae) are unusual in the Chrysomelidae, having lost the characteristic defensive glands that are distinctive in their close relatives in the Chrysomelinae (Pasteels *et al.*, 1984). Instead, the flea beetles, as their name suggests, escape by a quick jump. At least in *D. chelones*, there appears to be an additional defense of containing iridoid glycosides.

D. Comparative Processing of Iridoid Glycosides

Some insect species that specialize on iridoid glycoside-containing plants do not sequester these compounds through to the adult stage (Table V). For example, adult butterflies of *J. coenia* do not contain iridoid glycosides and are palatable (Bowers and Puttick, 1986; M. D. Bowers and S. Farley, unpublished data). Although they are present in the larva and pupa, they are eliminated in the meconium at adult eclosion and may function to deter invertebrate predators of the teneral adults (Bowers and Puttick, 1986). Another lepidopteran, *Platyptilia pica* (Wlsm.) (Pterophoridae), feeds on *Castilleja* species and some other species of Scrophulariaceae, but does not sequester iridoid glycosides (Roby and Stermitz, 1984a; Stermitz *et al.*, 1986b), whereas *E. anicia,* which also feeds on *Castilleja,* does sequester these compounds (Stermitz *et al.*, 1986a).

Iridoids ingested during larval feeding are also eliminated in the meconium of *C. catalpae* (Bowers and Puttick, 1986). The host plant *Catalpa bignonioides* Walt. (Bignoniaceae) contains an array of iridoid glycosides, possibly as many as 15 (Nayar and Fraenkel, 1963). We have been able to consistently distinguish five iridoid glycosides in *C. bignonioides* using TLC. The larva of *C. catalpae* contains only catalpol, while the larval feces contain the other four (Bowers and Puttick, 1986). There may also be some metabolic conversion of host-plant iridoids into catalpol, as was found in *E. anicia* (Stermitz *et al.,* 1986a). The adult of *C. catalpae* does not contain any iridoid glycosides and was palatable to gray jays, *Perisoreus canadensis* (Corvidae) (M. D. Bowers and S. Farley, unpublished data). Of note here is that the larvae of *C. catalpae* are gregarious and warningly colored black and yellow, while the adults are cryptic. Larvae of other *Ceratomia* species are solitary and cryptic, as are the adults (Hodges, 1971). This suggests that the ability to feed on *Catalpa,* with the sequestration of iridoids by the larvae and consequent unpalatability, has led to the evolution of gregarious behavior and aposematic coloration in *C. catalpae* larvae.

E. Insect Radiations onto Iridoid-Containing Plant Taxa

Several taxa of herbivorous insects appear to have radiated onto iridoid-containing plant species: *Euphydryas* (Lepidoptera: Nymphalidae), *Anartia* (Lepidoptera: Nymphalidae), and *Dibolia* (Coleoptera: Chrysomelidae). Where the host plants are known for North American *Dibolia,* all feed on plants containing iridoid glycosides (Table IV) (Parry, 1974). In *Euphydryas* and *Dibolia* these compounds are also sequestered and used as a defense against the predators of these insects, while individuals of the various *Anartia* species appear to be palatable (Silberglied *et al.,* 1979, 1980; Bowers *et al.,* 1987).

These genera are in interesting contrast to other iridoid specialists such as *C. catalpae, Papaipema nepheleptena* (Dyar) (Lepidoptera: Noctuidae), which is a stem borer in *C. glabra,* and the sawflies, *M. nigra* and *T. grandis,* whose close relatives use plant species which do not contain iridoid glycosides (Table VI, *P. nepheleptena*) (Hodges, 1971; Hessel, 1954; Stamp, 1984; Krombein *et al.,* 1979). Thus in some groups, the ability to feed on plants containing iridoid glycosides has led to a radiation

Table VI. Groups of Plants and Plant Families Fed on by Species of *Papaipema* (Noctuidae)

Plant Family	Number of Species of *Papaipema*
Ferns	4
Liliaceae	3
Araliaceae	1
Berberidaceae	1
Compositae	13
Labiatae	2
Leguminosae	1
Moraceae	1
Oleaceae	1
Primulaceae	1
Ranunculaceae	2
Sarraceniaceae	1
Scrophulariaceae	2
Umbelliferae	4

Note: The families Labiatae, Oleaceae, Sarraceniaceae, and Scrophulariaceae are known to contain iridoid glycosides.

onto a variety of iridoid-containing plant species, while in other groups this has not occurred.

In those insect taxa that have diversified, host-plant records indicate that although they are restricted to iridoid glycoside-containing plant species, single insect species may feed on plants in several families. In contrast, those insect taxa that have not speciated are monophagous: *C. catalpae* on *Catalpa*, *T. grandis* and *M. nigra* on *C. glabra*, and *P. nepheleptena* on *C. glabra*. For these latter insect taxa, specific cues other than iridoid glycosides may be restricting feeding to a single species or genus. Comparative studies of feeding behavior and dietary preference of insects that have and have not diversified onto iridoid-containing plant taxa would be rewarding.

Thus the iridoid glycosides are an example of a group of plant allelochemicals that play a critical role in the interactions of plants that contain them and insect herbivores on those plants. The information available about ecology, host-plant preferences, and population biology of the insect herbivores feeding on iridoid glycoside-containing plants, coupled with insight into the chemistry of the hostplants and its importance to insect host-plant preference and insect defense, suggest that this system may be particularly useful for studying short-term coevolutionary interactions.

IV. Chemistry and Coevolution

A coevolutionary interaction necessitates reciprocal responses on the part of all participants. With respect to plants and insects, this can be described in the following questions about a particular plant–insect–allelochemical system:

1. Has plant allelochemistry acted as a selective force on the insect population? And in turn,
2. Has the insect herbivore acted as a selective force on the plant population, and in particular to alter plant allelochemical makeup?

The answer to both of these questions must be yes for the critical feature of a coevolutionary relationship, reciprocity, to be met.

In trying to look at coevolution from a phylogenetic point of view, many factors may "conspire" to obscure the coevolutionary aspects of the relationship. Perhaps most crucial of these is the inability to distinguish between association by descent and colonization (Janzen, 1980; Mitter and Brooks, 1983). That is, have the insect and its host plant truly coevolved, or did the insect relatively recently invade a new habitat and find this host-plant species to be the one with the suite of defensive characters with which it was able to cope, both physiologically and behaviorally (Janzen, 1980; Futuyma, 1983)? A second important problem that becomes apparent in looking at chemistry and coevolution is that there may be a diverse guild of herbivores exerting selective pressures of very different kinds on a single plant population. For example, the effects of one herbivore species that feeds on the seeds may be obscured by the impact of the foliage feeders; and generalist herbivores may be very different as selective agents than specialist herbivores. Third, although the existence of crop cultivars and study of some natural population indicate that variation in plant allelochemistry is heritable, environmental effects may also affect plant chemistry. Age of the plant, the plant part or organ examined, nutrient levels, time of day, previous levels of herbivory, and microhabitat are only some of the factors that can affect variation in plant chemistry (Robinson, 1974; McKey, 1979; Woodhead, 1981; Krischik and Denno, 1983). These environmental effects may override the genotypic contribution to alleochemical variation, or at least obscure any clear patterns. Fourth, chemical defenses may be induced over relatively short periods of time in response to herbivory (Ryan, 1979; Carroll and Hoffmann, 1980; Schultz and Baldwin, 1982; Wratten *et al.*, 1984; Karban and Carey, 1984), potentially obscuring any indication of the genetic component of observed allelochemical variation.

Despite this impressive array of potentially confounding factors, and there are certainly others, certain systems may be less subject to some of these effects than others. Substantial evidence has accumulated that plant allelochemistry can and does act as a selective agent on insect populations, and/or that insect herbivores act as selective agents on plant populations. Evidence for a reciprocal interaction, that is, a system where there is evidence that both these processes are operating, is, however, a different matter. Evidence for the role of insects and plant allelochemicals as selective agents is discussed below, and is used to propose a scenario in which short-term coevolution is an integral part of the evolution of at least some insect–plant interactions.

A. Insects as Selective Agents on Plant Allelochemistry

For selection to operate, there must be genetic variation on which it can act. There is abundant evidence that there is substantial qualitative and quantitative variation in allelochemical content both within and between plant populations. In addition, there may be variation among individuals within a species (Dirzo and Harper, 1982a; Louda and Rodman, 1983a,b; Rodman and Louda, 1984; Mihaliak and Lincoln, 1985; Krischik and Denno, 1983), and even between different organs of an individual plant (Nelson *et al.,* 1981; Roby and Stermitz, 1984b; McKey, 1979; Krischik and Denno, 1983), in what particular compounds are present and in what amounts they occur. These elements of variation are not necessarily genetically determined, although they may be (see below). As discussed above, an important confounding factor is that the allelochemical composition may change with the time of the season that the plants are sampled (Rodman and Louda, 1984; Nelson *et al.,* 1981; Mooney and Chu, 1974) and the conditions under which the plant is growing (Lincoln and Mooney, 1984; Rhoades, 1979; Gershenzon *et al.,* 1978; Cooper-Driver *et al.,* 1977; McKey *et al.,* 1978), and even with the time of day that the plant is harvested (Robinson, 1974).

Despite the potential effects of all this variation, there are many examples that show that there is a genetic basis for at least some allelochemical variation (Lincoln and Langenheim, 1978, 1979; Woodhead, 1981; Krischik and Denno, 1983; discussion by Gould, this volume). In addition, certain genotypes have been shown to be more or less susceptible to herbivores. For example, a single gene has been shown to control the production of a bitter cucurbitacin in cucumber, *Cucumis sativus* L. (Cucurbitaceae) (DaCosta and Jones, 1971). Presence of this cucurbitacin was equivocally advantageous to the plant: it increased mortality of, and reduced damage by, the two-spotted mite, *Tetranychus urticae* Koch.; but at the same time it attracted cucumber beetles, which are cucurbitacin specialists, to the detriment of the plant (DaCosta and Jones, 1971). In a more complex system, the genetics of monoterpene production in the mint *Satureja douglasii* (Benth.) Briq. (Labiatae) have been worked out (Lincoln and Langenheim, 1981; Lincoln *et al.,* 1986). Different monoterpenoid compositional types are of different palatabilities to a molluscan herbivore, and there is differential predation on these compositional types in the field (Rice *et al.*, 1978).

One example where the genetic basis of allelochemical variation is well known is cyanogenesis in the legumes *Lotus corniculatus* L. and *Trifolium repens* L. (Leguminosae) (Hughes, 1981). Cyanide does not occur freely within the plant, but is released by the action of hydrolytic enzymes on a precursor glucoside (Hoesel, 1981). Cyanogenesis in *T. repens* and *L. corniculatus* is under the control of two genes (Hughes, 1981). Cyanogenesis only occurs in individuals with at least one dominant allele of both genes. The alleles of the gene *Ac* control the synthesis of the cyanogenic glucosides from their amino acid precursors. The hydrolysis of the cyanogenic glycosides to produce cyanide is determined by alleles of another, independently inherited gene, *Li*. Individuals heterozygous for *Ac* or *Li* have intermediate levels of glucosides or enzymes, respectively, when compared with homozygous individuals (Conn, 1979). The cyanogenic forms are well protected against herbivory by slugs and snails, the principal herbivores (Dirzo, 1980). The acyanogenic forms, in contrast, are quite tasty to these animals, are highly preferred in choice tests, but show some signs of being better competitors (Dirzo, 1980; Dirzo and Harper, 1982b). Dirzo and Harper (1982a) found that within a single population of *T. repens,* high slug and snail activity was correlated with a high density of cyanogenic individuals. In contrast, areas with little or no slug and snail activity contained predominately individual *T. repens* that were acyanogenic. Thus there is differential herbivory on different genotypes: that is, differential selection on allelochemical variation that is genetically based.

For glucosinolate-producing plants, it was found that herbivory by a guild of crucifer specialists was concentrated on individuals of *Cardamine cordifolia* Gray (Cruciferae) that were lowest in glucosinolate content (Fig. 3) (see also Louda and Rodman, 1983a,b; Rodman and Louda, 1984). Contrary to predictions of plant–herbivore theory (e.g., Blau *et al.*, 1978), these specialist insect species as a group were responding negatively to higher glucosinolate content of individual host plants and damaging plants with lower levels (Rodman and Louda, 1984). Thus in a population with sustained high levels of herbivory (Louda, 1984) concentrated on individuals low in glucosinolates and with consequent reduction in seed set by those individuals (Louda, 1984, 1986), selection by a specialist herbivore guild should lead to a population of plants relatively richer in glucosinolates than the present population. It is just such selection by the crucifer-feeding guild that is a critical component of the coevolutionary process, if there is a genetic component to glucosinolate production. Evidence from glucosinolate variation in different strains of cruciferous crops indicate that there is indeed a genetic basis to this variation (van Etten and Tookey, 1979).

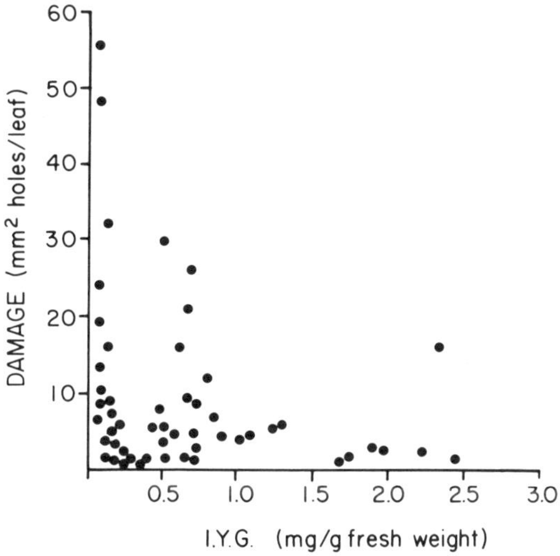

Fig. 3. Amount of damage to *Cardamine cordifolia* by a guild of crucifer specialist insects in relation to isothiocyanate-yielding glucosinolate (I.Y.G.) content of leaves. The regression coefficient (Kendall's tau) = 0.26, $p < 0.01$. Data from Louda and Rodman, unpublished.

For iridoid glycosides, there may be substantial interindividual and interpopulation variation in both the amounts and the kinds of iridoids present. Stermitz *et al.* (1986b) found substantial qualitative and quantitative variation in the iridoid glycoside content of *Castilleja sulphurea* Rydb. (Scrophulariaceae) and *C. occidentialis* Torrey from several populations. They examined seven different iridoid glycosides from various populations of these two species and found, for example, that the catalpol content of *C. sulphurea* varied from 21% of the total iridoid glycoside content in one population to 95% in another population. The population that contained 95% catalpol contained 5% of a second iridoid, aucubin, and none of the other iridoids that were found in varying amounts in the other eight populations. These different iridoid phenotypes may be perceived very differently by herbivores, and may have very different effects on their growth and survival, as has been found in experiments with iridoid glycosides in artificial diets (Bowers and Puttick, 1988; Puttick and Bowers, 1988).

Unfortunately, the genetics of iridoid glycoside production have not been studied. The genetic basis of at least some of the extensive variation found in these compounds must be established before their potential involvement in coevolutionary interactions can be completely assessed. Evidence from other groups of compounds, and the qualitative and quantitative variation found among different populations of *Castilleja* species described above, however, suggest a genetic basis for at least some of this variation.

In summary, insect herbivory can potentially be a potent selective force on the allelochemical makeup of plant populations. Insect herbivores may prefer certain allelochemical genotypes, or avoid individual plants with a certain allelochemical makeup. Such selection may be based on preference for, or enhanced growth on, these genotypes, or on avoidance of or reduced growth and survival on other genotypes. In a coevolutionary scenario, this constitutes one side of the interaction. The other side, that is, plant chemistry acting as a selective agent on the insect population, must also be a part of the interaction.

B. Plant Chemistry as a Selective Agent on Insect Populations

Many species groups of insects, such as *Euphydryas,* are tied to a particular suite of host plants by the allelochemicals of those host plants. Thus pierid butterflies specialize on crucifers containing mustard oils, danaines like the Monarch, *Danaus plexippus* L. (Danaidae), on milkweeds containing cardenolides, and *Euphydryas* on plants that contain iridoid glycosides. However, other closely related species groups may show no such importance of plant chemistry in their choice of host plants. For example, as discussed earlier, *Euphydryas phaeton* feeds primarily on *Chelone glabra* (Scrophulariaceae) and secondarily on other plant species that contain iridoid glycosides. Another specialist on *C. glabra, P. nepheleptena,* is monophagous on that plant and yet its relatives feed on a wide range of plant species (Table VI). *Papaipema* species are stemborers, however, and this habit may make host-plant chemistry less important in determining host range for them than for leaf-feeders. Thus plant allelochemicals may be important in the evolutionary diversification of some groups of insects and not others.

At the population level, host-plant races exist in many insect species of several orders, most notably the Lepidoptera (Fox and Morrow, 1981; Singer, 1971, 1983; Scriber, 1983; Bowers, 1986) and Coleoptera (Hsiao, 1978; Rausher, 1983), but also Diptera (Huettel and Bush, 1972; Jaenike and Grimaldi, 1983; Via, 1984a,b), tenthredinid sawflies in the Hymenoptera (Knerer and Atwood, 1973), and Homoptera (Alstad and Edmunds, 1983). Although substantial evidence of genetic divergence between such "host races" or populations exists in the form of differences in survival, growth rate, and utilization indices on different host-plant species (e.g., Tabashnik *et al.,* 1981; Rausher, 1982; Scriber, 1983), only rarely have these differences been linked directly to host-plant chemistry (but see Alstad and Edmunds, 1983).

For example, butterflies in the genus *Euphydryas* are well known for their diversification in host-plant utilization patterns (Ehrlich and Murphy, 1981; Singer, 1983). Different populations of *E. editha* and *E. chalcedona* may use different host plants and show behavioral and physiological adaptation to those host plants (Rausher, 1982; Singer, 1983; Bowers, 1986). However, in the one case where at least part of the allelochemical makeup of the host-plants was examined in addition to the performance of insects from each of two populations on their own host plant and that of the other population, both plant species contained the same iridoid glycosides (Bowers, 1986). This study involved two populations of *E. chalcedona,* one in the foothills of the Sierra, feeding on *Penstemon breviflorus* Lindl. (Scrophulariaceae), and one in the high Sierra, feeding on another *Penstemon* species, *P. newberryi* Gray. TLC showed that the profile of iridoid glycosides in these two host-plant species was the same, although possible quantitative differences in the relative amounts of the specific compounds were not determined and may have existed (Bowers, 1986). In addition, there may have been iridoids or other allelochemicals not detected by TLC. Although differences in insect herbivore performance and digestive efficiency on plants may be linked to other characteristics of the plant species under study, plant allelochemistry can affect insect growth and survival.

Differential response to different individual iridoid glycosides was found in larvae of the iridoid specialist, *Junonia coenia*. These larvae performed best on diets containing aucubin and catalpol, the two iridoids from one of their normal host-plants, *P. lanceolata,* but very poorly on artificial diets containing two other iridoid glycosides, loganin and asperuloside, which are biosynthetically related to aucubin and catalpol (Inouye, 1971). Thus different, but biosynthetically related, compounds, which may occur in the same host-plant species, population or individual, may have very different effects on herbivores. Quantitative, as well as qualitative, variation, such as that found by Stermitz *et al.* (1986b) in *Castilleja* populations, may affect larval feeding patterns (e.g., by avoidance of or preference for certain chemotypes) as well as growth and survival.

Differences in chemical structure, including relatively small changes in an R-group, can alter the "taste" (Kubota and Kubo, 1969) as well as toxicity of the compound, and perhaps determine whether or not the insect can perceive the compound, or how the compound is perceived (e.g., attractant, deterrent, or neutral). Thus the suite of iridoid glycosides (or other compounds) present in an individual plant may determine insect preference for, or survival on, plants with that genotype. Preferred plants or plant genotypes, or plants producing the fastest growth and/or highest survival of herbivores, may be selectively eaten, leading to a preponderance of unpreferred or more "toxic" plants in the population. Herbivore species and the consequent patterns of herbivory among individual plants may therefore be important in determining the allelochemical genotypes found in particular populations.

Despite the demonstrated importance of particular groups of allelochemicals in determining the host-plant range of many insect species, experiments have shown that selection can change host-plant preference or tolerance in only a few generations (Gould, 1979; Alstad and Edmunds, 1983; Gould, this volume). The ability of insect populations to adapt to novel chemicals in their environment has been the bane of the agricultural industry since the advent of insecticides (Georghiou, 1983), and suggests that they have a similar ability as regards to naturally occurring plant chemicals. Such an example is the Monarch butterfly, *D. plexippus* L., normally confined to feeding on milkweeds containing cardenolides. Brower (1969) was able to select a strain of *D. plexippus* that would feed on cabbage, which is in the Cruciferae and contains glucosinolates, a completely different group of compounds from those found in milkweeds. Although no data are given on the procedures or the number of generations involved in selecting this strain, it probably did not involve more than a dozen or so generations. Interestingly, sinigrin, a glucosinolate, was found to be highly toxic to another unadapted specialist herbivore, *Papilio polyxenes* Fab. (Papilionidae), the black swallowtail, a specialist on Umbelliferae (Blau *et al.*, 1978). The high mortality of *D. plexippus* described by Brower (1969) indicates that cabbage, containing glucosinolates, was initially toxic to *D. plexippus*. Nonetheless, despite the demonstrated toxicity of glucosinolates to nonadapted specialists, Brower (1969) was able to select a strain of *D. plexippus* that would feed and survive on cabbage.

Work with the iridoid glycoside specialist *J. coenia* showed that selection operating over a relatively short period of time, 30–40 generations, could alter the response of larvae to iridoid glycosides, the allelochemicals normally used as larval feeding cues (Bowers, 1984). Larvae reared from eggs from wild-collected females ("wild" strain) preferred artificial diets containing iridoid glycosides found in natural host plants over diets that did not, and grew significantly better on artificial diets containing dried leaf powder of the host plant, *P. lanceolata,* or on diets with pure iridoid glycosides from this host plant, than on a diet devoid of iridoid glycosides (Bowers, 1984; Bowers and Puttick, 1988). However, a strain of *J. coenia* that had been reared on artificial diets containing *P. lanceolata* leaf powder for 30–40 generations (designated the "lab" strain) grew equally well on artificial diets with and without iridoid glycosides (Fig. 4) and did not distinguish diets with and without iridoid glycosides in choice tests (M. D. Bowers and G. M. Puttick, unpublished data).

Newly hatched larvae of the "lab" and "wild" strains were reared for 15 days on one of five artificial diets: artificial diet with no additives (AD), with 3.6% *P. lanceolata* leaf powder (P.l.), with 0.18% aucubin (auc),

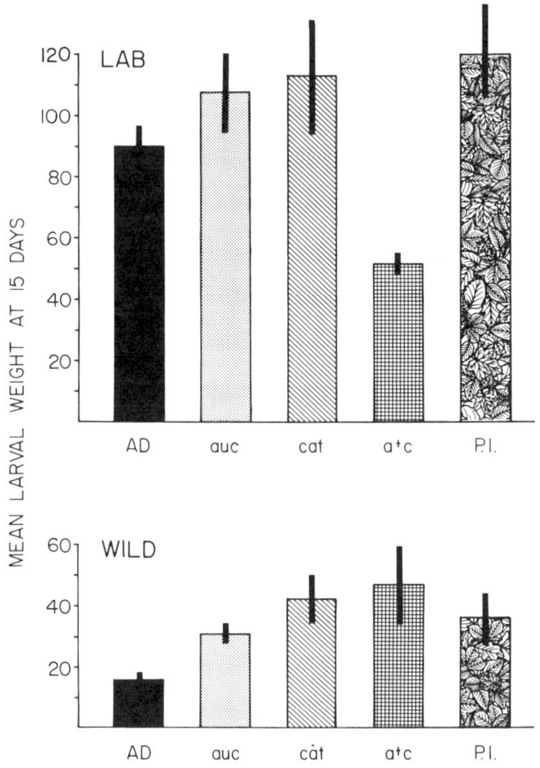

Fig. 4. Larval weight (mean ± SE) of *Junonia coenia* from the "lab" and "wild" strains, after feeding for 15 days on artificial diets containing no additives (AD), 0.18% aucubin (auc), 0.18% catalpol (cat), 0.09%, aucubin and 0.09% catalpol (a+c), or 3.6% dried *P. lanceolata* leaf powder (P.l.).

with 0.18% catalpol (cat), and with a combination of 0.09% aucubin and 0.09% catalpol, the two iridoid glycosides found in *P. lanceolata* (a+c). A comparison of the weight of larvae from the two strains after 15 days of growth on these diets (Fig. 4) showed several differences between these two strains. Two-way ANOVA showed that larvae from the "lab" strain grew significantly larger on the artificial diets than larvae from the "wild" strain ($F = 89.119$, df = 1, 30, $p < 0.001$). There was also a significant effect of diet; larvae from both strains grew differently on the five diets ($F = 3.414$, df = 4, 30, $p < 0.025$). Most importantly, there was a significant interaction of strain and diet ($F = 4.828$, df = 4, 30, $p < 0.005$), indicating that larvae from the two strains responded differently to the array of diets. Interestingly, larvae of the "wild" strain grew slightly, although not significantly, better on the artificial diet that contained a combination of the two iridoid glycosides found in the host plant (*P.*

lanceolata), aucubin and catalpol; the larvae of the "lab" strain grew poorly on this diet (Fig. 4).

Thus the "lab" strain had altered responses to the allelochemicals normally critical in the ability of larvae to distinguish appropriate host plants, the iridoid glycosides, after 30–40 generations of selection. And note that exposure of these larvae to iridoid glycosides had not been withdrawn: the females still oviposited on leaves of *P. lanceolata,* and the artificial diet contained about 4% dried *P. lanceolata* leaf powder. Thus larvae were exposed to the appropriate compounds, but in a lower dosage than in the host plant. Although the biochemical basis for the selection in this strain is unknown, the important point is that selection altered the response of these larvae to the allelochemicals normally encountered in the host plants, even though exposure to these chemicals had not been withdrawn. Another interesting facet of these experiments was that although larvae of the "wild" strain performed very poorly on artificial diets containing loganin or asperuloside, "lab" strain larvae survived and grew similarly on artificial diets with loganin and asperuloside as they did on the diets with aucubin and catalpol (M. D. Bowers and G. M. Puttick, unpublished data).

In summary, insects can evolve "resistance" to plant allelochemicals (or insecticides) in relatively few generations, and populations of at least some insect species have the ability to respond relatively quickly to changes in the allelochemical makeup of their host plant. Exposure to variation in the relative quantities of different individual allelochemicals, as well as different arrays of allelochemicals, can be important in the evolution of behavioral, ecological, and physiological responses by insect herbivores.

V. Conclusion—Chemical Coevolution in Plant–Insect Interactions

Short-term coevolutionary changes can be important in the interaction of plant and insect herbivore populations. As described in the previous sections, differential herbivory upon certain allelochemical phenotypes can alter the genotypic makeup of plant populations. Such changes in the host-plant population can in turn act as a selective force upon the insect population. Individual insects that can tolerate or perhaps even prefer the new genotype will have enhanced survival and reproductive output. The insect population may thus go through a bottleneck as the preferred plant genotypes are removed from the population and then recover as those individuals that could grow and survive on the new genotype reproduce. These coevolutionary changes can occur in decades and not necessarily over millennia. Short-term coevolution may thus be an important factor in determining the patterns of host-plant utilization observed in many insect species.

Certain systems of insects and their host plants are particularly amenable to an examination of these short-term reciprocal evolutionary changes. Characteristics of such systems are listed in Table VII. In general, insect and plant associations that can be profitably studied from the perspective of chemical coevolution consist of small, stable, sedentary, chemically well-characterized plant populations, with a limited number of specialist herbivores and little generalist herbivory.

The iridoid glycosides, plants that contain those compounds, and herbivorous insects specializing on those plants provide one such system. As described above, the iridoid glycosides are important at many levels to adapted specialist insects. They are critical as feeding cues and are the basis of the predator defense system of some species. Variation in iridoid glycoside makeup of host plants can affect the efficacy of this defense system. These compounds are deterrent and/or toxic to generalists, suggesting that there would be little generalist herbivory on plants containing these compounds. In addition, there is substantial qualitative and quantitative variation in iridoid glycoside content of individual host-plant species, populations, and individuals. Different iridoid glycosides may have very different effects on insect herbivores, and thus iridoid makeup of plants may be a critical factor in determining herbivory patterns.

There are several steps required for detection of the existence and importance of short-term coevolution in plant–insect interactions. These steps are described for systems involving iridoid glycosides, but would be similar for other systems. Documenting patterns of differential herbivory in relation to iridoid glycoside composition of individual plants within a population as has been done in the crucifer–pierid–glucosinolate system (Louda and Rodman, 1983b) is the first step. The second is to determine the genetic components of the observed variation in iridoid glycoside

Table VII. Characteristics of Interacting Plant and Insect Populations That Are Conducive to Using that System to Study Short-Term Chemical Coevolution

Characteristics of plant population
Plant species considered is an annual or short-lived perennial.
The plant population is discrete, persistent, and localized.
The plant species and the populations under study are well characterized chemically.
There is differential herbivory within or between plant populations.
Characteristics of herbivore population
There are no or few generalist herbivores, and one or a few specialist herbivores.
The specialist insects are good "chemists"; that is, the plant allelochemicals are important in their choice of host plant.
The herbivore population is stable, sedentary, and persistent.

content among individuals and populations of plants. The third is to assess what effects differential herbivory has on the fitness of individual plants. The fourth is to determine to what extent change in plant allelochemical content, as the result of this selection, affects insect feeding patterns, behavioral preferences, and physiological responses of insect herbivores. The data available on iridoid glycoside chemistry and occurrence in plants, insect herbivory patterns, and the importance of iridoid glycosides to associated insect herbivores make this system an exemplary one with which to begin to study the potential for such short-term coevolution in plant–herbivore systems.

Acknowledgments

This paper has benefited from discussion with Fred Gould, Gillian Puttick, Frank Stermitz, and Peter Stevens. I greatly appreciate permission to use unpublished data from Dale Gardner, Svata Louda, and Frank Stermitz. James Carpenter, Frances Chew, Svata Louda, Gillian Puttick, Kevin Spencer, Nancy Stamp, and Frank Stermitz provided helpful comments on the manuscript, which I very much appreciate. The author was supported by National Science Foundation grants BSR-8307353 and RII-8503816.

References

Alstad, D. N., and Edmunds, G. F., Jr. (1983). Adaptation, host specificity, and gene flow in the black pineleaf scale. *In* "Variable Plants and Herbivores in Natural and Managed Systems" (R. F. Denno and M. S. McClure, eds.), pp. 413–426. Academic Press, New York.

Berenbaum, M. (1983). Coumarins and caterpillars: A case for coevolution. *Evolution* **37,** 163–179.

Berkowitz, W. F., Sasson, I., Sampathkumar, P. S., Hrabie, J., Choudry, S., and Pierce, D. (1979). Chiral prostanoid intermediates from aucubin. *Tetrahedron Lett.* **19,** 1641–1644.

Bernays, E., and DeLuca, C. (1981). Insect anti-feedant properties of an iridoid glycoside: Ipolamiide. *Experientia* **37,** 1289–1290.

Bianco, A., Passacantilli, P., Righi, G., and Nicoletti, M. (1985). Iridoid glucosides from *Parentucellia viscosa. Phytochemistry* **24,** 1843–1845.

Blau, P. A., Feeny, P., Contardo, L., and Robson, D. (1978). Allylglucosinolate and herbivorous caterpillars: A contrast in toxicity and tolerance. *Science* **200,** 1296–1298.

Bobbitt, J. M., and Segebarth, K. P. (1969). Iridoid glycosides and similar substances. *In* "Cyclopentanoid Terpene Derivatives" (W. I. Taylor and A. R. Battersby, eds.), pp. 1–145. Dekker, New York.

Bowers, M. D. (1980). Unpalatability as a defense strategy of *Euphydryas phaeton* (Lepidoptera: Nymphalidae). *Evolution* **34,** 586–600.

Bowers, M. D. (1981). Unpalatability as a defense strategy of western checkerspot butterflies (*Euphydryas,* Nymphalidae). *Evolution* **35,** 367–375.

Bowers, M. D. (1983a). Mimicry in North American checkerspot butterflies: *Chlosyne harrisii,* a Batesian mimic of *Euphydryas phaeton* (Lepidoptera: Nymphalidae). *Ecol. Entomol.* **8,** 1–8.

Bowers, M. D. (1983b). Iridoid glycosides and larval hostplant specificity in checkerspot butterflies (*Euphydryas,* Nymphalidae). *J. Chem. Ecol.* **9,** 475–493.

Bowers, M. D. (1984). Iridoid glycosides and host-plant specificity in larvae of the buckeye butterfly, *Junonia coenia* (Nymphalidae). *J. Chem. Ecol.* **10,** 1567–1577.

Bowers, M. D. (1986). Population differences in larval hostplant use in the checkerspot butterfly, *Euphydryas chalcedona. Entomol. Exp. Appl.* **40,** 61–69.

Bowers, M. D., and Puttick, G. M. (1986). The fate of ingested iridoid glycosides in lepidopteran herbivores. *J. Chem. Ecol.* **12,** 169–178.

Bowers, M. D., and Puttick, G. M. (1988). The response of generalist and specialist insects to qualitative allelochemical variation. *J. Chem. Ecol.* **14,** 319–334.

Bowers, M. D., Brown, I. L., and Wheye, D. (1985). Bird predation as a selective agent in a butterfly population. *Evolution* **39,** 93–103.

Bowers, M. D., Crabtree, R. C., Harrison, S. P., Sobrevilla, C., Wells, M., and Wolfe, L. M. (1987). Predation on adults of *Anartia fatima* (Fab.). *J. Lepid. Soc.* **41,** 75–76.

Breedlove, D. E., and Ehrlich, P. R. (1968). Plant-herbivore coevolution: Lupines and lycaenids. *Science* **162,** 671–672.

Breedlove, D. E., and Ehrlich, P. R. (1972). Coevolution: patterns of legume predation by a lycaenid butterfly. *Oecologia* **10,** 99–104.

Brower, L. P. (1969). Ecological chemistry. *Sci. Am.* **220,** 22–29.

Brower, L. P. (1984). Chemical defense in butterflies. *R. Entomol. Symp.* **11,** 109–134.

Cavill, G. W. K., Ford, D. P., and Loksley, H. D. (1956). The chemistry of ants, Part I. Terpenoid constituents of some Australian *Iridomyrmex* species. *Aust. J. Chem.* **9,** 288–293.

Carroll, C. R., and Hoffmann, C. A. (1980). Chemical feeding deterrent mobilised in response to insect herbivory and counter adaptation by *Epilachna tredecemnotata. Science* **209,** 414–416.

Conn, E. E. (1979). Cyanide and cyanogenic glycosides. *In* "Herbivores: Their Interaction with Secondary Plant Metabolites" (G. A. Rosenthal and D. H. Janzen, eds.), pp. 387–412. Academic Press, New York.

Cooper-Driver, G., Finch, S., Swain, T., and Bernays, E. (1977). Seasonal variation in secondary plant compounds in relation to the palatability of *Pteridium aquilinum. Biochem. Syst. Ecol.* **5,** 177–183.

Cordell, G. A. (1974). The biosynthesis of indole alkaloids. *Lloydia* **37,** 219–298.

Cronquist, A. (1968). "The Evolution and Classification of Flowering Plants." Houghton-Mifflin, Boston.

DaCosta, C. P., and Jones, C. M. (1971). Cucumber beetle resistance and mite susceptibility controlled by the bitter gene in *Cucumis sativa* L. *Science* **172,** 1145–1146.

Dahlgren, R. (1977). A note on the taxonomy of the "Sympetalae" and related groups. *Publ. Cairo Univ. Herb.* **7, 8,** 83–102.

Dahlgren, R. (1980). A revised system of classification of the angiosperms. *Bot. J. Linn. Soc.* **80,** 91–124.

Dahlgren, R., Jensen, S. R., and Nielsen, B. J. (1981). A revised classification of the angiosperms with comments on the correlation between chemical and other characters. *In* "Phytochemistry and Angiosperm Phylogeny" (D. A. Young and D. S. Seigler, eds.), pp. 149–204. Praeger, New York.

Damtoft, S., Jensen, S. R., and Nielsen, B. J. (1983). The biosynthesis of iridoid glucosides from 8-*epi*-deoxyloganic acid. *Biochem. Soc. Trans.* **11,** 594–595.

Dirzo, R. (1980). Experimental studies on slug-plant interactions. I. The acceptability of thirty plant species to the slug *Agriolimax caruanae. J. Ecol.* **68,** 981–998.

Dirzo, R., and Harper, J. L. (1982a). Experimental studies on slug-plant interactions. III. Differences in the acceptability of individual plants of *Trifolium repens* to slugs and snails. *J. Ecol.* **70,** 101–118.

Dirzo, R., and Harper, J. L. (1982b). Experimental studies on slug-plant interactions. IV. The performance of cyanogenic and acyanogenic morphs of *Trifolium repens* in the field. *J. Ecol.* **70,** 119–138.

Dolinger, P. M., Ehrlich, P. R., Fitch, W. L., and Breedlove, D. E. (1973). Alkaloid and predation patterns in Colorado lupine populations. *Oecologia* **13,** 191–204.

Ehrlich, P. R., and Murphy, D. D. (1981). The population biology of checkerspot butterflies. (*Euphydras*). *Biol. Z.* **100,** 613–629.

Ehrlich, P. R., and Raven, P. H. (1964). Butterflies and plants: A study in coevolution. *Evolution* **18,** 586–608.

El-Naggar, L. J., and Beal, J. L. (1980). Iridoids: A review. *J. Nat. Prod.* **43,** 649–707.

El-Naggar, S. F., and Doskotch, R. W. (1980). Specioside: A new iridoid glycoside from *Catalpa speciosa. J. Nat. Prod.* **43,** 524–526.

Feeny, P., Rosenberry, L., and Carter, M. (1983). Chemical aspects of oviposition behavior in butterflies. *In* "Herbivorous Insects—Host-seeking Behavior and Mechanisms" (S. Ahmad, ed.), pp. 27–76. Academic Press, New York.

Fox, L. R., and Morrow, P. A. (1981). Specialization: Species property or local phenomenon? *Science* **211,** 887–893.

Franke, A., Rimpler, H., and Schneider, D. (1987). Iridoid glycosides in the butterfly *Euphydryas cynthia. Phytochemistry* **26,** 103–106.

Futuyma, D. J. (1983). Evolutionary interactions among herbivorous insects and plants. Ch. 10. *In* "Coevolution" (D. Futuyma and M. Slatkin, eds.), pp. 207–231. Sinauer Press, Sunderland, Massachusetts.

Gardner, D. R. (1987). Iridoid chemistry of some *Castilleja* and *Besseya* plants and their hosted checkerspot butterflies. Ph.D. thesis, Colorado State University, Fort Collins, Colorado.

Gardner, D. R., Narum, J. Zook, D., and Stermitz, F. R. (1987). New iridoid glycosides from *Castilleja* and *Besseya:* 6-Hydroxyadoxoside and 6-isovanilloylcatalpol. *J. Nat. Prod.* **50,** 485–489.

Gardner, D. R., and Stermitz, F. R. (1988). Hostplant utilization and iridoid

glycoside sequestration by *Euphydryas anicia* individuals and populations. *J. Chem. Ecol.* (in press).
Georghiou, G. P. (1983). Management of resistance in arthropods. *In* "Pest Resistance to Pesticides" (G. P. Georghiou and T. Saito, eds.), pp. 769–792. Plenum Press, New York.
Gershenzon, J., and Mabry, T. J. (1983). Secondary metabolites and the higher classification of angiosperms. *Nord. J. Bot.* **3,** 5–34.
Gershenzon, J., Lincoln, D. E., and Langenheim, J. H. (1978). The effect of moisture stress on monoterpenoid yield and composition in *Satureja douglasii*. *Biochem. Syst. Ecol.* **6,** 33–43.
Gould, F. (1979). Rapid host range evolution in a population of the phytophagous mite, *Tetranychus urticae* Koch. *Evolution* **33,** 791–802.
Gould, F. (1983). Genetics of plant-herbivore systems: interactions between applied and basic study. *In* "Variable Plants and Herbivores in Natural and Managed Systems" (R. F. Denno and M. S. McClure, eds.), pp. 599–653. Academic Press, New York.
Hansel, R., and Winde, E. (1959). *Arzneimittel-Forsch.* **9,** 189.
Hegnauer, R. (1966). Aucubin-like glucosides: Distribution and significance as a systematic criterion. *Pharm. Acta Helv.* **11,** 577–587.
Hegnauer, R. (1973). "Chemotaxonomie der Pflanzen," Vol. 6. Berkhauser, Basel.
Hegnauer, R., and Kooiman, P. (1978). Die systematische Bedeutung von iridoiden Inhaltstoffen im Rahmen von Wettstein's Tubiflorae. *Planta Med.* **33,** 1–37.
Hessel, S. A. (1954). A guide to collecting the plant-boring larvae of the genus *Papaipema* (Noctuidae). *Lepid. News* **8,** 57–63.
Hodges, R. (1971). "Moths of North America, Fascicle 21. Sphingoidea." Classey, London.
Hoesel, W. (1981). The enzymatic hydrolysis of cyanogenic glucosides. *In* "Cyanide in Biology" (B. Vennesland, Conn, E. E., Knowles, C. J., Westley, J., and Wissing, F., eds.), pp. 217–232. Academic Press, New York.
Hsiao, T. H. (1978). Host plant adaptations among geographic populations of the Colorado potato beetle. *Entomol. Exp. Appl.* **24,** 237–247.
Huettel, M. L. D., and Bush, G. L. (1972). The genetics of host selection and its bearing on sympatric speciation of *Procecidochares*. *Entomol. Exp. Appl.* **15,** 465–480.
Hughes, M. A. (1981). The genetic control of plant cyanogenesis. *In* "Cyanide in Biology" (B. Vennesland, Conn, E. E., Knowles, C. J., Westley, J., and Wissing, F., eds.), pp. 495–508. Academic Press, New York.
Inouye, H. (1971). Biosynthesis of iridoid and seco-iridoid glucosides. *In* "Pharmacognosy and Phytochemistry" (H. Wagner and L. Horhammer, eds.), pp. 290–313. Academic Press, New York.
Inouye, H., Takeda, Y., Uobe, K., Yamauchi, K., Nabuuchi, N., and Kuwano, S. (1974a). Purgative activities of iridoid glucosides. *Planta Med.* **25,** 285–288.
Inouye, H., Veda, S., Shingu, T., and Thies, P. W. (1974b). Die absolute Konfiguration von Valerosidatum und von Didrovaltratum. *Tetrahedron* **30,** 2317–2325.

Jaenike, J., and Grimaldi, D. (1983). Genetic variation for host preference within and among populations of *Drosophila tripunctata*. *Evolution* **37,** 1023–1029.

Janzen, D. H. (1980). When is it coevolution? *Evolution* **34,** 611–612.

Jensen, S. R., and Nielsen, B. J. (1985). Hygrophiloside, an iridoid glucoside from *Hygrophila difformis* (Acanthaceae). *Phytochemistry* **24,** 602–603.

Jensen, S. R., Nielsen, B. J., and Dahlgren, R. (1975). Iridoid compounds, their occurrence and systematic importance in the angiosperms. *Bot. Not.* **128,** 148–180.

Jermy, T. (1984). Evolution of insect/host plant relationships. *Am. Nat.* **124,** 609–630.

Kaplan, M. A. C., and Gottlieb, O. R. (1982). Iridoids as systematic markers in dicotyledons. *Biochem. Syst. Ecol.* **10,** 329–347.

Karban, R., and Carey, J. R. (1984). Induced resistance of cotton seedlings to mites. *Science* **225,** 53–54.

Knerer, G., and Atwood, C. E. (1973). Diprionid sawflies: Polymorphism and speciation. *Science* **179,** 1090–1099.

Krischik, V., and Denno, R. (1983). Individual, population, and geographic patterns in plant defense. *In* "Variable Plants and Herbivores in Natural and Managed Systems" (R. F. Denno and M. S. McClure, eds.), pp. 463–512. Academic Press, New York.

Krombein, K. V., Hurd, P. D., and Smith, D. R. (1979). "Catalog of Hymenoptera in America North of Mexico." Smithsonian Institution Press, Washington, D.C.

Kubota, T., and Kubo, I. (1969). Bitterness and chemical structure. *Nature (London)* **223,** 97–99.

Lincoln, D. E., and Langenheim, J. H. (1978). Effect of light and temperature on monoterpenoid yield and composition in *Satureja douglasii. Biochem. Syst. Ecol.* **6,** 21–32.

Lincoln, D. E., and Langenheim, J. H. (1979). Variation of *Satureja douglasii* monoterpenoids in relation to light intensity and herbivory. *Biochem. Syst. Ecol.* **7,** 289–298.

Lincoln, D. E., and Langenheim, J. H. (1981). A genetic approach to monoterpenoid variation in *Satureja douglasii. Biochem. Syst. Ecol.* **9,** 153–161.

Lincoln, D. E., and Mooney, H. A. (1984). Herbivory on *Diplacus aurantiacus* shrubs in sun and shade. *Oecologia* **64,** 173–177.

Lincoln, D. E., Murray, M. J., and Lawrence, B. M. (1986). Chemical composition and genetic basis for the isopinocamphone chemotype of *Mentha citrata* hybrids. *Phytochemistry* **25,** 1857–1863.

Louda, S. M. (1984). Herbivore effect on stature, fruiting, and leaf dynamics of a native crucifer. *Ecology* **65,** 1379–1386.

Louda, S. M. (1986). Insect herbivory in response to root-cutting and flooding stress on a native crucifer under field conditions. *Acta Oecol., Oecol. Gen.* **7,** 37–53.

Louda, S. M., and Rodman, J. E. (1983a). Ecological patterns in the glucosinolate content of a native mustard, *Cardamine cordifolia,* in the Rocky Mountains. *J. Chem. Ecol.* **9,** 397–421.

Louda, S. M., and Rodman, J. E. (1983b). Concentration of glucosinolates in relation to habitat and insect herbivory for the native crucifer *Cardamine cordifolia*. *Biochem. Syst. Ecol.* **11,** 199–207.

McKey, D. (1979). The distribution of secondary compounds within plants. *In* "Herbivores: Their Interaction with Secondary Plant Metabolites" (G. A. Rosenthal and D. H. Janzen, eds.), pp. 55–133. Academic Press, New York.

McKey, D., Waterman, P. J., Mbi, C. N., Gartlan, J. S., and Struhsaker, T. T. (1978). Phenolic content of vegetation in two African rainforests: Ecological implications. *Science* **202,** 61–64.

Marco, J. L. (1985). Iridoid glucosides of *Linaria clementei*. *Phytochemistry* **24,** 1609–1610.

Mihaliak, C., and Lincoln, D. E. (1985). Growth pattern and carbon allocation to volatile leaf terpenes under nitrogen limiting conditions in *Heterotheca subaxillaris* (Asteraceae). *Oecologia* **66,** 423–426.

Mitter, C., and Brooks, D. R. (1983). Phylogenetic aspects of coevolution. *In* "Coevolution" (D. J. Futuyma and M. Slatkin, eds.), pp. 65–98. Sinauer Press, Sunderland, Massachusetts.

Mitter, C., and Futuyma, D. J. (1983). An evolutionary–genetic view of host-plant utilization by insects. *In* "Variable Plants and Herbivores in Natural and Managed Systems" (R. F. Denno and M. S. McClure, eds.), pp. 427–459. Academic Press, New York.

Mooney, H. A., and Chu, C. (1974). Seasonal carbon allocation in *Heteromeles arbutifolia,* a California evergreen shrub. *Oecologia* **14,** 295–306.

Nayar, J. K., and Fraenkel, G. (1963). The chemical basis of host selection in the catalpa sphinx, *Ceratomia catalpae* (Lepidoptera: Sphingidae). *Ann. Entomol. Soc. Am.* **56,** 119–122.

Nelson, C. J., Seiber, J. N., and Brower, L. P. (1981). Seasonal and intraplant variation of cardenolide content in the California milkweed, *Asclepias eriocarpa,* and implications for plant defense. *J. Chem. Ecol.* **7,** 981–1010.

Parry, R. H. (1974). Revision of the genus *Dibolia* Latreille in America north of Mexico (Coleoptera: Chrysomelidae). *Can. J. Zool.* **52,** 1317–1354.

Pasteels, J. M., Rowell-Rahier, M., Braekman, J., and Daloze, D. (1984). Chemical defenses in leaf beetles and their larvae: The ecological, evolutionary, and taxonomic significance. *Biochem. Syst. Ecol.* **12,** 395–406.

Pavan, M. V. (1975). Gli iridoidi negli insetti. Pubblicazioni dell'instituto dei Entomologia Agraria dell'universita' de Pavia. No. 2.

Pavan, M. V., and Finzi, P. V. (1974). Chemically known constituents of arthropod defensive secretions. *Mem. Accad. Naz. Lincei. Sene VIII* **XII,** 109–146.

Pereyra, P. C., and Bowers, M. D. (1988). Iridoid glycosides as oviposition stimulants for the buckeye, *Junonia coenia* (Nymphalidae). *J. Chem. Ecol.* **14,** 917–928.

Petkov, V., and Manolov, P. (1972). Pharmacological analysis of the iridoid oleuropein. *Arzneimittel. Forsch.* **22,** 1476–1486.

Puttick, G. M., and Bowers, M. D. (1988). The effect of iridoid glycosides on growth, survival, and food choice of the armyworm, *Spodoptera eridania* (Lepidoptera: Noctuidae). *J. Chem. Ecol.* **14,** 335–351.

Rausher, M. D. (1978). Search image for leaf shape in a butterfly. *Science* **200,** 1071–1073.

Rausher, M. D. (1982). Population differentiation in *Euphydryas editha* butterflies: Larval adaptation to different hosts. *Evolution* **36,** 581–590.
Rausher, M. D. (1983). Ecology of host-selection behavior in phytophagous insects. *In* "Variable Plants and Herbivores in Natural and Managed Systems" (R. F. Denno and M. S. McClure, eds.), pp. 223–258. Academic Press, New York.
Rhoades, D. F. (1979). Evolution of plant chemical defense against herbivores. *In* "Herbivores: Their Interaction with Secondary Plant Metabolites" (G. A. Rosenthal and D. H. Janzen, eds.), pp. 4–54. Academic Press, New York.
Rice, R. L., Lincoln, D. E., and Langenheim, J. H. (1978). Palatability of monoterpenoid compositional types of *Satureja douglasii* to a generalist molluscan herbivore, *Ariolimax dolichophallus*. *Biochem. Syst. Ecol.* **6,** 45–53.
Robinson, T. (1974). Metabolism and function of alkaloids in plants. *Science* **184,** 430–435.
Roby, M. R., and Stermitz, F. R. (1984a). Pyrrolizidine and pyridine monoterpene alkaloids from 2 *Castilleja* plant hosts of the plume moth, *Platyptilia pica*. *J. Nat. Prod.* **47,** 846–853.
Roby, M. R., and Stermitz, F. R. (1984b). Penstemonoside and other iridoids from *Castilleja rhexifolia:* Conversion of penstemonoside to the pyridine monotepene alkaloid rhexifoline. *J. Nat. Prod.* **47,** 854–857.
Rodman, J. E., and Louda, S. M. (1984). Phenology of glucosinolate concentrations in roots, stems and leaves of *Cardamine cordofolia*. *Biochem. Syst. Ecol.* **12,** 37–46.
Rombouts, J. E., and Links, J. (1956). The chemical nature of the antibacterial substance present in *Aucuba japonica* Thunbg. *Experientia* **12,** 78–80.
Ryan, C. A. (1979). Proteinase inhibitors. *In* "Herbivores: Their Interaction with Plant Secondary Metabolites" (G. A. Rosenthal and D. H. Janzen, eds.), pp. 599–618. Academic Press, New York.
Schultz, J. D., and Baldwin, I. T. (1982). Oak leaf quality declines in response to defoliation by gypsy moth larvae. *Science* **217,** 149–151.
Scriber, M. (1983). Evolution of feeding specialization, physiological efficiency, and host races in selected Papilionidae and Saturniidae. *In* "Variable Plants and Herbivores in Natural and Managed Systems" (R. F. Denno and M. S. McClure, eds.), pp. 373–412. Academic Press, New York.
Sha'aban, F., El-Naggar, S. F., and Doskotch, R. W. (1980). Specioside: A new iridoid glycoside from *Catalpa speciosa*. *J. Nat. Prod.* **43,** 524–526.
Silberglied, R. E., Aiello, A., and Lamas, G. (1979). Neotropical butterflies of the genus *Anartia:* Systematics, life histories and general biology (Lepidoptera: Nymphalidae). *Psyche* **86,** 219–260.
Silberglied, R. E., Aiello, A., and Windsor, D. M. (1980). Disruptive coloration in butterflies, lack of support in *Anartia fatima*. *Science* **209,** 617–619.
Singer, M. C. (1971). Evolution of foodplant preferences in the butterfly *Euphydryas editha*. *Evolution* **25,** 383–389.
Singer, M. C. (1983). Determinants of multiple host use in a phytophagous insect population. *Evolution* **37,** 389–403.
Stamp, N. E. (1984). Effect of defoliation by checkerspot caterpillars (*Euphydryas phaeton*) and sawfly larvae (*Macrophya nigra* and *Tenthredo grandis*) on their hostplants (*Chelone* spp.). *Oecologia* **63,** 275–280.

Stephenson, A. G. (1981). Toxic nectar deters nectar thieves of *Catalpa speciosa*. *Am. Midl. Nat.* **105,** 381–383.

Stephenson, A. G. (1982). Iridoid glycosides in the nectar of *Catalpa speciosa* are unpalatable to nectar thieves. *J. Chem. Ecol.* **8,** 1025–1034.

Stermitz, F. S., Gardner, D. R., Odendaal, F. J., and Ehrlich, P. R. (1986a). *Euphydryas anicia* (Lepidoptera: Nymphalidae) utilization of iridoid glycosides from *Castilleja* and *Besseya* (Scrophulariaceae) host plants. *J. Chem. Ecol.* **12,** 1456–1468.

Stermitz, F. S., Harris, G. H., and Jing, W. (1986b). Iridoids and alkaloids from *Castilleja* (Scrophulariaceae) host plants for *Platyptilia pica* (Lepidoptera: Pterophoridae). Rhexifoline content of *P. pica*. *Biochem. Syst. Ecol.* **14,** 499–503.

Sticher, O. (1977). Plant mono-, di- and sesquiterpenoids with pharmacological or therapeutical activity. *In* "New Natural Products and Plant Drugs with Pharmacological, Biological, or Therapeutic Activity" (H. Wagner and P. Wolff, eds.), pp. 137–176. Springer-Verlag, Berlin.

Tabashnik, B., Wheelock, H., Rainbolt, J. D., and Watt, W. B. (1981). Individual variation in oviposition preference in the butterfly, *Colias eurytheme*. *Oecologia* **50,** 225–230.

Thomas, R. (1961). A possible biosynthetic relationship between the cyclopentanoid monoterpenes and the indole alkaloids. *Tetrahedron Lett.* **16,** 544–553.

Tixidre, A., Rolland, Y., Garnier, J., and Poisson, J. (1982). Aucubin, a source of prostanoid synthons—New hemisynthetic pathways. *Heterocycles* **19,** 253–256.

van Etten, C. H., and Tookey, H. L. (1979). Chemistry and biological effects of glucosinolates. *In* "Herbivores: Their Interaction with Secondary Plant Metabolites" (G. A. Rosenthal and D. H. Janzen, eds.), pp. 471–500. Academic Press, New York.

Via, S. (1984a). The quantitative genetics of polyphagy in an insect herbivore. I. Genotype–environment interaction in larval performance on different hostplant species. *Evolution* **38,** 881–895.

Via, S. (1984b). The quantitative genetics of polyphagy in an insect herbivore. II. Genetic correlations in larval performance within and among host plants. *Evolution* **38,** 896–905.

Wagner, H., and Vasirian, K. (1974). Desoxyamarogentin. Ein neuer Bitterstoff aus *Gentiana pannonica* Scop. *Phytochemistry* **13,** 615–617.

Woodhead, S. (1981). Environmental and biotic factors affecting the phenolic content of different cultivars of *Sorghum bicolor*. *J. Chem. Ecol.* **7,** 1035–1047.

Wratten, S. D., Edwards, P. J., and Dunn, I. (1984). Wound-induced changes in the palatability of *Betula pubescens* and *B. pendula*. *Oecologia* **61,** 372–375.

CHAPTER 7

Chemical Mediation of Coevolution in the *Passiflora–Heliconius* Interaction

KEVIN C. SPENCER
Department of Medicinal Chemistry and Pharmacognosy
University of Illinois
Chicago, Illinois

I. Introduction

The genus *Passiflora* is remarkable in producing plants of great structural complexity in both floral and vegetative parts. The latter have been proposed to represent evolutionarily advanced end products of an adaptive radiation of form derived from a multifaceted coevolutionary relationship with their primary herbivores, the heliconiine butterflies (Benson *et al.*, 1976).

Passiflora is a large and morphologically complex genus that was last monographed by Killip (1938) some 50 years ago. He divided the genus into 22 subgenera and further divided these into sections and series (Table I). Several workers have recently taken issue with this classification (Gentry, 1976; Gilbert, 1975; Escobar, 1980, 1986; MacDougal, 1988). Much recent taxonomic (Escobar, 1980; MacDougal, 1988; Gentry, 1980; Spencer, 1984, 1988; Spencer *et al.*, 1988) and chemical work (Seigler *et al.*, 1982; Spencer, 1984, 1987, 1988; Spencer *et al.*, 1986; Spencer and Seigler, 1983, 1984a–c, 1985a–c, 1987a,b) has been reported; a number of newly discovered taxa have been described (Gentry, 1980; Gilli, 1981; Holm-Neilsen, 1974).

The taxonomy of *Heliconius* is considered to be well understood (Turner, 1977). This genus has been intensively studied both in terms of systematics (Alexander, 1961a,b; Brown, 1972, 1975, 1976; Brown and

Table I. *Passiflora* L.

Passiflora L.	Number of Species
Subg. 1. *Apodogyne* Killip	1
Subg. 2. *Astephia* Killip	1
Subg. 3. *Tryphostemmatoides* (Harms) Killip	2
Subg. 4. *Deidamioides* (Harms) Killip	1
Subg. 5. *Plectostemma* Mast.	115
Sect. 1. *Cieca* (Medic.)	26
Sect. 2. *Mayapathanthus* Killip	1
Sect. 3. *Decaloba* (DC.) Mast.	70
Series 1. *Auriculatae*	3
Series 2. *Heterophyllae*	1
Series 3. *Sexflorae*	4
Series 4. *Apetalae*	2

Table I. (*continued*)

Passiflora L.	Number of Species
Series 5. *Luteae*	3
Series 6. *Organenses*	8
Series 7. *Miserae*	4
Series 8. *Punctatae*	45
Sect. 4. *Xerogona* (Raf.) Killip	8
Sect. 5. *Pseudodysosmia* (Harms) Killip	1
Sect. 6. *Pseudogranadilla* (Harms) Killip	6
Sect. 7. *Hahniopathanthus* (Harms) Killip	3
Subg. 6. *Chloropathanthus* (Harms) Killip	2
Subg. 7. *Murucuja* (Medic.) Mast.	4
Subg. 8. *Pseudomurucuja* (Harms) Killip	6
Subg. 9. *Psilanthus* (DC.) Killip	4
Subg. 10. *Adenosepala* Killip	1
Subg. 11. *Tacsoniopsis* (Tr. & Planch.) Killip	1
Subg. 12. *Rathea* (Karst.) Killip	1
Subg. 13. *Tacsonia* (Juss.) Tr. & Planch.	35
Subg. 14. *Granadillastrum* (Tr. & Planch.) Killip	6
Subg. 15. *Distephana* (Juss.) Killip	8
Subg. 16. *Calopathanthus* (Harms) Killip	1
Subg. 17. Tacsonioides (DC.) Killip	5
Subg. 18. *Granadilla* (Medic.) Mast.	97
Series 1. *Quadrangulares*	2
Series 2. *Digitatae*	1
Series 3. *Tiliaefoliae*	11
Series 4. *Marginatae*	1
Series 5. *Laurifoliae*	13
Series 6. *Serratifoliae*	4
Series 7. *Setaceae*	1
Series 8. *Pedatae*	1
Series 9. *Incarnatae*	6
Series 10. *Palmatisectae*	1
Series 11. *Kermesinae*	8
Series 12. *Imbricatae*	2
Series 13. *Simplicifoliae*	14
Series 14. *Lobatae*	28
Series 15. *Menispermifoliae*	4
Subg. 19. *Dysosmia* (DC.) Killip	11
Subg. 20. *Dysosmioides* Killip	5
Subg. 21. *Polyanthea* (DC.) Killip	1
Subg. 22. *Astrophea* (DC.) Mast.	47
Sect. 1. *Dolichostemma* Killip	2
Sect. 2. *Cirrhipes* Killip	1
Sect. 3. *Euastrophea* (Harms) Killip	15
Sect. 4. *Leptopoda* Killip	1
Sect. 5. *Pseudoastrophea* (Harms) Killip	17
Sect. 6. *Botryastrophea* (Harms) Killip	11

Source: Killip (1938).

Benson, 1974, 1975; Emsley, 1963, 1964, 1965; Turner, 1977; Brown and Mielke, 1972) and evolutionary ecology (Benson, 1978, 1982; Brown and Benson, 1974, 1975, 1977; Brown *et al.*, 1974; Ehrlich and Gilbert, 1973; Gilbert, 1972, 1976, 1977; Turner, 1971a,b, 1975, 1976, 1978, 1981). The most recent systematic treatment of *Heliconius* purports to reflect a true phylogenetic series of evolutionarily primitive through advanced species (Table II) (Brown, 1981). These insects are able to quickly defoliate their hosts, and probably represent the major biotic challenge to survival of individual *Passiflora* in the field (Benson *et al.*, 1976; Gilbert and Smiley, 1978).

The importance of chemistry in determining patterns of utilization by lepidopterans and other phytophagous insects has been demonstrated by Ehrlich and Raven (1964). Insect foodplant specialization is chemically primed, but in *Heliconius* the pattern of specialization is not obviously chemically cued in a one-to-one correspondence (Gilbert and Smiley, 1978). That toxicity of chemicals is not the only evolutionary determinant

Table II. Heliconiine Herbivores of *Passiflora* in an Advancing Phylogenetic Series

Heliconiine Species	Phylogenetic Group[a] (Brown, 1981)	Herbivore Group[a] (Benson *et al.*, 1976)
1 *Philaethria pygmalion*	1. *Philaethria* gr.	I. Primitive genera
2 *P. wernickei*		
3 *P. dido* (complex)		
4 *Dione juno*	2. *Dione* gr.	
5 *D. moneta*		
6 *D. glycera*		
7 *Agraulis vanillae*	3. *Agraulis* gr.	
8 *Dryadula phaetusa*	4. *Dryadula* gr.	
9 *Dryas julia*	5. *Dryas* gr.	
10 *Eueides vibilia*	6. *Eueides* gr.	II. Mature leaf radiation
11 *E. pavana*		
12 *E. isabella*		
13 *E. lybia*		
14 *E. tales*		
15 *E. aliphera*		
16 *Neruda godmani*	7. *Neruda* gr.	III. Early radiation
17 *N. metharme*		
18 *N. aoede*		
19 *Laparus doris*	8. *Laparus* gr.	
20 *Heliconius xanthocles*	9. *H. xanthocles* gr.	
21 *H. wallacei*	10. *H. wallacei* gr.	
22 *H. burneyi*		
23 *H. egeria*		
24 *H. astraea*		
25 *H. nattereri*	11. Silvaniform gr.	IV. Advanced radiation A
26 *H. numata*		
27 *H. ismenius*		

Table II. (*continued*)

Heliconiine Species	Phylogenetic Group[a] (Brown, 1981)	Herbivore Group[a] (Benson *et al.*, 1976)
28 *H. pardalinus*		
29 *H. hecale*		
30 *H. ethilla*		
31 *H. atthis*		
32 *H. cydno*	12. *H. cydno* gr.	
33 *H. pachinus*		
34 *H. heurippa*		
35 *H. timareta*		
36 *H. elevatus*		
37 *H. besckei*		
38 *H. melpomene*		
39 *H. charitonia*	13. *H. charitonia* gr.	V. Advanced radiation B
40 *H. hermathena*		
41 *H. erato*		
42 *H. clysonymus*		
43 *H. hortense*		
44 *H. hecalesia*		
45 *H. ricini*	14. *H. ricini* gr.	
46 *H. demeter*		
47 *H. leucadia*		
48 *H. sara*		
49 *H. antiochus*	15. *H. antiochus* gr.	
50 *H. hewitsoni*		
51 *H. sapho*		

Source: Brown (1981).
Note: An additional 14 species with unknown hosts are excluded from the present discussion.
[a] See text for explanation.

of host choice in this system is suggested by limited feeding trials (Smiley, 1978) in which several species of larvae exhibited successful growth on alternate food plants. The above data do not take into account possible chemical similarities among plants nor ultimate fitness, but they do show that searching ability, competition, resource availability, edaphic factors, and interactions with higher trophic levels must all be considered (Scriber, 1984; Miller and Strickler, 1984). Both the monophyletic *Heliconius* feeding guild and a polyphyletic flea beetle guild were able to differentiate between members of subgenus *Plectostemma* and subgenus *Granadilla*. Both diversified similarly onto *Passiflora*, and the determinants of this radiation were the characteristics of the host plants (Smiley, 1982).

It is proposed here that, despite the considerable influence of other factors, host-plant chemistry is central to the observed coevolutionary interaction between *Heliconius* and *Passiflora*, and accounts for much of the observed correlation between insect and plant taxa. Many plant–

insect interaction studies have dealt explicitly with chemical data but have only considered ecological parameters in passing. In this situation, the ecology of the system had been extensively studied, but the chemical parameters were mostly unknown. I have attempted to identify the types and distribution of cyanogenic compounds produced in many members of the genus *Passiflora*, and to determine whether any correlation exists between the chemical distributional pattern and host-plant specialization of *Heliconius*.

The presence in the insects of specialized searching abilities and feeding methods involving choice of plant part, age, and condition (Copp and Davenport, 1978a,b; Swihart and Swihart, 1970; Swihart, 1971; Mallet and Jackson, 1980; Young, 1978), the presence in *Passiflora* species of egg mimic structures (Williams and Gilbert, 1981), and variation of growth-rate patterns and leaf morphologies (Benson, 1978) and hooked trichomes (Gilbert, 1971) all provide evidence for coevolution between *Heliconius* and *Passiflora*. Host-plant abundance has been demonstrated to limit the number of *Heliconius* individuals attacking an individual plant (or plant species), thereby generating a high level of interspecific competition among insects for host plants (Benson, 1978; Gilbert, 1980). These insects are efficient defoliators of *Passiflora*, yet they experience very high egg and larval mortalities from ant and parasitoid predators (Benson, 1978) that are attracted by a variety of extrafloral nectaries (Gilbert and Smiley, 1978; J. T. Smiley, personal communication). These insects occupy unique niches involving different plant taxa, plant habitats, and plant parts or growth condition (Benson *et al.*, 1976; Benson, 1978; Gilbert, 1980). These pressures result in an evolutionary correlation between *Heliconius* species groups and *Passiflora* subgenera and morphotypes (Benson *et al.*, 1976). This has not been resolvable into clear ecological or morphological causes, yet forms the basis of a subdivision of *Heliconius*

Table III. *Passiflora* Species Used in Analyses, Listed by Subgenus, Section, and Series

Tryphostemmatoides			14. *tricuspis* Mast.
	1. *P. gracillima* Killip		15. *trifasciata* Lemaire
Plectostemma		3-8	16. *vespertilio* L.
1[a]	2. *coriacea* Juss.		17. *tuberosa* Jacq.
	3. *tenuiloba* Engelm.		18. *lancearia* Mast.
	4. *suberosa* L.		19. *cuspidifolia* Harms
	5. *trinifolia* Mast.		20. *alnifolia* HBK.
	6. *warmingii* Mast.		21. *cuneata* Willd.
	7. *truncata* Regel		22. *talamancensis* Killip
3-1	8. *auriculata* HBK.		23. *chelidonea* Mast.
3-4	9. *apetala* Killip		24. *punctata* L.
3-5	10. *lutea* L.		25. *biflora* Lam.
	11. *filipes* Benth.		26. *affinis* Engelm.
3-6	12. *organensis* Gardn.		27. *standleyi* Killip
3-7	13. *misera* HBK.		28. *pohlii* Mast.

Table III. (*continued*)

4	29. *costaricensis* Killip
	30. *capsularis* L.
	31. *rubra* L.
5	32. *adenopoda* DC.
6	33. *pulchella* HBK.
7	34. *hahnii* (Fourn.) Mast.
	35. *membranacea* Benth.
Chloropathanthus	
	36. *lancifolia* Desv.
Pseudomurucuja	
	37. *perfoliata* L.
	38. *tacsonioides* Griseb.
Psilanthus	
	39. *sanguiñolenta* Mast.
Tacsonia	
	40. *flexipes* Tr. & Planch.
	41. *pinnatistipula* Cav.
	42. *cumbalensis* (Karst.) Harms
	43. *molissima* (HBK.) Bailey
	44. *mixta* L.
Granadillastrum	
	45. *manicata* (Juss.) Pers.
Distephana	
	46. *glandulosa* Cav.
	47. *coccinea* Aubl.
	48. *quadriglandulosa* Rodschied
	49. *vitifolia* HBK.
	50. *speciosa* Gardn.
Calopathanthus	
	51. *racemosa* Brot.
Granadilla	
1	52. *quadrangularis* L.
	53. *alata* Dryand.
2	54. *serrato-digitata* L.
3	55. *ligularis* Juss.
	56. *seemanni* Griseb.
	57. *maliformis* L.
	58. *platyloba* Killip
5	59. *capparidifolia* Killip
	60. *ambigua* Hemsl.
	61. *laurifolia* L.
	62. *acuminata* DC.
	63. *riparia* Mart. ex Mast.
	64. *nitida* HBK.
	65. *guazumaefolia* Juss.
6	66. *bahiensis* Klotzsch
	67. *serratifolia* L.
7	68. *setacea* DC.
8	69. *pedata* L.
9	70. *recurva* Mast.
	71. *incarnata* L.
	72. *edulis* Sims
11	73. *kermesina* Link & Otto
	74. *miersii* Mast.
12	75. *sidaefolia* M. Roemer
13	76. *mucronata* Lam.
	77. *jileki* Wawra
	78. *actinia* Hook.
	79. *oerstedii* Mast.
14	80. *caerulea* L.
	81. *violacea* Vell.
	82. *amethystina* Mikan
	83. *picturata* Ker.
	84. *subpeltata* Ortega
	85. *elegans* Mast.
	86. *lonchophora* Harms
	87. *eichleriana* Mast.
	88. *resticulata* Mast. & Andre
	89. *cyanea* Mast.
	90. *garckei* Mast.
15	91. *menispermifolia* HBK.
Dysosmia	
	92. *clathrata* Mast.
	93. *foetida* L.
Dysosmioides	
	94. *villosa* Vell.
	95. *vellozii* Gardn.
Polyanthea	
	96. *cirrhiflora* Juss
Astrophea	
1	97. *pittieri* Mast.
3	98. *gigantifolia* Harms
5	99. *candida* (Poepp. & Endl.) Mast.
	100. *hexagonocarpa* Barb.
	101. *costata* Mast.
	102. *faroana* Harms
	103. *mansii* (Mart.) Mast.
	104. *pentagona* Mast.
	105. *rhamnifolia* Mast.
6	106. *skiantha* Huber
	107. *spinosa* (Poepp. & Engl.) Mast.
Related Species:	
	108. *Mitostemma glaziovii*
	109. *Dilkea parviflora*
	110. *Tetrastylis ovalis*
	111. *Tetrastylis lobata*

Source: Killip (1938).
[a] Margin numbers refer to sections and series given in Table I.

into five herbivore groups, each of which represents a radiation onto a cluster of *Passiflora* subgenera. These herbivore groups presage almost precisely (Table II) the phylogeny constructed by Brown (1981).

That an evolutionary radiation into diverse forms and many species has occurred in *Passiflora* is evident. An examination of the 22 subgenera described by Killip (1938) provides ample evidence for clustering of large numbers of species into relatively homogeneous groupings (Table III). Examples are the many enumerated and measurable character differences between the subgenera *Granadilla*, *Tacsonia*, and *Plectostemma*. It is quite reasonable to assume that extensive and rapid diversification has occurred within these subgenera, while more major differences exist between subgenera. The pattern is one of radiation from distinct set points of morphological and ecological characteristics.

Heliconius seems also to have radiated into a diverse array of species (Table IV). The degree of correlation between heliconiine taxa (species, phylogenetic groups (Brown, 1981) or herbivore groups (Benson *et al.*, 1976) and *Passiflora* taxa (species and subgenera) heretofore described is indicative of a coevolutionary process, that is, a reciprocal coadaptation (Janzen, 1980) of pairs of taxa. If this is indeed the case, we may wonder at the only small degree of correlation thus far explained.

It is the purpose of this chapter to report upon the intrusion of a third player into the coevolutionary *pas de deux* of these plants and insects, a highly diversified plant chemistry. As good choreography demands, the chemistry influences the evolutionary movements of both plant and insect, and turns of the latter necessitate revisions in the score.

Table IV. Heliconiine Utilization of *Passiflora* Species as Host Plants

Heliconiine Species (from Table II)	*Passiflora* Species (from Table III)
1	47, 100, 102, 103
2	4, 75, 76, 77, 78, 80, 82, 85, 105
3	46, 47, 49, 50, 60, 61, 66, 72, 76, 89, 101, 106
4	4, 6, 8, 40, 42, 43, 44, 45, 47, 48, 49, 50, 53, 54, 59, 62, 63, 64, 66, 67, 72, 75, 76, 80, 89
5	6, 30, 32, 34, 43, 72, 81, 111
6	20, 39, 40, 42, 43, 44, 45, 72
7	3, 4, 6, 8, 13, 26, 29, 30, 34, 47, 50, 52, 53, 55, 57, 64, 66, 68, 72, 73, 74, 75, 76, 77, 80, 81, 82, 89, 92, 93, 94, 103, 105, 110
8	4, 6, 13, 16, 17, 26, 28, 31, 34, 49, 72, 76, 77, 81, 103
9	1, 2, 4, 6, 7, 8, 12, 13, 15, 16, 17, 25, 28, 29, 30, 34, 36, 46, 52, 72, 74, 75, 80, 81, 101, 105
10	72, 101, 103, 108
11	105

Table IV. (*continued*)

Heliconiine Species (from Table II)	*Passiflora* Species (from Table III)
12	48, 53, 54, 56, 57, 58, 60, 61, 66, 67, 72, 73, 75, 76, 79, 80, 82, 89
13	30, 47, 49, 64, 65, 72
14	47, 49, 61, 63, 64
15	29, 30, 31, 47, 48, 49, 52, 61, 66, 68, 73, 75, 79, 80, 81, 82, 86, 89
16	109
17	58, 109
18	109
19	54, 60, 61, 63
20	64, 87, 91
21	46, 47, 48, 49
22	47, 49, 99
23	46, 99
24	46, 47, 53
25	110
26	8, 29, 46, 47, 48, 49, 53, 61, 62, 64, 66, 68, 71, 75, 77, 79, 86, 87, 95, 99, 101, 102, 105, 109, 110
27	47, 49, 52, 60, 67, 72
28	47, 52, 53, 64, 69, 107
29	8, 47, 48, 49, 52, 53, 58, 61, 62, 64, 69, 79, 107
30	51, 53, 66, 68, 70, 72, 73, 74, 75, 77, 79, 81, 82, 83, 84, 87, 89, 90, 95, 105, 110
31	87, 88
32	2, 8, 18, 25, 29, 49, 52, 60, 61, 65, 79, 97
33	49, 60, 79
34	61, 79
35	53
36	61, 72, 80
37	72, 78, 80, 94, 95
38	13, 14, 17, 46, 47, 48, 49, 53, 54, 55, 57, 59, 60, 61, 62, 64, 66, 72, 77, 79, 81, 87, 89, 91, 107, 110
39	4, 10, 26, 31, 32, 33, 34, 35, 37, 38, 111
40	100, 102
41	1, 2, 4, 6, 7, 8, 12, 13, 14, 15, 16, 17, 19, 21, 22, 23, 24, 25, 28, 30, 31, 33, 34, 53, 61, 71, 72, 74, 75, 77, 80, 81, 82, 88, 105, 109, 110
42	9, 11, 18, 21, 25, 27, 41, 49
43	5
44	1, 25, 111
45	59, 61
46	109
47	8
48	4, 7, 8, 72, 76, 77, 96, 99, 101, 102, 103, 104, 105, 107, 110
49	101, 106, 107
50	1, 25, 97
51	97, 98

II. Chemistry

Cyanogenesis is extremely widespread, occurring in bacteria and fungi (Knowles, 1976), millipedes (Duffey *et al.*, 1977; Eisner, 1970), insects (Lepidoptera: Jones *et al.*, 1962; Davis and Nahrstedt, 1979; Nahrstedt and Davis, 1981, 1986) ferns and gymnosperms (Hegnauer, 1977; Harper *et al.*, 1976), and more than 2000 species in over 110 families of angiosperms (Seigler, 1981). In the angiosperms, cyanogenesis is usually due to the presence of cyanogenic glycosides (Conn, 1981b). Outside the Passifloraceae, these constitute some 36 described structures, each derived from one of five amino acid precursors (Seigler, 1975; Conn, 1979b; 1980, 1981a) (Table V), following the biosynthetic pathway worked out by Conn (1979a). The derivation of acalyphin is not yet known (Nahrstedt *et al.*, 1982). Again excluding the Passifloraceae, in only one family, the Fabaceae (= Leguminosae), are all four noncyclopentenoid cyanogenic glycoside pathways expressed (10 compounds). Three (Phe, Leu, Iso/Val) occur together within the genus *Acacia*. Four other families produce cyanogenic glycosides by two pathways: the Asteraceae, Euphorbiaceae, Poaceae, and Rosaceae (Saupe, 1981).

The remaining 103 families apparently elaborate only one structure. Within the families possessing more than one biogenetic pathway, the

Table V. Derivations of Cyanogenic Glycosides in Plants

Amino Acid Precursor	Cyanogenic Glycosides (Number and Names)
Leucine	8—Proacacipetalin, epiproacacipetalin, heterodendrin, epiheterodendrin, cardiospermin, cardiospermin sulfate, cardiospermin-*p*-hydroxybenzoate, 3-hydroxyheterodendrin
Isoleucine/valine	4—Linamarin, lotaustralin, linustatin, neolinustatin
Phenylalanine	10—Prunasin, sambunigrin, amygdalin, vicianin, lucumin, holocalin, zierin, epilucumin, epilucumin-4″-*p* (β-D-glycopyranosyloxyl)-(Σ)-cinnamate, epilucumin-4″-*p* (β-primeverosyloxy)-(Σ)-cinnamate
Tyrosine	7—Taxiphyllin, dhurrin, proteacin, *p*-glycosyloxymandelonitrile, triglochinin, triglochin methyl ester[a], isotriglochinin[a]
2-Cyclopentenylglycine	6—Gynocardin, tetraphyllin A, tetraphyllin B, epi-tetraphyllin B, deidaclin, epigynocardin[b]
Cyanopyridone	1—Acalyphin[c]

Source: Spencer (1988).

[a] Structure possibly an artifact of isolation. Acacipetalin (leucine) has been determined to represent such an artifact and is not included here.

[b] Structure report in preparation.

[c] Of unknown biosynthetic origin.

distribution of these follows subfamilial or tribal divisions, without exception. Even in *Acacia*, the distribution of pathways follows subgeneric and biogeographic divisions.

The overall picture is of widespread distribution of noncyclopentenoid cyanogenic glycoside pathways, with each family or subfamilial taxon producing generally only one compound, a few producing several related compounds. Diversification in glycosylation has not been extensive. All but eight of the known cyanogenic glycosides are monoglucosides, four are gentiobiosides, two are primeverosides, and one contains a vicianose moiety (Conn, 1981b).

In terms of systematic distribution within plants, cyanogenic glycoside biogenetic pathways seem to have arisen either early, or many times. Tyrosine-derived compounds occur in gymnosperms, many monocots, and primitive dicots, and are present in many more advanced families as well. Phenylalanine-derived cyanogenic glycosides are elaborated by 13 families in several orders, and by certain ferns. The leucine pathway and the isoleucine/valine pathway are more restricted in occurrence (five families), but these families tend to be advanced and unrelated.

Passiflora and other genera in the Passifloraceae have long been recognized as cyanogenic (Rosenthaler, 1919, 1929; Steyn, 1929), but the Passifloraceae is unusual in producing cyanogenic glycosides with a cyclopentene moiety (Paris *et al.*, 1969; Russell and Reay, 1971; Tantisewie *et al.*, 1969; Clapp *et al.*, 1970; Fischer *et al.*, 1982; Gondwe *et al.*, 1978; Seigler *et al.*, 1982; Spencer, 1984, 1987; Spencer and Seigler, 1984b, 1985a–c, 1987a,b; Spencer *et al.*, 1988). This trait is shared only with members of five closely related families, the Turneraceae (Spencer and Seigler, 1980, 1981; Spencer *et al.*, 1985), Malesherbiaceae (Spencer and Seigler, 1985e), Flacourtiaceae (Coburn and Long, 1966; Kim *et al.*, 1970; Spencer *et al.*, 1981; Spencer and Seigler, 1985d), Achariaceae (Jensen and Nielsen, 1986), and Caricaceae (Spencer and Seigler, 1984c).

Compounds that possess a cyclopentene ring structure occur naturally only as cyclopentenoid cyanogens or lipids, a few prostaglandins, some pyrethrins, and a few terpenoids. The cyanogenic compounds are probably biosynthesized from the nonprotein amino acid 2-cyclopentenylglycine via oxidation, oxidative decarboxylation, dehydration, and glycosylation (Conn, 1979a). This is a canalized process involving several genes. The ability to elaborate cyanogenic glycosides from 2-cyclopentenylglycine has apparently arisen only once in the evolution of plants.

Previous studies (Seigler *et al.*, 1982; Spencer, 1984; Spencer and Seigler, 1983; 1984a–c; 1985a–c, 1987a,b) indicate that a great diversity of cyclopentenoid cyanogens exist in the genus *Passiflora*, and that this diversity might be sufficient to permit taxonomic inferences to be made and to correlate these with data on insect distribution. Recently, the distribution of only four cyclopentenoid cyanogens was demonstrated to be a good indicator of taxonomic relationships within the related genus *Adenia* (Spencer, 1984; Spencer *et al.*, 1988) by correlating chemical distributions a posteriori with the results of numerical analysis of a large morphological data set. Other studies have confirmed the usefulness of cyclopentenoid cyanogen distributional patterns in studies of systematic and phylogenetic problems in other families of plants (Spencer and Seigler, 1984c, 1985d,e; Spencer *et al.*, 1985). It was reasonable to expect therefore, that a distributional analysis of *Passiflora* chemicals would yield information on the evolution of taxa within the genus.

Previous chemical work has shown *Passiflora* to contain simple harmane alkaloids (Schilcher, 1968; Lutomski, 1960) and an unusual triterpene glycoside (Bombardelli *et al.*, 1975). The unique flavonoid chemistry of *Passiflora* has been explored in a series of papers (McCormick and Mabry, 1981; Ulubelen *et al.*, 1982; Ulubelen and Mabry, 1980). None of these compound classes has been demonstrated to correlate in distribution to host-plant specialization in *Heliconius* nor to other known special-

Table VI. Quantitative Cyanide Assay of Plant Parts of *Passiflora* × *alatocaerulea*

1	Stigma	32[a]
2	Anther	35
3	Filament	31
4	Gynoecium	45
5	Stalk of Gynoecium	42
6	Outer corona filaments	43
7	Petal	36
8	Sepal	44
9	Bract	41
10	Peduncle	46
11	Tendril	45
12	Base of inner corona	33
13	Outer receptacle	35
14	Inner receptacle	41
15	Stem	39
16	Leaf	40
17	Root	60
18	Pericarp, unripe fruit	45
19	Aril, unripe fruit	48
20	Seed, unripe fruit	35
21	Pericarp, ripe	5
22	Aril, ripe	5
23	Seed, ripe	0

Source: Spencer (1988).
[a] Measured in μmol/g HCN, average of five samples. All ranges ± 5%.

ists including flea beetles (*Altica*, *Disonycha*, *Monomacra*, *Strabala*) and locally important coreid bugs (*Anisocelis* and *Holymenia*) (Gilbert and Smiley, 1978; Pemberton, 1983). In terms of both metabolic cost and biosynthetic diversity, cyanogenic glycosides undoubtedly represent the major chemical defense of *Passiflora*, and provide the best indicators of taxonomic relationships.

A survey for cyanogenesis in over 570 accessions of *Passiflora* and an additional 100 species from other genera of the Passifloraceae was carried out (Spencer, 1988). The results were consistent with previous reports of cyanogenesis in *Passiflora* (Tjon Sie Fat, 1979). Two-thirds (210/306) of the species of *Passiflora* were found to be cyanogenic; these were distributed through 20 of Killip's 22 subgenera. Lack of material from two other subgenera precluded testing of those groups. Enzymatic discrimination showed that most species contained cyclopentenoid cyanogenic glycosides, although other types of cyanogenic glycosides were encountered in some instances. Plant parts were tested separately, but little difference was found. Quantitative analysis of HCN production in *P.* × *alatocaerulea* showed that all parts of the plant usually produce similar amounts of cyanogenic compounds, although more may be accumulated in the roots (Table VI). None is present in mature seeds and very little in ripe fruits. These findings are consistent with previous work (Spencer and Seigler, 1983).

The cyanogenic principles of 37 species of *Passiflora* in 11 subgenera have been isolated (Fischer *et al.*, 1982; Seigler *et al.*, 1982; Spencer and Seigler, 1983, 1984a,b, 1985a–c, 1987a,b; Spencer, 1984, 1988; Spencer *et al.*, 1986) (Table VII). A tremendous diversity in cyanogen type was found in *Passiflora* (Table VIII and Fig. 1). Not only are a number of cyclopentenoid cyanogenic glycosides elaborated, but isoleucine/valine-derived cyanogenic compounds, cyanohydrins, and aromatic cyanogens are also found. Many of these compounds had not been previously described. The compounds can be logically assigned to six structural classes: sulfates, aliphatics, aromatics, simple cyclopentenoid, complex cyclopentenoid, and cyanohydrin (Table VIII). The distribution of cyanogenic compounds in *Passiflora* has no parallel. No other genus previously has been known to produce more than eight structurally related types, and only one other genus (*Acacia*) was previously known to elaborate cyanogens by three distinct biosynthetic pathways (Seigler, 1981). Comparison of the distribution of cyclopentenoid cyanogens in the genus *Passiflora* with those found in related families underscores this diversification (Table IX).

The distribution of identified cyanogenic compounds within Killip's phylogenetic interpretation of *Passiflora* (Killip, 1938; 1960) is nonrandom and taxonomically significant (Spencer, 1988). Members of subgenus *Granadilla* elaborate sulfated cyanogens. Most members of subgenus

Table VII. Isolation of Cyanogenic Glycosides from *Passiflora*

Subgenus	Section	Series	Species[a]	Compound[b]
Plectostemma	Cieca		*coriacea*	17–20
			suberosa	17–20
			warmingii	6, 8
			bryonoides	27
			lobbii	6
			pendens	6–9
	Decaloba	Apetalae	*apetala*	**15**
		Luteae	*lutea*	6, 7, 22, 23
		Miserae	*trifasciata*	24
		Punctatae	*talamancensis*	22
			biflora	22, 23
	Xerogona		*conzattiana*	21
			capsularis	21
	Pseudodysosmia		*adenopoda*	6, 7
Pseudomurucuja			*perfoliata*	29
			cuprea	**12, 21**
Psilanthus			*sanguinolenta*	**15**
Tacsonia			*molissima*	1, 2
Granadillastrum			*manicata*	1, 12–14
			antioquiensis	12–14
Distephana			*coccinea*	3, 4
			vitifolia	3, 4
Calopathanthus			*racemosa*	12, 13
Granadilla		Quadrangulares	*quadrangularis*	1, 2, 28
			alata	1, 2
		Tiliaefoliae	*platyloba*	**11**
		Laurifoliae	*laurifolia*	1, 2, 5
			nitida	1, 2
		Incarnatae	*incarnata*	15
			edulis	10
		Simplicifoliae	*oerstedii*	**15**
		Lobatae	× *alatocaerulea*	1, 2
			caerulea	1, 2
Dysosmia			*foetida*	2, 26
Astrophea	Dolichostemma		*pittieri*	2, **22**
	Euastrophea		*lindeniana*	**11, 25**
	Pseudoastrophea		*sclerophylla*	**17,18,19**

Source: Spencer (1988).
[a] Nomenclature follows Killip (1938).
[b] Compound reports in boldface are based on preliminary nuclear magnetic resonance (NMR), HPLC, and enzyme specificity data only. Numbers refer to Table VIII.

Plectostemma, section *Decaloba*, produce tetraphyllin B diglycosides. Both members of subgenus *Plectostemma*, section *Xerogona*, produce passicapsin. Members of subgenus *Distephana* produce derivatives of sulfated cyanogenic compounds, suggesting a relationship with subgenus

Table VIII. Cyanogens of the Genus *Passiflora* and Relatives

Compound Class	Compound Name
A—Sulfates	1 Tetraphyllin B sulfate
	2 Epitetraphyllin B sulfate
	3 Passicoccin
	4 Epipassicoccin
	5 Passilaurifolin[a]
B—Aliphatics	6 Linamarin
	7 Lotaustralin
	8 Linustatin
	9 Neolinustatin
C—Aromatics	10 Prunasin
D—Simple cyclopentenoid	11 Tetraphyllin A
	12 Deidaclin
	13 Tetraphyllin B
	14 Taraktophyllin[a]
	15 Gynocardin
	16 Epigynocardin[a]
	17 Epivolkenin[a]
	18 Volkenin[a]
E—Complex cyclopentenoid	19 Passisuberosin
	20 Epipassisuberosin
	21 Passicapsin
	22 Passibiflorin
	23 Epipassibiflorin
	24 Passitrifasciatin
	25 Unknown cyclopentenoid
F—Complex cyanohydrins	26 Passifoetidin[b]
	27 Passibryonoidin[b]
	28 Passiquadrangularin[b]
	29 Unknown α-hydroxynitrile

Source: Spencer (1988).
[a] Nomenclatural and structural reassignments after Jaroszewski and Olafsdóttir (1986,1987), Jaroszewski *et al.* (1987a,b), and Seigler and Spencer (1988).
[b] Structure report in preparation.

Granadilla or subgenus *Tacsonia*. Putatively primitive *Passiflora* species (subgenus *Astrophea*) have simple compounds, while the supposedly most advanced species have complex compounds or have noncyclopentenoid cyanogens (subgenus *Plectostemma*). Most members of subgenus *Plectostemma,* section *Cieca*, produce leucine and isoleucine/valine-derived cyanogenic glycosides and may be thus more closely related to section *Pseudodysosmia* (*P. adenopoda*) than has previously been recognized.

The putatively more primitive members of the Passifloraceae (Table X) elaborate cyanogens typical of the Turneraceae and Malesherbiaceae.

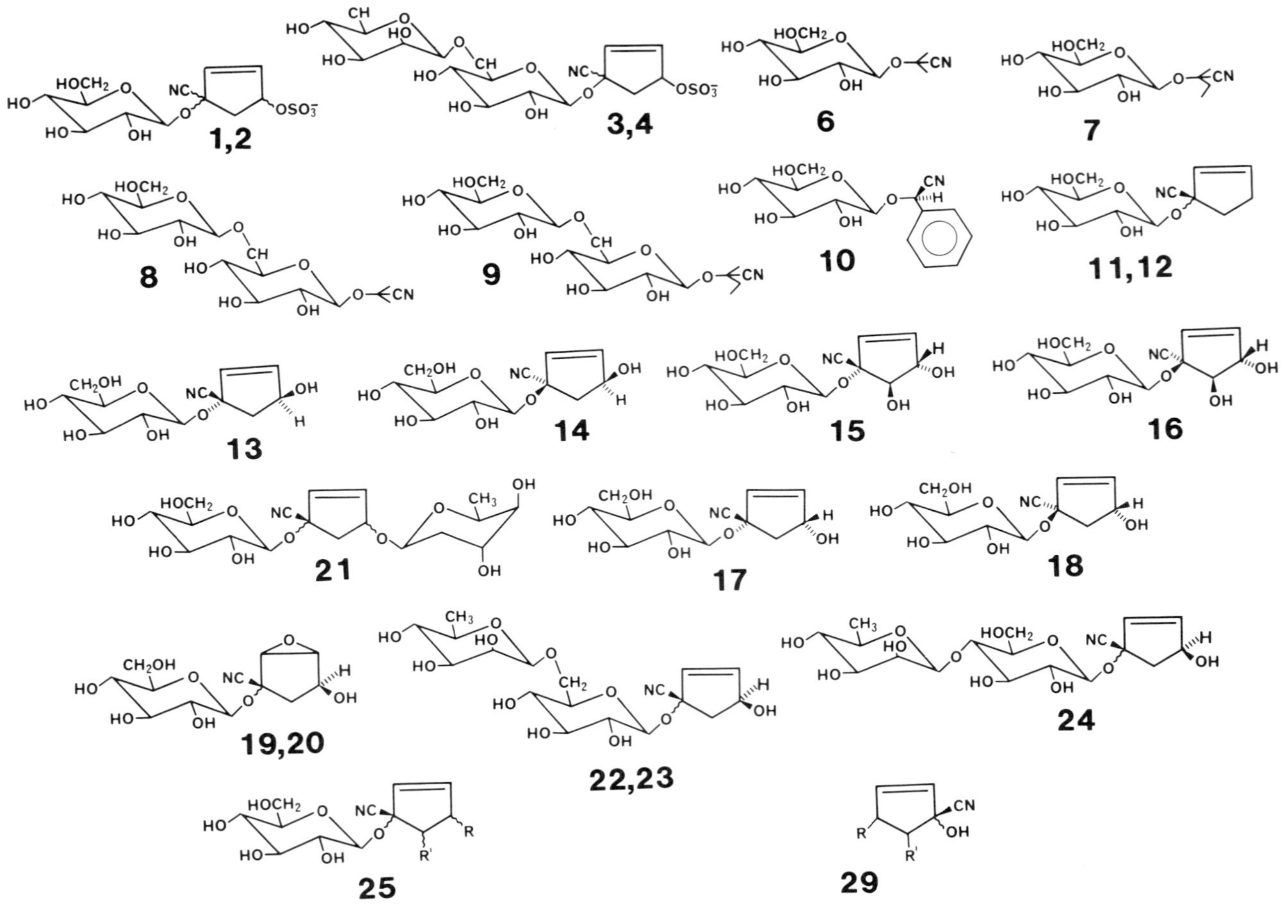

Fig. 1. Cyanogenic compounds isolated from *Passiflora* and related genera.

Table IX. Distribution of Cyanogens in the Violales

	Compound Numbers[a]																												
Taxa	1	2	3	4	5	6	7	8	9	10	11	12	13	14	15	16	17	18	19	20	21	22	23	24	25	26	27	28	29
Passifloraceae	X	X	X	X	X	X	X	X	X	X	X	X	X	X	X		X	X	X	X	X	X	X	X	X	X	X	X	X
Other genera in the Passifloraceae											X	X	X	X															
Turneraceae											X	X	X	X															
Malesherbiaceae											X	X	X	X															
Flacourtiaceae							X							X	X	X	X												
Achariaceae															X														
Caricaceae										X			X																

Source: Spencer (1988).
[a] Compound numbers refer to Table VIII. An X beneath a compound number denotes its occurrence in the taxon named.

Table X. Cyanogenic Glycosides of Other Genera of the Passifloraceae

Species	Compound (Numbers Refer to Table VIII)
Androsiphonia adenostegia Stapf.	11, 13
Barteria fistulosa Mast.	14
Crossostemma laurifolium Planch.	11–14
Deidamia clematoides (C. H. Wright) Harms	12
Dilkea parviflora Killip	11, 13
Mitostemma brevifilis Gontsch.	11
Paropsia brazzeana Baill.	11
Smeathmannia pubescens Soli ex R. Br.	11, 13
Tetrapathea tetrandra Cheesm.	11–14
Tryphostemma friensii Norlindh.	13, 14
Adenia spp. (26)	One or more of 11–14, 17, 18

Source: Spencer (1988).

III. Enzymes

Glycosides themselves are generally regarded as representing simple storage products, accumulable derivatives of the aglycone moiety (Hösel, 1981a). Glycosidation is often disregarded as irrelevant in terms of potential biological activity, except in reducing this activity as part of a storage process (Pridham, 1965). It is certainly appreciated that glycosidase-mediated enzymatic hydrolysis of glycosides (Hösel, 1981b) releases the aglycone whereby these glycosides may become biologically active toxins, but this process is also regarded as essentially a storage–release phenomenon.

β-Glucosidases and other glycosidases often show an exceedingly high degree of specificity toward a particular substrate (Hösel and Nahrstedt, 1975; Nahrstedt *et al.*, 1979). In recent work (Hösel and Conn, 1982; Dale *et al.*, 1985) this specificity was attributable to the structure of the aglycone. Several recent reviews (Nisizawa and Hashimoto, 1970; Spencer and Seigler, 1984a; Conn, 1979a,b) have indicated that a specific glycosidase corresponding to a specific glycoside structure is generally present in the same plant. However, this necessary coelaboration has been frequently overlooked as a functional mechanism of evolutionary change in diversified systems of coevolved plants and insects.

I will argue in this chapter that the evolution of the process of hydrolysis is an event under the selection of the competing interests of plant and insect species, and that "toxic" plant glycosides should be regarded as such only in terms of an inseparable, targeted glycoside–glycosidase system.

Table XI. Specificity of β-Glucosidases of *Passiflora* and Related Species

	Compound										
Enzyme Preparation	Linamarin	Prunasin	Tetraphyllin B sulfate	Deidaclin	Tetraphyllin B	Gynocardin	Passibiflorin	Passitrifasciatin	Passisuberosin	Passicoriacin	Amygdalin
Turnera ulmifolia	−	−	−	+++	+	−	+	−	−	−	−
Passiflora foetida	−	−	+++	+++	+++	−	+++	−	−	−	−
P. × *alatocaerulea*	−	−	+++	−	+	−	+	−	−	−	−
P. caerulea	−	−	+++	−	+	−	+	−	−	−	−
P. trifasciata	−	−	−	−	−	−	−	+++	−	−	−
P. suberosa	−	−	−	−	−	−	−	−	+++	+	−
P. coriacin	−	−	−	−	−	−	−	−	+	+++	−
Emulsin (Sigma)	−	+++	−	−	−	−	−	−	−	−	+++
Linamarase	+++	−	−	−	−	−	−	−	−	−	−
Gynocardia odorata	−	−	−	−	−	+++	−	−	−	−	−

Notes: Progress of hydrolysis: no reaction (−), slow reaction (+), fast reaction (+++). From Spencer (1987).

I will extend this argument to encompass the diversification of such systems in lineages of plants in an effort to correlate enzyme-mediated glycoside toxicity with the evolution of host-plant specificity in insects and the coevolution of plants and insects.

Cyanogenic glycosides yield free cyanide and an aglycone through the hydrolytic action of a β-glucosidase (Hösel, 1981a,b). These β-glucosidases are very specific in action, so a series of different enzyme preparations may be used to classify the cyanogen as aromatic, aliphatic, or cyclopentenoid, and, in the case of cyclopentenoid cyanogens, to indicate the probable substitutional isomer and epimer (Spencer and Seigler, 1984a). The specificity of β-glucosidases toward their substrates has been confirmed in the production of cyanide from cyanogenic glycosides of *Passiflora* and its relatives (Table XI). The specificity is such that even a crudely purified enzyme preparation (Spencer, 1984; Spencer and Seigler, 1984a) can be used to identify biosynthetic and structural types of cyanogens.

The observation that particular combinations of β-glucosidases from

Table XII. Inhibition of Natural β-Glucosidase Activity by Addition of Similar Glucosidases of Competing Substrates

Combinations of Enzyme Preparations and Cyanogenic Compounds	Compound: Deicalin	Tetraphyllin B sulfate	Prunasin	Linamarin	Gynocardin	Passibiflorin	Passitrifasciatin	Passisuberosin	Passicoriacin	Amygdalin
T. ulmifolia + *P.* × *alatocaerulea*	−	−	−	−	−	−	−	−	−	−
T. ulmifolia + emulsin	+	−	−	−	−	−	−	−	−	−
T. ulmifolia + linamarase	+	−	−	+	−	−	−	−	−	−
T. ulmifolia + *P. foetida*	−	−	−	−	−	−	−	−	−	−
P. × *alatocaerulea* + *P. foetida*	−	+	−	−	−	−	−	−	−	−
P. foetida + emulsin	−	+	−	−	−	−	−	−	−	−
P. biflora + *P. trifasciata*	−	−	−	−	−	−	−	−	−	−
P. suberosa + *P. biflora*	−	−	−	−	−	−	−	−	−	−
P. coriacea + *P. trifasciata*	−	−	−	−	−	−	−	−	−	−
T. ulmifolia + amygdalin	−	−	−	−	−	−	−	−	−	−
Emulsin + tetraphyllin B	−	−	−	−	−	−	−	−	−	−
G. odorata + linamarin	−	−	−	−	−	−	−	−	−	−
P. × *alatocaerulea* + prunasin	−	−	−	−	−	−	−	−	−	−

Notes: Progress of hydrolysis: no reaction (−), very slow reaction (+). From Spencer (1987).

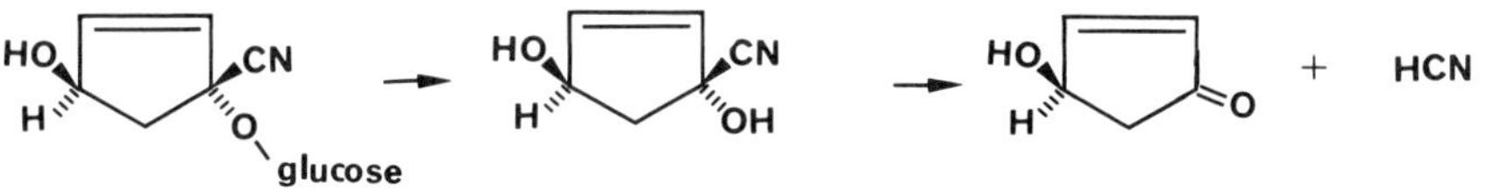

Fig. 2. Enzymatic hydrolysis of a cyclopentenoid cyanogenic glycoside. The first hydrolysis is catalyzed by a plant β-glucosidase. The second reaction occurs either spontaneously or through the action of an α-hydroxynitrile lyase.

various related plant species, or combinations of plant β-glucosidases and substrates, frequently prevented expected hydrolysis (Table XII), which gave the first indication that the interaction of plant enzymes could be of importance in the *Passiflora–Heliconius* system. The same inhibitory interaction has since been observed in insect–plant glucosidase combinations (Spencer, 1987). The cyclopentenoid cyanogenic glycosides undergo hydrolysis according to the reaction illustrated in Fig. 2. This two-step process is consistent with that determined for other cyanogenic glycosides (Hösel, 1981a). The second step is thermodynamically favored, and occurs rapidly at normal cell pH even in the absence of α-hydroxynitrile lyase. The ketone produced is toxic (Fig. 3) (see Section IV).

Recent work (Goldstein and Spencer, 1985) has shown that plant tannins are capable of inactivating plant β-glucosidases *in vitro* (Fig. 4). Insect β-glucosidases are also inactivated by plant tannins at field concentrations (Fig. 5). *Heliconius* β-glucosidase fractions inhibit *in vitro Passiflora* β-glucosidase hydrolysis of cyclopentenoid cyanogenic glycosides, as determined through quantitative measurement of HCN release in the test reaction (Spencer, 1987). It is possible that these ecoregulatory enzymes, used by the insect to inhibit targeted plant toxification systems, may themselves be the principal target of the production of tannins by plants. The discovery of other examples of secondary chemical interac-

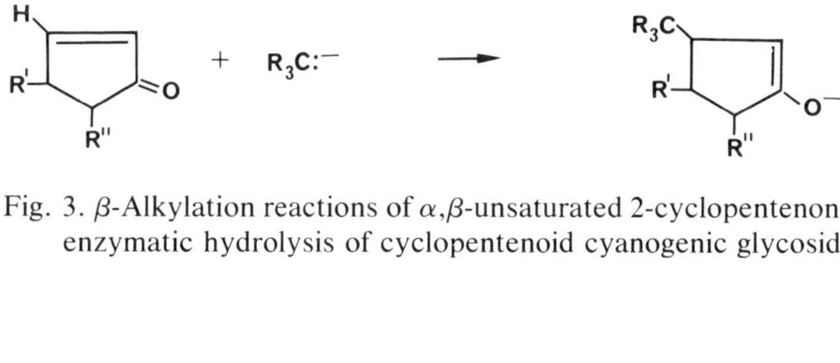

Fig. 3. β-Alkylation reactions of α,β-unsaturated 2-cyclopentenones produced by enzymatic hydrolysis of cyclopentenoid cyanogenic glycosides.

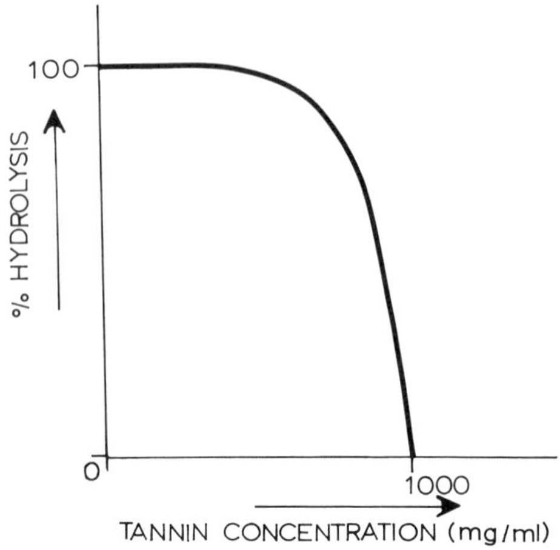

Fig. 4. Inhibition by tannin of hydrolysis of cyclopentenoid cyanogenic glycosides by *Passiflora* β-glucosidase preparations. Percent hydrolysis is measured over a standard time interval (see Goldstein and Spencer, 1985).

tions with enzymes dedicated to the regulation of toxification and detoxification mechanisms may be expected as our knowledge of chemically mediated plant–insect interaction expands.

There are additional indications that *Heliconius* has evolved enzymological competence toward *Passiflora* compounds, possibly as a coevolu-

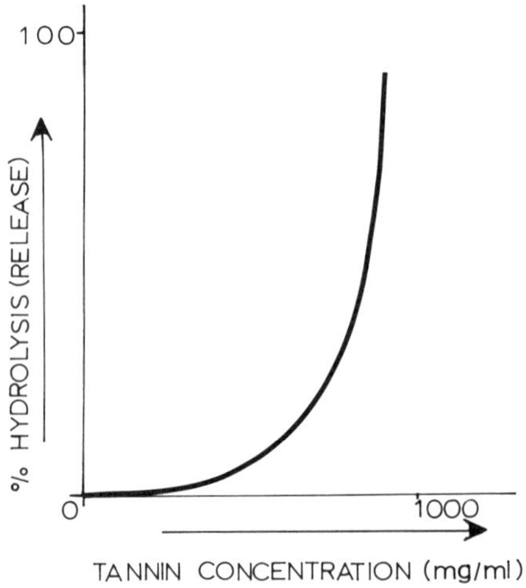

Fig. 5. Release from inhibition by *Heliconius* inhibitory enzymes of plant-mediated hydrolysis of cyanogenic glycosides upon deactivation of *Heliconius* enzymes with tannin. Percent hydrolysis is measured over a standard time interval (see Spencer, 1987).

tionary response. *Heliconius* caterpillars, pupae, and adults are themselves cyanogenic and contain linamarin and lotaustralin (Nahrstedt and Davis, 1981), as do relatives (Davis and Nahrstedt, 1979); these are simple cyanogens common in various plants but thus far found only in a few *Passiflora* species (Table VII) and are synthesized via a pathway different from the majority of *Passiflora* cyanogens. Further work suggests that *Heliconius* and relatives synthesize their own cyanogenic glycosides (Wray *et al.*, 1983) yet also may sequester them (Nahrstedt and Davis, 1986). Whether or not the cyanogenic glycosides are synthesized in the butterfly, derived from the plant, or both, it is clear that *Heliconius* possess their own metabolism for processing cyanogenic compounds.

IV. Biological Defense

Cyanogenic glycosides are known toxins (Jones, 1962, 1972, 1981) that have been demonstrated to be ecologically important deterrents to herbivory (Angseesing and Angseesing, 1973; Cooper-Driver and Swain, 1976; Bernays *et al.*, 1977; Woodhead and Bernays, 1977, 1978; Brower, 1984). In all of these cases, their deterrency is manifested only in the release of hydrogen cyanide upon enzymatic hydrolysis, and is generally not regarded as being particularly effective (i.e., Moertel *et al.*, 1981, 1982).

The hydrolytic enzyme is localized in mesophyll tissues and the cyanogenic glycoside in epidermal tissues of sorghum (Kojima *et al.*, 1979). This separation of glucosidase from glucoside is probably of general occurrence, and protects the substrate from hydrolysis until leaf damage occurs, as during herbivory. The toxic manifestation of these compounds to herbivores as HCN has not been adequately explored in the field nor *in vivo*, but HCN toxicity has been quantitatively determined in many systems (Solomonson, 1981; Way, 1981).

What must be stressed is that although HCN is a fairly toxic compound ($LDO_{50} \simeq 0.7$ mg/kg for rats), it is entirely nonspecific in its action, and hence cannot account for variations in distribution of different cyanogenic glycoside structures. Metabolically expensive diversification of glycoside structures should not occur if their selective value lies merely as a carrier of a cyano entity. In fact, very many animal and plant enzyme systems are capable of rapidly and efficiently detoxifying HCN (Volini and Alexander, 1981; Brattsten *et al.*, 1983; Miller and Conn, 1980), and while selection by herbivores of acyanogenic versus cyanogenic plants has been demonstrated (for reviews see Jones, 1981; Scriber, 1978), actual selective reduction of fitness by HCN has not been quantified in plant–insect systems.

Thus, as stated previously, hydrogen cyanide evolution is only half the story in regards to cyanogenic glycoside hydrolysis. The aglycone moiety generated along with HCN may also possess biological activity. As an example, the repellent activity of benzaldehyde, the aglycone product of hydrolysis of prunasin and other aromatic cyanogenic glycosides, has been well studied and a review is given by Blum (1981). The compound is toxic [LDO_{50} = 0.10% (weight basis, normal meal) in rats (Jenner *et al.*, 1964)] and is a feeding repellent, deterrent, or irritant to many insects including moths (Hirai *et al.*, 1978; Hirai, 1982), aphids (Dreyer *et al.*, 1981), ants (Blum *et al.*, 1969), and millipedes and other arthropods (Eisner, 1970; Eisner *et al.*, 1978). HCN is also produced by arthropods (Duffey, 1981; Duffey *et al.*, 1977), and Eisner and Eisner (1965) and Eisner *et al.* (1978) have concluded that both HCN and benzaldehyde are biologically active in field situations (see also Brower, 1984), a statement I find to be prescient on first principles of consideration of chemistry and parsimonious expenditure. If cyanide were the only defensive portion of a cyano compound, it should be delivered on a structure that is least expensive biosynthetically. Benzaldehyde is energetically too costly to waste as a delivery compound.

For other cyanogenic glycoside parental structures, hydrolysis generates a variety of ketones: linamarin yields acetone, and lotaustralin yields 2-butanone, for example. These compounds have a relatively high toxicity to mammals [LDO_{50} = 0.05–0.06% (weight basis, normal meal) for rats (Smyth *et al.*, 1962)] and are irritating and repellent in high concentrations. Their toxicity to some insects is considerably greater due at least partly to their ability to denature flexible insect cuticular integuments (Spencer, unpublished).

The 2-cyclopentenones (Fig. 2) produced by enzymatic hydrolysis of cyclopentenoid cyanogenic glycosides are ketones but are chemically very different in nature from the aliphatic and aromatic ketones mentioned above. Because the 2,3-diene ring unsaturation is conjugated with the carbonyl at C-1, these ketones possess an α,β-unsaturated system, which is very prone to nucleophilic addition reactions (Fig. 3) (Spencer, 1987; Hall *et al.*, 1977). These Michael additions can involve a wide variety of biologically important molecules, including enzymes, nucleic acids, and essential metabolites, which may become alkylated by the cyclopentene moiety and inactivated. Preliminary data indicate that the alkylation reaction is so rapid and nonspecific on *in vitro* alkylations of nucleic acids, proteins, and membranes that development of a specific rather than a general posthydrolysis resistance mechanism in an herbivore is unlikely. The insect species should therefore be highly induced to produce a defensive capability that prevents hydrolysis altogether. This selective pressure, targeted against the ketone, is not at all necessarily the same as would induce the development of a defense against HCN.

Cyclopentenones can therefore be quite toxic (Hall *et al.*, 1977) by a mode of action quite different from other ketones.

Changes in the structure of the cyclopentenoid glycoside influence the specificity of the nucleophilic attack in two ways. First, alteration of either the sugar moiety or the aglycone changes the susceptibility of the molecule to hydrolysis by a specific β-glucoside and hence, the availability of the alkylating ketone in the reaction milieu. Second, alteration of the structure of the aglycone moiety by ring substitution or changes in symmetry will alter the electronic state of the α,β-unsaturated system with concomitant alteration of its ability to accept a particular nucleophile. Within the *Passiflora/Heliconius* interaction, therefore, different cyclopentenoid cyanogenic glycosides should have different toxicities toward different insects. Moreover, toxicity can be due to HCN and/or the aglycone. The latter can be a generally toxic ketone or a specifically toxic alkylating ketone, and the entire process is regulated by specificity in action of hydrolytic enzymes of both the plant and the insect.

Confirmation of the above hypothesis requires demonstration of differential toxicity of different glycoside structures, and this must be related to enzyme activity. Further, this test should utilize an organism lacking cyclopentenoid-specific glycosidases, so as to be free of the compensatory effects of specialist adaptation. These results should then be related to the *Heliconius/Passiflora* interaction by finding evidence of similar effects.

A bioassay was designed to determine the independent toxicity to lepidoptera of HCN and aglycone moieties of a variety of cyanogenic glycosides: aliphatic, aromatic, and cyclopentenoid. In addition, a number of cyanohydrins (α-hydroxynitriles), cyanogenic compounds retaining both CN and aglycone but lacking a glycoside moiety and so requiring no enzyme to hydrolyze, were tested.

An assay of the toxic effects of various cyanogenic glycosides, cyanohydrins, and cyanide was performed using *Heliothis zea* (Boddie), which does not utilize cyclopentenoid-producing host plants, nor does it prefer cyanogenic host plants. Negative controls (no cyanide) and controls using serial concentrations of NaCN and KCN (LDO_{50} = 0.001% in rats) (Hayes, 1967; Smyth *et al.*, 1969) and glucose were also run. All compounds were tested at concentrations of 0.1%, 1.0%, and 2.0% w/v, bracketing the normal concentrations in living plants of 0.5–1.0% (Conn, 1980, 1981b; Spencer, 1984; Spencer and Seigler, 1979–1987). Materials were weighed and mixed with 5.0 g of cooling *H. zea* media (Bioserv, Inc., number 939R4) in 1-oz rearing cups. Two-day-old (to reduce mortality due to handling) first instar larvae were introduced into these cups singly, and kept under standard conditions. Ten replicates of each concentration were run, and results were scored according to criteria of larval growth rate (weight, size, and stage), pupation, and eclosion. Rearing procedures followed Waldbauer *et al.* (1984).

It can be concluded from these initial experiments (Table XIII) that (1) cyanide below lethal concentrations retarded larval development, (2) a standard lethal concentration (LC_{50}) of cyanide was reached at about 1.0%, the same level reported by Brattsten *et al.* (1983), (3) cyclopentenoid cyanogenic glycosides were more toxic than aromatic or aliphatic cyanogenic glycosides, (4) toxicity due to cyanogenic glycosides is not due solely to their titer of cyanide, as given amounts of cyanide carried on cyanogenic glycosides frequently differed in toxicity from identical molar amounts of free cyanide, and (5) complex cyclopentenoid cyanogens, cyanohydrins, and substituted cyclopentenoid cyanogens possess differing levels of toxicity toward *H. zea*. In a further demonstration of specific toxicity of cyclopentenoid cyanogens, topical application of tetraphyllin B to *Heliconius ismenius* fifth instar larvae as a 1.0% solution in dimethyl sulfoxide (DMSO) caused death within 3 days, while controls dosed with DMSO or an equivalent titer of cyanide (0.1%) or with linamarin or amygdalin pupated normally.

A digestive enzyme preparation of *Heliothis* larvae was shown to contain β-glucosidase activity that hydrolyzed many cyanogenic compounds inefficiently. Simple Feigl–Anger quantitative assay (Feigl and Anger, 1966) of HCN released during hydrolysis revealed large differences in the rate of hydrolysis of various compounds. This enzyme preparation caused complete or partial inhibition of hydrolysis in plant glycoside–plant glucosidase fraction pairs, entirely consistent with the data obtained from the *Heliconius*/*Passiflora* and *Pieris*/crucifer interactions (Spencer, 1987). Inducibility of cyanide resistance was not observed to occur as it did in *Spodoptera eridania* Cramer (Brattsten *et al.*, 1983). Pupal weight of *H. zea* does tend to increase with cyanogen dose, but survival and emergence time generally decrease. The data presented herein on cyanogenic glycoside toxicity to *H. zea* shows this organism to be relatively unable to tolerate these compounds, as opposed to *S. eridania*, which has as a preferred host plant a cyanogenic species. The confusing data pertaining to the general toxicity of cyanogenic glycosides toward insects (see Bernays, 1983, for a review) indicates strongly that assessments of toxicity made independent of ecological considerations are not meaningful. Toxicity is not an innate characteristic of chemical compounds: it is a biological effect that has ecological meaning because of the history of evolutionary interaction between the chemical source and its target.

It should be clear that the only adequate bioassay to determine the toxicity of *Passiflora* cyanogens toward *Heliconius* will require laboratory and field testing using *Heliconius* species for which host-plant specialization patterns are known. Initial *in vitro* studies in this laboratory have established that specific resistances to toxification by cyclopente-

Table XIII. Toxicity of Cyanogenic Compounds to *Heliothis zea*

	0.5%				1.0%				2.0%			
	GR	PW	T to E	E	GR	PW	T to E	E	GR	PW	T to E	E
Control	100	100	100	100	100	100	100	100	100	100	100	100
Glucose	100	100	100	100	100	100	100	100	100	100	100	100
Sodium cyanide[a]	60	105	108	40	60	—	—	0	0	—	—	0
Potassium cyanide	60	98	119	40	60	—	—	0	0	—	—	0
Amygdalin + β-glucosidase	60	98	115	20	60	—	—	0	0	—	—	0
Linamarin + β-glucosidase	60	102	136	40	50	110	—	40	0	—	—	0
Tetraphyllin B sulfate + β-glucosidase	60	102	112	40	50	101	131	40	0	—	—	0
Tetraphyllin B sulfate + sulfatase	50	115	100	40	50	106	112	40	0	—	—	0
Tetraphyllin B sulfate[a]	100	109	118	60	100	114	131	40	50	88	140	20
Amygdalin	100	110	108	80	100	118	115	40	80	86	120	40
Linamarin	100	99	104	60	100	116	108	60	100	68	120	40
Tetraphyllin B	100	102	138	20	80	101	110	60	60	89	135	20
Volkenin	100	89	119	40	80	107	121	40	80	103	135	40
Deidaclin	100	106	123	60	100	108	115	60	80	111	125	40
Gynocardin	100	100	123	40	80	101	127	20	50	97	140	20
Cyanogenic glycoside from *P. costaricensis*	80	102	127	60	80	101	131	20	40	80	—	0
Cyanohydrin from *P. alata*	80	107	115	20	20	—	—	0	65	93	—	0
Cyanohydrin from *P. bryonoides*	50	101	108	40	20	—	—	0	5	—	—	0
Cyanohydrin from *P. foetida*	100	113	123	20	50	101	119	40	20	111	—	0

Note: GR, larval growth rate (% of control); PW, pupal weight (% of control); T to E, time to eclosure relative to controls; E, successful eclosure (%).

[a] Pupal weights vary from ±SE = 0.60–17.67%, $\overline{X}$ = ±SE = 8.46%.

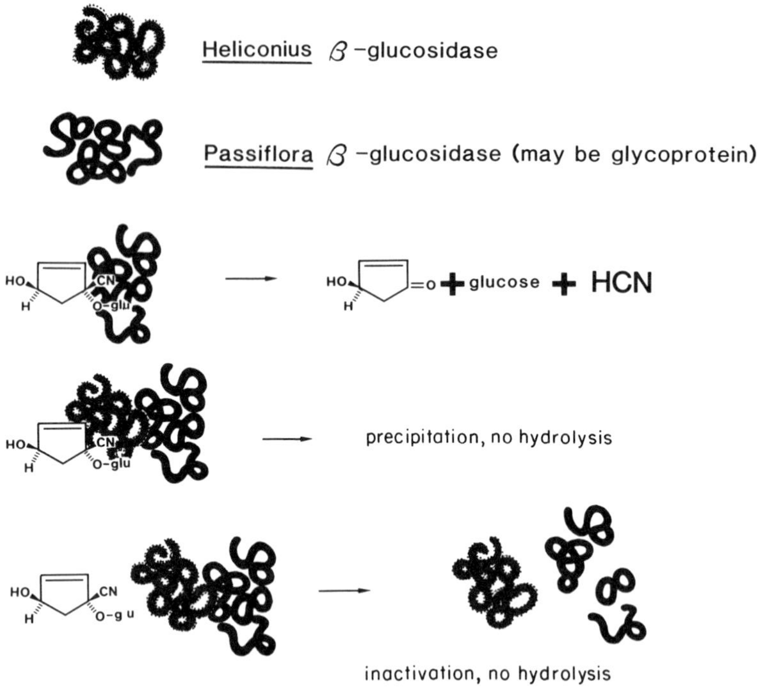

Fig. 6. Two hypothetical mechanisms for inactivation of plant-mediated hydrolysis of cyclopentenoid cyanogenic glycosides by *Heliconius* β-glucosidases.

noid cyanogenic glycosides exist in *Heliconius* and its relatives, and that the process involves the specific inhibition of the hydrolysis of the compounds present in the host plant. While the actual mechanism has not been satisfactorily quantitatively demonstrated, our data are consistent with the models illustrated in Fig. 6. In the first reaction, the *Heliconius* β-glucosidase binds to the plant substrate–enzyme complex, either during or after complex formation, in a competitive manner. The one-substrate, two-enzyme complex precipitates out of solution. In the second reaction, the *Heliconius* β-glucosidase and/or other glycosidases actually attack and hydrolyze the plant β-glucosidase. Both reactions have been measured *in vitro* and appear to occur *in vivo*. Both reactions are highly substrate-specific, and both result in the inactivation of the plant β-glucosidase and prevent hydrolysis. The specificity of the reaction is such that the production of *Heliconius* inhibitory enzymes may be regarded as a coevolved response to the specific *Passiflora* toxification system (cyanogenic glycoside plus β-glucosidase) which is targeted against it. Such an interaction is illustrated in Fig. 7.

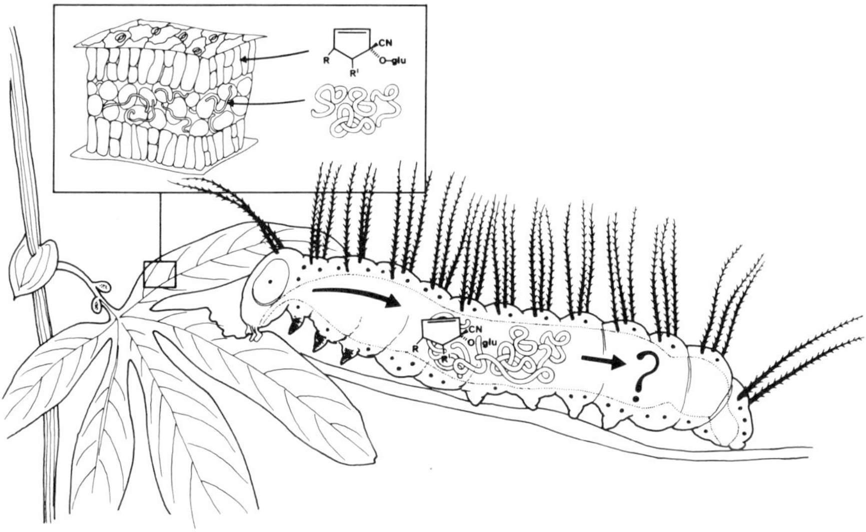

Fig. 7. Illustration of *in situ* experience of phenomena described in Figs. 2–6. During feeding of *Heliconius* larvae upon *Passiflora*, cyclopentenoid cyanogenic glycosides sequestered in the epidermis and specific β-glucosidases sequestered in the mesophyll are mixed. Ordinarily, this results in hydrolysis. However, *Heliconius* produces β-glucosidases of its own that competitively inhibit hydrolysis, at least temporarily. The stoichiometric fate of these three molecules remains unknown.

V. Analyses of Correlation between *Heliconius* and *Passiflora* and Plant Chemistry

In order to evaluate the possible importance of chemistry in shaping the evolutionary interaction between *Heliconius* and *Passiflora*, it is necessary to describe the interaction quantitatively. To conclude that *Passiflora* and *Heliconius* have coevolved in a way responsive to changes in plant chemistry, it is necessary to show (1) that specialization of heliconiine taxa upon *Passiflora* taxa has indeed occurred, (2) that heliconiines are correlated with *Passiflora* and with *Passiflora* chemicals, (3) that heliconiine taxa are restricted in the breadth of their ability to use hosts and tolerate chemicals, (4) that this correlation has a taxonomic (evolutionarily historical) component, yet (5) is not entirely the result of nonselective taxonomic association, (6) that plant taxa are correlated with specific chemistries, and finally, that the correlation of heliconiine taxa with plant taxa is explained by correlation of heliconiine taxa with chemistry, and (7) specialization of heliconiines upon *Passiflora* is mirrored in chemical specialization, as is (8) generalization in host-plant use.

Correlation analyses were carried out to test the validity of these statements. Pearson product moment correlation coefficients (for ranked data, Spearman rank correlation coefficients) were generated using all available data for identified variables (see below), and are subject to error factors described below. Ordinary least-squares regression analyses provided a *t*-statistic for each coefficient. Other analyses, such as multiple regressions, were carried out as described. All analyses were performed on a Digital Equipment Corp. VAX 11/780 computer with VMS operations system using the IDA interactive data analysis program package (Roberts and Ling, 1982; Ling and Roberts, 1982).

Field host-plant data (Table IV) were assembled from the literature, beginning with the detailed compilations of Benson *et al.* (1976) and Brown (1981) and encompassing published reports to date. Insect variables treated include 51 herbivore species (HS) for which host-plant records were available; 15 heliconiine phylogenetic groups (HPG), defined by Brown (1981) as logical subdivisions of the heliconiine lineage based on evolutionary relatedness as determined by cladistic analysis; and five heliconiine herbivore groups (HHG), defined by Benson *et al.* (1976) as logical subdivisions of the heliconiine lineage representing four major evolutionary radiations in host-use specialization from ancestral stock. These are listed in Table II, and are ranked from evolutionarily most primitive (HS 1, HPG 1, HHG 1) to most advanced (HS 51, HPG 15, HHG 5) (Brown, 1981). As HHG and HPG represent ranked linear subdivisions of HS, all three insect variables are entirely correlated with one another.

Plant variables treated (Table III) include 111 *Passiflora* species (PS) and 15 *Passiflora* subgenera (PSG) for which host-use reports are known, ranked following the taxonomic scheme of Killip (1938) (Table I). This tends to be a representation of an evolutionary linage of most advanced (PS 1, PSG 1) to most primitive (PS 111, PSG 15), although this has been questioned as described previously. Also included are the 29 chemical species or types (CS), so far isolated from *Passiflora* and given in Table VIII. These are ranked from 1 to 29 in an order that allows them to be grouped in six chemical classes (CC) based on biosynthetic type and complexity. All analyses tested heliconiine species, phylogenetic groups, and herbivore groups independently. Additionally, tests were made using heliconiine species ranked from most primitive to most advanced according to phylogeny, and by wholly randomizing the ranks of species. In this way, it is possible to determine (1) absolute correlation of plant and insect species with test variables, (2) ecological sense of purported herbivore groupings, (3) rationality of purported phylogeny with respect to radiation upon host-plant taxa, and (4) level of coevolutionary association with regard to heliconiine taxon and ecological type.

A. Error Sources

It should be noted throughout the following section that several sources of error exist. First, these analyses rely wholly on literature reports of field host plant use. These represent some of the most carefully gathered and thoroughly investigated host-plant data extant but may by no means be complete for either plant or insect. They include only occurrence (host use in feeding and oviposition) reports and do not include field determinations of degree of specialization, nor ranking of host plants by choice. Further, such reports may be biased by any of a number of ecological constraints on host-plant selection, which have not been completely investigated in the field (e.g., habitat restriction, mimicry ring participation, predation rates, stochastic events, phenology). Information was available for 51 species of the 65 known heliconiines; the remaining have unknown host-plant preferences. All groups are represented. Most of the missing species are phylogenetically advanced specialists, probably with restricted host-plant ranges (Brown, 1981; Benson *et al.*, 1976). If so, the specialization correlations reported herein will be highly conservative.

Second, the chemical data obtained, while substantial, are neither the result of random sampling of species of *Passiflora*, nor have they been accumulated for an adequate statistical sample of any population of a species. Indeed, I will later suggest that the rapidity of evolution in this system is such that population dynamics and differentiation are of paramount importance. Also, only a small portion of the *Passiflora* species known to support *Heliconius* has been studied chemically.

Finally, it should be clearly stated that these are preliminary analyses designed to detect a pattern of interaction consistent with available host-plant data, taxonomic information, chemical data and proposed biochemical modes of action as obtained in the laboratory. These will point the way for experimental confirmation and permit more detailed study in the field, but do not now adequately incorporate such data.

B. Restriction in Distribution of the Herbivores

If significant host-plant selection is taking place (as observed by Gilbert, 1971, 1975; Gilbert and Smiley, 1978; Benson, 1978; Benson *et al.*, 1976; Spencer, unpublished), there must be noticeable specialization of *Heliconius* upon *Passiflora* species. Thus, the 51 heliconiines investigated should not range freely across the possible choices among some 450 species of *Passiflora* but each should utilize only a few species. Further, it would be expected that a restriction might be present in the number of plant subgenera employed (Ehrlich and Raven, 1964; Singer, 1975), although this will depend strongly on the evolutionary correctness of the taxonomic subdivision of *Passiflora*. If plant chemistry is an important factor in the determination of host-plant choice by whatever means, *Heliconius* should be limited in its tolerance of *Passiflora* chemicals in a way parallel to its limitation on host plants. Finally, if the laboratory rationale for a biochemical mode of specialization is correct as presented, restriction in tolerance among chemical classes must also be evident. Ranges of tolerance will be discussed separately.

The average number of *Passiflora* species and subgenera utilized by each heliconiine was calculated from the host plant data (Table XIV). On average, each heliconiine species (HS) utilizes only eight *Passiflora* species (PS), representing less than 8% of all *Passiflora* species that are utilized as host plants by any heliconiid and less than 2% of all *Passiflora* species. For convenience and to retain pertinence to the variables under analysis, I will henceforth consider only *Passiflora* species and subgenera (PSG) for which heliconiine use is known. On average, each HS utilizes no more than 20% (3 of the 15) subgenera that the heliconiines are known to utilize. Therefore, HS are indeed restricted in their apparent ability to utilize individual *Passiflora* species, and this restriction has a taxonomic focus at the subgeneric level.

If we take the reasonable step of assuming that an herbivore that is reported to utilize a host-plant species can tolerate the chemicals produced by that species, subject to the inadequacies of field population and species sampling admitted to previously, we may calculate the average number of chemical types and classes that are tolerated by the herbivores. By substituting chemical profiles (Table VII) for plant species identity in the above calculations, we find that, on average, a given heliconiine can tolerate five chemical species (CS) or about 19% of those presently known from *Passiflora* (Table XIV). Note that this is approximately the same

Table XIV. Plants and Chemicals Utilized or Tolerated by Heliconiines in the *Passiflora–Heliconius* Interaction

	HS		HPG		HHG	
	$\overline{X}$	%	$\overline{X}$	%	$\overline{X}$	%
Passiflora species utilized	8.4	7.6	22.5	20.3	48.2	43.4
Passiflora subgenera utilized	2.6	17.3	4.5	30.0	7.4	49.3
Chemical species tolerated	5.4	18.6	8.1	27.9	14.6	50.3
Chemical classes tolerated	2.3	38.3	3.5	58.3	5.0	83.3

Notes: Figures represent average number and percentage of total of plants and chemicals utilized or tolerated. HS, Heliconiine species; HPG, phylogenetic group; HHG, herbivore group.

percentage discrimination as was found above for *Passiflora* species utilization. When we reorganize the chemical species data into chemical class (CC) groupings, we find that an average heliconiine can tolerate but two classes, 38% of all classes tolerated by heliconiines. Again, we may note with interest the similarity between this level of discrimination and that found for *Passiflora* subgenera. Independent of range of tolerance data, we might be tempted to hypothesize immediately that chemical factors are completely responsible for host-plant choice in this system. This would be, needless to say, premature.

In calculating the same averages for heliconiine phylogenetic groups (HPG) and herbivore groups (HHG) (Table XIV) we obtain similar relative results (with a slight degenerate trend in restriction–heliconiine herbivore groups are apparently least discriminating). Results of this first analysis plainly confirm an actual restriction of distribution of heliconiine herbivores upon available host plants. A similar restriction in apparent chemical tolerance exists, again suggesting the possibility that chemical factors are important in the process of distribution of the insects on the plants.

If the observed host-plant restriction is caused by factors under the control of the plant, whatever they may be, then a similar restriction in the number of heliconiines supported by plants should be evidenced. On average, field reports (Table IV) show that a *Passiflora* species supports less than four HS, which is less than 8% of the HS known to utilize *Passiflora* (Table XV). This is the same percentage found for HS utilization of PS above. In addition, a given PS supports only 20% of the available HPG and only 43% of the available HHG. This restriction holds at the level of *Passiflora* subgenera (PSG), which on average support 17% of the available HS and 30% of the HPG. These figures are similar to those obtained for HS utilization of PS. These results indicate that the insect mode of specialization on plants is identical to the mode of restriction by plants.

Table XV. Heliconiines Supported by *Passiflora* Species or Subgenera, or Tolerating a Chemical Species or Class

	PS		PSG		CS		CC	
	$\bar{X}$	%	$\bar{X}$	%	$\bar{X}$	%	$\bar{X}$	%
HS	3.86	7.6	8.80	17.2	8.03	15.7	16.50	32.4
HPG	3.04	20.3	4.53	30.2	5.03	33.5	8.50	56.7
HHG	2.18	43.6	2.47	49.4	2.41	48.2	4.0	80.0

Notes: Figures represent average number and percentage of total of heliconiines. PS, *Passiflora* species; PSG, *Passiflora* subgenera; CS, chemical species; CC, chemical class; HS, heliconiine species; HPG, heliconiine phylogenetic group; HHG, heliconiine herbivore group.

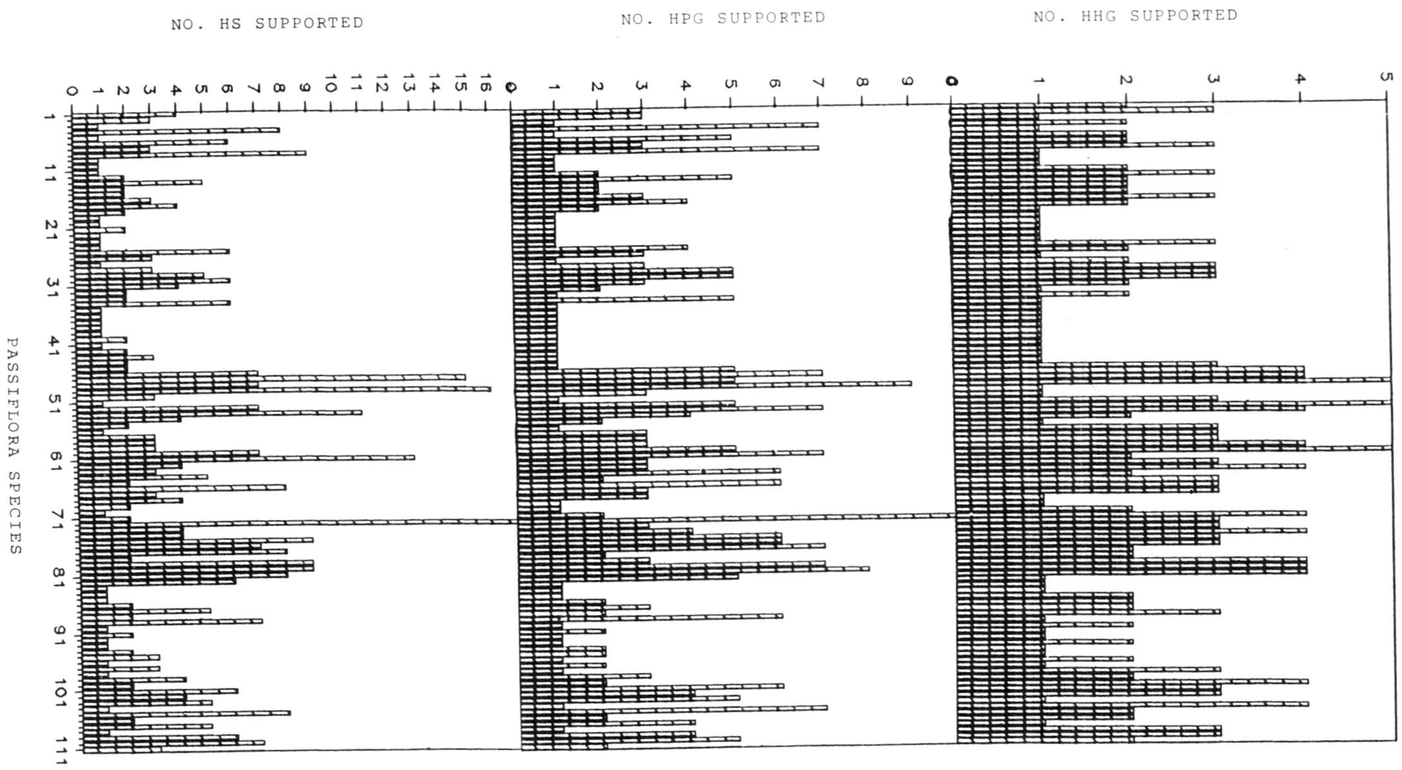

Fig. 8. Summary of *Passiflora* support of heliconiines. Numbers of heliconiine species (HS), phylogenetic groups (HPG), and herbivore groups (HHG) are plotted for each species of *Passiflora* (PS). *Passiflora* species numbers refer to Table III.

There are very many factors that could account for such specialization; one might be plant chemistry. If so, a first requirement is that a similar degree of restriction be evident in tolerance of CS and CC. We find this to be so: on average, 16% of HS and 34% of HPG tolerate a given CS, and 32% of HS and 57% of HPG tolerate a given CC (Table XV). We will return to a discussion of the degree and ranges of tolerance, but some limitations on individual heliconiines' competence with plant chemistry seem to be present along with a limited ability to utilize *Passiflora* species. Figure 8 shows the restricted distribution of heliconiines upon *Passiflora* species, and Fig. 9 shows the distribution of heliconiines upon *Passiflora* subgenera.

C. Correlation between Heliconiines and Plant and Chemical Variables

An analysis of correlation between randomized heliconiine species data and each of the 4 plant and chemical variables taken from Tables II, IV, and VII, showed significant negative correlation (-0.18, $p \leq 0.01$) with *Passiflora* species, and a similar correlation (-0.11, $p \leq 0.05$) with chemicals and chemical class (-0.12, $p \leq 0.05$). Analysis of randomized data for heliconiine phylogenetic and herbivore groups showed no correlation trends. For nonrandomized data (HS 1-51, ranked) a significant correlation of 0.12, $p \leq 0.05$, was detected with chemicals only.

A significant correlation (-0.30, $p \leq 0.01$) was obtained between *Passiflora* species and chemical species, and an associated lesser correlation with chemical class (-0.26, $p \leq 0.05$). Chemicals also correlate (-0.39, $p \leq 0.01$) with *Passiflora* subgenera, which in turn showed a smaller correlation to chemical class (-0.33, $p \leq 0.02$). Randomization of *Passiflora* species data (ranked from 1 to 111) improved these figures slightly. Plant and chemical variables are correlated more with each other than with insect variables. A simple perfect correlation reflecting the cause of host-plant specialization obviously does not exist. Some factor other than taxonomic concatenation is responsible for the distribution of heliconiines upon *Passiflora*, as has been suspected (Benson *et al.*, 1976; Gilbert and Smiley, 1978; Gilbert, 1975, 1984; Smiley, 1978). The detection of a significant correlation between HS and PS randomized data (-0.18, $p \leq 0.001$), despite the enormous sampling errors alluded to previously, indicates clearly that this distribution is nonrandom, and that an actual species-specific host-plant/herbivore association does indeed exist. This is of course an obvious requirement of any coevolutionary hypothesis in this system.

Having failed to find a simple explanatory correlation between insects and chemical species, but recognizing that persistent correlations among

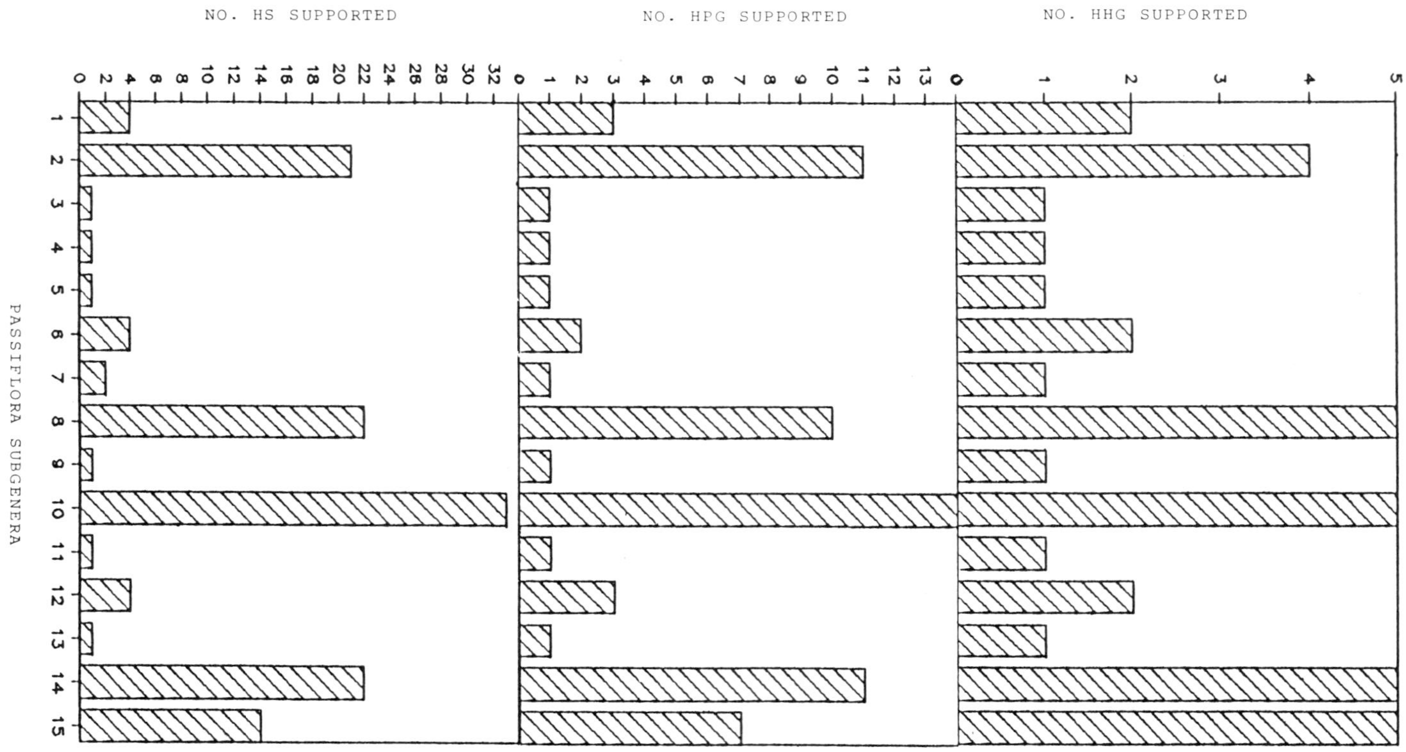

Fig. 9. See Fig. 8. Data presented are for *Passiflora* subgenera (PSG).

variables exist, we must look for other modes of association that could form the basis for specialization in this system. It is possible to deduce that, on average, insects are avoiding most chemical structures and are not obviously associated with particular chemicals. This is at first glance discouraging to any hypothesis that requires that *Heliconius* species simply choose chemical types during specialization, but upon further reflection it will be appreciated (a) that considerations of avoidance as well as attraction are components of choice and (b) that active choice may be a very different process than an evolved preference or association (involving volatile chemical recognition, and adult choice versus deterrence of larval feeding). We do see a trend ($p \leq 0.10$) toward association with particular classes of chemicals. It becomes essential to realize that the basis of chemical association of insects to plants can depend on any relative chemical difference as accommodated by an appropriate internal system of molecular recognition. An association between *Heliconius* and chemical class would therefore be a necessary and sufficient initial event in the evolution of host-plant choice in this system.

Further discrimination adequate to account for the host-plant restriction of heliconiines upon *Passiflora* requires another variable (one of a higher level of diversity). The significant correlation (-0.30, $p \leq 0.05$) between *Passiflora* species and chemical species variables indicates that a plant's chemical complement contains information sufficient to permit identification or selection of the plant species. The two variables are to some degree equivalent, and it is therefore possible that specialization upon *Passiflora* species is dependent on, or caused by, specializations upon chemical type. This indication is contradicted in the above discussion; hence we must look for an explanation consistent with the observation that *Heliconius* both avoids *and* is positively associated with plant chemical variables, in further treatment of the data.

The lack of correlation of heliconiine phylogenetic and herbivore groupings with *Passiflora* subgenera would seem to indicate that no important phylogenetic association between the plant and animal lineages exists. That it is easily recognizable only at higher levels of insect taxa indicates again that a simple taxonomic coordination has not occurred. Rather, we must again anticipate, upon further evaluation of the interaction, a theory that can accommodate the influence of a nonlinear phylogenetic component.

Hence, while phylogeny has undoubtedly contributed to the construction of the system, it cannot alone have caused the specialization that we now observe. This supposition is strengthened by the trends in correlation between *Passiflora* species and chemical class, and between subgenera and chemical species. It seems likely that all four variables are important determinants of the structure of the *Heliconius*/*Passiflora* interaction.

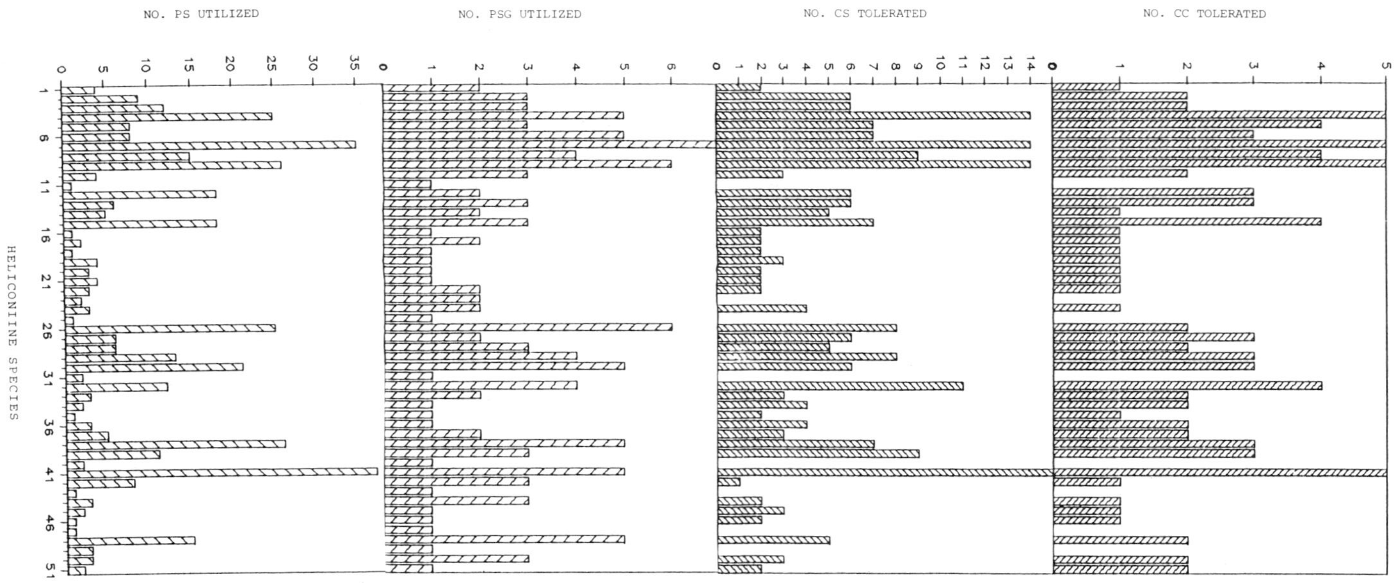

Fig. 10. Summary of heliconiine plant and chemical restriction. Number of *Passiflora* species (PS) and subgenera (PSG) utilized as hosts and chemical structures (CS) and classes (CC) tolerated are plotted for each heliconiine species (HS). Heliconiine species numbers refer to Table II.

D. Ranges in Plant Utilization and Chemical Tolerance by Heliconiines

Evolutionary processes depend as much on variability of response as they do on selection of an adaptive response; indeed, one makes possible the other. If *Passiflora* and *Heliconius* are involved in a coevolutionary interaction, and especially if the rate of diversification and specialization is rapid as herein implied, there must exist a variability of response in host-plant selection, as is typical for such interactions (Ehrlich and Raven, 1964; Lawton and Strong, 1981; Lawton, 1978; Futuyma and Gould, 1979; Gilbert, 1977). As Figs. 10, 11, and 12 illustrate, the *Passiflora/Heliconius* interaction is no exception. Ranges of utilization/tolerance of each of the four variables tested in the previous analysis are plotted against the phylogenetic ranking of heliconiines from purportedly most primitive to most advanced. Within this overall representation of variable ranges of response in the insects are hidden some more discrete mathematical structures. We will examine each in turn.

First, ranges are indeed restricted. No insect species can tolerate all chemical classes nor all chemical species. However, a few insects can tolerate a large portion of them (80% and 52%, respectively, data from Tables II, IV, and VII). No herbivore species utilizes more than one-third of all *Passiflora* species utilized by heliconiines, nor more than one-half of plant subgenera. Restriction in range therefore obtains, as well as restriction in average concentration of usage upon groups of taxa, consistent with a system operating under ongoing selective pressure manifested in the test variables. We see also a variation in ranges across the heliconiine phylogeny. I will return to the actual ranges in a separate analysis of generalization versus specialization, but let us focus upon the range across phylogeny first.

We can see (Figs. 10–12) that more variability in response exists in "most primitive" and "most advanced" heliconiines. This enhanced range (above the average) of the proximal and distal part of the phylogenetic lineage of *Heliconius* is significantly greater than that of the central, and prompted me to conduct a broken regression analysis (Pindyck and Rubinfeld, 1981), which will form a separate section. Suffice it to say here that simple linear correlations of plant and animal species indicative of wholly congruent phylogenies will not obtain.

Note also that the V-shaped distribution of ranges is comparable in each of the four variables. Are then PS, PSG, CS, and CC correlated? Indeed, they must be to some degree if (a) the plant subgeneric divisions as proposed have some basis in evolutionary reality, (b) specialization has indeed occurred, and (c) chemistry is an important determinant of host-plant choice in this system. Pearson product moment correlation analysis (Table XVI) shows the four variables to be highly correlated ($p \leq 0.001$)

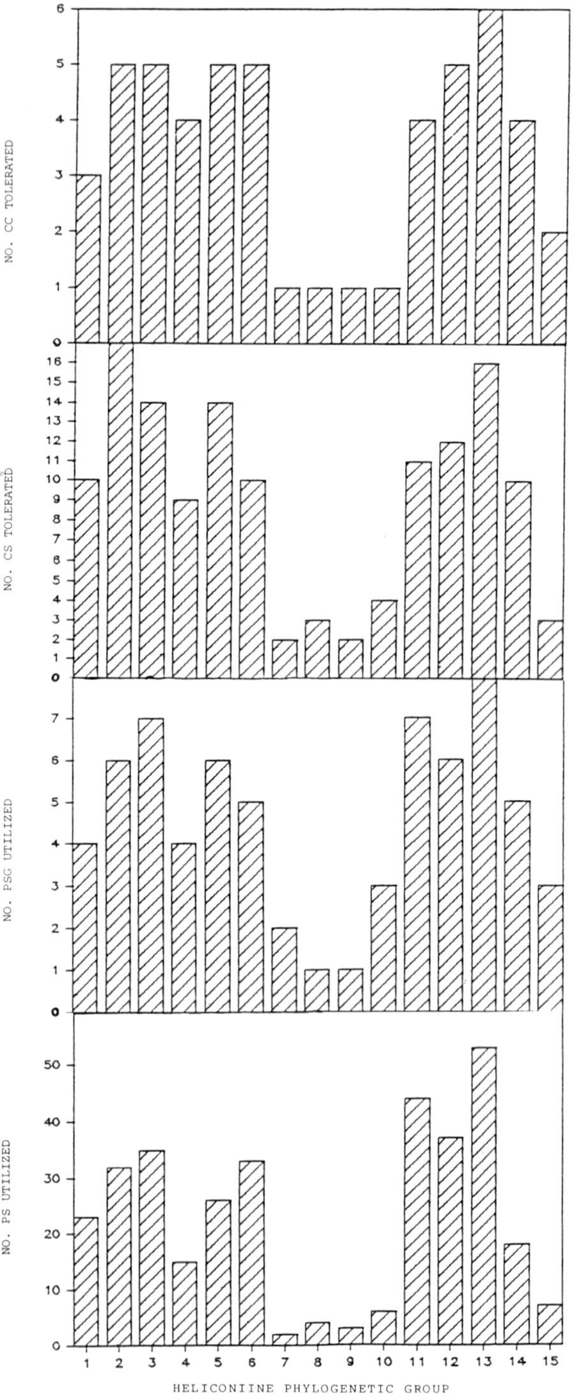

Fig. 11. See Fig. 10. Data presented are for heliconiine phylogenetic groups (HPG).

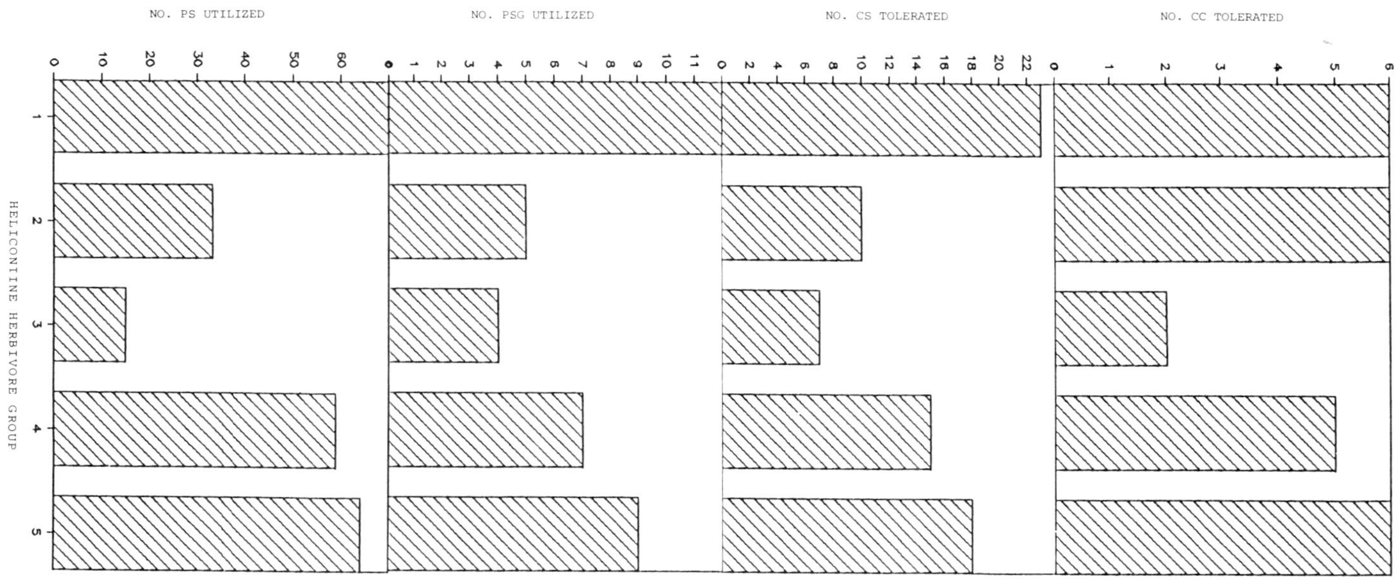

Fig. 12. See Fig. 10. Data presented are for heliconiine herbivore groups (HHG).

Table XVI. Correlations Between Numbers of *Passiflora* Species, Subgenera, Chemical Species, and Chemical Classes: Comparison of Data for Heliconiine Species Versus Heliconiine Phylogenetic Groups Versus Heliconiine Herbivore Groups

		HS		HPG		HHG	
		Rank[a]	Corr.	Rank[a]	Corr.	Rank[a]	Corr.
No. PS	vs no. CS	2	0.86	4	0.87	2	0.95
	vs no. CC	4	0.79	3	0.89	4	0.76**
	vs no. PSG	2	0.86	1	0.94	3	0.90*
No. CS	vs no. CC	1	0.94	1	0.94	—	—
	vs no. PSG	3	0.82	2	0.92	1	0.99
No. CC	vs no. PSG	5	0.77	2	0.92	—	—

Notes: All $p \leq 0.001$ except $^*p \leq 0.01$, $^{**}p \leq 0.10$. Abbreviations as in Table XV.
[a] Rank of strength of correlation within a particular heliconiine data set.

with one another, as expected. An analysis of correlation of heliconiine species to ranges of each of the variables is given in Table XVII. Remember that we are dealing with numbers of, for instance, plant species utilized, not actual species–species correlations. We find that heliconiine species are significantly correlated in a negative way with chemical type and class. This is consistent with restricted distribution being caused by chemistry. That specialization on plants is dependent on chemistry is further indicated by the lesser correlation trend with plant species and subgenera ($p \leq 0.10$). Once again, significance of observation declines with sample size (ranked data) for higher taxa of the insects. The negative sign of the correlation reveals that on average the range of tolerance and utilization in all variables declines with increasing phylogenetic advancement. But it is obvious from Figs. 10–12 that there is a nonlinear structure that has not been accommodated by linear regression analysis. The problem becomes more apparent as we lump the 51 herbivore species into 15 phylogenetic groups and further into five herbivore groups (Table II). The basic difficulty is that the heliconiine phylogeny is not necessarily related to individual host-plant selection events. In other words, *Heliconius* did not radiate upon *Passiflora* in a manner consistent with simple colinear

Table XVII. Correlation of Heliconiines to Numbers of Chemical Species, Chemical Classes, and Host-Plant Species and Subgenera

	No. CS	No. CC	No. PS	No. PSG
Heliconiine herbivore group	−0.12	−0.9	0.9	−0.20
Heliconiine phylogenetic group	−0.28	−0.13	−0.2	−0.1
Heliconiine species	−0.33**	−0.32*	−0.21	−0.25

Notes: $^{**}p \leq 0.02$, $^*p \leq 0.05$. CS, chemical species; CC, chemical classes; PS, host-plant species; PSG, subgenera.

coevolution. Some selective factor is at work that is more important than phylogeny. This can be further tested in two ways. The first is by broken regression analysis (BRA; see following section), which will test the approximate value of the insect phylogeny in this system and correct for the imposition of mathematical structure in variable distribution by the phylogenetic ranking (V-curve). This will be carried out in the next section. Second, we can simply randomize the ranking of the 51 heliconiines in the correlation analyses, which will remove the influence of insect phylogeny entirely.

E. Broken Regression Analysis of Heliconiine Host-Plant Restriction

The V-shaped distribution of restriction data over the phylogenetic ranking of heliconiines (Figs. 10–12) may strongly influence linear correlation analyses in the absence of data randomization. Randomization will correct for an artificial imposition of structure by the phylogenetic ranking, but will prevent us from recognizing evolutionarily and ecologically important trends in host-plant specialization that may have a phylogenetic component.

Such a V-distribution represents a spline function (Suits *et al.*, 1978) in which a major shift in slope in an otherwise continuous function confounds ordinary linear regression approaches. By separating the function into two separate linear portions, a piecewise linear regression (Pindyck and Rubinfeld, 1981) can be carried out that retains much of the power and estimation characteristics of ordinary least-squares linear regression (Poirier, 1976).

Therefore, a broken regression analysis (BRA) was carried out to separate the downward from the upward trends and thus to linearize the regression comparison of heliconiine phylogeny with host-plant specialization across the four host-plant restriction variables (PS, PSG, CS, CC). This treatment removes the major error caused by deformation from the linear of the correlation function of insect species with plant restriction variables as determined by linear regression. This introduces no error into either our assessment of variable correlations nor insect phylogeny, but necessitates the separate analysis of the two directional halves of the phylogeny (see below). Conclusions drawn from this analysis in isolation will be incomplete if the phylogeny is absolutely correct, and the results of the analysis are constrained in probabilistic significance by a decreased total number of observations.

The results of a BRA treatment of the data used to derive Table XVII (from Tables II, IV and VII) are given in Table XVIII. The heliconiine line was partitioned at the point of demarcation of downward and upward trends (species 22/23, see Figs. 10–12), but the treatment is robust at a

Table XVIII. Correlation of Heliconiines to Numbers of Chemical Species, Chemical Classes, Host-Plant Species, and Subgenera as Revealed by Broken Regression Analysis

	No. CS	No. CC	No. PS	No. PSG
Heliconiine species (1–22)	−0.51**	−0.48***	−0.43***	−0.58*
Heliconiine phylogenetic group (1–10)	−0.80*	−0.72*	−0.74*	−0.70***
Heliconiine species (23–51)†	−0.19	−0.10	−0.11	−0.17
Heliconiine phylogenetic group (11–15)†	−0.60	−0.53	−0.78	−0.74

Notes: $*p \leq 0.01$, $**p \leq 0.02$, $***p \leq 0.05$; † all $p \geq 0.50$. Abbreviations as in Table XVII.

considerable distance in either direction from that point. The first half of the overall data matrix (species 1–22) is shown to be strongly and negatively correlated across all variables to the same degree (~0.50). This correlation is improved by 50% when phylogenetic groups are used. The second half of the data (species 23–51) shows both poorer correlation and greater discrimination; phylogenetic grouping increases the strength of the correlation. The negative correlation indicates increasing specialization (decrease in range of utilization/tolerance) of heliconiines with purported increasing evolutionary advancement within each half of the phylogenetic ranking. The most primitive insect species are the most generalized, while species in the center and terminus of the phylogeny are most specialized. The relationship has some linearity (strength of correlation) and hence some dependence on phylogenetic (evolutionary) events as is required by our hypothesis. The irregularity of occurrence of generalists, that is, the heterogeneity of the relationship, must yet be explained. Herbivore group BRA was not possible as the number of data points was too small.

Table XIXa. Percent Variance Accounted for by Broken Regression Analysis of Each of Five Variables

Heliconiine species 1–22		Species 23–51	
HS	20	HS	0
No. PS	84	No. PS	82
No. CS	94	No. CS	90
No. CC	86	No. CC	86
No. PSG	82	No. PSG	74
Heliconiine phylogenetic group 1–10		Group 11–15	
HPG	41	HPG	0
No. PS	86	No. PS	79
No. CS	89	No. CS	86
No. CC	86	No. CC	71
No. PSG	86	No. PSG	89

Note: See text for explanation.

Taken at face value, these data indicate that much more specialization upon host plants occurs in putatively advanced *Heliconius* species, entirely consistent with the conclusions of Benson *et al.* (1976). Further, it would seem that advanced *Heliconius* phylogenetic groups are more strongly cued to specific plant subgenera than are primitive groups. This is confirmed by examination of the variance explained by each variable in a multivariate analysis (Table XIXa). In this analysis [perfectly collinear variables (HS, HPG, HHG) were treated in independent analyses], heliconiine data were treated as the dependent variable, and numbers of CS, CC, PS, and PSG as independent variables. Though multicollinearity is present, the model is assumed to rely on the inclusion of all variables, and the values generated thus adequately represent the importance of each variable. It may be seen that for all heliconiine species data, specialization occurs at the level of chemical species, this variable being equivalent to the *Passiflora* species variable. For phylogenetic groups, all variables are equivalently discriminatory, as was the case for the unbroken data matrix (Table XIXb). This indicates that individual species specialization events take place at the level of chemical species (= plant species), but that for phylogenetic groups, coevolutionary events take place with a strong dependence on larger chemical classes and plant subgenera as well. These data are consistent with an interpretation of initial coevolutionary focus upon a given chemical class and/or subgenus, followed by specialization at a finer level during subsequent evolution.

Table XIXb. Percent Variance Accounted for by Each of Five Variables for Unbroken Data in Ranked and Randomized Form

Heliconiine species 1–51		Randomized Data
HS	6	3
No. PS	81	81
No. CS	92	92
No. CC	87	88
No. PSG	75	75
Heliconiine phylogenetic group 1–15		
HPG	25	47
No. PS	85	90
No. CS	92	93
No. CC	88	91
No. PSG	92	91
Heliconiine herbivore group 1–5		
HPG	0	0
No. PS	92	93
No. CS	98	98
No. CC	0	0
No. PSG	92	92

Table XX. Correlation Between Numbers of *Passiflora* Species, Subgenera, Chemical Species, and Chemical Classes as Revealed by Broken Regression Analysis of Data for Heliconiine Species and Phylogenetic Group

		HS 1–22		HS 23–51		HPG 1–10		HPG 11–15	
		Rank	Corr.	Rank	Corr.	Rank	Corr.	Rank	Corr.
No. PS	vs No. CS	2	0.93	3	0.81	3	0.93	4	0.91*
	vs No. CC	4	0.88	4	0.73	1	0.95	6	0.87**
	vs No. PSG	6	0.85	2	0.88	3	0.93	2	0.98
No. CS	vs No. CC	1	0.94	1	0.93	3	0.93	1	0.99
	vs No. PSG	3	0.91	4	0.73	2	0.94	3	0.95
No. CC	vs No. PSG	5	0.87	5	0.67	3	0.93	5	0.89

Notes: All $p \leq 0.001$, except $^{*}p \leq 0.02$, $^{**}p \leq 0.05$. Abbreviations as in Tables XV and XVI.

A BRA treatment of data used to compile Table XVI (correlations between numbers of plant and chemical variables utilized or tolerated by heliconiines) was also carried out and revealed few differences, as would be expected for independent variables. These data are presented in Table XX to emphasize that (a) plant subgenera are much less the focus of specialization for advanced heliconiines than for those more primitive and (b) chemicals and chemical classes are elaborated selectively by individual *Passiflora* species, and are not associated only with higher taxonomic units as would be expected if biosynthesis were determined by phylogeny alone.

F. Correlation Analysis after Randomization of Proposed Heliconiine Phylogeny

The results of an analysis of correlation between ranges of tolerance of heliconiines to chemical species and classes, and ranges of utilization of plant species and subgenera are given in Table XXI. This analysis utilized the same data that produced Table XVII, but here the order of heliconiine species 1–51 was completely randomized. This treatment permits evaluation of host-plant restriction factors independent of any constraints due to insect phylogeny. In other words, events that take place at the species level (microevolution) should become more apparent, while events deriving from historically coordinated species group interaction with plants should be diminished. The amount of the observed difference will depend on both the usual sampling error and the artificiality of the proposed heliconiine phylogeny. The sign of the correlation will also now be a significant representation of acceptance or avoidance. Several differences between Table XXI (randomized data) and Table XVII (ranked data) can be noted. First, the randomized data show a significantly greater difference in correlation of insect with chemical species versus class. As the change in correlation is positive and retained in the phylogenetic and

Table XXI. Correlation of Randomized Heliconiine Herbivore Groups, Phylogenetic Groups, and Heliconiine Species to Numbers of Chemicals, Chemical Classes, Host-Plant Species, and Subgenera

	No. CS	No. CC	No. PS	No. PSG
HHG	0.52	0.18	0.68	0.44
HPG	0.10	−0.11	−0.23	−0.7
HS	0.28[a]	0.20	0.22	0.20

[a] $p \leq 0.05$.

herbivore group data, the conclusion may be drawn that each *Heliconius* species avoids (proximately or ultimately determined) most chemical classes strongly, and avoids most chemicals and plant species as well. The sum of the correlations across all four variables accounts for much of the variance in the system. The insects range rather more freely across plant subgenera, however. These correlations change when heliconiines are treated as pooled phylogenetic groups (HPG and HHG: these groups were formed, then randomized also). Here it is seen (Table XXI) that related groups of heliconiines avoid individual chemical species less than they avoid given classes of chemicals. It is clear that larger groups of insects are correlated with larger numbers of chemicals and plants; resolution of specialization becomes obscured. The change in the values of HPG and HHG correlations with variables between Table XVII and Table XXI is a measure of the amount of specialization facilitated by linear radiation along taxonomic branches. This is consistent with the radiation of real phylogenetic groups of insects upon plants' individual chemical profiles *after* the insect group has managed to overcome the chemical class elaborated by that plant group.

The difference between restriction in assumed chemical tolerance and plant species utilization may accommodate other untested ecological variables, lag time in evolutionary response, and production of multiple chemicals or shifts in production of classes of compounds. The positive correlation of heliconiine herbivore groups with all factors reflects a loss in discrimination with increasing group size and increasing ranges of subgenera and chemical class, as is expected mathematically.

G. Correlations between Host-Plant Choice Variable Numbers

If heliconiines are recognizing host plants by the cyanogenic compounds they produce, whether in real or evolutionary time frames, there should exist a direct correlation between plant taxon and chemical type.

This should be present at all levels (plant species and subgenera, chemical type and class) to some degree. Additionally, if chemical profile is the primary determinant of host-plant choice in this system, there will exist a high degree of correlation between the number of host-plant species utilized and the number of chemical types tolerated. The same should be true to varying degrees for chemical classes and plant subgenera. Referring again to Table XVI, the correlations obtained between all combinations of variables as derived from data sets prepared for heliconiine species, phylogenetic groups, and herbivore groups (see Figs. 10–12) are shown. Strong and significant positive correlations were obtained between all variables for the first two data sets. Herbivore group data were numerically limited and failed to yield significant values in half the comparisons, although the strong positive trends are still present.

Rankings of the strengths of correlation are also given, and reveal much about the importance of chemicals in the host-plant choice of *Heliconius*. Restriction in use of plant species is equivalent to restricted tolerance of chemical types. Thus, the primary focus of the insect in choosing a host plant is the avoidance of many specific chemical structures and the acceptance of a very few. The insect possesses another level of competence in the avoidance of unacceptable chemical classes. The strong correlation between numbers of chemical species tolerated and numbers of chemical classes tolerated (0.94, $p \leq 0.001$) indicates that the mechanism of chemical tolerance is at once highly structure-specific and yet has developed in such a way as to allow recognition of structural classes. In other words, the insect must be able to recognize both the macromolecular structure of an acceptable chemical class, *and* the micromolecular structure of an acceptable chemical species.

In evolutionary terms, this ability would require that an insect lineage first acquire the ability to overcome the toxicity of a chemical class and then acquire tolerance to specific chemical structures, while having developed and retained throughout an ability to recognize tolerated chemicals.

The poorer correlations (0.86, $p \leq 0.001$) of numbers of plant species utilized with numbers of plant subgenera utilized, and especially of subgenera utilized with chemical species and chemical classes tolerated (0.82, 0.87, $p \leq 0.001$), indicate that *Passiflora* subgenera are not the primary focus of heliconiine tolerance (Table XVI). Although this conclusion may be influenced by taxonomic error in delineation of subgenera, it is in fact conservative upon closer examination of the data. Certain chemical classes are very highly and significantly correlated (Tables VII, XXVII, XXVIII) with certain subgenera. Subgenus *Granadilla*, comprising one-third of all species considered here, is typified by the production of class A sulfated cyanogenic glycosides. Subgenus *Plectostemma* is typified by the production of class E complex cyanogenic glycosides or class B aliphatics. These expressions of taxonomic chemical conformity are almost entirely responsible for the strengths of correlations between chemical species or classes and *Passiflora* subgenera shown in Table XVI. The discrimination in utilization of subgenera is in fact obscured by the presence of generalists capable of tolerating these widely distributed compounds, as will be discussed in the next section. Finally, it should be noted that pooling of heliconiine species into phylogenetic groups alters the rank and degree of correlation, indicating that the phylogenetic ranking of species 1–51 does not reflect their host plant affinities in a simple way. The dissection of this problem is handled by broken regression analysis (Section V,E). The construction of herbivore groups, which should improve correlations with host-plant choice variables [as this is the basis of their delineation (Benson *et al.*, 1976)], lowers strength or significance of correlation (Table XVII) in every case except for one. In this case, numbers of *Passiflora* species utilized group strongly into numbers of subgenera utilized. This is an artifact of the construction of herbivore groups, which was strongly biased to expose subgeneric specialization. These data indicate that heliconiine herbivore groups are not constructed using variables most important in host plant choice.

H. Specialization of *Heliconius* upon *Passiflora* through Plant Chemistry

The correlative data presented so far are indicative of an evolutionary interaction involving a selective pressure leading to host-plant specialization on the part of *Heliconius* by means of (1) development by ancestral insect groups of tolerance for a given chemical class, followed by (2) radiation of species upon plants characteristically producing that chemical class. Plants then (3) produce altered structures of compounds within that parental class, which (4) the specializing heliconiines overcome with a specialized biosynthetic pathway, investment in which makes it unlikely that other structures may be overcome.

Initial observations of this system pointed out a relationship between *Passiflora* subgenera and *Heliconius* distribution. Restriction in use of subgenera by heliconiines is real, as has been shown. Yet no clear correlation is obtained which would tie particular heliconiines to particular subgenera. Data given in Table XXII demonstrate clearly that the range of subgenera utilized is, on average, small: 35% of all heliconiine species are restricted to a single subgenus. This is, however, a species-specific phenomenon. Two related insects may use unrelated subgenera. Only one species, *A. vanillae*, is able to exploit seven of the 15 subgenera included in our analysis, two others being able to exploit six. Thus, most heliconiines are indeed restricted in their distribution upon *Passiflora* subgenera, but the cause, as has been seen, is not nonselective association

Table XXII. Ranking of Heliconiines by Degree of Specialization upon Subgenera of *Passiflora*

Number of Subgenera Utilized as Host Plants	Number of Heliconiines	Percent of Heliconiines
7	1—*A. vanillae*	2
6	2—*D. julia, H. numata*	4
5	6—(4, 6, 30, 38, 41, 48)[a]	12
4	3—(8, 29, 32)[a]	6
3	11	22
2	10	20
1	18	35

[a] Number of heliconiines species using this number of subgenera, followed by identifying species numbers as given in Table II.

between plant and insect taxa. Also, some heliconiines are generalists across a range of *Passiflora* subgenera.

Let us next examine real distribution of heliconiines among *Passiflora* species. Table XXIII shows that a high degree of specialization characterizes most heliconiines, yet six insect species are able to utilize more than 20% of the 111 plant species investigated, and another five to utilize

Table XXIII. Ranking of Heliconiine Host-Plant Specialization

Number of Reported Host-Plant Species	Heliconiine Species
37	*H. erato*
35	*A. vanillae*
26	*D. julia, H. melpomene*
25	*D. juno, H. numata*
21	*H. ethilla*
18	*E. isabella, E. aliphera*
15	*D. phaetusa, H. sara*
13	*H. hecale*
12	*P. dido, H. cydno*
11	*E. pavana*
9	*P. wernickei*
8	3—(5, 6, 42)[a]
6	3—(13, 27, 28)[a]
5	2—(14, 37)[a]
4	4—(1, 10, 19, 21)[a]
3	8—(20, 22, 24, 33, 36, 44, 49, 50)[a]
2	7—(17, 23, 31, 34, 40, 45, 51)[a]
1	8—(11, 16, 18, 25, 35, 43, 46, 47)[a]

[a] Number of heliconiine species using this number of host plants, followed by identifying species numbers as given in Table II.

Table XXIV. Ranking of Heliconiine Tolerance of *Passiflora* Compounds

Number of Compounds Tolerated	Heliconiine Species
15	*H. erato*
14	*D. juno, A. vanillae, D. julia*
11	*H. cydno*
9	*D. phaetusa, H. charitonia*
8	*H. numata, H. hecale*
7	4—(5, 6, 15, 38)[a]
6	6—(2, 3, 12, 13, 27, 30)[a]
5	3—(14, 28, 48)[a]
4	3—(24, 34, 36)[a]
3	6—(10, 19, 33, 37, 45, 50)[a]
2	2—(1, 16, 17, 18, 20, 21, 22, 35, 44, 46, 51)[a]
1	1—(42)[a]

[a] Number of heliconiine species tolerating this number of compounds, followed by identifying species as given in Table II.

nearly twice the average number. Seven of these 10 heliconiines are included as the major subgenera generalists in Table XXII. I have stated that chemical species is nearly equivalent to plant species, so a ranking of heliconiine specialization upon chemistry should be similar. Table XXIV shows that five of the top nine chemical generalists are included among the 10 best *Passiflora* generalists and are well represented as subgenera generalists, the fit being less perfect as the correlation analysis indeed predicts.

It is crucial to realize that, while restriction in range across all three parameters (PS, PSG, CS) has been shown, and a relative ranking of importance of these three variables may be deduced, no discriminatory power can be attributed to any of these three variables that would provide a basis for host-plant choice in individual insect species. For that we turn to our fourth variable, chemical class. A ranking of heliconiines by relative chemical specialization is given in Table XXV. No single insect species is capable of tolerating all six chemical classes, and only four, the extreme generalists determined from the previous three rankings, are capable of tolerating five chemical classes. In fact, over two-thirds of heliconiine species are restricted in tolerance to only one or two chemical classes. Most of the insects that can tolerate three or more classes of chemicals were also found to be generalists in terms of the other three variables. These results are better interpreted by using Tables IV and VII to reassign chemical class names to the restriction data of Table XXV. Every group of compound classes tolerated by an insect includes class A, the sulfated cyanogenic glycosides. The most restricted group includes

Table XXV. Ranking of Heliconiines by Degree of Specialization upon *Passiflora* Cyanogen Compound Classes

Number of Compound Classes Tolerated	Number of Heliconiines	Percent of Heliconiines
6	0	0
5	4—*D. juno, H. erato, D. julia, A. vanillae*	9
4	4—*D. moneta, D. phaetusa, E. aliphera, H. cydno*	9
3	8—(6, 12, 13, 27, 29, 30, 38, 39)[a]	19
2	12	28
1	15	39

[a] Number of species tolerating this number of compound classes, followed by the numbers of these species as given in Table II.

nine of the 19 insects known only to tolerate sulfates; the second most restrictive group includes all of the insects with sulfates as the second tolerated class. In fact, we find that group A compounds are tolerated by the majority of heliconiines. If toxicity of sulfates was overcome early by heliconiines, or if sulfates represent a different mode of toxicity, specialization would depend on only the remaining chemical classes and we would then observe it to be much more focused indeed.

The breadth of tolerance of compound classes is summarized for higher groupings of heliconiines in Table XXVI. Both groupings reflect a degeneration in discrimination caused by lumping specialists and generalists together. All of the increased tolerance in groups is due to inclusion of generalist species. Discrimination between phylogenetic groups is still possible as each tolerates a unique set of chemical classes. The heliconiine herbivore groupings are not discriminating.

Table XXVI. Breadth of Tolerance of Classes of *Passiflora* Cyanogenic Compounds by *Heliconius*

Number of Compound Classes Tolerated	Number of Herbivore Species	Percent[a]	Number of Heliconiine Phylogenetic Groups	Percent[a]	Number of Heliconiine Herbivore Groups	Percent[a]
6	0	0	1	7	3	60
5	4	9	5	33	1	20
4	4	9	3	20	—	—
3	8	19	1	7	—	—
2	12	28	1	7	1	20
1	15	35	4	27	—	—

[a] Percentage of heliconiines using *Passiflora*.

If the heliconiines are so restricted in their ability to handle chemical classes, this should be evident at every level of specialization. An insect species, having evolved the (apparently quite difficult to obtain) capability to tolerate a given chemical class, should find it far simpler to adapt to structural changes within that class then to drop its tolerance to that class and develop another (Feeny, 1976).

I. Generalization versus Specialization—Tests

We have found that some heliconiines are apparently able to tolerate many compounds, and that these generalists are all able to tolerate sulfated cyanogenic glycosides (CC group A). In fact, 79% of all heliconiine species, all herbivore groups, and all but one monotypic phylogenetic group tolerate sulfates (Tables II, IV, VII). Removing these generalist insects and chemicals from the analysis, it was found that the remaining correlation of specialist insects with remaining chemical classes (randomized data) yielded a strongly positive correlation (+0.35, $p \leq 0.01$), whereas a negative correlation of only 0.12 was previously detected. The correlation between the generalist heliconiines, as identified from Tables XXII–XXV, and sulfates is 1.0 ($p \leq 0.001$), indicating an evolutionary cause-and-effect relationship between generalism and a plant chemical response (production of sulfates). Similar analysis of the same heliconiine specialist species as above and nonsulfated chemical species yielded zero correlation. This finding indicates that in this system specialization is actually extreme and is based on avoidance of nontolerated chemicals. Nonrandomized heliconiine species data correlated with "specialized" chemicals at a level of 0.19 ($p \leq 0.01$). This correlation is therefore derived in its entirety from phylogenetic trends in chemical specialization, indicating that insect species radiation occurs after specialized chemicals have been encountered and overcome.

Specialized insects are more closely associated with individual chemicals than with their host plants (correlation of 0.20 versus 0.08, $p \leq 0.001$), and are associated with chemical class to the same degree as chemical type.

As it has been postulated that radiation of chemical structure is caused by *Heliconius* host-plant specialization within *Passiflora*, it should be expected that this radiation will occur within any other level of plant taxon. Table XXVII lists host-plant utilization data within subgenus *Plectostemma*. This subgenus comprises the purportedly most evolutionarily advanced group in the genus, and produces a wide range of chemical structures. These fall into four chemical classes distributed throughout the subgenus. Three sections produce two or three different classes, and several plant species produce more than one class. Members of the subgenus experience herbivory and specialization from a large number of heli-

Table XXVII. Utilization as Host Plants of Subgenus *Plectostemma* by Heliconiines

Section	Series	Number of Heliconiine Species	Number of Heliconiine Phylogenetic Groups	Number of Heliconiine Herbivore Groups	Number of Chemicals	Chemical Classes[a]
Cieca		11	8	3	9	B,E,F
Decaloba		16	9	3	6	B,D,E
	Auriculatae	9	7	3	—	—
	Apetalae	1	1	1	1	D
	Luteae	2	1	1	4	B,E
	Organenses	2	2	2	—	B,E
	Miserae	5	5	3	1	E
	Punctatae	11	7	3	2	E
Xerogona		10	8	3	1	E
Pseudodysosmia		2	2	2	2	B
Pseudogranadilla		2	1	1	—	—
Hahniopathanthus		6	5	2	—	—

[a] See Table VIII.

coniine species from many different groups. The diversification in chemical structure and class here observed mimics that of the genus as a whole. Comparing Table XXVII with data presented previously, the process of plant/insect interaction does not appear to be fundamentally different for lower taxonomic levels than it is for the level that includes the whole of the interaction. In other words, the primary mode of determination of host-plant choice appears to take place at the species level or below.

Table XXVIII. Utilization as Host Plants of Subgenus *Granadilla* by Heliconiines

Series	Number of Heliconiine Species	Number of Heliconiine Phylogenetic Groups	Number of Heliconiine Herbivore Groups	Number of Chemicals	Chemical Class[a]
Quadrangulares	15	8	5	3	A,F
Digitatae	4	3	4	—	—
Tiliaefoliae	5	5	4	1	D
Laurifoliae	19	10	5	3	A
Seratifoliae	9	6	3	—	—
Setaceae	4	3	3	—	—
Pedatae	2	1	1	—	—
Incarnatae	18	10	4	2	C,D
Kermesianae	6	5	4	—	—
Imbricatae	9	7	4	—	—
Simplicifoliae	18	9	4	1	D
Lobatae	17	10	5	2	A
Menispermifoliae	2	2	2	—	—

[a] See Table VIII.

Table XXVIII shows similar data for subgenus *Granadilla*, which is regarded as a more primitive group. In this subgenus, typified by sulfate elaboration, insect specialization is much lower (see Tables IV and VII). All generalists, as previously defined, can consume most of these plant species. Where specialization does occur, plants tend to be defined chemically by having developed chemical structures and classes different from the typical sulfated cyanogenic glycosides. Compared to subgenus *Plectostemma*, the chemical diversification here is lower, and so is the observed insect specialization. The development of sulfate chemistry represents a fundamentally different way of dealing with heliconiine herbivory, and will be discussed in the next section.

VI. Ecological Action

The cyanogenic glycosides so far isolated from *Passiflora* may be grouped into six structural classes (Fig. 13). Of these, the aliphatic and aromatic types are quite widespread throughout the plant kingdom, as we have

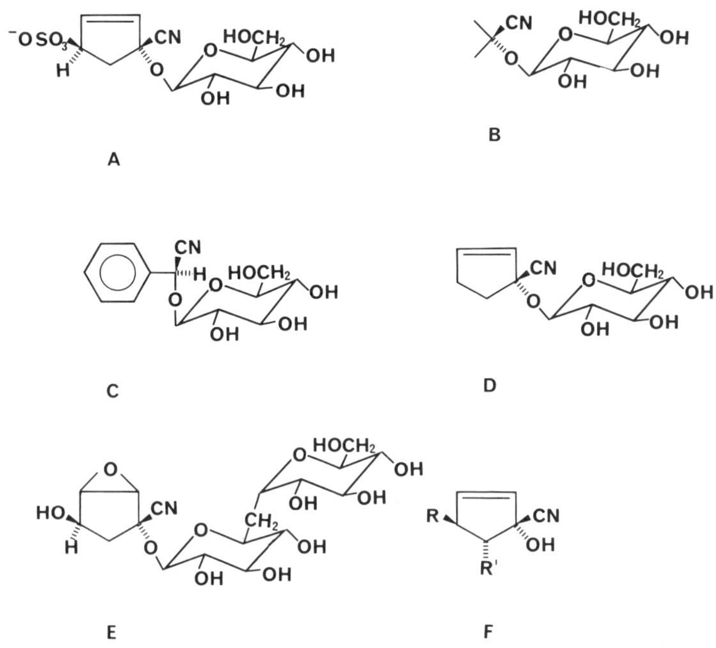

Fig. 13. The six structural classes of cyanogenic compounds produced by *Passiflora*. A, Sulfated cyclopentenoid cyanogenic glycosides. B, Aliphatic. C, Aromatic. D, Simple cyclopentenoid cyanogenic glycosides. E, Complex cyclopentenoid cyanogenic glycosides. F, Cyclopentenoid cyanohydrins.

seen. It seems rather a contradiction that they should be found only in the putatively more advanced species and subgenera of *Passiflora*. We must wonder why cosmopolitan compounds of relatively low toxicity are produced as late derivatives in evolutionarily advanced plants.

It should be emphasized here again that, in terms of cyanide production, the compounds can differ in toxicity toward *Heliconius* only (1) in the amount of cyanide that may be delivered per unit molecular weight, (2) in the rate at which cyanide is delivered, which depends on substrate structure and enzyme activity and (3) on innate resistance to inhibition of these processes by *Heliconius*, which depends on compound structure as well as plant and insect enzyme structure and interaction. The relative toxicity of the aglycone has a greater range, major biosynthetic differences in chemical class representing one basis of differentiation of toxicity, substitutional diversification of cyclopentenoid structure another. In altering its chemistry, either in terms of changes in molecular structure or in shifts in biosynthetic pathway to change the chemical class produced, *Passiflora* resists the development by *Heliconius* of competence in defeating the delivery of toxification.

The theoretical coevolutionary interaction of the *Passiflora* β-glucosidase and cyanogenic glycoside with a specialized *Heliconius* herbivore is summarized in Fig. 14 (Spencer, 1987). Glucose is assumed as the model

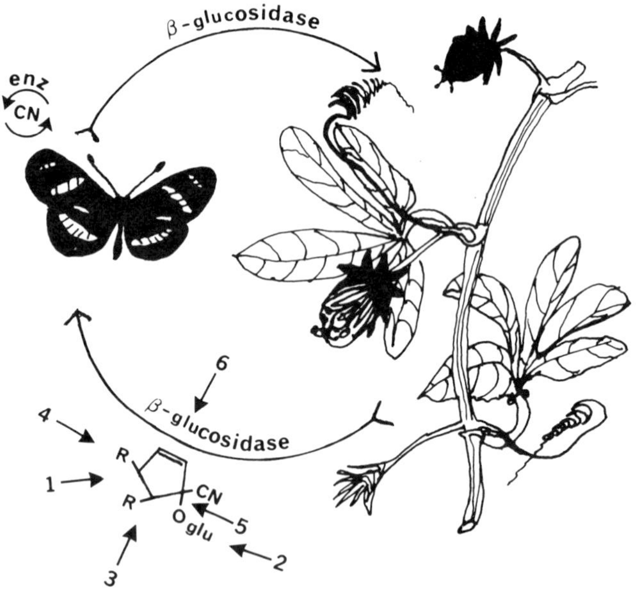

Fig. 14. Summary of the chemical responses of *Passiflora* to coevolutionary interaction with *Heliconius*. For explanation, see text. (From Spencer, 1987).

sugar moiety. Here, in response to the development of an insect β-glucosidase capable of inactivating the plant toxification syndrome, the plant species may evolve any one or more of the following changes:

1. Modification of aglycone structure. This occurs in *Passiflora* through the attachment of different substituents or replacements of the double bond with a single bond or epoxide, or through changes in symmetry.
2. Modification of the sugar moiety, through a change in number, type or linkage of sugar substituents.
3. Change in cyanogenic glycoside skeletal type through an alteration in the biosynthetic pathway to another precursor to yield an alternate type cyanogenic glycoside (i.e., cyclopentenoid becomes aromatic as 2-cyclopentenylglycine is replaced with phenylalanine as precursor).
4. Production of ionically destabilized cyanogenic glycosides; in *Passiflora* by attachment of a sulfate at C-4.
5. Production of cyanohydrins through the omission of the final glycosylation step in biosynthesis. This results in an exceedingly unstable form of cyanogenic compound (α-hydroxynitrile), which, having no sugar moiety, no longer requires a β-glucosidase for hydrolysis.
6. Change in structure of the plant β-glucosidase complement (a) to facilitate hydrolysis of an altered structure, (b) to resist binding and inactivation by a given insect β-glucosidase, or (c) to hydrolyze insect β-glucosidases.

A. Sulfated Cyclopentenoid Cyanogenic Glycosides and Simple Cyanohydrins

The large, forest plants typical of subgenus *Granadilla* contain sulfated cyanogens that occur in high concentrations. These compounds are easily and rapidly hydrolyzed to release large amounts of cyanide but are not highly specific in action (Table XI), showing little structure diversification and as substrates being generally cleaved by most β-glucosidases. It seems possible that in these plants, the primary purpose of cyanogenesis is to provide a general, "quantitative" mode of defense, as has been suggested for tannins in oaks (Rhoades and Cates, 1976; Feeny, 1976).

Many species of subgenus *Granadilla* are large and long-lived and hence are less able to escape herbivory in space or time, fitting the model of "apparent" plants (Feeny, 1976). *Passiflora alata* is such a species. *Passiflora alata* is strongly deterrent to a general feeder (the grasshopper *Osmilia flavolineata* L.) (Denno and Donnelly, 1981). This plant also produces quantities of tetraphyllin B sulfate, a molecule that may have been constructed in order to facilitate the production of large quantities of the general toxin HCN. This idea is supported by the finding that the β-glucosidase fraction present in this plant is particularly efficient in hydrolyzing this compound, and difficult to inhibit in activity compared to enzyme fractions from other species. A simple aliphatic cyanohydrin is produced by a related plant, *P. quadrangularis* (subgenus *Granadilla*).

This compound is unstable, and when the plant is damaged, large amounts of hydrogen cyanide are emitted. This compound is unlikely to possess toxicity outside of its general cyanogenic nature.

The reported host-plant specialization of heliconiine species on subgenera of *Passiflora* reveals that many more herbivore species and species groups occur on plant species that are larger resources for larval development and more reliable food sources (see Gilbert and Smiley, 1978) (subgenus *Granadilla, Astrophea, Distephana*). Greater specialization as measured by restriction of host plant utilization to only a few *Heliconius* species is seen on ostensibly more advanced *Passiflora* species (subgenus *Plectostemma* versus *Granadilla*).

B. Complex Cyclopentenoid Cyanogenic Glycosides and Alternative Pathways

In subgenus *Plectostemma* and other small, fugitive plants, more complex compounds occur, which seem to be tolerated by few heliconiines. These plants also utilize alternative biosynthetic routes of cyanogenic glycoside production most frequently. This suggests the presence of specifically toxic, "qualitative" defenses and makes this an interesting system for studying the "apparency" theory. A recent study showed that *Heliconius* species that specialize on plants of subgenus *Plectostemma* respond differently to host plant chemistry than do specialists on members of subgenus *Granadilla* (Smiley, 1978). Specialists on *Granadilla* are able to feed on some species of *Plectostemma*, while the reverse is not true. The former are therefore more generalist and the latter specialist feeders.

Generalist heliconiines are found to be distributed upon plant species that we observe to have developed a few general, quantitative toxins in response to generalized herbivory. These insects include the ostensibly most primitive genera (i.e., *Agraulis, Dryas*) with representation by only a few *Heliconius* species from higher in the proposed phylogeny (i.e., *H. erato*). More advanced *Heliconius* species are distributed upon the plants with many different complex cyanogenic compounds, which represent qualitative defenses. However, some extreme specialists are found on purportedly advanced plants that elaborate simple aliphatic cyanogenic compounds commonly produced by many plant families.

This distribution may be explained if one considers that elaboration of cyanogens derived from noncyclopentenoid pathways counteracts the evolved tolerance of *Heliconius* species to cyclopentenoid cyanogens. Within species pairs and within clades, *Heliconius* species continuously track *Passiflora* species and seem to be chasing different plant groups into corners of morphological advancement (Benson, *et al.*, 1976). For example, the small, fugitive, and rapidly growing members of subgenus *Plec-*

tostemma, regarded as advanced based on their derived, reduced floral morphology (Killip, 1938; Gilbert, 1975), may have evolved partly in response to heavy herbivore pressure upon larger, longer-lived growth forms, such as species of subgenus *Granadilla*. Putatively more advanced (Gilbert, 1975; Brown, 1981) *Heliconius* species have followed them. This tracking depended on the development of advanced modes of searching behavior over a long period of directional evolution (Benson, 1978; Gilbert and Smiley, 1978). As *Heliconius* species track the chemistry of *Passiflora*, a sudden shift to another biosynthetic pathway of cyanogenic glycoside production could defeat the *Heliconius* counteradaptations while retaining cyanogenesis as an effective general deterrent to other herbivores. Differences in the apparent tolerance for different epimers of a given structure (viz., *P. platyloba* versus *P. lindeniana*) are observed among insects. In subgenus *Granadilla*, different ratios of epimers of tetraphyllin B sulfate are produced among the various species. These differences may explain some of the host choice exhibited by insects, although in this group morphological variation must be considered significant. In *P. incarnata* (subgenus *Granadilla*), the type species of the genus, we find the singular production of gynocardin, typical of the so-called primitive Flacourtiaceae. This temperate American species is not subject to heliconiine attack in most of its range and perhaps represents a preselection, ancestral, chemotype. *Passiflora foetida* (subgenus *Dysosmia*) is a successful, wide-ranging species with dozens of known varieties (Benson *et al.*, 1976; Killip, 1960). This species produces a unique and complex cyclopentenoid cyanohydrin compound and is well protected against attack. *Passiflora lindeniana* of the "primitive" woody subgenus *Astrophea* produces a simple cyclopentenoid compound. Reversion to the production of simple cyclopentenoid glycosides may protect this species from most *Heliconius* species. This last argument may best be considered in light of the production of a simple nonhydroxylated cyclopentenoid by the "primitive" passifloraceous genus *Mitostemma*, which is apparently protected against attack by most *Heliconius* species.

Gilbert (1979) has pointed out that mechanical defenses can be more selective than chemical defenses, and that a specialized mechanical defense may evolve only after a series of chemical changes has been overcome. In *P. adenopoda*, it seems possible that a series of progressively more complex compounds (in this instance, cyclopentenoid cyanogens) has been eliminated and replaced with another series of cyanogens, namely, linamarin. This shift may have freed the species from most coevolved herbivores. Timely production of trichomes (which are capable of tearing the insects' integuments) by the plant may have further limited the ability of any remaining species of *Heliconius* to survive on the host (Gilbert, 1971). In any case, *Passiflora* species with special (derived) morphological characteristics often possess chemistry that is otherwise

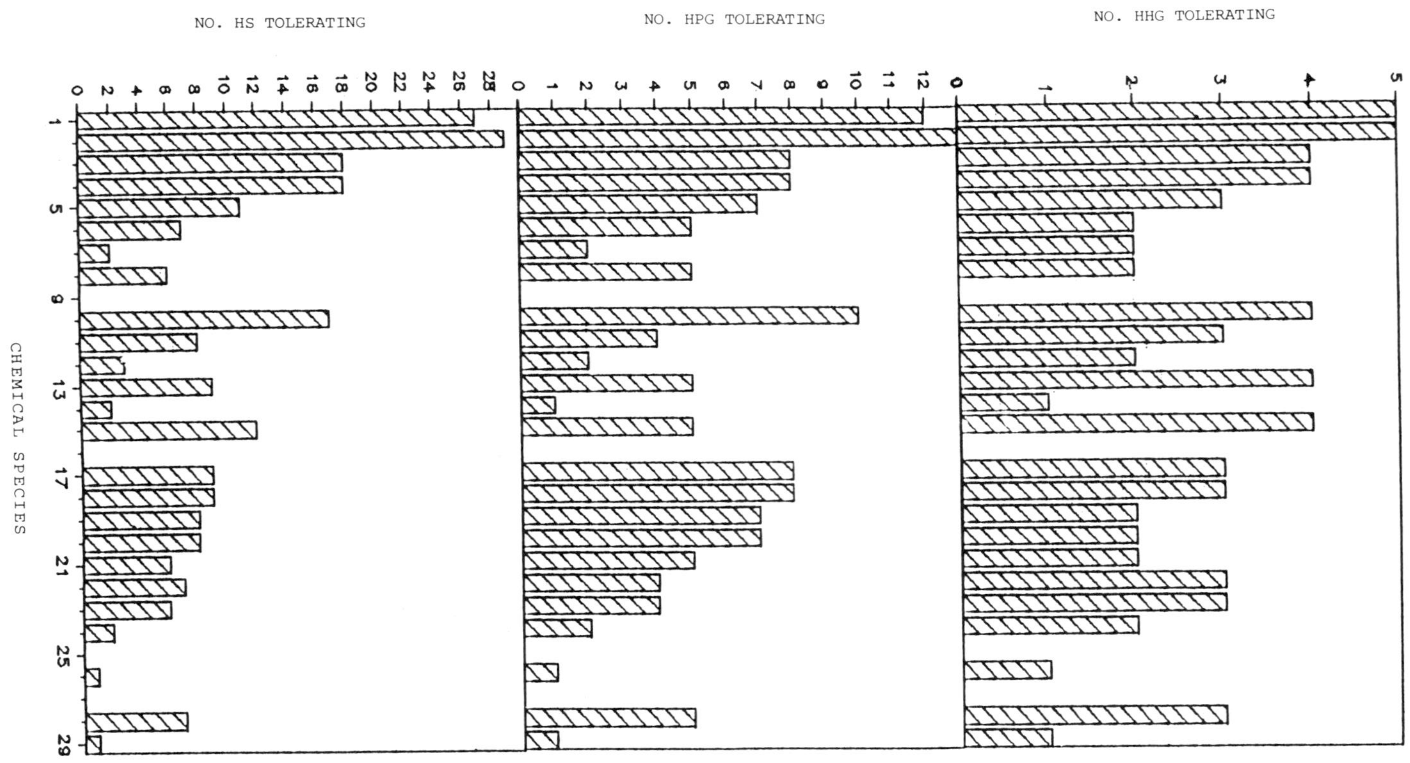

Fig. 15. Frequency of tolerance of chemical species by heliconiines. Numbers of heliconiine species (HS), phylogenetic groups (HPG), and herbivore groups (HHG) tolerating each structure are plotted against the structure number. Chemical structure numbers refer to Table VIII.

unusual in the genus (e.g., *P. adenopoda*, trichomes; *P. foetida*, glandular trichomes; *P. perfoliata*, modified perfoliate leaves; *P. warmingii*, trichomes, egg mimics, and petiolar glands). These data support an interpretation of the genus as a rapidly evolving system in which chemistry and morphology are linked. Because of the apparent selection for variation in *Passiflora* [see Janzen (1968) for a description of one mechanism that ensures outcrossing in the genus], the morphology is highly variable, and taxonomic decisions based on morphological characters alone will result in an artificial classification.

Passiflora species with distinct ranges (see Benson *et al.*, 1976; Killip, 1938) often have similar chemistry (e.g., subgenus *Granadilla* and *Tacsonia*), whereas others with similar ranges have distinct chemistry (e.g., sections *Xerogona* and *Decaloba* in *Plectostemma*). The diversification of chemistry in *Passiflora* seems to be correlated with habitat and associated heliconiine species (see Benson, 1978, 1982; Benson *et al.*, 1976, for reviews of the latter). It is not surprising to find that the Old World species *P. moluccana* of subgenus *Plectostemma*, section *Decaloba*, series *Polyanthea* (De Wilde, 1972), elaborates only tetraphyllin A as is true for primitive *Passiflora* species, while other species of section *Decaloba* characteristically make more complex cyclopentenoid diglycosides. Herbivory by *Heliconius* species may have provided the selective pressure responsible for this difference. It has been observed by Janzen (Brown, 1981) that one *Heliconius* species feeds on *Turnera ulmifolia* L. [which contains deidaclin and other simple cyclopentenoid cyanogenic glycosides (Spencer and Seigler, 1980, 1981; Spencer *et al.*, 1985)]. Another heliconiid, *Eueides procula*, also specializes on members of the Turneraceae (Waage *et al.*, 1981). Some of these insects have probably retained the ancestral ability to resist toxification by the simple types of cyanogens present in the Turneraceae and other primitive relatives of *Passiflora*. Figure 15 shows that increasingly complex cyanogenic compounds are tolerated by progressively fewer heliconiines.

VII. Summary

The phylogenies of heliconiine butterflies and their larval host plants of the genus *Passiflora* are correlated at several taxonomic levels, but are not wholly congruent. The distribution of the insects upon the plants is seen to be a nonrandom specialization upon individual *Passiflora* species. Some correlation also exists between higher phylogenetic groups of *Heliconius* species and *Passiflora* subgenera, indicative of an important historical factor in determining the present state of the observed interaction.

Passiflora species are characterized by the production of cyanogenic glycosides. Species are limited in the number of chemical types each produces. These compounds are diverse in structure and fall within six structural classes. The plant species produce unique structures and combinations of structures within a very limited number of structural classes. The correlation between *Passiflora* subgenera and chemistry is fairly good at the level of chemical class, but the overall pattern within the genus is one of representation of species as unique chemotypes.

Heliconius species cannot apparently tolerate many individual *Passiflora* cyanogenic compounds, nor can they tolerate many structural classes of these compounds. The observed correlation between heliconiine specialization and chemistry is good, consistent with a pattern of avoidance of most structures and restriction in acceptance to one or a very few structural classes.

It is concluded that *Passiflora* species produce unique chemical structures, and shifts in chemical classes elaborated, as an evolutionary response to heliconiine selective pressure. The distribution of heliconiines upon *Passiflora* species and subgenera may have been ultimately determined by differences in the cyanogenic compounds elaborated by the plants. These serve as the medium of information exchange in a coevolved system, where plant chemical changes reciprocate insect digestive enzyme changes.

Heliconius is responding to a selection pressure to specialize upon host plants, probably as a result of interspecific competition for food plants that are rare or difficult to locate. *Passiflora* species appear to shift their chemistry during the course of evolution, in response to the resulting herbivore pressure, in order to defeat the efficient destruction of plants which accompanies specialization.

Originally, ancestors to *Passiflora* may have developed cyclopentenoid cyanogenic glycosides as general toxins and deterrents to insect herbivory. Heliconiine ancestors probably began exploitation of *Passiflora* due to the presence of or development of an ability to process cyanide and tolerate cyanide in food plants. This may have been a direct result of the presence of sequestration and cyanide-processing ability developed in *Heliconius* in order to increase their distastefulness and toxicity under the selection pressures of fitting aposematic and mimicry regimes of predator avoidance.

Heliconius historically developed an ability to tolerate ancestral plant chemistry, which probably consisted of simple cyclopentenoid cyanogenic glucosides, possibly through the evolutionary development of β-glucosidases which inhibited the normal release of cyanide and α,β-ketone toxins when plant compound and plant β-glucosidase were mixed during the biting and chewing process. Although cyclopentenoids probably were the initial cyanogenic glycosides of the family and evolved because of a preexisting cyclopentenoid biochemistry, they represent an improved type of cyanogenic compound in terms of toxicity as compared to other biosynthetic types.

Under increasingly efficient herbivore pressure, *Passiflora* individuals that possessed altered chemistries that defeated specialized *Heliconius* inhibitory enzymes would be at a selective advantage. *Heliconius* individuals possessing altered enzyme activities that maintained inhibition of cyanogenesis would be successful at exploiting the new forms. This process is envisaged as being basically a gene-for-gene exchange in alteration of chemical and protein structure. Specialization of both plant and insect would be an expected result, and should proceed from many different sources in a variety of directions. When substrate structural variation had reached a certain level, the plant biosynthetic pathway would be long, complex, and costly enough that an advantage would exist in dropping the pathway entirely and substituting another. The former process, dropping the pathway, can be a one-gene event, but the latter, adding a pathway, would be more involved and should occur more rarely. Addition of a second pathway need not require elimination of a preexisting pathway but would be that much more costly, and varying combinations of pathways might be expected to coexist, subject to biochemical constraints at the regulatory level.

As in other systems of chemically mediated plant/insect interaction, a fundamental difference is here detected between qualitative and quantitative chemical defenses. Maintenance of efficient cyanide production appears to have been selected as a generalist defense in many *Passiflora* species, especially those constituting less ecologically ephemeral, more predictable resources. This would seem to be necessary, as a number of heliconiines exploit large numbers of *Passiflora* species. This perhaps represents a selectively advantagenous avoidance of extreme specialization, which might in turn result from avoidance of interspecific competition or avoidance of dependence on very limited and unpredictable resources. It is also possible that such a process proceeded nonselectively from the release of resources due to successful chemical defense of plants against specialists, causing extinction of the latter. Alternatively, this process may have proceeded from an exaptation in derived *Heliconius* species of an enzyme complement inhibitory to toxification by complex cyanogenic compounds, which is somehow resistant to gene-for-gene development of counter-inhibition as described.

A biochemical expression of generalized feeding habits could be accomplished through the development by the insects of glycosidases that are structurally limited in their specificity, such that specificity exists sufficient to cause partial inhibition of hydrolysis of toxic substrates, yet is insufficient to permit development of chemical structures of high inhibitory efficiency by the plant.

Throughout, the observed trends toward specialization of generalization have been more complex than a simple collinear phylogenetic adaptive cospeciation. However, the persistence of a small degree of correlation between higher taxa of insects and plants, and the retention of correlation between chemical class and subgenera of plants, indicate that a historical component of association has been important in the development of the present pattern of interaction.

It is proposed that the initial association of the Heliconiinae with *Passiflora* was exploitative and facilitated by insect exaptation or adaptation to an undiversified plant chemistry. Initial specialization was probably driven by insect interspecific competition, and manifested in sensory discrimination of morphology or sensory or passive discrimination of habitat. Several initial associations may therefore have existed prior to chemical diversifications.

Adaptive radiation of plant chemistry, followed by plant speciation, would therefore have occurred from several centers of intense herbivore pressure. This multicentered process would be exacerbated and repeated as *Passiflora* lineages were "cornered" into inefficient and ineffective overly complex chemical developments necessitating very major and abrupt shifts into whole new biosyntheses. The alternative would be extinction of the plant species.

The development of the present observed coevolutionary interaction between *Heliconius* and *Passiflora* is therefore interpreted as representing a "punctuated equilibrium" (Eldredge and Gould, 1972) of chemical diversification. The resultant form of the interaction, the structure of the system, and the biochemical bases of its development are proposed to be of common and frequent occurrence.

Acknowledgments

I am grateful to M. Berenbaum, M. D. Bowers, L. Brower, R. Cates, F. Gould, and J. M. Scriber for their many helpful comments and suggestions.

References

Alexander, A. J. (1961a). A study of biology and behavior of caterpillars, pupae, and emerging butterflies of the subfamily Heliconiinae in Trinidad, West Indies. Part I. Some aspects of larval behavior. *Zoologica* **46,** 1–24.

Alexander, A. J. (1961b). A study of the biology and behavior of caterpillars, pupae, and emerging butterflies of the subfamily Heliconiinae in Trinidad, West Indies. Part II. Molting and the behavior of pupae and emerging adults. *Zoologica* **46,** 105–124.

Angseesing, J. P., and Angseesing, W. J. (1973). Field observations on the cyanogenesis polymorphism in *Trifolium repens*. *Heredity* **31,** 276–282.

Benson, W. W. (1978). Resource partitioning in passion vine butterflies. *Evolution* **32,** 493–518.

Benson, W. W. (1982). Alternative models for infrageneric diversification in the humid tropics: Tests with passion vine butterflies. *In* "Biological Diversification in the Tropics" (G. T. Prance, ed.), Columbia Univ. Press, New York.

Benson, W. W., Brown, K. S., Jr., and Gilbert, L. E. (1976). Coevolution of plants and herbivores: Passion flower butterflies. *Evolution* **29,** 659–680.

Bernays, E. A. (1983). Nitrogen in defense against insects. *In* "Nitrogen as an Ecological Factor" (J. A. Lee, S. McNeill, and I. H. Rorison, eds.), Ch. 15, pp. 321–344. Blackwell, Oxford.

Bernays, E. A., Chapman, R. F., Leather, E. M., Mccafferty, A. R., and Modder, W. W. D. (1977). The relationship of *Zonocerus variegatus* (L.) (Acridoidea: Pyrgomorphidae) with cassava (*Manihot esculenta*). *Bull. Entomol. Res.* **67,** 391–404.

Blum, M. S. (1981). "Chemical Defenses of Arthropods." Academic Press, New York.

Blum, M. S., Padovani, F., Curley, A., and Hawk, R. E. (1969). Benzaldehyde: Defensive secretion of a harvester ant. *Comp. Biochem. Phys.* **29,** 461–465.

Bombardelli, E., Bonati, A., Gabetta, B., Martinelli, E. M., Mustich, G., and Danieli, B. (1975). Passiflorine, a new glycoside from *Passiflora edulis*. *Phytochemistry* **14,** 2661–2665.

Brattsten, L. B., Samuelian, J. S., Long, K. Y., Kincaid, S. A., and Evans, C. K. (1983). Cyanide as a feeding stimulant for the southern armyworm, *Spodoptera eridania*. *Ecol. Entomol.* **8,** 125–132.

Brower, L. P. (1984). Chemical defense in butterflies. *Symp. R. Entomol. Soc. London* **VII,** 109–134.

Brown, K. S., Jr. (1972). The heliconians of Brazil (Lepidoptera: Nymphalidae). III. Ecology and biology of *Heliconius nattereri,* a key primitive species near extinction, and comments on the evolutionary development of *Heliconius* and *Eueides*. *Zoologica* **57,** 41–69.

Brown, K. S., Jr. (1975). Geographical patterns of evolution in neotropical Lepidoptera. Systematics and derivation of known and new Heliconiini (Nymphalidae: Nymphalinae). *J. Entomol. B* **44,** 201–242.

Brown, K. S., Jr. (1976). An illustrated key to the silvaniform *Heliconius* (Lepidoptera: Nymphalidae) with descriptions of new subspecies. *Trans. Am. Entomol. Soc.* **102,** 373–484.

Brown, K. S., Jr. (1981). The biology of *Heliconius* and related genera. *Annu. Rev. Entomol.* **26,** 427–456.

Brown, K. S., Jr., and Benson, W. W. (1974). Adaptive polymorphism associated with multiple Mullerian mimicry in *Heliconius numata* (Lepid. Nymph.). *Biotropica* **6,** 205–228.

Brown, K. S., Jr., and Benson, W. W. (1975). West columbian biogeography. Notes on *Heliconius hecalesia* and *H. sapho* (Lepidoptera: Nymphalidae). *J. Lepid Soc.* **29,** 199–212.

Brown, K. S., Jr., and Benson, W. W. (1977). Evolution in modern Amazonian non-forest islands: *Heliconius hermathena*. *Biotropica* **9,** 95–117.

Brown, K. S., Jr., and Mielke, O. H. H. (1972). The Heliconians of Brazil (Lepidoptera: Nymphalidae). Part II. Introduction and general comments, with a supplementary revision of the tribe. *Zoologica* **57,** 1–40.

Brown, K. S., Jr., Sheppard, P. M., and Turner, J. R. G. (1974). Quaternary refugia in tropical America: Evidence from race formation in *Heliconius* butterflies. *Proc. R. Soc. London Ser, B* **187,** 369–378.

Clapp, R. C., Ettlinger, M. G., and Long, L., Jr. (1970). Deidaclin: A natural glucoside of cyclopentenone cyanohydrin. *J. Am. Chem. Soc.* **92,** 6378–6379.

Coburn, R. A., and Long, L., Jr. (1966). Gynocardin. *J. Org. Chem.* **31,** 4312–4315.

Conn, E. E. (1979a). Biosynthesis of cyanogenic glycosides. *Naturwissenschaften* **66,** 28–34.

Conn, E. E. (1979b). Cyanide and cyanogenic glycosides. *In* "Herbivores: Their Interaction with Secondary Plant Metabolites" (G. A. Rosenthal and D. H. Janzen, eds.), pp. 387–412. Academic Press, New York.

Conn, E. E. (1980). Cyanogenic glycosides. *Encycl. Plant Physiol.* **8,** 461–492.

Conn, E. E. (1981a). Biosynthesis of cyanogenic glycosides. *In* "Cyanide in Biology" (B. Vennesland, E. E. Conn, C. J. Knowles, J. Westley, and F. Wissing, eds.), pp. 183–196. Academic Press, New York.

Conn, E. E. (1981b). Cyanogenic glycosides. *In* "The Biochemistry of Plants" (E. E. Conn and P. K. Stumpf, eds.), Vol. 7, pp. 479–500. Academic Press, New York.

Cooper-Driver, G. A., and Swain, T. (1976). Cyanogenic polymorphism in bracken in relation to herbivore predation. *Nature* (*London*) **260,** 604.

Copp, N. H., and Davenport, D. (1978a). *Agraulis* and *Passiflora* I. Control of specificity. *Biol. Bull.* **155,** 98–112.

Copp, N. H., and Davenport, D. (1978b). *Agraulis* and *Passiflora* II. Behavior and sensory modalities. *Biol. Bull.* **155,** 113–124.

Dale, M. P., Emsley, H. E., Kern, K., Sastry, K. A. R., and Byers, L. D. (1985). Reversible inhibitors of β-glucosidase. *Biochemistry* **24,** 3530–3539.

Davis, R. H., and Nahrstedt, A. (1979). Linamarin and lotaustralin as the source of cyanide in *Zygaena filipendulae L.* (Lepidoptera). *Biochem. Physiol.* **64B,** 395–397.

Denno, R. F., and Donnelly, M. A. (1981). Patterns of herbivory on *Passiflora* leaf tissues and species by generalized and specialized feeding insects. *Ecol. Entomol.* **6,** 11–16.

De Wilde, W. J. J. O. (1972). The indigenous old world *Passifloras. Blumea* **20,** 227–250.

Dreyer, D. L., Reese, J. C., and Jones, K. C. (1981). Aphid feeding deterrents in sorghum. Bioassay, isolation, and characterization. *J. Chem. Ecol.* **7,** 273–284.

Duffey, S. S. (1981). Cyanide and arthropods. *In* "Cyanide in Biology" (B. Vennesland, E. E. Conn, C. J. Knowles, J. Westley, and F. Wissing, eds.), pp. 385–414. Academic Press, New York.

Duffey, S. S., Blum, M. S., Fales, H. M., Evans, S. L., Roncadori, R. W., Tiemann, D. L., and Nakagawa, Y. (1977). Benzoyl cyanide and mandelonitrile benzoate in the defensive secretions of millipedes. *J. Chem. Ecol.* **3,** 101–113.

Ehrlich, P., and Gilbert, L. E. (1973). Population structure and dynamics of the tropical butterfly *Heliconius ethilla. Biotropica* **5,** 69–82.

Ehrlich, P., and Raven, P. H. (1964). Butterflies and plants: A study in coevolution. *Evolution* **18,** 586–629.

Eisner, T. (1970). Chemical defense against predation in arthropods. *In* "Chemical Ecology" (E. Sondheimer and J. B. Simeone, eds.), pp. 157–217. Academic Press, New York.

Eisner, T., and Eisner, H. E. (1965). Mystery of a millipede. *Nat. Hist.* **74,** 30–37.

Eisner, T., Alsop, D., Hicks, K., and Meinwald, J. (1978). Defensive secretions of millipedes. *Exp. Pharmakol.* **48,** 41–72.

Eldredge, N., and Gould, S. J. (1972). Punctuated equilibria: An alternative to phyletic gradualism. *In* "Models in Paleobiology" (T. J. M. Schopf, ed.), pp. 82–115. Freeman, San Francisco.

Emsley, M. G. (1963). A morphological study of imagine Heliconiinae (Lep: Nymphalidae) with a consideration of the evolutionary relationships within the group. *Zoologica* **48,** 85–130.

Emsley, M. G. (1964). The geographical distribution of the color-pattern components of *Heliconius erato* and *Heliconius melpomene* with genetical evidence for the systematic relationship between the two species. *Zoologica* **49,** 245–286.

Emsley, M. G. (1965). Speciation in *Heliconius:* Morphology and geographic distribution. *Zoologica* **50,** 191–254.

Escobar, L. K. (1980). Interrelationships of the edible species of *Passiflora* centering around *Passiflora mollissima* (H. B. K.) Bailey, subgenus Tacsonia. Ph.D. thesis, University of Texas, Austin.

Escobar, L. K. (1986). New species and varieties of *Passiflora* (Passifloraceae) from the Andes of South America. *Syst. Bot.* **11,** 88–97.

Feeny, P. (1976). Plant apparency and chemical defense. *Recent Adv. Phytochem.* **10,** 1–40.

Feigl, A., and Anger, V. A. (1966). Replacement of benzidine by copper ethylacetoacetate as spot-test reagent for hydrogen cyanide and cyanogen. *Analyst* **91,** 282–284.

Fischer, F. C., Fung, S. Y., and Lankhorst, P. P. (1982). Cyanogenesis in Passifloraceae. II. Cyanogenic compounds from *Passiflora capsularis, P. warmingii* and *P. perfoliata. Planta Med.* **45,** 42–45.

Futuyma, D. J., and Gould, F. (1979). Associations of plants and insects in a deciduous forest. *Ecol. Monogr.* **49,** 33–50.

Gentry, A. H. (1976). Additional Panamanian Passifloraceae. *Ann. Mo. Bot. Gard.* **63,** 341–345.

Gentry, A. H. (1980). New species of Apocynaceae, Bignoniaceae, Passifloraceae, and Piperaceae from Coastal Columbia and Ecuador. *Phytologia* **47,** 97–107.

Gilbert, L. E. (1971). Butterfly-plant coevolution: Has *Passiflora adenopoda* won the selectional race with heliconiine butterflies? *Science* **172,** 585–586.

Gilbert, L. E. (1972). Pollen feeding and reproductive biology of *Heliconius* butterflies. *Proc. Natl. Acad. Sci. U.S.A.* **69,** 1403–1407.

Gilbert, L. E. (1975). Ecological consequences of a coevolved mutualism between butterflies and plants. *In* "Coevolution of Animals and Plants" (L. E. Gilbert and P. H. Raven, eds.), pp. 108–240. Univ. of Texas Press, Austin.

Gilbert, L. E. (1976). Postmating female odor in *Heliconius* butterflies: A male-contributed aphrodisiac? *Science* **193,** 419–420.

Gilbert, L. E. (1977). The role of insect-plant coevolution in the organization of ecosystems. *Colloq. Int. C.N.R.S.* **265,** 399–413.
Gilbert, L. E. (1980). Food web organization and the conservation of neotropical diversity. *In* "Conservation Biology" (M. E. Sonte and B. A. Wilcox, eds.), pp. 11–33. Sinauer, Suderland, Massachusetts.
Gilbert, L. E., and Smiley, J. T. (1978). Determinants of local diversity in phytophagous insects: Host specialists in tropical environments. *In* "Diversity of Insect Faunas" (L. A. Mound and N. Waloff, eds.). Blackwell, Oxford.
Gilli, A. (1981). Beiträge zur Flora von Ecuador 2. Choripetalae. *Feddes Rep.* **92,** 676–687.
Goldstein, W. S., and Spencer, K. C. (1985). Inhibition of cyanogenesis by tannins. *J. Chem. Ecol.* **7,** 847–858.
Gondwe, A. T. D., Seigler, D. S., and Dunn, J. E. (1978). Tetraphyllin B and *epi*-tetraphyllin B from *Adenia volkensii. Phytochemistry* **17,** 271–274.
Hall, I. H., Lee, K. H., Mar, E. C., Starnes, C. O., and Waddell, T. G. (1977). Antitumor agents. 21. A proposed mechanism for inhibition of cancer growth by tenulin and helanalin and related cyclopentenones. *J. Med. Chem.* **20,** 333–337.
Harper, N. L., Cooper-Driver, G. A., and Swain, T. (1976). A survey for cyanogenesis in ferms and gymnosperms. *Phytochemistry* **15,** 1764–1767.
Hayes, W. J. (1967). The 90-dose LD_{50} and a chronicity factor of measures of toxicity. *Toxicol. Appl. Pharmacol.* **11,** 327–335.
Hegnauer, R. (1962–1973). "Chemotaxonomie der Pflanzen," Vols. 1–6. Birkhauser, Basel.
Hegnauer, R. (1977). Cyanogenic compounds as systematic markers in Tracheophyta. *Plant. Syst. Evol. Suppl.* **1,** 191–209.
Hirai, K. (1982). Directional flow of male scent released by *Pseudaletia separata* Walker (Lepidoptera: Noctuidae) and its repellent effect on adults and larvae of four noctuid and one phycitine moth. *J. Chem. Ecol.* **8,** 1263–1270.
Hirai, K., Shorey, H. H., and Gaston, L. K. (1978). Competition among courting male moths: Male-to-male inhibitory pheromone. *Science* **202,** 644–645.
Holm-Neilsen, L. B. (1974). Notes on Central Andean Passifloraceae. *Bot. Not.* **127,** 338–351.
Hösel, W. (1981a). Glycosylation and glycosidases. *In* "The Biochemistry of Plants" (P. K. Stumpf and E. E. Conn, eds.), Ch. 23, pp. 725–754. Academic Press, New York.
Hösel, W. (1981b). The enzymatic hydrolysis of cyanogenic glycosides. *In* "Cyanide in Biology" (B. Vennesland, E. E. Conn, C. J. Knowles, J. Westley, and F. Wissing, eds.), pp. 217–232. Academic Press, New York.
Hösel, W., and Conn, E. E. (1982). The aglycone specificity of plant β-glucosidases. *Trends Biochem. Sci.* **7,** 219–221.
Hösel, W., and Nahrstedt, A. (1975). Specifische Glucosidasen für das Cyanglucosid Triglochinin: Reinigung und Charakterisierung von α-Glucosidasen aus *Alocasia macrorrhiza* Schott. *Hoppe-Seylers Z. Physiol. Chem.* **356,** 1265–1275.
Janzen, D. J. (1968). Reproductive behavior in the Passifloraceae and some of its pollinators in Central America. *Behavior* **32,** 34–48.
Janzen, D. J. (1980). When is it coevolution? *Evolution* **34,** 611–612.

Jaroszewski, J. W., Andersen, J. V., and Billeskov, I. (1987a). Plants as a source of chiral cyclopentenes: Taraktophyllin and epivolkenin, new cyclopentenoid cyanohydrin glucosides from Flacourtiaceae. *Tetrahedron* **43,** 2349–2354.

Jaroszewski, J. W., and Olafsdóttir, E. S. (1986). Natural glycosides of cyclopentenone cyanohydrines: Revised structure of so-called epitetraphyllin B. *Tetrahedron Letters* **27,** 5297–5300.

Jaroszewski, J. W., and Olafsdóttir, E. S. (1987). Monohydroxylated cyclopentenone cyanohydrin glucosides of Flacourtiaceae. *Phytochemistry* **26,** 3348–3349.

Jaroszewski, J. W., Olafsdóttir, E. S., Cornett, C., and Schaumburg, K. (1987). Cyanogenesis of *Adenia volkensii* Harms and *Tetrapathaea tetrandra* Cheeseman (Passifloraceae) revisited: Tetraphyllin B and Volkenin. Optical rotary power of cyclopentenoid cyanohydrin glucosides. *Acta Chem. Scand. B* **41,** 410–421.

Jenner, P. M., Hagan, E. C., Taylor, J. M., Cook, E. L., and Fitzhugh, O. G. (1964). Food flavorings and compounds of related structure. I. Acute oral toxicity. *Food Cosmet. Toxicol.* **2,** 327–343.

Jensen, S. R., and Nielsen, B. J. (1986). Gynocardin in *Ceratiosicyos laevis* (Achariaceae). *Phytochemistry* **25,** 2349–2350.

Jones, D. A. (1962). Selective eating of the acyanogenic form of the plant *Lotus corniculatus* L. by various animals. *Nature* (*London*) **193,** 1109–1110.

Jones, D. A. (1972). Cyanogenic glycosides and their function. *In* "Phytochemical Ecology" (J. B. Harborne, ed.). Academic Press, New York.

Jones, D. A. (1981). Cyanide and coevolution. *In* "Cyanide in Biology" (B. Vennesland, E. E. Conn, C. J. Knowles, J. Westley, and F. Wissing, eds.), pp. 509–516. Academic Press, New York.

Jones, D. A., Parsons, J., and Rothschild, M. (1962). Release of hydrocyanic acid from crushed tissues of all stages in the lifecycle of species of Zygaeninae (Lepidoptera). *Nature* (*London*) **193,** 52–53.

Killip, E. P. (1938). The American species of Passifloraceae. *Field Mus. Nat. Hist. Bot. Ser.* **19.**

Killip, E. P. (1960). Supplemental notes on the American species of Passifloraceae with description of new species. *Contrib. U.S. Nat. Herb.* **35,** 1–23.

Kim, S., Jeffrey, G. A., Panke, D., Clapp, R. C., Coburn, R. A., and Long, L., Jr. (1970). The X-ray crystallographic determination of the structure of gynocardin. *Chem. Commun.* 381–382.

Knowles, C. J. (1976). Microorganisms and cyanide. *Bacteriol. Rev.* **40,** 652–680.

Kojima, M., Poulton, J. E., Thayer, S. S., and Conn, E. E. (1979). Tissue distributions of dhurrin and of enzymes involved in its metabolism in leaves of *Sorghum bicolor*. *Plant Physiol.* **63,** 1022–1028.

Lawton, J. H. (1978). Host-plant influences on insect diversity: The effects of space and time. *Symp. R. Entomol. Soc. London* **9,** 41–55.

Lawton, J. H., and Strong, D. R., Jr. (1981). Community patterns and competition in folivorous insects. *Am. Nat.* **118,** 317–338.

Ling, R. F., and Roberts, H. V. (1982). "IDA: A User's Guide to the IDA Interactive Data Analysis and Forecasting System." Scientific Press/McGraw-Hill, New York.

Lutomski, J. (1960). Isolation of the major alkaloids from *Passiflora incarnata*. *Biul. Inst. Roslin Leczniczych*. **6,** 209–219.

Macdougal, J. M. (1988). Revision of Section Pseudodysosmia (Harms) Killip (Passifloraceae). In preparation.

McCormick, S., and Mabry, T. J. (1981). Flavonoids of *Passiflora pavonis*. *J. Nat. Prod.* **44,** 623–624.

Mallet, J. L. B., and Jackson, D. A. (1980). The ecology and social behavior of the neotropical butterfly *Heliconius xanthocles* Bates in Colombia. *J. Linn. Soc. London Zool.* **70,** 1–13.

Miller, J. M., and Conn, E. E. (1980). Metabolism of hydrogen cyanide by higher plants. *Plant Physiol.* **65,** 1199–1202.

Miller, J. R., and Strickler, K. L. (1984). Finding and accepting host plants. *In* "Chemical Ecology of Insects" (W. J. Bell and R. T. Cardé, eds.), pp. 127–158. Sinauer, Sunderland, Massachusetts.

Moertel, C. G., Ames, M. M., Kovach, J. S., Moyer, T. P., Rubin, J. R., and Tinker, J. H. (1981). A pharmacologic and toxicological study of amygdalin. *J. Am. Med. Assoc.* **245,** 591–594.

Moertel, C. G., Fleming, T. R., Rubin, J., Kvols, L. K., Sarna, G., Koch, R., Currie, V. E., Young, C. W., Jones, S. E., and Davignon, J. P. (1982). A clinical trial of amygdalin (laetrile) in the treatment of human cancer. *N. Engl. J. Med.* **306,** 201–206.

Nahrstedt, A., and Davis, R. H. (1981). The occurrence of the cyanoglucosides, linamarin and lotaustralin, in *Acrea* and *Heliconius* butterflies. *Comp. Biochem. Physiol. B* **68,** 575–577.

Nahrstedt, A., and Davis, R. H. (1983). Occurrence, variation and biosynthesis of the cyanogenic glycosides linamarin and lotaustralin in species of the Heliconiini (Insecta: Lepidoptera). *Comp. Biochem. Physiol. B.* **75,** 65–73.

Nahrstedt, A., and Davis, R. H. (1986). Uptake of linamarin and lotaustralin from their foodplant by larvae of *Zygaena trifolii*. *Phytochemistry* **25,** 2299–2302.

Nahrstedt, A., Hösel, W., and Walther, A. (1979). Characterization of cyanogenic glycosides and β-glucosidases in *Triglochin maritima* seedlings. *Phytochemistry* **18,** 1137–1141.

Nahrstedt, A., Kant, J.-D., and Wray, V. (1982). Acalyphin, A cyanogenic glucoside from *Acalypha indica*. *Phytochemistry* **21,** 101–105.

Nahrstedt, A., Wray, V., Grotjahn, L., Fikenscher, L. H., and Hegnauer, R. (1983). New acylated cyanogenic diglycosides from fruits of *Anthemis cairica*. *Planta Med.* **49,** 143–148.

Nisizawa, K., and Hashimoto, Y. (1970). Glycoside hydrolases and glycosyl transferases. *In* "The Carbohydrates" (W. Pigman and D. Horton, eds.), 2nd Ed., Vol. IIA, pp. 241–300. Academic Press, New York.

Paris, M., Bouquet, A., and Paris, R. (1969). Sur le barctérioside, nouvel hétéroside cyanogénétique des écorces de racine du *Barteria fistulosa* Mast. *C. R.* **268,** 2804–2808.

Pemberton, R. W. (1983). "Exploration for Natural Enemies of *Passiflora mollissima* in the Andes." USDA Biological Control of Weeds Laboratory. Albany, California.

Pindyck, R. S., and Rubinfeld, D. L. (1981). "Econometric Models and Economic Forecasts," 2nd Ed., pp. 126–127. McGraw-Hill, New York.

Poirier, D. J. (1976). "The Econometrics of Structural Change." North-Holland Publ., Amsterdam.

Pridham, J. B. (1965). Low molecular weight phenols in higher plants. *Annu. Rev. Plant Physiol.* **16,** 13–36.

Rhoades, D. F., and Cates, R. G. (1976). Toward a general theory of plant antiherbivore chemistry. *Recent Adv. Phytochem.* **10,** 168–213.

Roberts, H. V., and Ling, R. F. (1982). "Conversational Statistics with IDA: An Introduction to Data Analysis and Regression." Scientific Press/McGraw-Hill, New York.

Rosenthaler, L. (1919). Beiträge zur Blausäure-Frage. Die Verbreitung der Blausäure im Pflanzenreich. *Schweiz. Apoth. Z.* **57,** 279–283, 295–297, 307–313, 324–329, 341–346.

Rosenthaler, L. (1929). Beiträge zur Blausäure-Frage. Neue Blausäure-Vorkommen. *Pharm. Acta Helv.* **4,** 196–199.

Russell, R. B., and Reay, P. F. (1971). The structure of tetraphyllin A and B, two new cyanoglucosides from *Tetrapathea tetrandra*. *Phytochemistry* **10,** 1373–1377.

Saupe, S. G. (1981). Cyanogenic compounds and angiosperm phylogeny. *In* "Phytochemistry and Angiosperm Phylogeny" (D. A. Young and D. S. Seigler, eds.), pp. 80–116. Praeger, New York.

Schilcher, H. (1968). Zur Kentniss der Flavo-C-Glykoside in *Passiflora incarnata* L. *Z. Naturforsch. B* **23,** 1393.

Scriber, J. M. (1978). Cyanogenic glycosides in *Lotus corniculatus*. *Oecologia* **34,** 143–155.

Scriber, J. M. (1984). Host-plant suitability. *In* "Chemical Ecology of Insects" (W. J. Bell and R. T. Carde, eds.), pp. 159–204. Sinauer, Sunderland, Massachusetts.

Seigler, D. S. (1975). Isolation and identification of naturally occurring cyanogenic glycosides. *Phytochemistry* **14,** 9–29.

Seigler, D. S. (1981). Cyanogenic glycosides and lipids: Structural types and distribution. *In* "Cyanide in Biology" (B. Vennesland, E. E. Conn, C. J. Knowles, J. Westley, and F. Wissing, eds.), pp. 133–143. Academic Press, New York.

Seigler, D. S., and Spencer, K. C. (1988). The structures of passicoriacin, epipassicoriacin, epitetraphyllin B, taraktophyllin, volkenin and epivolkenin, and the distribution of these cyanogenic glycosides. *Phytochemistry* (submitted).

Seigler, D. S., Spencer, K. C., Statler, W. S., Conn, E. E., and Dunn, J. E. (1982). Tetraphyllin B and Epitetraphyllin B. Sulfates: Novel cyanogenic glycosides from *Passiflora caerulea* and *P. alatocaerulea*. *Phytochemistry* **21,** 2277–2285.

Smiley, J. T. (1978). Plant chemistry and the evolution of host specificity: New evidence from *Heliconius* and *Passiflora*. *Science* **201,** 745–747.

Smiley, J. T. (1985). Are chemical barriers necessary for evolution of butterfly-plant associations? *Oecologia* **65,** 580–583.

Smyth, H. F., Jr., Carpenter, C. P., Weil, C. S., Pozzani, U. C., and Striegel, J. E. (1962). Range-finding toxicity data: List VI. *Am. Ind. Hyg. Assoc. J.* **23,** 95–107.

Smyth, H. F., Jr., Carpenter, C. P., Weil, C. S., Pozzani, U. C., Striegel, J. E., and Nycum, J. S. (1969). Range-finding toxicity data: List VII. *Am. Ind. Hyg. Assoc. J.* **30,** 470–476.

Solomonson, L. P. (1981). Cyanide as a metabolic inhibitor. *In* "Cyanide in Biology" (B. Vennesland, E. E. Conn, C. J. Knowles, J. Westley, and F. Wissing, eds.), pp. 11–28. Academic Press, New York.

Spencer, K. C. (1984). Cyclopentenoid cyanogens: Their chemistry, importance in plant systematics and their role in the coevolution of plants and insects. Ph.D. thesis, University of Illinois, Urbana.

Spencer, K. D. (1987). Specificity of action of allelochemicals: Diversification of glycosides. *Am. Chem. Soc. Symp. Ser.* **330,** 275–288.

Spencer, K. C. (1988). Cyanogenic glycosides and the systematics of the Passifloraceae. *Biochem. Syst. Ecol.* (submitted).

Spencer, K. C., and Seigler, D. S. (1980). Deidaclin from *Turnera ulmifolia. Phytochemistry* **19,** 1863–1864.

Spencer, K. C., and Seigler, D. S. (1981). Tetraphyllin B from *Turnera diffusa. Planta Med.* **43,** 175–178.

Spencer, K. C., and Seigler, D. S. (1983). Cyanogenesis of *Passiflora edulis. J. Agric. Food Chem.* **31,** 794–796.

Spencer, K. C., and Seigler, D. S. (1984a). Isolation and identification of cyclopentenoid cyanogens. *Phytochem. Bull.* **16,** 13–21.

Spencer, K. C., and Seigler, D. S. (1984b). Gynocardin from *Passiflora. Planta Med.* **51,** 356–357.

Spencer, K. C., and Seigler, D. S. (1984c). Cyanogenic glycosides of *Carica papaya* and its phylogenetic position with respect to the Violales and Capparales. *Am. J. Bot.* **71,** 1444–1447.

Spencer, K. C., and Seigler, D. S. (1985a). Passicoccin, a novel cyclopentenoid cyanogen from *Passiflora coccinea. Phytochemistry* **24,** 2615–2617.

Spencer, K. C., and Seigler, D. S. (1985b). Passibiflorin, epipassibiflorin and passitrifasciatin: Novel cyclopentenoid cyanogenic glycosides from *Passiflora. Phytochemistry* **24,** 981–986.

Spencer, K. C., and Seigler, D. S. (1985c). The co-occurrence of valine/isoleucine derived and cyclopentenoid cyanogenic glycosides in *Passiflora* species. *Biochem. Syst. Ecol.* **13,** 303–304.

Spencer, K. C., and Seigler, D. S. (1985d). Cyanogenic glycosides and systematics of the Flacourtiaceae. *Biochem. Syst. Ecol.* **13,** 421–431.

Spencer, K. C., and Seigler, D. S. (1985e). Cyanogenic glycosides in *Malesherbia. Biochem. Syst. Ecol.* **13,** 23–24.

Spencer, K. C., and Seigler, D. S. (1987a). Passisuberosin and epipassisuberosin: Two cyclopentenoid cyanogenic glycosides from *Passiflora suberosa. Phytochemistry* **26,** 1665–1667.

Spencer, K. C., and Seigler, D. S. (1987b). Passicoriacin and epipassicoriacin: C-4 epimers of tetraphyllin B and epitetraphyllin B from *Passiflora coriacea. Phytochemistry* **26,** 1661–1663.

Spencer, K. C., Seigler, D. S., Fikenscher, L. H., and Hegnauer, R. (1981). Gynocardin and tetraphyllin B from *Carpotroche brasiliensis. Planta Med.* **44,** 28–29.

Spencer, K. C., Seigler, D. S., and Fraley, S. W. (1985). Cyanogenic glycosides of the Turneraceae. *Biochem. Syst. Ecol.* **13,** 433–435.

Spencer, K. C., Seigler, D. S., and Nahrstedt, A. (1986). Linamarin, lotaustralin, linustatin and neolinustatin from *Passiflora* species. *Phytochemistry* **25,** 645–649.

Spencer, K. C., Hauser, L. A., and Seigler, D. S. (1988). Systematics of *Adenia* (Passifloraceae): Numerical analysis and chemistry. *Biochem. Syst. Ecol.* (submitted).

Steyn, D. G. (1929). *Annu. Rep. Director Vet. Serv., South Afr., 15th,* 777–803.

Suits, D., Mason, A., and Chan, L. (1978). Spline functions fitted by standard regression methods. *Rev. Econ. Stat.* **60,** 132–139.

Swihart, C. A. (1971). Colour discrimination by the butterfly, *Heliconius charitonius* Linn. *Anim. Behav.* **19,** 156–164.

Swihart, C. A., and Swihart, S. L. (1970). Colour selection and learned feeding preferences in the butterfly, *Heliconius charitonius* Linn. *Anim. Behav.* **18,** 60–64.

Tantisewie, B., Ruijgrok, H. W. L., and Hegnauer, R. (1969). Die verbreitung der blausäure bei den Cormophyten, 5. Mitteilung: über cyanogene Verbindingen bei den Parietales und bei einigen weiteren Sippen. *Pharm. Weekblad* **104,** 1341–1355.

Tjon Sie Fat, L. A. (1979). Contribution to the knowledge of cyanogenesis in angiosperms. Ph.D. dissertation, Rijksuniversiteit, Leiden.

Turner, J. R. G. (1971a). Studies of Mullerian mimicry and its evolution in Burnet moths and Heliconiid butterflies. *In* "Ecological Genetics and Evolution" (E. R. Creed, ed.), pp. 224–260. Blackwell, Oxford.

Turner, J. R. G. (1971b). Two thousand generations of hybridisation in a *Heliconius* butterfly. *Evolution* **25,** 471–482.

Turner, J. R. G. (1975). Communal roosting in relation to warning colour in two heliconiine butterflies (Nymphalidae). *J. Lepid. Soc.* **29,** 221–226.

Turner, J. R. G. (1976). Mullerian mimicry: Classical "bean bag" evolution and the role of ecological islands in adaptive race formation. *In* "Population Genetics and Ecology" (S. Karlin and E. Nevo, eds.), pp. 185–218. Academic Press, New York.

Turner, J. R. G. (1977). A bibliography of *Heliconius* and the related genera. *Prog. Ecol. Evol. Contrib.* (151), 1–21.

Turner, J. R. G. (1978). Butterfly mimicry: The genetical evolution of an adaptation. *In* "Evolutionary Biology" (M. K. Hecht, W. C. Steere, and B. Wallace, eds.), pp. 163–206. Plenum, New York.

Turner, J. R. G. (1981). Adaptation and evolution in *Heliconius:* A defense of neoDarwinism. *Annu. Rev. Ecol. Syst.* **12,** 99–121.

Ulubelen, A., and Mabry, T. J. (1980). C-glycosylflavonoids of *Passiflora serratifolia. J. Nat. Prod.* **43,** 162–163.

Ulubelen, A., Kerr, R. R., and Mabry, T. J. (1982). Two new neoflavonoids and C-glycosylflavones from *Passiflora serratodigitata. Phytochemistry* **21,** 1145–1147.

Volini, M., and Alexander, K. (1981). Multiple forms and multiple functions of the Rhodaneses. *In* "Cyanide in Biology" (B. Vennesland, E. E. Conn, C. J. Knowles, J. Westley, and F. Wissing, eds.), pp. 77–91. Academic Press, New York.

Waage, J. K., Smiley, J. T., and Gilbert, L. E. (1981). The *Passiflora* problem in

Hawaii: Prospects and problems of controlling the forest weed *P. mollissina* (Passifloraceae) with heliconiine butterflies. *Entomophaga* **26,** 275–284.

Waldbauer, G. P., Cohen, R. W., and Friedman, S. (1984). An improved procedure for laboratory rearing of the corn earworm, *Heliothis zea* (Boddie) (Lepidoptera: Noctuidae). *Great Lakes Entomol.* **17,** 113–118.

Way, J. L. (1981). Pharmacologic aspects of cyanide and its antagonism. *In* "Cyanide in Biology" (B. Vennesland, E. E. Conn, C. J. Knowles, J. Westley, and F. Wissing, eds.), pp. 29–49. Academic Press, New York.

Williams, K. S., and Gilbert, L. E. (1981). Insects as selective agents on plant vegetative morphology: Egg mimicry reduces egg laying by butterflies. *Science* **282,** 467–469.

Woodhead, S., and Bernays, E. (1977). Changes in release rates of cyanide in relation to palatability of *Sorghum* to insects. *Nature* (*London*) **270,** 235–236.

Woodhead, S., and Bernays, E. (1978). The chemical basis of resistance of *Sorghum bicolor* to attack by *Locusta migratoria*. *Entomol. Exp. Appl.* **24,** 123–144.

Wray, V., Davis, R. H., and Nahrstedt, A. (1983). Biosynthesis of cyanogenic glycosides in butterflies and moths: Incorporation of valine and isoleucine into linamarin and lotaustralin by *Zygaena* and *Heliconius* species (Lepidoptera). *Z. Naturforsch. Sect. C* **38,** 583–588.

Young, A. M. (1978). "Disappearances" of eggs and larvae of *Heliconius* butterflies (Nymphalidae: heliconiinae) in northeastern Costa Rica. *Entomol. News* **89,** 811–887.

CHAPTER 8

Tale of the Tiger: Beringial Biogeography, Binomial Classification, and Breakfast Choices in the *Papilio glaucus* Complex of Butterflies

J. MARK SCRIBER
Department of Entomology
Michigan State University
East Lansing, Michigan

I. Introduction

Most concisely defined as "reciprocal evolutionary change in interacting species" (Thompson, 1982), coevolution has been interpreted variously by numerous researchers subsequent to the coining of the term by Ehrlich and Raven (1964). Janzen (1980) suggested that a trait in each of two species must evolve in response to one another in order for the process of coevolution to occur. A more generalized definition would include the evolution of a specific trait in one or more species in response to a trait or group of traits in a number of other species (or "diffuse coevolution"; Gilbert and Raven, 1975; Fox, 1981; Futuyma and Slatkin, 1983). Even in tightly coadapted systems of insects and plants, constraints on coevolution are significant.

This chapter will address the adaptation of the North American tree-feeding *Papilio* (Lepidoptera: Papilionidae) species to their various foodplant species, and the genetically based differentiation of these butterflies from each other. The Papilionidae nomenclature in this chapter (see Table I) will follow Scriber (1984b) and Munroe (1961). I will attempt to construct a likely biogeographical scenario using current host-plant affiliations and the historical distributions of their foodplants. A hypothetical phylogeny of foodplant use abilities will be presented as a framework for future hypotheses testing mechanisms of speciation.

In this system, plant chemistry certainly shapes patterns of host-plant choice and host-plant suitability for the insects. It is assumed, but not known, that insects exert a significant selection pressure upon their hosts (cf. Coley *et al.*, 1985; Thompson, 1986). Hence they may not even qualify for consideration under the process of "coadaptation" (reciprocal evolutionary adaptation) let alone "cospeciation" (reciprocal speciation; see Thompson, 1986). With at most a few solitary *Papilio* larvae occurring on large and abundant tree species such as wild black cherry (*Prunus serotina* Ehrh.), white ash (*Fraxinus americana* L.), quaking aspen (*Populus tremuloides* Michx.), and tulip tree (*Liriodendron tulipifera* L.), it is difficult to imagine that a significant physiological impact, and even less likely a significant ecological/evolutionary impact, would be incurred by these plants. While we cannot rule out "parallel evolution" (Brues, 1920), "diffuse coevolution," or "sequential evolution" (Jermy, 1976, 1984), interactions of the North American tree-feeding *Papilio* and their foodplants may not really involve coevolution at all.

Host switches among closely related plants are not uncommon, but the mechanisms of insect speciation (e.g., sympatric versus allopatric considerations) are poorly known for a variety of reasons (see reviews by Diehl and Bush, 1984; Strong *et al.*, 1984; Futuyma and Peterson, 1985). This chapter will attempt to address some biochemical, behavioral, and ecological mechanisms of plant-mediated (and chemically mediated) speciation in the insects, and the potential effects of insects on their food plants. My conclusion is that coevolution between tree-feeding swallowtails and their host plants in North America is unlikely except where specific chemical toxins may be involved. Local specialization in the *Papilio glaucus* group is more likely to be due to several ecological factors, rather than

Table I. Species Richness by 10° Belts of Latitude of the 563 Species of World Papilionidae (Lepidoptera) and the Latitudinal Distribution of the 62 Generalist Species

Global Latitude	Percent Generalist Species	Total Number of Species Present	Number of Species of Generalists Present by Tribe[a]						
			Baroniini	Parnassiini[b]	Zerynthiini	Graphiini[c]	Papilionini[d]	Troidini[e]	Total
60–70°N	50.0	8	—	2	—	—	2	—	4
50–60°N	55.0	20	—	4	—	1	6	—	11
40–50°N	31.6	57	—	4	0	1	11	2	18
30–40°N	26.0	123	—	3	0	6	19	4	32
20–30°N	19.3	135	—	0	0	8	13	5	26
10–20°N	15.3	176	0	—	0	11	11	5	27
0–10°N	12.8	234	—	—	—	13	11	6	30
0–10°S	10.6	264	—	—	—	11	13	4	28
10–20°S	18.5	157	—	—	—	14	12	3	29
20–30°S	23.9	92	—	—	—	12	8	2	22
30–40°S	42.9	28	—	—	—	5	6	1	12
40–50°S	50.0	2	—	—	—	1	—	—	1

Note: Nomenclature follows that of Munroe (1961) and Scriber (1984b).

[a] "Generalist" includes any species reported on two or more foodplant families. (0) indicates members of the tribe present but no generalists; (—) indicates no species present for that tribe and latitude. (Modified from Scriber, 1984b; updated with Ackery, 1975, and personal communication.) Some unverified literature records (likely to be incorrect) still exist (K. Brown, personal communication).

[b] Generalist species of Parnassiini include *Parnassius apollonius*, *P. honrathi*, *P. phoebus*, and *P. clodius*.

[c] Generalist species of Graphiini include *Lamproptera curius*, *Eurytides branchus*, *E. lysithous*, *E. asius*, *E. protesilaus*, *E. helios*, *E. dolicaon*, *E. leosthenes*, *Iphiclides podalirius*, *Graphium macleayanus*, *G. sarpedon*, *G. doson*, *G. eurypylus*, *G. agamemnon*, *G. antheus*, *G. policenes*, *G. leonidas*, *G. pylades*, *G. antiphates*, and *G. eurous*.

[d] Generalist species of the Papilionini include *Papilio clytia*, *P. aegeus*, *P. fuscus*, *P. ambrax*, *P. ophidicephalus*, *P. demoleus*, *P. demodocus*, *P. hospiton*, *P. machaon*, *P. oregonius*, *P. zelicaon*, *P. bairdi*, *P. polyxenes*, *P. indra*, *P. rudkini*, *P. paris*, *P. arcturus*, *P. dardanus*, *P. constantinus*, *P. nobilis*, *P. glaucus*, *P. rutulus*, *P. multicaudatus*, *P. eurymedon*, *P. troilus*, *P. palamedes*, *P. thoas*, *P. cresphontes*, *P. hectorides*, *P. scamander*, and *P. homerus*.

[e] Generalist species of Troidini include *Parides lysander*, *P. neophilus*, *P. philoxenus*, *P. hector*, *Battus philenor*, *B. polydamus*, and *B. crassus*.

specific phytochemicals and a chemical arms race between the plant defense and insect detoxication systems.

A new hypothesis is proposed herein that suggests that isolation of the proto "canadensis" (i.e. the pre *Papilio glaucus canadensis*) group in the Beringial refuge during the entire 36,000 years of Pleistocene glaciation is feasible, based on recent studies of pollen/fossil records. Quaking aspen and/or balsam poplar of the Salicaceae could have supported these *Papilio*, leading to further adaptations in oviposition behavior and detoxication. The evolution of photoperiodically obligate diapause (see Rockey *et al.*, 1986b) as well as the inability to utilize foodplants of the Magnoliaceae and Lauraceae would be a logical development during this time frame and suggests that allopatric speciation is as viable an explanation for differences in the *Papilio glaucus* group as are parapatric and sympatric processes. We are currently in the process of testing these hypotheses

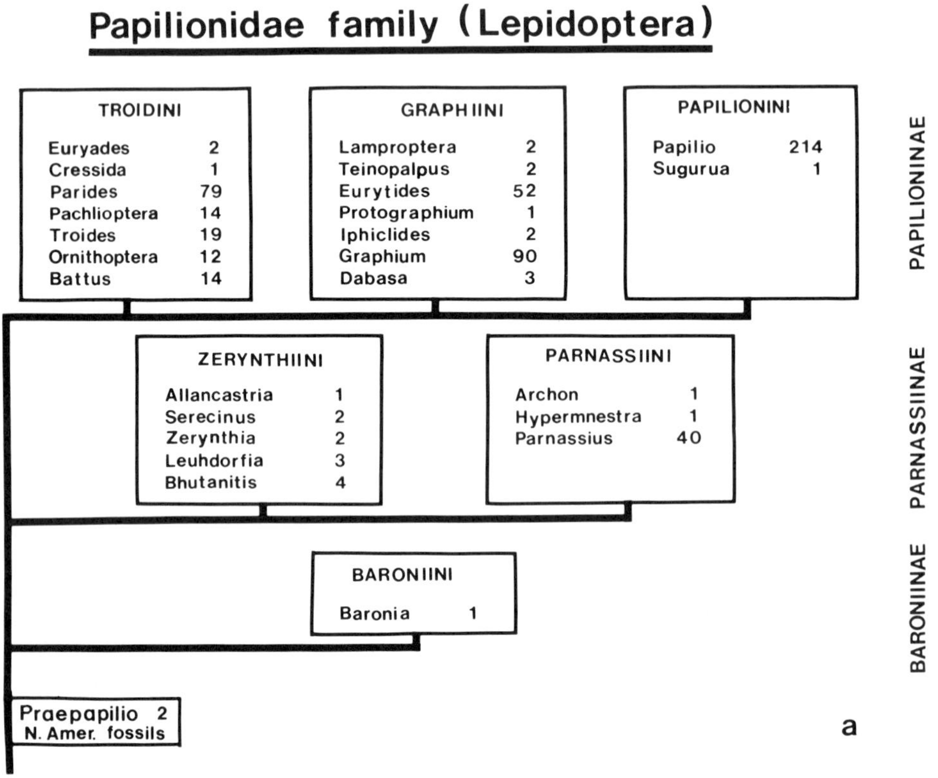

Fig. 1. Genera and number of swallowtail butterflies of the world, arranged horizontally by tribes and vertically by subfamilies. (a) Data from Scriber (1984b); after Munroe (1961), Okano (1983). (b) data of Hancock (1983).

using morphometrics and allozymes as independent tools for establishing a phylogeny for comparison with adaptive traits described herein.

II. Swallowtails in North America

The swallowtail butterflies of the Papilionidae consist of 563 species, divided among three subfamilies (Munroe, 1961; Scriber, 1984). Few undescribed species are thought to exist, and few changes at the species level have been reported within the primary tribes (Leptocircini, Troidini, and Papilionini), since the revisions of Munroe (1961) and Munroe and Ehrlich (1960). The recent transfer of *Meandrusa* Moore from the Leptocircini to the Papilionini (Igarashi, 1984) and separation of *Teinopalpus* from the Leptocircini (Miller, 1986) are the only major changes in the relatively stable taxonomy of the Papilionidae. Figures 1a and b adapted from Munroe (1961) and Hancock (1983), summarize earlier taxonomic treatments of the family.

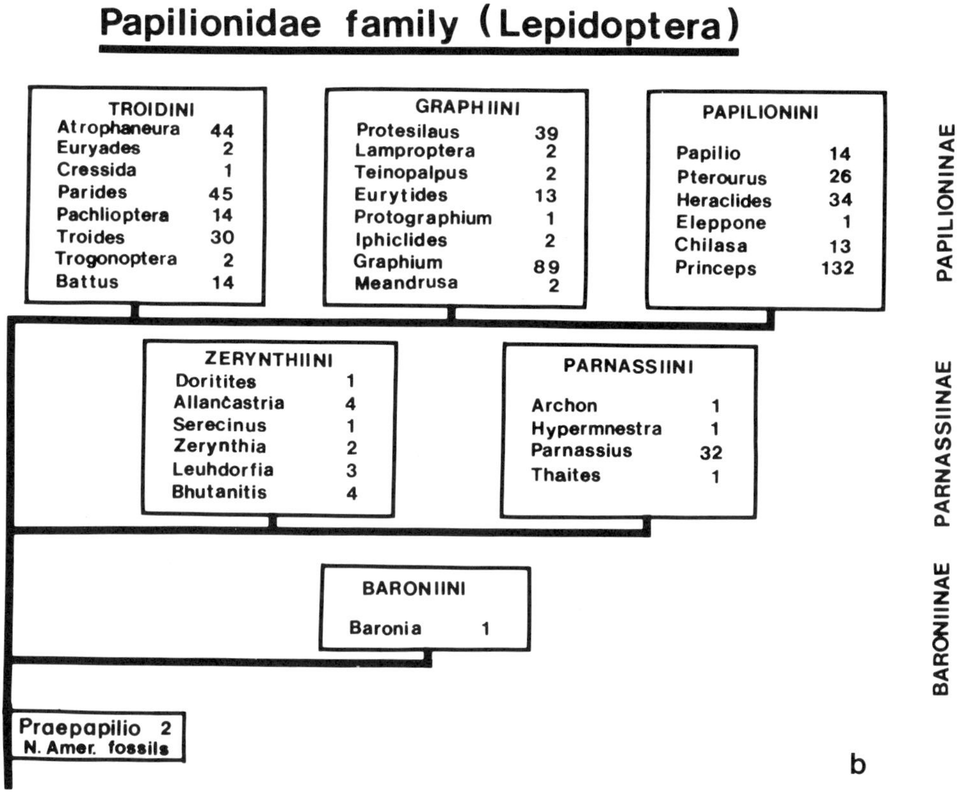

Fig. 1. (*continued*)

The swallowtail butterflies have figured prominently in investigations of the evolutionary relationships between herbivorous insects and plants (Brues 1920; Dethier, 1941, 1954; Ehrlich and Raven, 1964; Wiklund, 1975; Brown *et al.*, 1980; Rausher, 1981; Feeny *et al.*, 1983; Berenbaum, 1983; Hancock, 1983; Okano, 1983; Scriber, 1973, 1983, 1984a; Miller and Feeny, 1983; Richard and Guedes, 1983; Hagen, 1986; Miller, 1986; Sperling, 1986). Most swallowtail species are dietary specialists as larvae: only 62 of the 563 species of Papilionidae are reported to use hosts in more than one plant family (Table I) (Scriber, 1984b). Within the family, there are distinct preferences of lineages associated with plant families possessing distinctive secondary chemistry: the Troidini feed primarily on Aristolochiaceae, many of the Leptocircini on Annonaceae, and the majority of Papilionini on plants in the families Lauraceae, Rutaceae, and Umbelliferae (reviews by Feeny *et al.*, 1983; Scriber, 1984b). For several reasons, errors in foodplant records are innate in literature from the tropics (K. Brown, personal communication) and are often perpetuated in secondary references. More detailed field work is needed to resolve some of these concerns (see e.g., DeVries, 1986, 1987).

Evidence suggesting a North American ancestry of the Papilionidae [rather than a neotropical or Indo-Australian origin; see Shields and Dvorak (1979), Brown and Gibson (1983)], is provided by a mid-Eocene fossil genus *Praepapilio* Durden & Rose (Praepapilioninae) in Colorado (Durden and Rose, 1978). These authors consider these fossils to be more primitive than the ancestor of the Parnassiinae/Papilioninae and perhaps more primitive than the Leguminoseae-feeding *Baronia brevicornis* Salvin of the Baroniinae (see Munroe and Ehrlich, 1960), with the origin of the Praepapilioninae possibly before the late Cretaceous. Hancock (1983) suggests that the three subfamilies had differentiated during the Cretaceous before the splitting of the Laurasian land mass, with the Baroniinae in the western North American block, the Papilioninae in the eastern North American/European block, and the Parnassiinae in the Asian block. According to Hancock (1983), the ancestral species of Papilionini, from a distribution center in eastern North America, appears to have dispersed to the Palearctic and South America before the end of the Cretaceous. The ancestral North American *Pterourus* (presently comprised of the *Papilio glaucus* group and the *P. troilus* group) was likely to have originally been a Rutaceae feeder with an early switch to the Lauraceae and the Magnoliaceae. (See evidence in Munroe, 1961; Scriber, 1972; Ae, 1979; Hancock, 1983).

A. Plant Chemistry as a Factor in Foodplant Choice

The chemical mechanism underlying conservatism of *Papilio* butterfly foodplant affiliations is an important factor that can be investigated experimentally. The tropical host-plant families such as the Aristolochiaceae, Lauraceae, Annonaceae, Magnoliaceae, and Rutaceae account for approximately 90% of the choices of these 563 species of Papilionidae (Scriber, 1984b; Feeny *et al.*, 1983). I earlier hypothesized (Scriber, 1973) that there was a necessary or fortuitous alteration of allelochemical specializations either as species extended their geographic ranges beyond those of these tropical ancestral hosts or as they remained in areas from which these hosts were eliminated (e.g., as a result of climatic changes or agricultural practices). This idea remains difficult to evaluate because we still do not understand whether polyphagy in general arose from stenophagy or vice versa (Dethier, 1954; Feeny, 1975; Smiley, 1978).

It is rare to find any species of the Papilionidae naturally using more than three plant families. In fact, this is known to occur in only 23 species, and the majority of these are North American (Scriber, 1984b). Therefore, the occurrence of swallowtail species with broad host ranges or atypical host plant families is of particular interest against this pervasive background pattern of feeding specialization (Table II, Fig. 2).

Among the most significant of the reported generalists are the members of the *Papilio glaucus* species group. The *glaucus* species group of *Papilio* consists of six North American species with diverse feeding habits (Brower, 1958a,b; Scriber, 1984a) (Fig. 3). The *troilus* species group is probably a sister taxon to the *glaucus* group (Hancock, 1983), and consists of three exclusively North American species (Fig. 4). Together these comprise section III of the Papilionidae (Forbes, 1951; Munroe, 1961).

The relationships of species within the *glaucus* and *troilus* groups are still not clearly resolved. Brower (1959a) suggested that *P. rutulus* Lucas, *P. eurymedon* Lucas, and *P. multicaudatus* Kirby were more closely related to one another than to *P. glaucus*, with *P. rutulus* being most similar to *P. glaucus* (Klots, 1951; Brower, 1959a,b; Scott and Shepard, 1976). The positions of *P. alexiares* Hopffer and *P. experanza* Beutelspacher (Beutelspacher, 1975; see also Frances and Elvira, 1978) remain unclear, although *alexiares* appears to be very close to the *P. glaucus glaucus* and *P. g. australis* subspecies in a number of features (Scriber, 1986b,c). Relationships among the three *troilus* group species (*P. troilus* L., *P. palamedes* Drury, and *P. pilumnus*) have not been studied at all.

Most of the foodplant families of the Papilionini, including the Magnoliaceae, Lauraceae, and Rutaceae, share phytochemical features with each other and with the Annonaceae and Aristolochiaceae [the primary foodplants of the Graphinii and Troidini (Dethier, 1954; Ehrlich and Raven, 1964; Feeny *et al.*, 1983)]. For example, the Magnoliales has several families (e.g. the Lauraceae, Magnoliaceae, Annonaceae, Furmariaceae, and Aristolochiaceae) that share benzylisoquinoline alkaloids with the Rutaceae and Rhamnaceae (see Feeny *et al.*, 1983, for a review). In addition to the above-mentioned families, the terpenoid and phenylpropa-

Table II. Observed (Literature Reports) Ecological Partitioning of Foodplant Use (at the Family Level) by North American Tree-Feeding *Papilio*

Foodplant Family	*P. glaucus glaucus*	*P. g. canadensis*	*P. rutulus*	*P. australis*	*P. alexiares*	*P. multicaudatus*	*P. eurymedon*	*P. pilumnus*	*P. palamedes*	*P. troilus*	*P. cresphontes*	*P. thaos*
Piperaceae	0	0	0	0	0	0	0	0	0	0	++	+++
Rutaceae	+++	+	0	0	0	+++	0	0	0	++	+++	++
Lauraceae	+	0	0	0	0	+	0	++	+++	+++	+	+
Magnoliaceae	+++	0	0	+++	0	0	0	0	++	++	0	0
Rhamnaceae	0	0	0	0	0	0	+++	0	0	0	0	0
Rosaceae	+++	+++	+++	0	+++	+++	++	0	0	0	0	0
Oleaceae	+++	++	+	0	0	+++	0	0	0	0	0	0
Plantanaceae	0	0	+++	0	0	0	0	0	0	0	0	0
Salicaceae	0	+++	+++	0	0	+	0	0	0	0	+	0
Betulaceae	++	+++	++	0	0	0	+	0	0	0	0	0
Tiliaceae	+++	+	0	0	0	0	0	0	0	0	0	0
Carpinaceae	+	0	0	0	0	0	0	0	0	0	0	0

Notes: Nomenclature follows Munroe (1961). Foodplants of *P. alexiares* were recently discovered (Scriber *et al.*, 1988a). The intensity (frequency) of literature reports is reflected by (+, ++, or +++) with (0 = no verified reports); see Scriber (1984b).

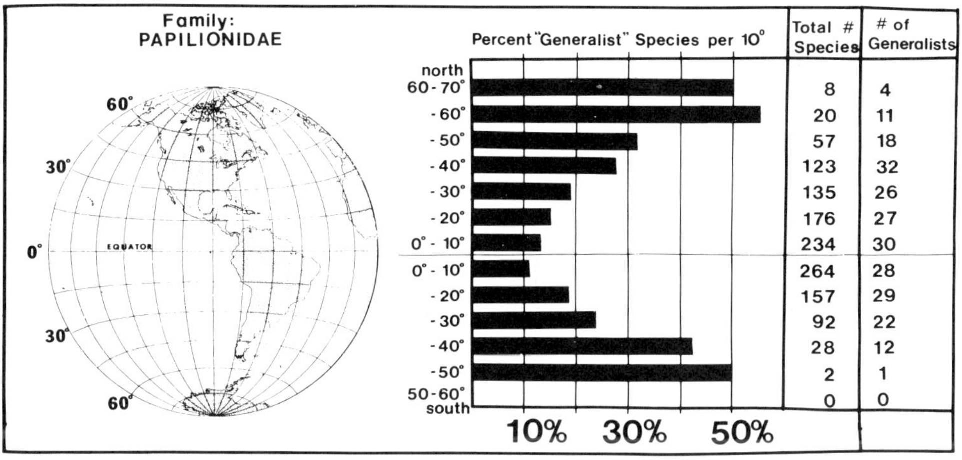

Fig. 2. Latitudinal gradients in species richness and percent ''generalist'' (feeding upon more than one family) of the 563 species of world Papilionidae. (After Scriber, 1973, 1984b; see Table I).

noid essential oils provide biochemical links with a number of other favored foodplants of the Papilionidae, including Piperaceae, Verbenaceae, Rubiaceae, Asteraceae, Leguminosae, Betulaceae, Saxifragaceae, and Umbelliferae. Hydroxycoumarins and furanocoumarins are shared by at least eight of the above-mentioned plant families and are known to be of potentially critical importance for the evolutionary interactions between these plants and the Papilionidae (Berenbaum, 1983; Berenbaum and Feeny, 1981).

One logical hypothesis is that the Lauraceae and/or Magnoliaceae (hosts of the *troilus* group) represent the ancestral hosts from which the host ranges of the *glaucus* group were derived (see Forbes, 1932, 1958; Munroe, 1961), in which case polyphagy involving unusual host families (e.g., Oleaceae, Salicaceae, Betulaceae, Tiliaceae, or Plantanaceae) might be considered apomorphies arising within the group (Hagen, 1986). Alternatively, it is possible that polyphagy arose from specialized feeding habits only in the lineage leading to the current *Papilio glaucus*, with the current relatively specialized ranges of the other *glaucus* group species arising from host shifts from ancestors that fed on more typical swallowtail foodplants (e.g., Lauraceae, Magnoliaceae, Rutaceae).

Our research on host use in the *glaucus* and *troilus* groups suggests that larval ability to feed on *Prunus* species (Rosaceae), *Fraxinus* species (Oleaceae), *Betula* (Betulaceae), and *Ptelea* (Rutaceae) may be synapomorphies for the *glaucus* group as a whole, in spite of the fact that not all species are known to use them in the field. This supports the hypothesis of a polyphagous common ancestor. However, since larval ability to feed on *Populus* species (Salicaceae) appears to be restricted to a few taxa (Fig.

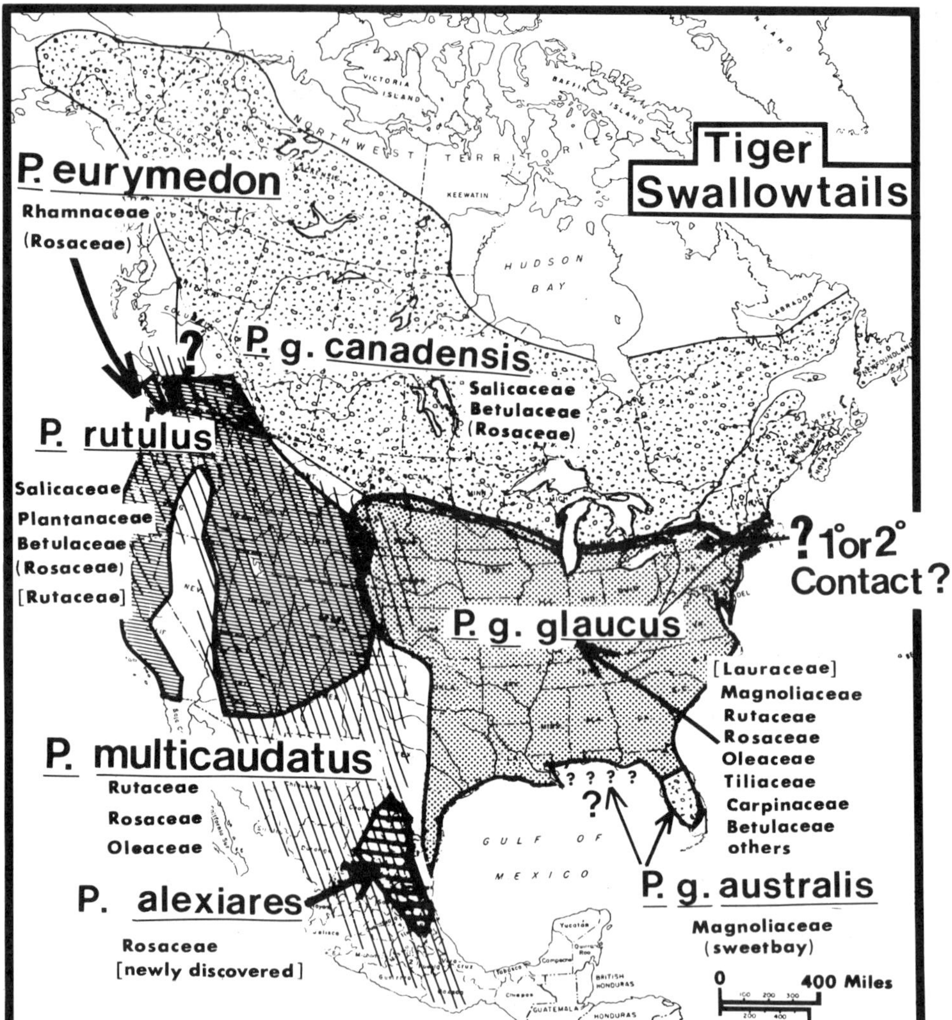

Fig. 3. North American distribution of tiger swallowtail butterflies of the *Papilio glaucus* complex, and their primary foodplant families. (After Scriber *et al.*, 1987. Copyright 1987 American Chemical Society).

3), it is feasible that a single evolutionary origin for *Populus*-feeding arose from the *glaucus* group and that the inability of some species to feed upon Salicaceae (Scriber, 1982b; Scriber *et al.*, 1982) is due to the fact that the ability never existed, rather than a loss of the ability (i.e. negative genetic correlations) with specialization elsewhere (Rausher, 1983; Scriber, 1983; Via, 1984). Thus, a complex set of hypotheses including all of these possi-

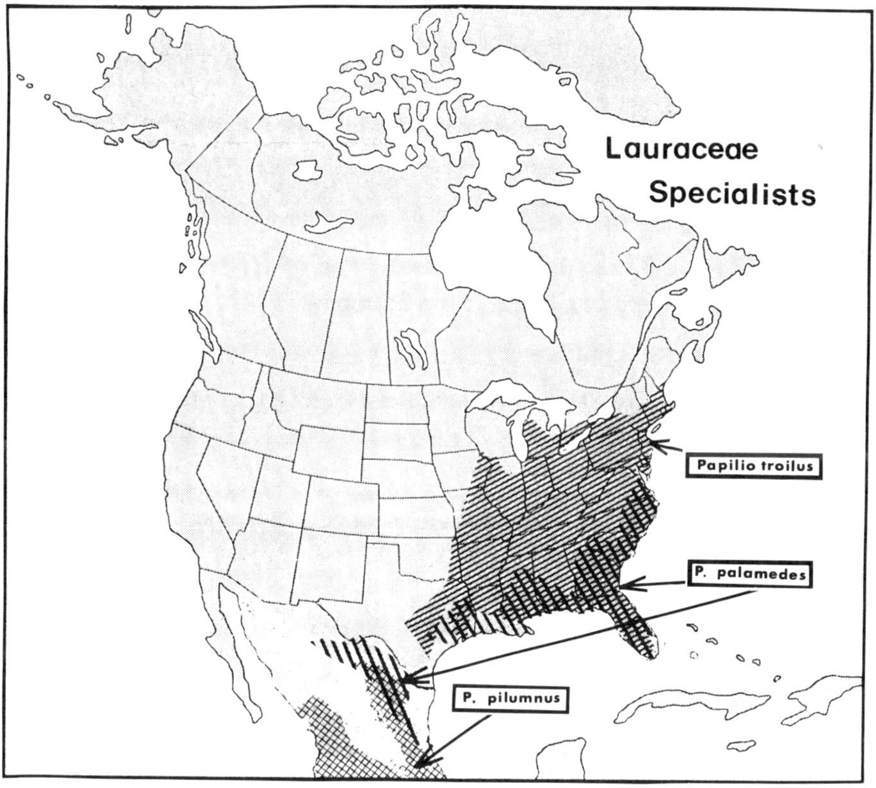

Fig. 4. North American distribution of swallowtail butterflies of thc *Papilio troilus* species group. (After Opler and Krizek, 1984; Beutelspacher and Howe, 1984).

bilities may be required to explain the evolutionary pattern of host use in these North American tree-feeding swallowtails (see also Hagen, 1986). The simple dichotomy between a specialist or a generalist ancestor may prove to be an inappropriate hypothesis (see also Dethier, 1954; Benson *et al.*, 1975; Chemsak, 1963; Wood, 1980; Guttman *et al.*, 1981; Berlocher and Bush, 1982; Mitter and Futuyma, 1983; Diehl and Bush, 1984; Via and Lande, 1985).

Discriminating between the influence of host-plant chemistry, environmental factors, and insect adaptations (particularly feeding specialization) as factors influencing postingestive growth performance is the first important step in resolving the above-mentioned hypotheses regarding the evolution of feeding specialization. It is important *not* to make conclusions about the degree of evolutionary association or adaptation of herbivores with particular hosts simply based on insect oviposition, survival, growth, and reproduction, unless careful consideration has been given to the nutritional and allelochemical quality of the plant species (see Scriber, 1983; Scriber and Feeny, 1979; Hare and Kennedy, 1986). The evolutionary duration of host affiliation cannot be deduced from larval growth performance and survival, since variation in nutritional quality can mask true, genetically based adaptations (cf. Worth *et al.*, 1982; Rausher, 1982; Scriber, 1984a).

B. The Tiger Swallowtail as a Test Species

The section III tiger swallowtail species present an excellent opportunity to study the evolutionary mechanisms and processes that lead to particular patterns of host use, both of which are the subject of much current interest and debate (Gilbert, 1979; Mitter and Futuyma, 1983; Mitter and Brooks, 1983; Diehl and Bush, 1984; Futuyma and Peterson, 1985). Both host shifts and changes in diet breadth appear to have occurred within this section.

Remington (1951) and Brower (1958a,b, 1959b) discuss one potential scenario for the evolutionary history of the *Papilio glaucus* group. Implicit in this model is the assumption that the common ancestor for the group was polyphagous and that this characteristic is retained by the eastern (allopatric) *P. glaucus*. The more specialized host use of the three western (sympatric) species are assumed to be derived from this ancestral *glaucus*-type polyphagy. It was further presumed that the Pleistocene glaciations aided this reproductive isolation between the eastern and western ancestral components. Subsequent subdivision of foodplant choices (i.e., feeding specialization) may have been reinforced by interspecific competition for escape from predation in the sympatric western populations (Brower, 1958b). In contrast, the same predation pressures on tiger swallowtails might favor the use of varied types of foodplants in the east, where only a single species remained (Brower, 1958b).

The post-Pleistocene pattern of host-plant and insect distribution is generally presumed to be similar to that observed today even in areas that were close to the glacial front (e.g., see Morgan *et al.*, 1985). The three western species of swallowtails in the glaucus group (i.e., *P. rutulus*, *P. eurymedon* and *P. multicaudatus*) might be considered to have ecologically subdivided the broad range of *P. glaucus* foodplants such that each normally uses only one to three mutually exclusive plant families (Table II) (Brower, 1958a). This is consistent with well-known resource-partitioning interpretations (e.g., Benson, 1978). Presumably, other reproductive isolating mechanisms evolved in the sympatric western group, including different flight periods, different altitudinal (or habitat) preferences in the mountains, and different wing color (yellow or white) patterns (Brower, 1959a,b). All of these possibilities warrant additional research, especially with regard to mating preferences in the group (see Platt *et al.*, 1984; Scriber *et al.*, 1985; R. C. Lederhouse and J. M. Scriber, unpublished results).

In the East, the situation is even less clear. Three subspecies have been described within *P. glaucus* (*P. g. canadensis* Rothschild and Jordan, *P. g. australis* Maynard, and *P. g. glaucus* L.). The evidence for the distinctness of these subspecies has been disputed (Clark and Clark, 1951; Remington, 1951; Forbes, 1960). Our studies and others, however, indicate that *P. g. glaucus* and *P. g. canadensis* are well differentiated in (1) diapause response (Scriber, 1982, 1983; Hainze and Scriber, 1985; Ritland and Scriber, 1985; Hagen and Lederhouse, 1985; Scriber and Hainze, 1986; Rockey *et al.*, 1987a,b), (2) occurrence of melanic female wing color polymorphism (Rothschild and Jordan, 1906; Shapiro, 1974; Scriber, 1985; Scriber *et al.*, 1987a), (3) wing pattern and morphology as determined via multivariate discriminant analyses (Scriber, 1982b; Luebke, 1985; Luebke *et al.*, 1988), and (4) larval feeding abilities (Scriber, 1982, 1983, 1984a, 1986c; Scriber *et al.*, 1982, 1987; Lindroth *et al.*, 1986a,b; Hagen, 1986). The status of *P. g. australis* in Florida and the southeastern United States and the Mexican species *P. alexiares* are also unclear at the present time, although they are each much more similar to *P. g. glaucus* in most features than to *P. g. canadensis* or any other species in these above-mentioned characteristics (Scriber *et al.*, 1988a).

C. Tiger Swallowtail Foodplant Choice

To understand the relative role of foodplants as ecological factors affecting the present distribution and abundance of the tiger swallowtail butterflies (e.g., Scriber and Hainze, 1987) and the potential evolutionary implications of these host affiliations in the historical biogeography across North America, we must have a reliable base of information about the particular plant species used (or usable at numerous locations throughout the range of the taxa in the group) (Gilbert, 1979; Fox and Morrow, 1981; Scriber, 1983). Most of our laboratory's effort has been focused upon the polyphagous and wide-ranging *Papilio glaucus*, including all three subspecies. The following pages will describe our efforts at deciphering the evolution of host use patterns. We were initially restricted by difficulties with collecting sufficient field observations, by difficulties in inducing oviposition in laboratory females, and by problems in locating adequate mass-rearing facilities and in developing efficient hand-pairing techniques. However, we are now successfully compiling the needed data base to permit the initial unraveling of the biogeographical/phylogenetic story.

D. Experimental Precautions (Physiological versus Genetic Effects)

The ability of a particular larva to feed successfully upon a specific host depends on various evolutionary, ecological, physiological, and behavioral adaptations. These abilities reflect different degrees of flexibility, depending on the very recent experience of the larva as well as the feeding adaptations of its ancestors. It is interesting that all three of the most important life-history (i.e., "fitness") characters (i.e., survival probability, developmental time, and fertility/fecundity) appear to be significantly and directly related to foodplant quality (see below). Therefore, to understand better the genetic basis and mode of expression of these characteristics, we must also carefully assess the differential capacities for induction of behavioral, physiological, and detoxication abilities in the larvae.

Proper interpretation of the significance of differences in various nutritional indices of growth (i.e., rates and efficiencies) also requires knowledge of the nutritional and allelochemical status of the host plant, which is more difficult to determine and experimentally control than various abiotic factors (see Scriber, 1984c). The fundamental and dominant influence of phytochemical quality (especially nitrogen, water, and allelochemics) upon the nutritional ecology of immature insects has recently been reviewed (Scriber and Slansky, 1981; Slansky and Scriber, 1985; Mattson and Scriber, 1987).

A few studies have been described recently in which attempts have been made to assess the physiological advantages of feeding specialization at the species level (Schroeder, 1976, 1977; Scriber and Feeny, 1979; Futuyma and Wasserman, 1981; Worth *et al.*, 1982) and at the population level (Blau *et al.*, 1978; Hsiao, 1978; Smiley, 1978; Gould, 1979, 1983; Rausher, 1982; Service and Lenski, 1982; Tabashnik, 1983; Scriber, 1983; Jaenike and Grimaldi, 1983; Via, 1984). At the individual level there is even less known regarding the rates and/or efficiency of foodplant processing with respect to the diet experience of the previous generations or previous instars (but see Schoonhoven and Meerman, 1978; Scriber, 1979a, 1982a; Grabstein and Scriber, 1982a,b). In addition to significant food plant-mediated alterations in larval growth performance that can occur between instars (see Scriber, 1982a), it has also been shown that large within-instar differences in larval growth performance can be induced by the particular sequence or amounts in which food is offered to a larva (Scriber, 1981; Grabstein and Scriber, 1982a). Furthermore, subtle physiological differences between sexes of Lepidoptera may result in different efficiencies and/or growth rates, with significant ecological implications (e.g., Lederhouse *et al.*, 1982; Oliver, 1983). These considerations should not be overlooked in the interpretation of physiological adaptations and genetic covariances of characters (Yamada, 1962; Cheverud *et al.*, 1983; Riska *et al.*, 1984).

In the last few years, a "physiological efficiency model" has been designed (Scriber, 1984a) to make it possible to assess more realistically insect adaptations as measured by efficiency and/or rate of assimilation, conversion, and growth. This is an empirically derived two-dimensional insect response surface based on leaf water and nitrogen content. This descriptive model provides a needed mechanism for interpretation of experimental data in regard to the "feeding specialization hypothesis" because it aids in factoring out environmentally modulated effects (e.g., variable plant chemistry), making genuine insect adaptations more easily discernable (Scriber, 1984a). Here I will discuss primarily the crucial neonate larval stage survival differences and not the more subtle aspects of differential larval growth and reproductive potential on various plants.

E. Local Specialization

Papilio glaucus L. (with its three subspecies) is easily the most polyphagous of all of the 563 species of the Papilionidae in the world. It is reported to feed upon at least 17 foodplant families, 30 genera, and dozens of species of plants (Scudder, 1889; Tietz, 1972; Scriber, 1973, 1984b) (Table I). We have tested 120 different plant species from 34 different plant families (Table III). Twenty-seven additional species have been reported in the literature (see Table I), but have not yet been verified in our own feeding studies. However, this insect species does have particular foodplant preferences in different parts of its geographic range (Fig. 5). This fact, combined with the erroneous foodplant records that occur in the literature (see Brower, 1958a; Shields *et al.*, 1970), mandates a comprehensive and careful analysis of food plant use at a local population level (Dethier, 1954; Fox and Morrow, 1981; Scriber, 1983, 1986b). It is important to determine whether local feeding specialization is the cause or the result of genetic divergence among insects.

This concept of local preferences for (regional specialization on) particular foodplants in the *Papilio glaucus* group is at least a century old. In reference to Scudder's (1889) report of willow as a foodplant for *Papilio glaucus* in London, Ontario (Gosse, 1840), Edwards (1885a) noted that neither he in Coalburgh, West Virginia, nor John Akhurst in Brooklyn, New York, had ever found larvae of *Papilio glaucus* on willow (*Salix* spp.). In addition, his attempts to force *P. glaucus* larvae to eat willows met with no success (Edwards, 1884). Allowing *P. glaucus* to achieve the fifth larval stage on a preferred food plant and subsequently switching them onto willow did not improve the results, as all larvae died (Edwards, 1885b). However, in response to Edwards's discussion, several reports of

Table III. Foodplant Species Reported as Hosts for *Papilio glaucus* or Determined in the Laboratory

Confirmed Larval Survival[a] (First Instar)					
P.g. glaucus	*P.g. canadensis*	Latin Name of Plant	Common Name	Plant Family	Literature References for Each Foodplant Record[a,b]
		Northern group			
–	+	*Populus tremuloides* Michx.	Quaking aspen	Salicaceae	1, 2, 3, 5, 6, 8, 16, 31
–	+	*Populus balsamifera* L.	Balsam poplar	Salicaceae	2, 3, 6, 16, 31
+	+	*Betula papyrifera* Marsh.	Paper birch	Betulaceae	1, 2, 3, 5, 6, 18, 30, 31
+	+	*Prunus virginiana* L.	Choke cherry	Rosaceae	1, 2, 3, 4, 5, 6, 7, 9, 12, 14, 16, 18, 25, 31
+	+	*Prunus pennsylvanica* L.	Pin cherry	Rosaceae	1, 2, 27, 30, 31
+	+	*Sorbus americana* Marsh.	Mountain ash	Rosaceae	2, 5, 6, 18, 31
?	+	*Amelanchier alnifolia* Nutt.	Serviceberry	Betulaceae	2, 3
–	+	*Populus grandidentata* Michx.	Big-toothed aspen	Salicaceae	3, 5, 6, 16, 31
+	+	*Betula alleghaniensis* Britt.	Yellow birch	Betulaceae	2, 3, 18, 31
?	?	*Salix alaxensis* (Anderss.) Cov.	Felt-leaf willow	Salicaceae	1, 2, 3, 11, 15, 18
?	–	*Salix bebbiana* Sarg.	Belt willow	Salicaceae	1, 2, 3, 11, 15, 18
?	?	*Salix discolor* Muhl.	Pussy willow	Salicaceae	1, 2, 3, 11, 15, 18
		Central group			
–	+	*Alnus rugosa* Regel	Speckled alder	Betulaceae	1, 2, 3, 12, 31
+	+	*Fraxinus pennsylvanica* Marsh.	Green ash	Oleaceae	2, 3, 5, 6, 9, 12, 16, 31
+	+	*Fraxinus americana* L.	White ash	Oleaceae	1, 2, 3, 4, 5, 6, 7, 12, 16, 18, 26, 31
?	+	*Fraxinus nigra* Marsh.	Black ash	Oleaceae	1, 3, 4, 7, 8, 18, 31
+	+	*Prunus serotina* Ehrh.	Black cherry	Rosaceae	1, 2, 3, 4, 5, 6, 7, 8, 9, 12, 14, 16, 17, 18, 19, 31

?	?	*Prunus americana* Marsh.	American plum	Rosaceae	1, 3, 10
?	?	*Betula lenta* L.	Sweet birch	Betulaceae	1, 3, 31
–	+	*Populus deltoides* Marsh.	Cottonwood	Salicaceae	3, 6, 16, 18, 31
?	?	*Salix exigua* (Rowlee) Crong.	Coyote willow	Salicaceae	1, 2, 3, 11, 15, 18
?	?	*Salix amygdaloides* Anderss.	Peachleaf willow	Salicaceae	1, 2, 3, 11, 15, 18
?	?	*Salix nigra* Marsh.	Black willow	Salicaceae	1, 2, 3, 11, 15, 18
+	+	*Plantanus occidentalis* L.	American sycamore	Plantanaceae	5, 6
+	–	*Carpinus caroliniana* Walt.	American hornbeam	Betulaceae	4, 6, 7, 31
+	+	*Tilia americana* L.	Basswood	Tiliaceae	1, 2, 5, 6, 9, 14, 18, 31
		Southern group			
+	+	*Ptelea trifoliata* L.	Hoptree	Rutaceae	1, 4, 5, 6, 7, 17, 20, 31
+	+	*Sassafras albidum* (Nutt.) Nees	Sassafras	Lauraceae	1, 3, 5, 6, 31
+	–	*Lindera benzoin* (L.) Blume	Spicebush	Lauraceae	3, 5, 6, 7, 21, 31
+	–	*Liriodendron tulipifera* L.	Tulip tree	Magnoliaceae	1, 3, 5, 6, 7, 10, 16, 18, 19, 27, 31
+	+	*Magnolia acuminata* L.	Mountain magnolia	Magnoliaceae	1, 3, 4, 6, 16, 17, 18, 31
+	–	*Magnolia virginiana* L.	Sweetbay	Magnoliaceae	5, 6, 7, 16, 31
?	–	*Catalpa bignonioides* Walt.	Indian bean	Bignonaceae	1, 3, 31
?	–	*Carya ovata* (Mill.) Koch.	Shagbark hickory	Juglandaceae	1, 3, 13, 31
		Acer spp.	Maple	Aceraceae	2, 3, 31
?	?	*Quercus tinctoria* Bartr. (= *Q. velutina* Lam.)	Black oak	Fagaceae	1, 3
?	–	*Humulus lupulus* L.	Hop	Moraceae	1, 3
?	?	*Styrax americana* Lam.	Snowbell	Styracaceae	1, 3, 31
?	–	*Ulmus americana* L.	Elm	Ulmaceae	3
?	?	*Vaccinium* spp.	Blueberries		3
?	–	*Vitus* spp.	Grape	Vitaceae	3
–	+	*Rhamnus cathartica* L.	Buckthorn	Rhamnaceae	
–	?	*Rhamnus rubra* Greene	—	Rhamnaceae	
–	–	*Ceanothus integerrimus* Hook *et* Arn.	—	Rhamnaceae	

(*continued*)

Table III. (continued)

Confirmed Larval Survival[a] (First Instar)					
P.g. glaucus	*P.g. canadensis*	Latin Name of Plant	Common Name	Plant Family	Literature References for Each Foodplant Record[a,b]
?	–	*Eleagnus angustifolia* L.	Russian olive	Eleagnaceae	
–	–	*Angelica atropurpurea* L.	Angelica	Umbelliferae	22
+	–	*Citrus limon* (L.) Burmi	Orange	Rutaceae	
–	–	*Zanthoxylum americanum* Mill.	Northern prickly ash	Rutaceae	31
–	?	*Zanthoxylum fagara* (L.) Sarg.	Lime prickly ash	Rutaceae	
+	?	*Dictamnus fraxinella* (= *D. albus* L.)	Gas plant	Rutaceae	
–	–	*Liquidambar styraciflua* L.	Sweet gum	Hamamelidaceae	
?	?	*Magnolia fraseri* Walt.	Fraser magnolia	Magnoliaceae	5, 16
?	–	*Magnolia tripetala* L.	Umbrella magnolia	Magnoliaceae	16
?	–	*Magnolia* sp.	Magnolia sp[a]	Magnoliaceae	16
?	–	*Magnolia* sp.	Magnolia sp[b]	Magnoliaceae	16
?	–	*Magnolia* sp.	Magnolia sp[c]	Magnoliaceae	
–	?	*Magnolia* sp.	Magnolia sp	Magnoliaceae	
?	?	*Cinnamomum camphora* L.	Camphor tree	Lauraceae	3, 16, 31
–	–	*Persea borbonia* (L.) Spreng.	Red Bay	Lauraceae	
–	?	*Asimina triloba* (L.) Dunal.	Paw paw	Annonaceae	
?	–	*Acer negundo* L.	Box elder	Aceraceae	
?	–	*Acer saccharinum* L.	Silver maple	Aceraceae	
?	–	*Acer saccharum* Marsh.	Sugar maple	Aceraceae	
?	–	*Acer rubrum* L.	Red maple	Aceraceae	
?	–	*Acer spicatum* Lam.	Mountain maple	Aceraceae	
?	–	*Quercus macrocarpa* Michx.	Bur oak	Fagaceae	
?	–	*Quercus bicolor* Willd.	Swamp white oak	Fagaceae	
?	–	*Quercus alba* L.	White oak	Fagaceae	

?	–	*Quercus rubra* L.	Red oak	Fagaceae
–	?	*Castanea dentata* (Marsh.) Borkh.	American chestnut	Fagaceae
?	–	*Ailanthus altissima* (Mill.) Swingle	Tree of heaven	Simaroubaceae
?	–	*Morus rubra* L.	Red mulberry	Moraceae
–	?	*Maclura pomifera* (Raf.) Schneid.	Osage orange	Moraceae
?	–	*Juglans nigra* L.	Black walnut	Juglandaceae
?	–	*Rhus typhina* L.	Sumac	Anacardiaceae
?	–	*Ginko biloba* L.	Maidenhair tree	Ginkoaceae
?	–	*Comptonia peregrina* L.	Sweet fern	Myricaceae
?	–	*Hamamelis virginiana* L.	Witch hazel	Hamamelidaceae
?	–	*Gleditsia triacanthos* L.	Honey locust	Leguminoseae
?	–	*Cercis canadensis* L.	Red bud	Leguminoseae
?	–	*Araucaria excelsa* R. Br.	Norfolk Island pine	Araucariaceae
?	–	*Juniperus virginiana* L.	Eastern red cedar	Cupressaceae
?	–	*Thuja occidentalis* L.	Northern white cedar	Cupressaceae
?	–	*Taxodium distichum* (L.) Rich.	Bald cypress	Taxodiaceae
?	–	*Pseudotsuga menziesii* Franco	Douglas fir	Pinaceae
?	–	*Pinus strobus* L.	White pine	Pinaceae
?	–	*Pinus resinosa* Ait.	Red pine	Pinaceae
?	–	*Picea pungens* Engelm.	Blue spruce	Pinaceae
–	–	*Larix laricina* (DuRoi) Koch.	Larch, tamarack	Pinaceae
–	–	*Larix decidua* Mill.	European larch	Pinaceae
?	–	*Taxus* spp.	Yew	Taxaceae
?	–	*Sambucus canadensis* L.	American elderberry	Caprifoliaceae
?	–	*Viburnum acerifolium* L.	Mapleleaf viburnum	Caprifoliaceae
?	–	*Viburnum lentago* L.	Nannyberry	Caprifoliaceae
?	–	*Viburnum* spp.	Viburnum #1	Caprifoliaceae
?	–	*Lonicera* spp.	Honeysuckle	Caprifoliaceae
?	–	*Celtis occidentalis* L.	Hackberry	Ulmaceae
?	–	*Ulmus* spp.	Elm	Ulmaceae
?	–	*Ulmus pumila* L.	Siberian elm	Ulmaceae
+	+	*Cornus stolonifera* Michx.	Red-osier dogwood	Cornaceae
–	?	*Cornus racemosa* Lam.	Grey dogwood	Cornaceae

(*continued*)

Table III. (continued)

Confirmed Larval Survival[a] (First Instar)					
P.g. glaucus	*P.g. canadensis*	Latin Name of Plant	Common Name	Plant Family	Literature References for Each Foodplant Record[a,b]
+	+	*Prunus nigra* Ait.	Canada plum	Rosaceae	1, 2, 3
+	+	*Malus malus* L.	Apple, cult.	Rosaceae	1, 2, 3, 6, 9, 14, 31
?	?	*Prunus cerasus* L.	Cherry, cult.	Rosaceae	31
?	?	*Prunus domestica* L.	Plum, cult.	Rosaceae	3, 31
?	?	*Prunus persica* L.	Peach, cult.	Rosaceae	3, 25, 31
–	–	*Pyrus communis* L.	Pear, cult.	Rosaceae	3
+	+	*Amelanchier canadensis* (L.) Medic.	Serviceberry shadbush	Rosaceae	2, 3, 31
+	?	*Prunus emarginata* Dougl.	Western chokeberry	Rosaceae	
?	?	*Prunus caroliniana* (Mill.) Ait.	Carolina laurelcherry	Rosaceae	
?	?	*Cotoneaster acutifolius* Turez.	Peking cotoneaster	Rosaceae	28
–	+	*Crataegus* spp.	Hawthorn spp.[a]	Rosaceae	1, 3, 5, 14, 31
+	–	*Crataegus* spp.	Hawthorn spp.[b]	Rosaceae	
+	+	*Syringa vulgaris* L.	Lilac	Oleaceae	1, 2, 3, 4, 6, 8, 9, 10, 25, 29
?	?	*Fraxinus caroliniana* Mill.	Carolina ash	Oleaceae	3, 16, 31
+	–	*Ligustrum vulgare* L.	Privet	Oleaceae	
?	?	*Fraxinus trifoliata* (Torr.) H. Lewis *et* Epling	Ash	Oleaceae	1, 3, 13, 16
?	?	*Fraxinus platycarpa* Michx.	Swamp ash	Oleaceae	1, 13, 16
+	–	*Ostrya virginiana* (Mill.) Koch	Ironwood	Betulaceae	

?	+	*Alnus rhombifolia* Nutt.	White alder (Idaho)	Betulaceae	
?	?	*Corylus americana* Walt.	Hazel	Betulaceae	2
–	+	*Betula nigra* L.	Riverbirch	Betulaceae	16
–	+	*Betula populifolia* Marsh.	Grey birch	Betulaceae	
+	–	*Betula jacquemonti* Spach.	Jacquemonti birch	Betulaceae	
+	+	*Betula turkestanica* Litw.	Turkestan birch	Betulaceae	
–	+	*Betula ermanii* Chami	Russian rock birch	Betulaceae	
–	?	*Betula albo-sinensis* Burkill	China paper birch	Betulaceae	
–	+	*Betula platyphylla* Sukaczew	Japanese white birch	Betulaceae	
+	+	*Salix babylonica* L.	Weeping willow	Salicaceae	11, 15, 16
?	–	*Salix alba* L.	White willow	Salicaceae	
–	+	*Salix* spp.	Willow #1	Salicaceae	
–	+	*Salix* spp.	Willow #2	Salicaceae	
–	+	*Salix* spp.	Red-stemmed white willow	Salicaceae	
–	+	*Populus alba* L.	White poplar	Salicaceae	16, 24
–	+	*Populus nigra* L. (var. italica)	Lombardy poplar	Salicaceae	
–	+	*Populus maximowiczii* Henry	Japanese poplar	Salicaceae	
–	–	*Populus tremula* L.	European poplar	Salicaceae	

Notes: Plant species after Gleason and Cronquist (1963), Preston (1976), and Dirr (1983). Laboratory in Madison, Wisconsin (1978–1986).

[a] A first-instar survival of greater than 10% (+); survival of 0–9% (–); not yet tested (?).

[b] (1) Scudder, 1889 (2) McGugan, 1958 (3) Tietz, 1972 (4) Shapiro, 1974 (5) Scriber, 1975 (6) Scriber *et al.*, 1982 (7) Opler and Krizek, 1984 (8) M. Nielson, personal communication MI (9) J. Mertens, personal communication IA (10) J. Luhman, personal communication MN (11) Gosse, 1840 (12) Durban, 1857, 1860 (13) Couper, 1874 (14) Saunders, 1874 (15) Edwards, 1884, 1885a, 1886 (16) Clark and Clark, 1951 (17) Levin and Angleberger, 1972 (18) Tyler, 1975 (19) Makielski, 1972 (20) Scriber, 1972 (21) Scriber *et al.*, 1975 (22) Berenbaum, 1981 (23) Kimball, 1965 (24) Clarke and Shepard, 1955 (25) Wm. Houtz, personal communication PA (26) J. Troubridge, personal communication Ontario (27) Shapiro, 1963 (28) A. M. Young, personal communication WI (29) L. Kass, personal communication (N.Y.) (30) R. Hagen and R. Lederhouse (northern Wisconsin) (31) Scott, 1986.

[c] A number of additional foodplant survival bioassays have recently been reported by Hagen (1986); however, the New York populations are likely to reflect introgression between the *P. g. canadensis* and *P. g. glaucus* subspecies, and therefore cannot be properly categorized at this time.

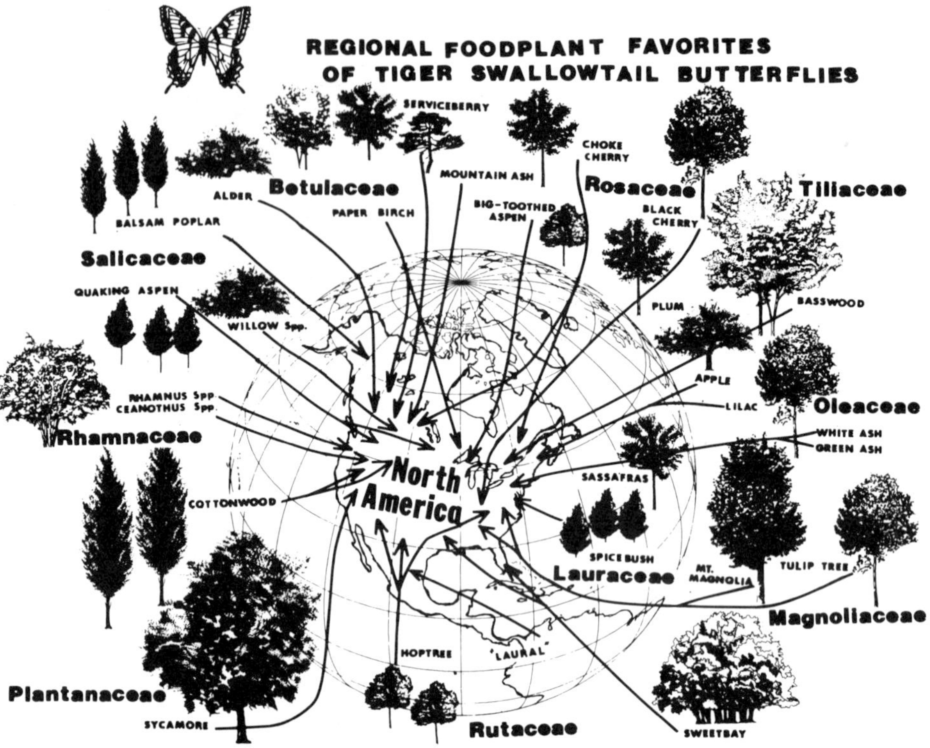

Fig. 5. Selected regionally favored foodplant species of the *Papilio glaucus* complex of Papilionidae. (After Scriber, 1986b).

willow as a natural foodplant of *Papilio glaucus* were received from J. D. Sherman in Peekskill, New York; Carolin G. Soule of Boston, Massachusetts; and W. Bentenmuller of New York (Edwards, 1886). Edwards's conclusion is that "larval habits as to food may differ decidedly in different localities" (see also Clark and Clark, 1951; Scriber, 1972).

No single foodplant can suffice throughout the range of *Papilio glaucus* in North America, because no foodplant species has a range that large. Specialization is, therefore, a local necessity. In addition to restricted availability (Fig. 6), a low abundance/density of physiologically suitable foodplants at any location and numerous behavioral and ecological factors may further restrict the realized feeding niche (Gilbert, 1979). In fact, one might conclude that local foodplant specialization or "ecological monophagy" is likely to be the rule in Lepidoptera rather than the exception. The degree to which such feeding specialization mediates an ecologically or metabolically more efficient utilization of food resources than polyphagy generally remains an interesting unknown for insect her-

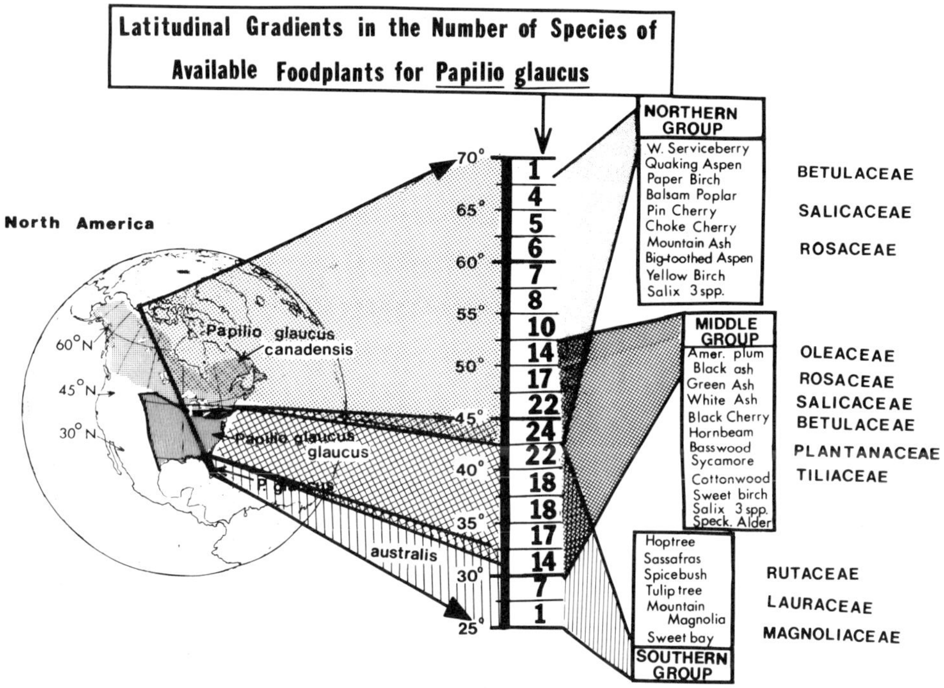

Fig. 6. Latitudinal distribution (availability) of 32 most frequently reported foodplants of *Papilio glaucus* along a 6000-km transect in North America. Note the overlap and transition of plant families at 40–45°N. (After Scriber, 1986b).

bivores. It is also interesting to know whether negative genetic correlations evolve with this specialization such that the ability to survive and grow upon allopatric/ancestral food plants is lost (see Smiley, 1978; Scriber, 1983).

F. The Southern Populations

While *Papilio glaucus* is generally the most polyphagous species of Papilionidae in the world (Scriber, 1984b), in Florida it might be considered one of the most monophagous (i.e., feeding on a single plant species). Throughout the southern half of Florida the only foodplant that is available for use by the eastern tiger swallowtail species, *Papilio glaucus*, is sweetbay, *Magnolia virginiana*. Ecological specialization of Florida in *Papilio glaucus* on sweetbay appears to have resulted in biochemical/physiological adaptations to this foodplant that are not realized by individuals of other allopatric populations of the eastern tiger swallowtail butterfly (Scriber, 1986b; Lederhouse and Scriber, 1987a). For example,

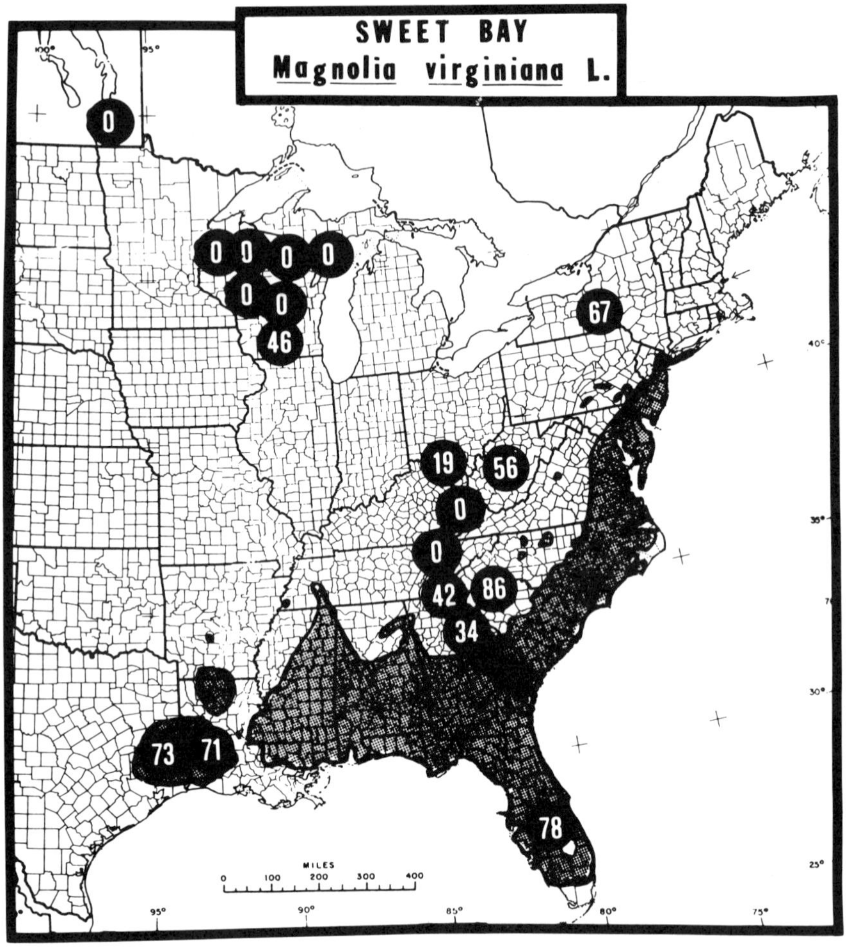

Fig. 7. First-instar (neonate) larval survival (indicated as a percentage) of various populations of *Papilio rutulus, P. glaucus canadensis, P. g. glaucus,* and *P. g. australis* on sweetbay (*Magnolia virginiana*) in controlled environment studies. Larvae tested from each location: 10, California; 4, Manitoba; 31, northern Wisconsin; 50, central Wisconsin; 28, southern Wisconsin; 86, Ohio; 20, West Virginia; 8, Kentucky; 15, Tennessee; 9, New York; 7, South Carolina; 545, southern Georgia; 12, northern Georgia; 161, Louisiana; 638, Texas; and 523, Florida. The shaded region represents the current geographic distribution of sweetbay. (After Scriber, 1986b).

none of the *P. g. canadensis* larvae (and very few of the Ohio, Kentucky, Tennessee, and Georgia *P. g. glaucus*) survived the first instar on sweetbay (Fig. 7). Furthermore, these Florida populations of the putative *Papilio glaucus australis* exhibit poor neonate survival on most other foodplants such as the Salicaceae and Betulaceae used to the north by *P. glaucus glaucus* and *P. glaucus canadensis* (Tables III and IV) (Scriber, 1986b).

The number of larvae obtained in the laboratory from Florida females (all Highlands County) totalled 1712 (Table IV). On sweetbay, 78% of the neonate larvae from a total of 16 different females survived to the second instar. The range in survival on other plants reported for *P. glaucus* (Scriber, 1984b) is from 0% on three species of the Salicaceae to more than 57% on all of the eight species tested of the Oleaceae, Rosaceae, and Magnoliaceae (Table IV). It may not be surprising based on phytochemical similarities in the Magnoliaceae that tulip tree and mountain magnolia appear to be generally as suitable larval foodplants as is sweetbay; however, neonate larvae also have excellent abilities to use the three allopatric species of Rosaceae (i.e., black cherry, choke cherry, and mountain ash) and the two allopatric species of Oleaceae (i.e., white ash and lilac) as well. It has likely been a long time since any South Florida populations of *P. g. australis* have seen anything but sweetbay as a host (Scriber, 1986b).

From these results (Table IV) it might be concluded that any population of *Papilio glaucus australis* that strayed in oviposition from the primary host (sweetbay) where other suitable foodplants occurred (e.g., northern Florida and the coastal regions of the southeastern United States), would not *necessarily* be doomed to certain acute toxicity or starvation. This would depend on the plant selected by the ovipositing female. Nonetheless, toxicity for neonate larvae might be virtually assured with certain oviposition mistakes (cf. Wiklund, 1975; Berenbaum, 1981) by *P. g. australis* upon several of the unsuitable plants.

It is intriguing that we find that the western populations of *P. rutulus* and the northern populations of *P. glaucus glaucus* and *P. g. canadensis* in no-choice situations apparently lack the behavioral acceptance responses or detoxication abilities and/or physiological capacities to effectively process sweetbay (at least during the first instars; Fig. 7). None of the neonate larvae of *P. rutulus* from California nor any of the *P. g. canadensis* subspecies from Manitoba or northern Wisconsin survived through the first instar. These populations have not had the opportunity to use this plant species for thousands of years (see Section III).

In contrast, some neonate larvae from every population of the *P. g. glaucus* subspecies tested survived the first instar on sweetbay. Similarly, larvae of some females the Mexican swallowtail, *P. alexiares* Hoppf., survive on sweetbay (Scriber, 1986c) (Table V). In eastern Texas females from Jasper, Tyler, and Newton Counties are considered together because of the geographic proximity of collection sites. It is noteworthy that the adults captured were generally large in wingspan, and many were ochreous (burnt orange) in coloration (much as the *P. g. australis* from

Table IV. No-Choice Feeding Bioassays of *Papilio glaucus* from Highlands County, Florida, on Foodplants Reported for Other Populations in the Literature. (After Scriber, 1986b).

Plant Family	*Genus Species* (Common Name)	Number of Fertile Females	Number of Larvae	First-Instar Surviva
Salicaceae	*Populus deltoides* Marsh. (cottonwood)	3	15	0.0%
Salicaceae	*Populus grandidentata* Michx. (big-toothed Aspen)	3	11	0.0%
Salicaceae	*Populus balsamifera* L. (balsam poplar)	1	2	0.0%
Salicaceae	*Populus tremuloides* (Michx.) (quaking aspen)	16	124	5.6%
Betulaceae	*Betula papyrifera* Marsh. (paper birch)	12	66	27.3%
Rhamnaceae	*Rhamnus cathartica* L. (buckthorn)	2	6	0.0%
Hamamelidaceae	*Liquidambar styraciflua* L. (sweetgum)	2	16	0.0%
Platanaceae	*Platanus occidentalis* L. (sycamore)	4	22	18.2%
Betulaceae	*Carpinus caroliniana* Walt. (hornbeam)	5	27	0.0%
Tiliaceae	*Tilia americana* L. (basswood)	3	16	50.0%
Rutaceae	*Ptelea trifoliata* L. (hoptree)	2	14	35.7%
Lauraceae	*Sassafras albidum* (Nutt.) Ness (sassafras)	4	27	25.9%
Lauraceae	*Lindera benzoin* (L.) Blume (spicebush)	4	30	16.6%
Lauraceae	*Persea borbonia* (L.) Spreng. (redbay)	2	41	0.0%
Oleaceae	*Fraxinus americana* L. (white ash)	5	34	71.2%
Oleaceae	*Syringa vulgaris* L. (common lilac)	1	7	57.1%
Rosaceae	*Sorbus americana* (Marsh.) (mountain ash)	2	19	63.2%
Rosaceae	*Prunus pennsylvanica* L. (pin cherry)	1	10	80.0%
Rosaceae	*Prunus virginiana* L. (choke cherry)	2	12	91.7%
Rosaceae	*Prunus serotina* Ehrh. (black cherry)	20	300	83.0%
Magnoliaceae	*Magnolia acuminata* L. (mountain magnolia)	2	21	61.9%
Magnoliaceae	*Liriodendron tulipifera* L. (tulip tree)	17	276	75.4%
Magnoliaceae	*Magnolia virginiana* L. (sweetbay)	16	616	77.6%

Note: Standard errors are not provided because survival is presented as a population of larvae and not evenly weighted by individual adult females (see text).

Florida). In addition, we did find larvae in this region feeding on sweetbay as well as black cherry. It is entirely possible that some or all of these *Papilio* could represent host-specialized subpopulations and belong to the *P. g. australis* subspecies; however, such conclusions must await intensified field studies and additional information obtained from multivariate discriminant analyses of morphometric traits, electrophoresis studies, and studies of developmental compatabilities of various hybrid crosses (Berlocher, 1984; Collins, 1984; J. M. Scriber and R. C. Lederhouse, unpublished). In any case, the Texas larval survival is excellent and is generally comparable to that observed for the Florida larvae. Most of our eastern populations (see Fig. 7) have too few larvae bioassayed to permit any rigorous conclusions about the degree of differential biochemical adaptation to sweetbay at this time.

While only a single foodplant is available for tiger swallowtail populations in Florida, many other choices are available to the north. Of the 32 most frequently reported foodplants for the species, the largest number of potential host species along a transect (which bisects the geographical range of *P. glaucus*) occurs at latitudes between 40 and 45° N (Fig. 6). This area is of interest because it also closely corresponds with the known zone of suspected hybrid interaction between *P. g. glaucus* and *P. g. canadensis* and represents a "tension zone" for many plant species (Curtis, 1959; Scriber *et al.*, 1982). The degree to which these populations are structured into host-associated subdemes is unknown, although we are expanding our data bases and also assessing the relative contribution of environmental versus genetic factors as causal mechanisms for the observed geographic and intrapopulation differences in host utilization traits (Scriber, 1986b). As with the population from Highlands County, Florida, we have conducted controlled studies of neonate larval survival for a number of populations from 20 states and Canadian provinces across the eastern half of North America.

G. Survival and Foodplant Origin (General Methodology)

Adult females of *P. glaucus* from Florida to New England and from Texas to Manitoba were collected and taken in envelopes and on ice for oviposition in the laboratory. Females were individually allowed to oviposit on foodplant leaves (generally black cherry and choke cherry) supported in water-filled Aquapics in clear plastic boxes (12 × 20 × 30 cm) under heat from an incandescent light bulb placed at a distance approximately 0.5 m from the boxes. Eggs were removed from leaves after oviposition, and neonate larvae were subsequently distributed to a controlled environmental growth chamber (at 16 : 8 photo/scotophase with a corresponding temperature regime of 23.5/19.5°C), where they were reared on excised leaves of various species of plants (in 4 × 15 cm petri dishes with

Table V. First-Instar (Neonate) Larval Survival for (North American) *Papilio*

	Lauraceae		Magnoliaceae		Salicaeae	
	Redbay	Spicebush	Sweet Bay	Tulip Tree	Quaking Aspen	Balsam Poplar
P. troilus	92	70	0	0	0	0
P. palamedes	45	—	0	—	—	—
P. pilumnus	—	—	—	—	—	—
P. g. glaucus	0	16	52	81	7	0
P. g. australis	0	19	78	75	6	0
P. alexiares	0	20	11	20	0	0
P. g. canadensis	0	1	1	1	80	70
P. rutulus	—	0	0	0	78	50
P. eurymedon	—	0	0	0	72	33

Notes: Madison, Wisconsin (1979–1986). All values are expressed as percent survival to the data), 0, no survival, —, untested.
[a] California population of *P. rutulus* use Sycamore as a local host.

screened ventilation) through to pupation and eclosion as adults. Foodplant leaves were kept turgid by use of water-filled plastic vials with rubber caps (see Scriber, 1977), and changed as needed. We used a mixture of mostly healthy-looking mature (fully expanded) leaves and some younger leaves for these studies.

Only the survival of larvae through completion of the first instar (i.e., to the molting stage) and not for subsequent instars is reported, because for some of our mass-rearing procedures we were forced to switch larger larvae to other foodplant species that were more abundant/available. Our laboratory rearing studies were carried out during the field seasons of 1981–1986 in Madison, Wisconsin, using different groups of *P. glaucus* from various locations (1986 data on larvae are not included here, but see Fig. 14). Neonate larvae from eggs of each individual adult were distributed evenly across as many foodplants as possible. Where few eggs of a female were obtained, we attempted to distribute the larvae primarily across several key foodplants (e.g., quaking aspen, paper birch, black cherry, sweetbay, and tulip tree).

All foodplants used in our survival bioassays were field collected in Madison, Wisconsin except for (1) balsam poplar, *Populus balsamifera* L., which was collected from various northern Wisconsin counties, and (2) sweetbay, *Magnolia virginiana*, which was obtained from Highlands County, Florida, and the Big Thicket region of Texas (Tyler, Jasper, and Newton Counties) (as well as from the University of Wisconsin Arboretum). These leaves were shipped weekly and stored in refrigeration at 6–8°C until use.

Taxa on Nine Selected Plant Families

Rosaceae						
Black Cherry	Choke Cherry	Rutaceae, Hop Tree	Betulaceae, Paper Birch	Oleaceae, White Ash	Plantanaceae, Sycamore	Rhamnaceae, *Rhamnus* spp.
0	0	0	0	0	0	0
0	—	0	—	—	—	—
—	—	—	—	—	—	—
81	67	66	33	74	35	3
82	92	36	27	71	18	0
89	100	45	40	100	0	0
74	61	70	73	75	37	9
81	50	67	67	17	93[a]	0
92	100	100	50	—	0	40

second instar. More than 26,000 larvae have been used in these bioassays (J. M. Scriber, unpublished

Larvae used in our study were obtained from females field captured (or hand-mated with a male from the same population no more than a single generation removed from the field). The primary locations or counties of populations used for our bioassays were: Sacramento County, California; Columbia and Garfield Counties, Washington; near Sandilands Provincial Park, Manitoba, Canada; Bayfield, Clark, Langlade, Lincoln, Oneida, Taylor, and Price Counties, northern Wisconsin; Juneau, Green Lake, Adams, and Marquette Counties, central Wisconsin; Gogebic and Ontonogan Counties, Michigan; Dane and Richland Counties, southern Wisconsin; Rock Island and Champaign Counties, Illinois; Adams, Preble, and Scioto Counties, Ohio; St. Lawrence and Tompkins Counties, New York; Clay County, West Virginia; Wolfe County, Kentucky; Campbell County, Tennessee; Schuykill and Dauphin Counties, Pennsylvania; Spartanburg and Greenville Counties, South Carolina; Jefferson County, Alabama; Clark, Habersham, Henry, Oglethorpe, and Rabun Counties, Georgia; Beauregard County, Louisiana; and Jasper, Newton, and Tyler Counties, Texas.

We noted tremendous variability in larval survival among various populations of *P. glaucus* for any given foodplant. Variation in foodplant quality locally is likely to be as large as geographic differences, and both should be investigated in much more detail. We know that there is a genetic basis to some of these differential detoxication/biochemical processing abilities, and that these abilities can be transferred from one taxon to another via a hand-pairing technique (Scriber, 1982, 1986a). The key phytochemical compounds responsible for the differential toxicity in the Salicaceae have recently been identified (Lindroth *et al.*, 1988), but several additional candidates are still under investigation (see Lindroth *et al.*,

1986a,b; Scriber *et al.*, 1987). The importance of these genetic differences in enzyme-mediated detoxication abilities to the understanding of reproductive isolation and speciation processes is of general interest (Diehl and Bush, 1984; Futuyma and Peterson, 1985; Scriber, 1986a) and will be the focus of continued research.

H. A Wisconsin Hybrid Zone and Northern Limits to Gene Flow

Dramatic differences in host use abilities (as well as other biological traits) across the presumed zone of hybridization of *P. g. glaucus* and *P. g. canadensis* suggest that host race formation and/or host shifts near this known transition (blend) zone of vegetation may be significant predisposing factors to speciation. While this possibility has been under investigation at a county and local population level across the state of Wisconsin for several years, the actual pattern of reproductive isolation and of foodplant choice by ovipositing females in the field has yet to be properly documented (see Scriber, 1975; Grossmueller and Lederhouse, 1985).

The numbers of oviposition choices (of the 32 most frequently reported foodplants) feasible for each county in Wisconsin are presented in Fig. 8. It can be seen that at least 17 different plant species (potential foodplants) exist in each county of Wisconsin for an ovipositing tiger swallowtail female. We also know that many of these foodplants are acceptable for larval survival and growth of both *P. g. glaucus* and *P. g. canadensis* and their hybrids. For example, black cherry (*Prunus serotina*) is an excellent foodplant (although the only one) that has been documented in the field as a natural choice of both Wisconsin subspecies (see Scriber *et al.*, 1982). However several other foodplants exist for which each *Papilio* subspecies exhibits a reciprocal inability to survive on the others' favorite choice (see Fig. 9). (Scriber, 1986b; Scriber *et al.*, 1987).

The situation described above for tiger swallowtails provides a unique opportunity to delve experimentally into the mechanisms of speciation. The capabilities of hand-paired hybridization, mass-rearing of all parental stock and hybrids upon commonly acceptable foods such as black cherry, and the genetically based reciprocal inabilities in foodplant detoxication traits offer a chance to manipulate experimentally the laboratory "populations." Opportunities to observe gene flow throughout a fairly well-defined hybrid zone in the Great Lakes region (see Fig. 3) are also possible with the use of genetic markers (nearly fixed allele differences at two different loci for each subspecies) (Hagen, 1986; Scriber *et al.*, unpublished).

The Wisconsin differentiation zone between *P. g. glaucus* and *P. g. canadensis* is sharply delineated geographically at the 1444 degree-day line (Scriber and Hainze, 1987). This line delineates the northernmost

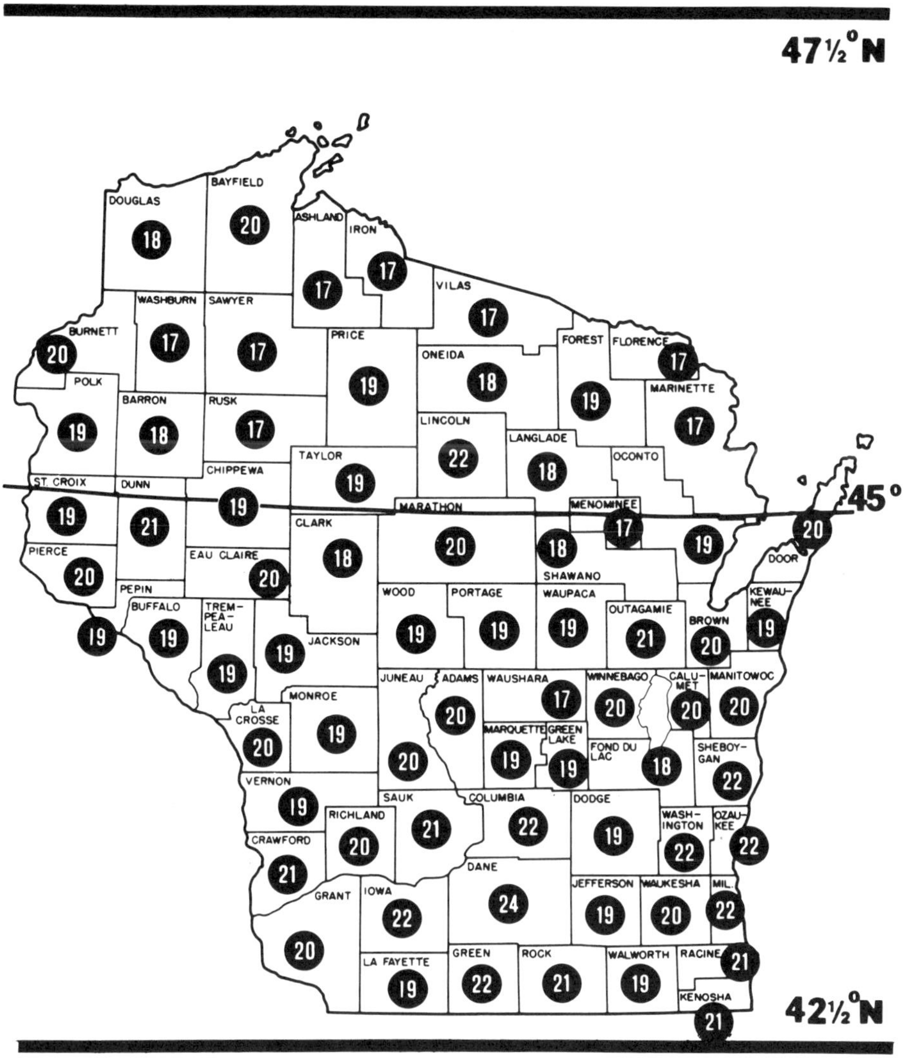

Fig. 8. Number of available *P. glaucus* foodplant species in each county of Wisconsin (of the 32 most frequently reported hosts). This figure provides detail of the pattern portrayed in Fig. 6.

limits of the potential for successful bivoltinism (see Scriber, 1982; Scriber and Lederhouse, 1983; Ritland and Scriber, 1985). The selection of foodplants in this area may be especially critical, and appears to have exerted selection pressure for the "best" host (see Fig. 10) (Scriber and Hainze, 1987; Grossmueller and Lederhouse, 1986). This line also repre-

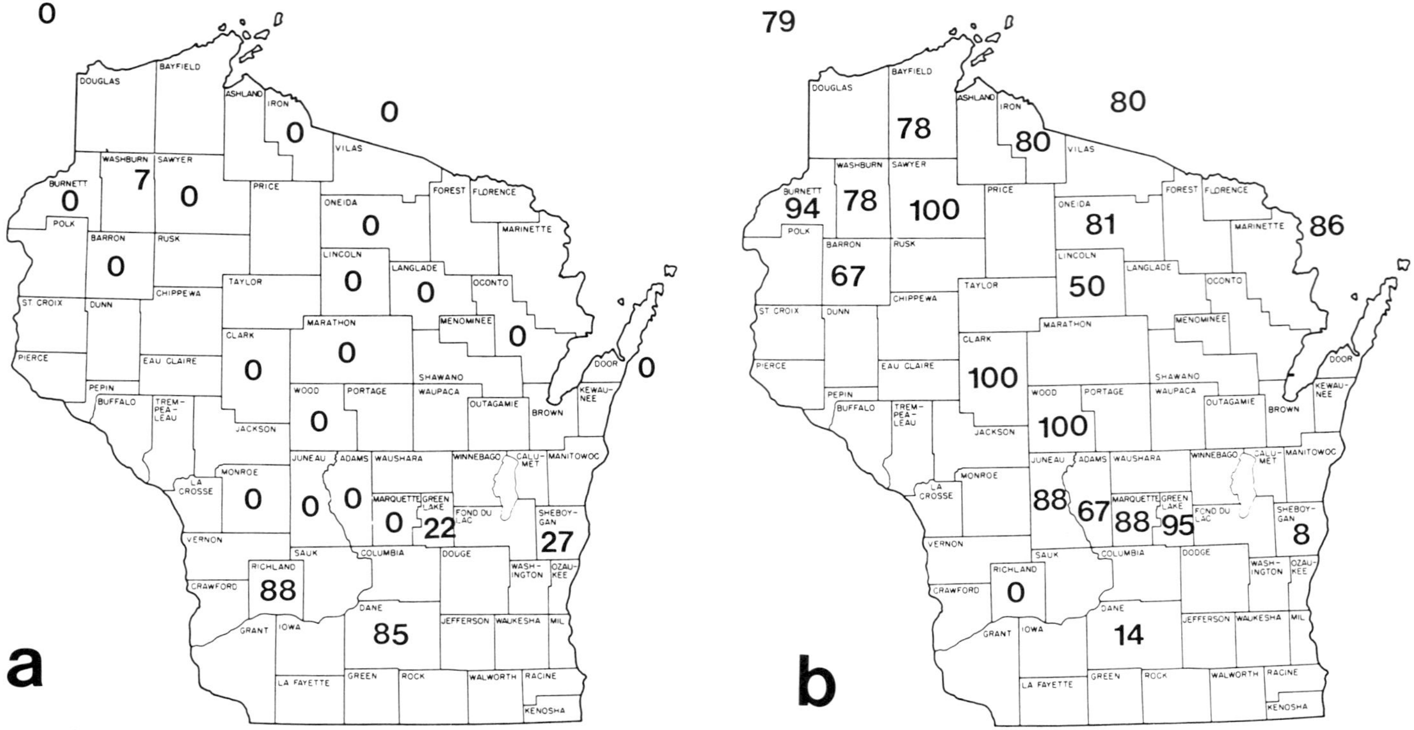

Fig. 9. Neonate survival (to second larval instar) of various Wisconsin populations of *Papilio glaucus* under controlled environmental conditions (16 : 8 photo/scotophase; corresponding temperatures of $23\frac{1}{2}$, $19\frac{1}{2}$°C) on (a) tulip tree, *Liriodendron tulipifera*, and (b) quaking aspen, *Populus tremuloides*.

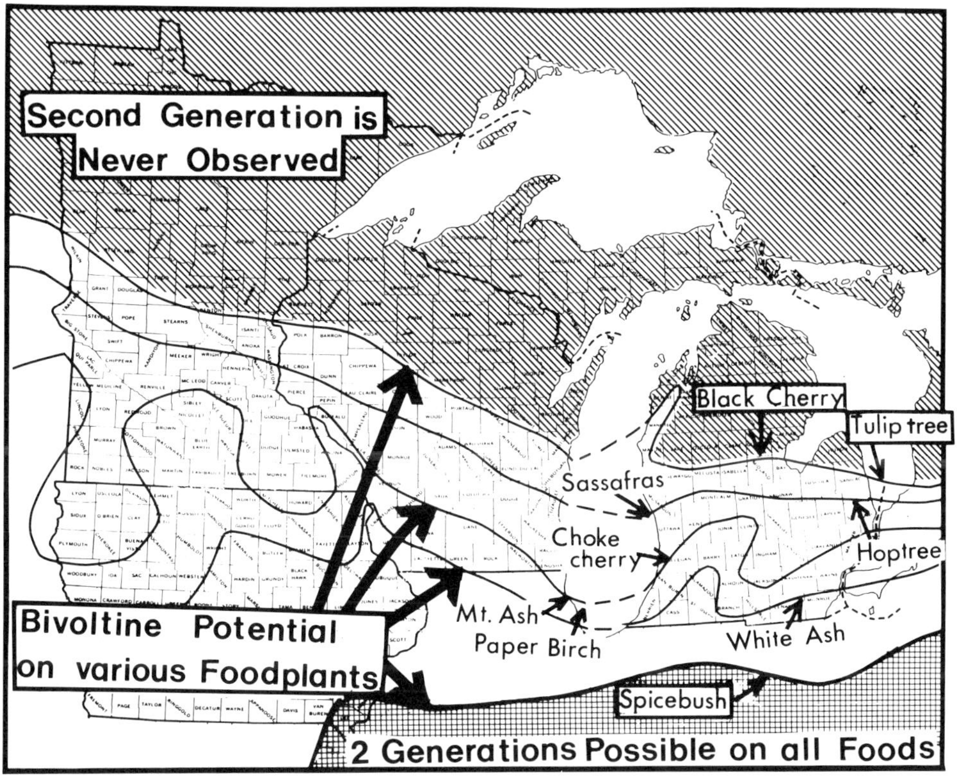

Fig. 10. The northernmost limits of bivoltine potential of *P. glaucus* upon various potential host plants calculated from various rearing studies and the 20-year mean thermal accumulations across Wisconsin and Michigan. (After Scriber and Hainze, 1987. Copyright 1987 Academic Press).

sents the northernmost collection records for dark morph females (Fig. 11) (Scriber, 1985; Scriber *et al.*, 1982b, 1987) and has been independently elucidated with univariate (Fig. 12) and multivariate morphometric techniques (Luebke *et al.*, 1988). The genetic capacity for dark morph polymorphism is *Y*-linked (Clarke and Sheppard, 1962) and apparently does not exist in *P. g. canadensis* females (but see Scriber and Evans, 1986a). The genetic basis of diapause control has recently been determined to be *X*-linked (Fig. 13) (Hainze and Scriber, 1985; Rockey *et al.*, 1987a), and appears to be photoperiodically obligate in *Papilio glaucus canadensis* (Fig. 14) (Rockey *et al.*, 1987b).

All of these data indicating significant restriction in gene flow between *P. g. glaucus* and *P. g. canadensis* and our findings of dramatic differences in abilities of the putative subspecies to detoxify the Magnoliaceae

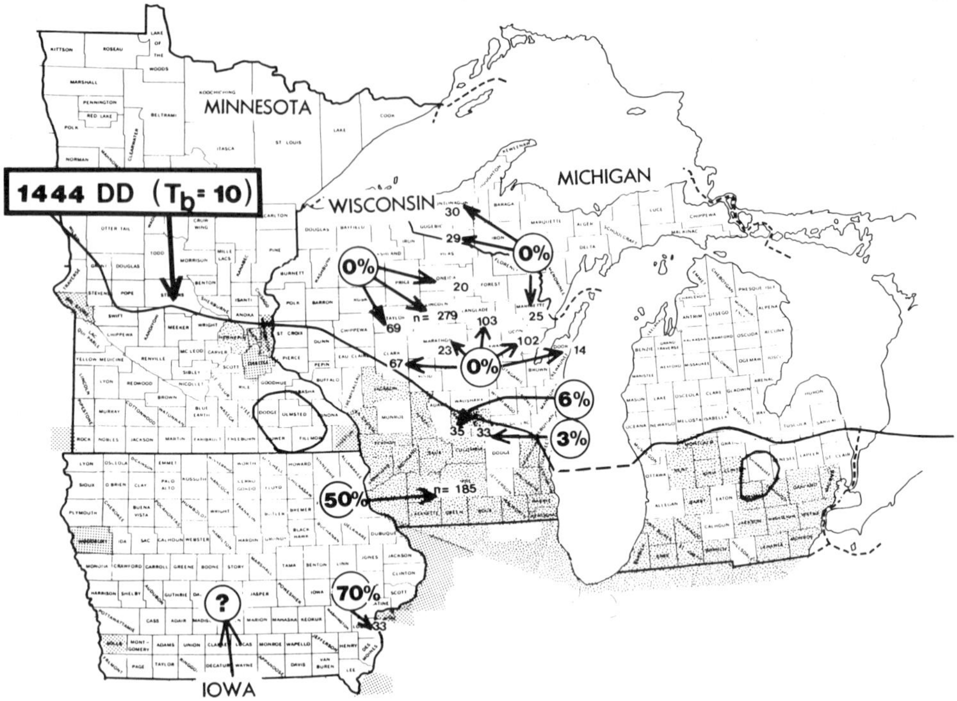

Fig. 11. Northernmost county records for captures of the dark morph females in Minnesota, Wisconsin, and Michigan (J. M. Scriber, M. Nielsen, M. H. Evans, R. C. Lederhouse, J. Luhmen, and W. Warfield, unpublished) in relation to isoline of thermal unit accumulation of 1444 (°C) degree-days. Sample size is indicated by county.

and Salicaceae (Scriber, 1982b, 1986a; Scriber *et al.*, 1987) suggest that considerable genetic divergence has occurred. We suspect asymmetrical gene flow across this thermally defined zone of potential hybridization. Northern flow of *P. g. glaucus* type traits (e.g., facultative diapause, dark morph females, inability to detoxify Salicaceae, and distinctive wing characters and larval characters) seems to stop sharply at this zone in Wisconsin. Whether *P. g. canadensis* genes spread southward and the extent of linkage disequilibrium are unknown at this time, but are under investigation. It remains unclear whether this divergence occurred in parapatry or whether these differences accumulated in proto-*canadensis* stock in allopatric isolation (or both). I would suggest that recent evidence from the glacial age may provide important evidence in this regard.

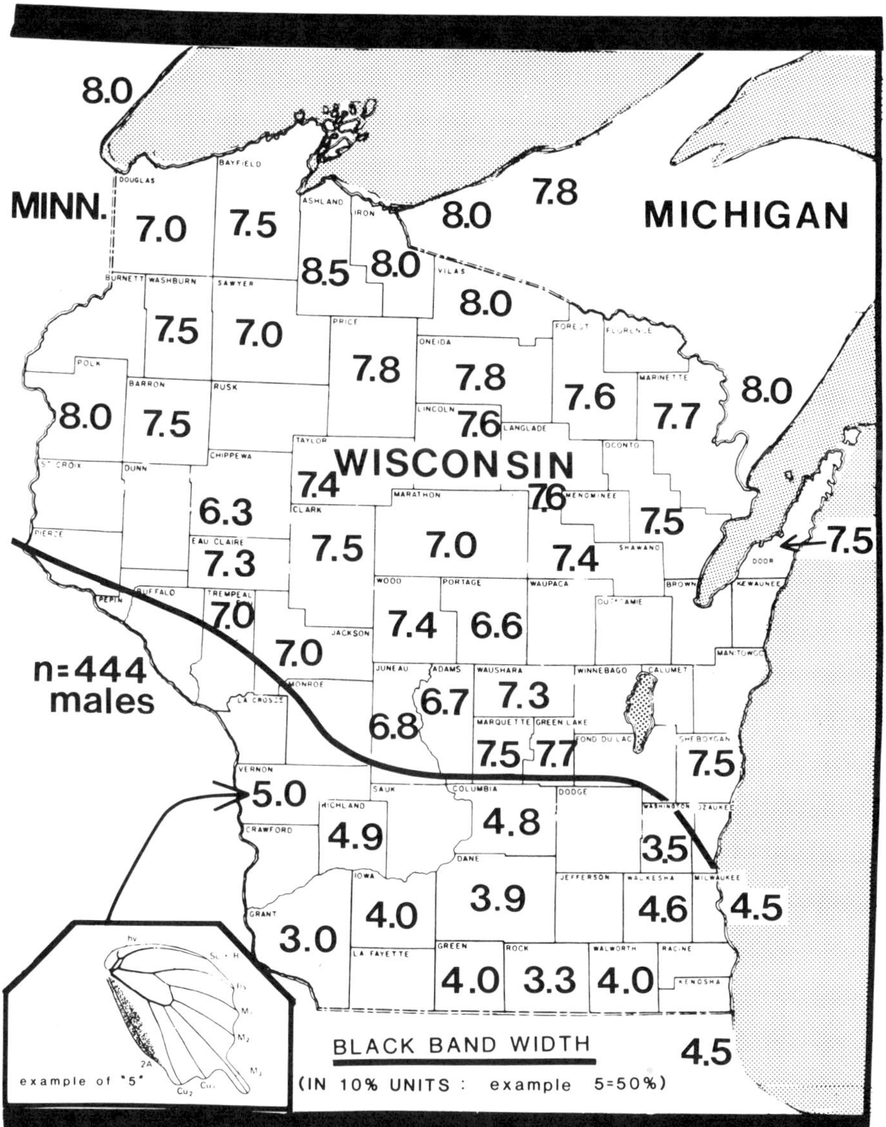

Fig. 12. Width of the black band along hindwing anal margin of field-collected male *P. glaucus*, averaged by county. Heavy line indicates the approximate position of the 1444 degree-day isotherm using 10°C base threshold for development.

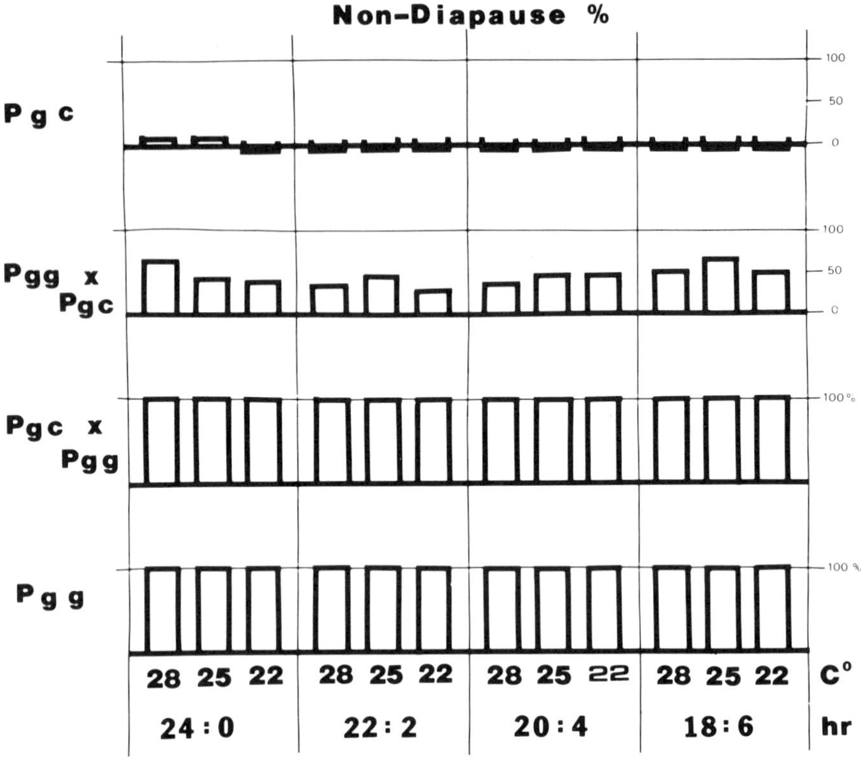

Fig. 13. The proportions of larvae of *Papilio glaucus canadensis* (Pgc), *P. g. glaucus* (Pgg), and the reciprocal hybrids (female listed first) that produce pupae that direct develop (i.e., do not diapause) when reared under various photoperiod and constant temperature regimes. (After Rockey *et al.*, 1987a).

III. Glaciation Effects and Biogeography

Pleistocene glaciations undoubtedly played a major role in determining the current biogeographical distribution of the swallowtails (Figs. 3 and 4). This current pattern of distribution in *Papilio* was mediated directly by the effect the glaciers had upon the plant communities and the climatic effects on voltinism patterns of the *Papilio*. For example, it is not surprising that the Magnoliaceae- and Lauraceae-feeders currently occur only in the southeastern part of North America, since there is general lack of plant species in this family elsewhere (north of Mexico). Similarly, Betulaceae- and Salicaceae-feeding populations are currently found primarily from the Rocky Mountains and across the northern United States and Canada where these plants predominate (see Figs. 5 and 6).

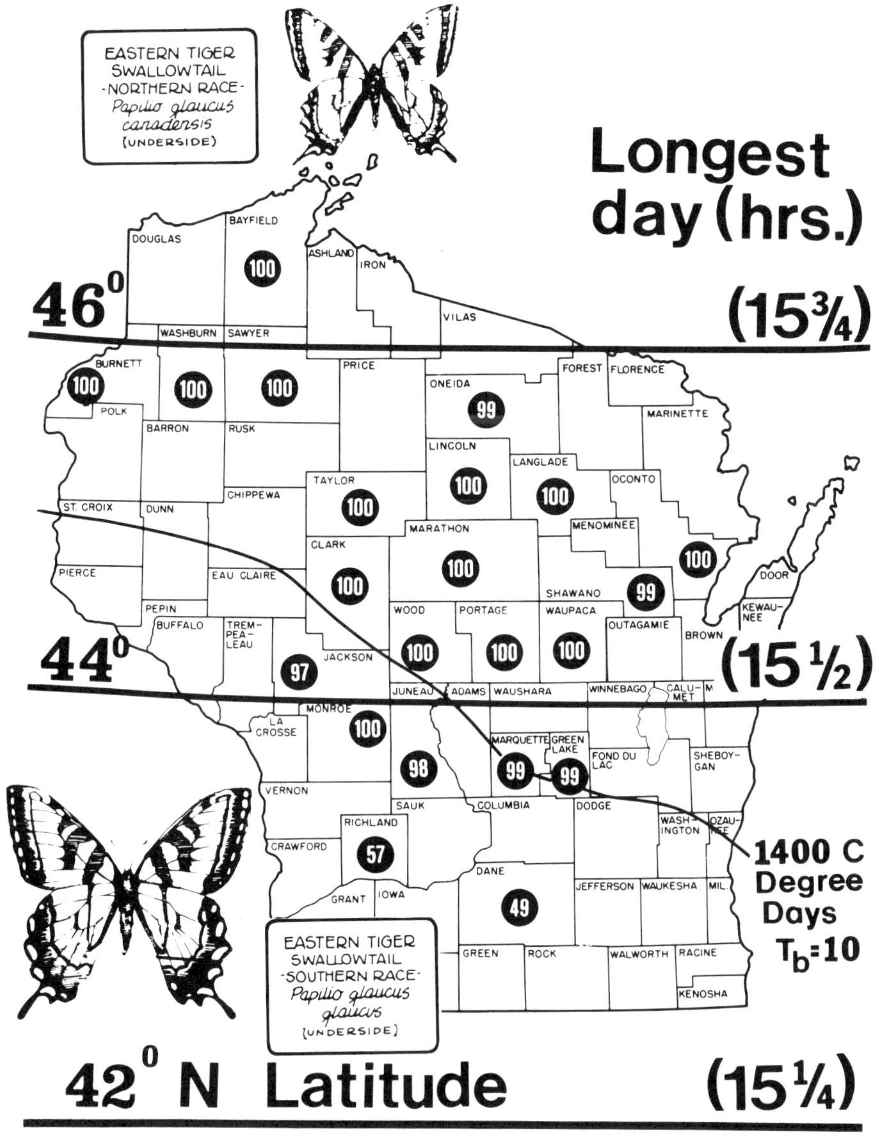

Fig. 14. Wisconsin latitude and longest day length with the percent diapause response observed for various collection sites for *Papilio glaucus* (n = 993) reared under 16 : 8 photophase. (After Rockey *et al.*, 1987b). A second generation is possible only south of the 1444°C degree-day line (see also Ritland and Scriber, 1985).

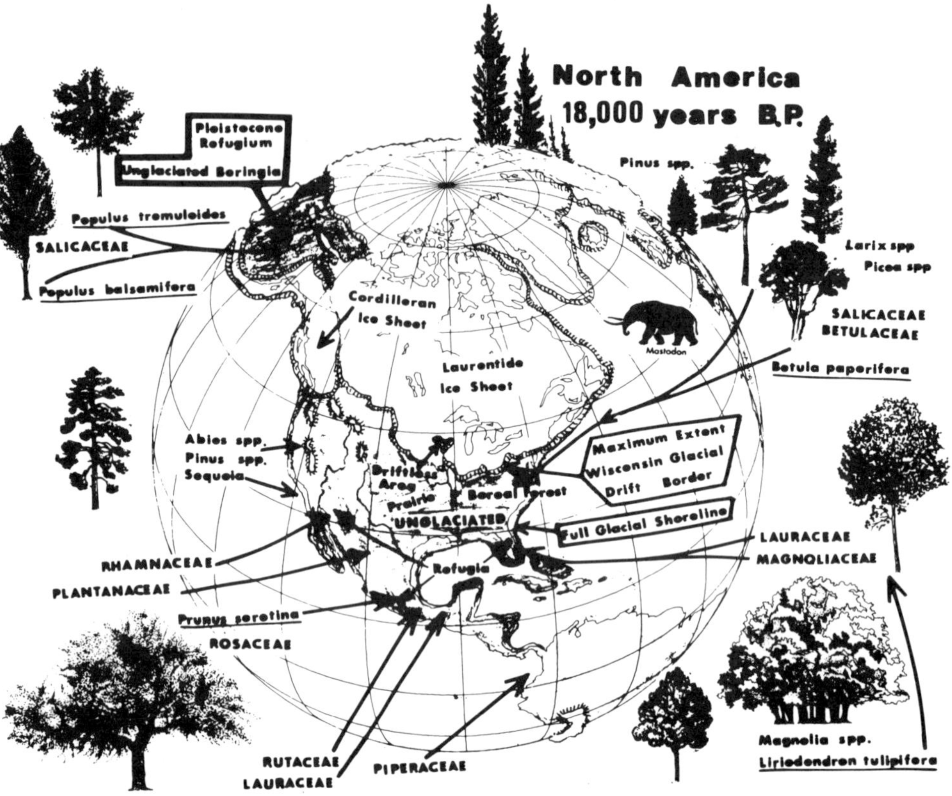

Fig. 15. North America 18,000 years B.P. as reconstructed from paleontological data and fossils (see Braun, 1955; Axelrod, 1958; Graham, 1964; Wolfe, 1978; Peterson *et al.*, 1979; Pielou, 1979; Watts, 1979, 1980; Colinvaux, 1981; Hopkins *et al.*, 1981; Murray, 1981; Brown and Gibson, 1983).

Assuming that these differential foodplant utilization abilities or host plant affiliations were in existence during the glacial maxima [e.g., 18,000 years B.P. (before present); see Fig. 15], we might be better able to reconstruct geographic histories as well as phylogenetic relationships of the taxonomic groups in the manner of vicariance biogeography (Wiley, 1981). Platnick and Nelson (1978) suggest that explanations for disjunct distributions of related individuals/taxa are generally placed in two classes: (1) dispersal biogeography, in which migration across a previous barrier occurs, and (2) vicariance biogeography, in which barriers fragment previously continuous distributions of a taxon. Assuming that the rate of genetic divergence of populations is proportional to the duration of isolation, then a logically constructed cladistic phylogeny can provide

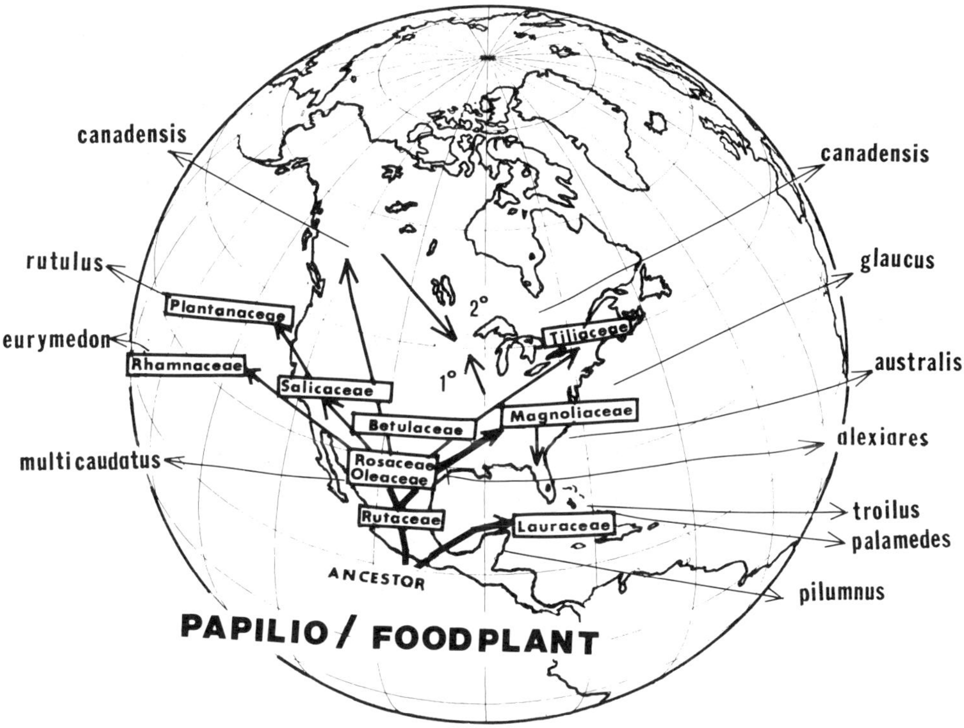

Fig. 16. Hypothetical cladograms (*Papilio*/foodplant affiliations) superimposed upon the area of natural occurrence. Negative genetic correlations in foodplant use abilities are observed for the species on the Lauraceae, Magnoliaceae, Salicaceae, and Rosaceae (for the *troilus* group) (see text).

hypotheses about both the historical relationships among the taxa and also their biogeographical affinities (see Brown and Gibson, 1983; but cf. Farris, 1981; Felsenstein, 1982, 1984, 1985).

Based on known patterns of foodplant affiliations and our data regarding ability/inability for detoxication/processing of plant species by neonate larvae of the various taxa, I propose one scenario for the evolutionary divergence of the North American tiger swallowtail butterflies (see Fig. 16). This hypothetical evolutionary schematic is based upon both the current foodplant use abilities (Fig. 16) and the current geographical distribution of the insect taxa (Fig. 17); while *P. alexiares* in Mexico will also be important, the northern *P. g. canadensis* is crucial to the interpretations of this phylogeny (Fig. 16) and will be described in more detail below.

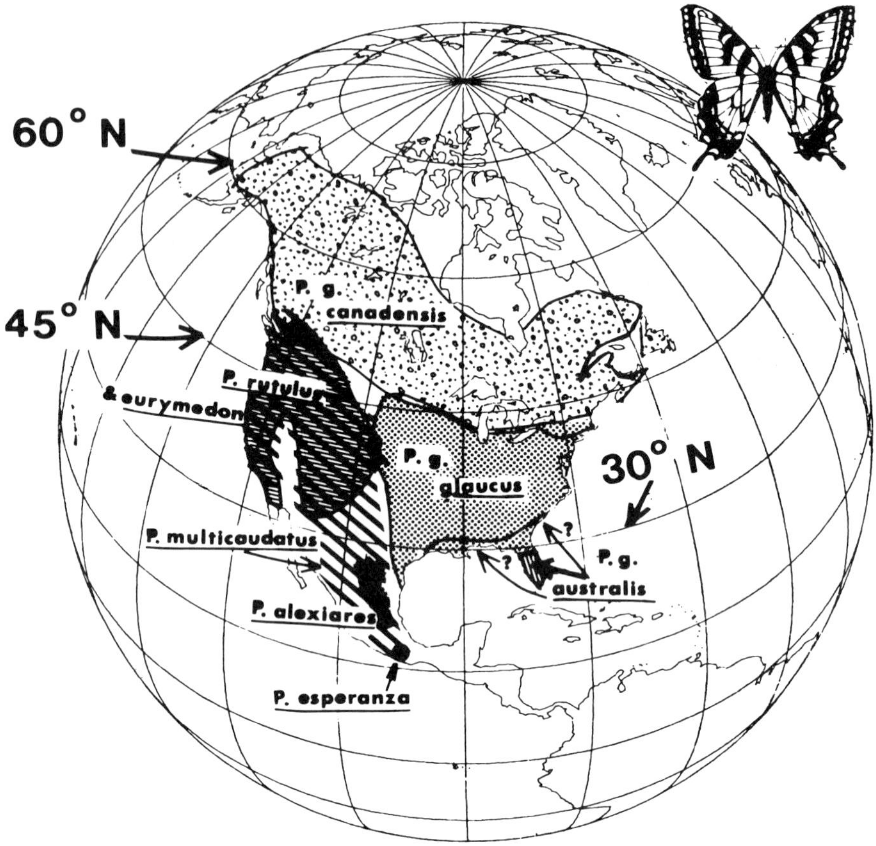

Fig. 17. The geographic ranges of the *P. glaucus* complex for comparison with the underlying foodplant use patterns (see Fig. 16). (After Scriber, 1986b).

A. A *Papilio*/Plant/Geographic Phylogeny

Since the Rutaceae, Rosaceae, Oleaceae, and Betulaceae can sustain survival and growth of neonate larvae of all taxa tested thus far (Table V) (Scriber, 1986c), it seems feasible that these plant families may have contained a chemical common denominator (see Feeny *et al.*, 1983), and perhaps served as hosts, for the common ancestor of the *P. glaucus* group. The key here may be with oviposition (Lederhouse and Scriber, unpublished) rather than larval use, however (see Hagen, 1986). Nonetheless, a major split appears to have occurred between the known Magnoliaceae-feeders (*P. g. glaucus* and *P. g. australis*) and the Salicaceae-feeders (*P. rutulus*, *P. g. canadensis*, and *P. eurymedon*). While *P. eurymedon* has specialized on Rhamnaceae, larvae can eat species of Salicaceae suc-

Table VI. Survival of Four Different *Papilio* Taxa on Redbay and Sweetbay

Taxa	Total Number Females	Total Number Larvae (*N*)	Redbay (Lauraceae) First-Instar Survival (%)	(*N*)	Sweetbay (Magnoliaceae) First-Instar Survival %
P. palamedes (Florida)	4	(68)	64.2 ± 15.3	(97)	0.0 ± 0.0
P. troilus (Florida)	2	(24)	95.7 ± 4.4	(34)	0.0 ± 0.0
P. g. glaucus (Georgia)	9	(190)	0.0 ± 0.0	(375)	28.0 ± 6.8
P. g. australis (Florida)	2	(41)	0.0 ± 0.0	(93)	69.7 ± 4.1

Note: Madison, Wisconsin (1985). Data are presented as mean ± SE.

cessfully (Fig. 16, Table V). Nothing is known about the *P. multicaudatus* ability/inability to detoxify and grow on the Salicaceae and Magnoliaceae at this time. Current foods include the Rutaceae, Oleaceae, and Rosaceae, and the current geographic distribution of *P. multicaudatus* is throughout Mexico north through the Rocky Mountains in the western United States. Similarly, *P. alexiares* and *P. esperenza* in Mexico may hold the key to the foodplant phylogeny. Recent discovery of black cherry, *Prunus serotina*, and ability to use both the Magnoliaceae and Betulaceae by the Mexican swallowtail *P. alexiares garcia* strengthens this suggestion (see Scriber, 1986b,c).

In the eastern half of the United States, we have seen that *Papilio glaucus glaucus* and *P. g. australis* are abundant. Their choice of and ability to use the Magnoliaceae is unique; however, unlike the members of the sympatric and specialized Lauraceae-feeding *P. troilus* group (*P. troilus*, *P. palamedes*, and *P. pilumnus*), the *glaucus* group can utilize foodplants of other families (see Fig. 15). Although literature records abound for use of sweetbay (Magnoliaceae) by the Lauraceae-feeding *troilus* group, we have shown that this is not likely to be correct, and in our experiments larvae of *P. troilus* and *P. palamedes* all died on sweetbay (Table VI; cf. Table 2).

The preference for and ability to use quaking aspen and balsam poplar of the Salicaceae by the northern *P. g. canadensis* and the western *P. rutrulus* is coupled with a virtually total inability to use sweetbay and tulip tree of the Magnoliaceae (Table V) or spicebush and red bay of the Lauraceae (Tables V and VI). The reciprocal pattern of use abilities/inabilities is observed for *P. g. glaucus* and *P. g. australis* on the Salicaceae (Table V; see also Scriber *et al.*, 1987). In other words, fitness (as measured by survival and growth rates) on one host appears to be negatively correlated with fitness on other hosts. However, if coadapted preference/viability gene complexes exist, they would be more likely to contribute significantly to incipient host races or species (Wiklund, 1975; Bush and Diehl, 1982; Feeny *et al.*, 1983; Scriber, 1983; Rausher, 1984; Futuyma and Peterson, 1985).

Determining the range of foodplants that can be successfully detoxified and processed by various populations of each taxon of tiger swallowtails across North America will help determine which populations (or taxa) possess the "preadaptations" for use of novel hosts and which lack (or have lost) the capacity to successfully process an otherwise potential foodplant. While the appropriate information on the genetic basis of the differential host use abilities of various populations will be difficult to accumulate rapidly, we have enough hybrid studies initiated to generate a few hypotheses concerning the evolutionary scenarios that are most likely to have occurred in North American tiger swallowtails. Investigation of gene flow in current zones of parapatry (or suspected areas of hybridization) may provide a useful mechanism of approach (see Remington, 1968; Endler, 1977, 1982; McDonnell *et al.*, 1978; Jackson and Pounds, 1979; Hafernik, 1982; Harrison, 1983; Collins, 1984; Thorpe, 1984).

In summary, we have experimentally detected significant phytochemically based differences (Lindroth *et al.*, 1987a,b; Lindroth and Pajutee, 1987, unpublished) in survival and growth of *Papilio* between the Lauraceae, Magnoliaceae, and Salicaceae that effectively correlate with reproductive barriers among the tree-feeding swallowtails of North America. While some plant families may serve as common foods (e.g., Rutaceae, Rosaceae, Oleaceae, and Betulaceae), we have observed dramatic reciprocal inabilities of the various *Papilio* taxa to successfully accept, consume, and process plants of the three key families (Lauraceae, Magnoliaceae, and Salicaceae). The implications in speciation of North American tree-feeding *Papilio* are significant, and a likely foodplant use phylogeny can be constructed for additional hypothesis testing (Fig. 16).

For example, the Mexican swallowtail, *P. alexiares*, could represent a Pleistocene relict population, and understanding its physiological/ detoxication capacities on various plant families (Fig. 16) would provide clues to the systematic affiliation of this insect taxon (Scriber *et al.*, 1988a). Also, we know that the northern subspecies *P. glaucus canadensis* has been a taxonomic enigma (see below). Whether or not the current parapatry between *P. g. canadensis* and *P. g. glaucus* throughout the Great Lakes region is a result of primary or secondary contact is of interest evolutionarily, and it may be possible to determine by assessing its genetically based physiological/ecological capabilities across its current geographic distribution (Fig. 17). Discriminating among alternatives A, B, C, and D of Fig. 18 may be difficult and a variety of systematic tools must be employed (see below).

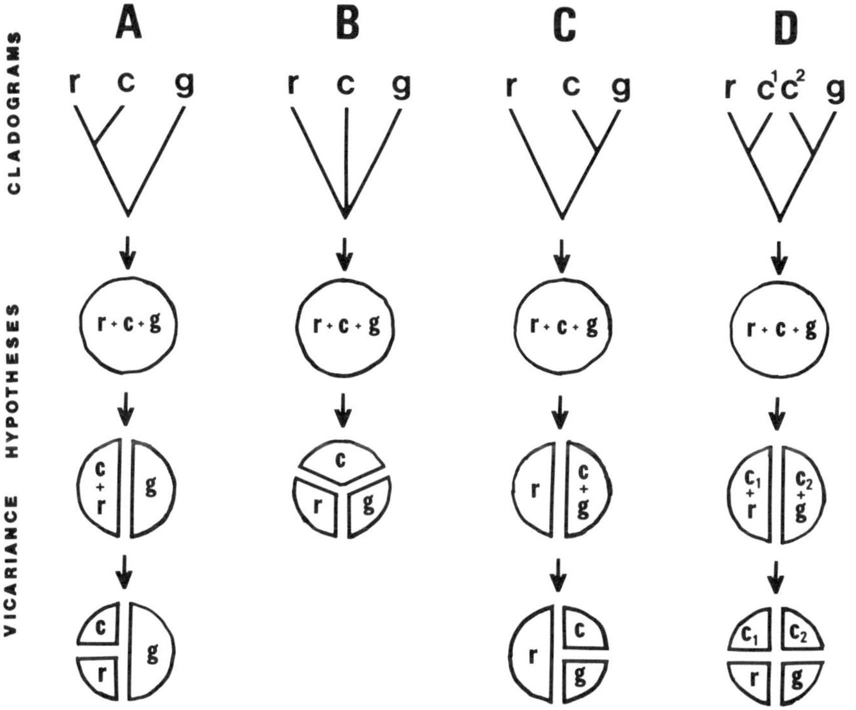

Fig. 18. Alternative cladograms and vicariance biogeography scenarios for the relationships between *Papilio glaucus glaucus* (g), *P. g. canadensis* (c), and *P. rutulus* (r).

B. A Northern Key to the Puzzle

The purported Canadian tiger swallowtail subspecies, *P. glaucus canadensis*, is of particular value in our systematic studies because of its unique biology relative to the other two purported subspecies (*P. g. glaucus* and *P. g. australis*). Its parapatric distribution (suspected narrow hybrid zones occur) in regard to the contact with both the western *P. rutulus* and the eastern *P. glaucus glaucus* is distinct and important (see Fig. 3). The taxonomic status of these northern populations of the rubric *canadensis* is unique in many ways, but uncertain because of the fact that it apparently shares some traits with both *P. glaucus* and *P. rutulus*. One of the most significant, of course, is the feeding preference and ability to use the Salicaceae (as *rutulus* does) and inability to use the Magnoliaceae (as *glaucus* does). On the other hand, they share morphological characteristics with *glaucus* [e.g., the single "eyespot" pattern of fifth-instar larvae is as *glaucus*, not as the three western species with two-component eyespots; see Brower (1959a)]. As is the case with *P. rutulus*, *canadensis* lacks the genetic capacity for dark morph females that both *P. g. glaucus* and *P. g. australis* possess (Scriber and Evans, 1986; but see Scriber *et al.*, 1987a). Adult wing pattern morphology/color as determined by multivariate discriminant analyses (Reyment *et al.*, 1984; Dixon and Brown, 1979) are unique and different from both *P. rutulus* and *P. glaucus* (Luebke, 1985; Scriber *et al.*, unpublished), but some characters are shared with only *P. rutulus*, and others shared with only *P. glaucus*.

In summary, the correct cladogram is, at this point, uncertain and deserving of intensified study (Fig. 18). Recent hybridization studies involving reciprocal crosses among all of these taxa (Fig. 19) suggest that the rubric *canadensis* examined thus far are genetically/developmentally more compatible with *P. glaucus* than with *P. rutulus* (Scriber and Lederhouse, unpublished; Oliver, 1979). For example, we very rarely obtain an F_1 hybrid female from a *rutulus* × *glaucus* or *glaucus* × *rutulus* cross; however, we obtain F_1 females from both *canadensis* × *glaucus* and *glaucus* × *canadensis* crosses. However, it should be noted that all *canadensis* thus far hand-paired are from the eastern half of the continent, and that crosses with northwestern (Alaska) *canadensis* may show more in common developmentally with *rutulus* (and *eurymedon*). In other words, perhaps two types of *canadensis* exist (e.g., "eastern" and "western"; see Fig. 18).

This vicariance biogeographical approach assumes that barriers divided a once-contiguous range of a taxon/taxa (Pielou, 1979; Wiley, 1981; Brown and Gibson, 1983). The Pleistocene glaciations would have provided such a mechanism, especially if the northern refugia (Beringia) is included in the scenario (Fig. 10) (see also discussions in Braun, 1955; Graham, 1964; Peterson *et al.*, 1979; Watts, 1979, 1980; Colinvaux, 1981; Danks, 1981; Hoffman, 1981; Murray, 1981). The most significant feature about this Beringial refugium for *P. g. canadensis* is that populations of balsam poplar, *Populus balsamifera*, and of quaking aspen, *Populus tremuloides*, are believed to have existed in this area throughout the entire period of glaciation [from 36,000 years B.P. to the present postglacial (Hopkins *et al.*, 1981)], although aspen may have been differentiated in the southern Rocky Mountains (Barnes, 1975). While "shrub birch" (e.g., *Betula nana*, *B. glandulosa*, and *B. kenaica*) may have also persisted in this refugium, tree paper birch (*B. papyrifera*) and spruce were apparently eliminated (absent from 30,000 to 11,500 years ago), with a reintroduction via dispersal from a refugium south of the Laurentide ice sheet only after the glaciation retreat in the Holocene (Hopkins *et al.*, 1981). Eastern *P. tremuloides* more closely resembles the northern (Beringial) form than the southwestern Rocky Mountain *P. tremuloides* (Barnes, 1975), and it would be interesting to compare the phytochemistry of the aspen used by

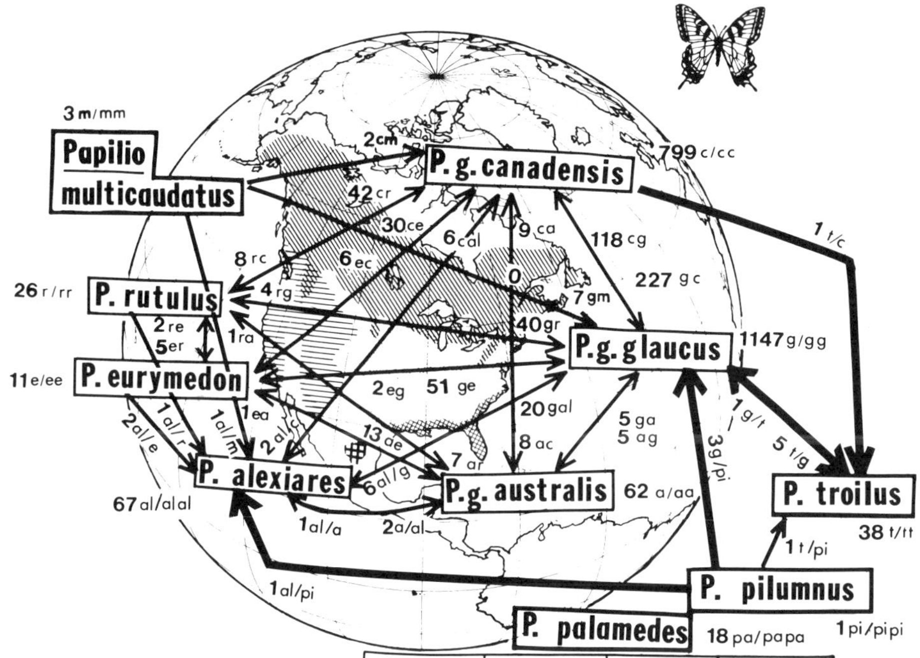

	1983	1984	1985	1986
PRIMARY & F_1	n= 232	n= 905	n= 858	n= 922
BACKCROSS	111	109	72	163
F_2 HYBRID	3	38	3	6
TOTAL FEMALES	346	1052	933	1087
TOTAL EGGS	22,449	38,674	29,365	37,106

Fig. 19. Number and type of interpopulation, intersubspecific, and interspecific laboratory hand-pairings for which copulation was of a sufficiently long duration (> 30 min) to permit transmission of a spermatophore. Reciprocal crosses are indicated by the arrows (female is listed first).

P. g. canadensis (northern and east) versus that used by *P. rutulus* in the Rocky Mts.

Should the ancestral *Papilio* stock have been isolated in this Beringial region (now Alaska) for the period 36,000 to 9500 years B.P., it would have been without the common foodplants enjoyed by their relatives remaining south of the Cordilleran and Laurentide ice sheets [e.g., cherry, paper birch, and ash would have been available even as far north as in the spruce forests (Watts, 1980)]. Among the Arcto-tertiary plant communities in southeastern North America (Graham, 1964; cf. Axelrod, 1958; Wolfe, 1978), the bay forests with sweetbay, *Magnolia virginiana*, did not become predominant in the southeast until 5000 B.P. (Watts, 1980). This may represent the earliest opportunity for the rubic *P. g. australis* to differentiate in association with sweetbay, its favorite (only?) foodplant (Scriber, 1986b).

Such isolation in Beringia with basically only the Salicaceae as foodplant options for more than 25,000 years could easily account for the genetic differentiation observed in "canadensis" today. Dispersal from this refugium southward and eastward could also explain the striking difference in voltinism currently observed between the *P. g. canadensis* (obligate diapause) and *P. g. glaucus* (facultative diapause) across the narrow blend zone in central Wisconsin (Figs. 9–14) (Rockey *et al.*, 1987a,b). Determination of the extent and direction of gene flow across the zone of parapatry (delineated very precisely by seasonal thermal unit accumulations above the basal temperature for larval development; see Scriber and Lederhouse, 1983; Ritland and Scriber, 1985; Hagen and Lederhouse, 1985; Scriber and Hainze, 1987) will aid in subsequent interpretations regarding the primary/secondary (sympatric/ allopatric) mechanisms of genetic divergence. The "spring brood" of *Papilio glaucus* in the eastern United States is very "canadensis" in many respects, yet exists in the midst of predominantly "glaucus" populations (Clark and Clark, 1951; Scriber *et al.*, unpublished). Perhaps the Appalachian "boreal refuge" has harbored *P. g. canadensis* types, much as presumed in the "Beringial refuge." Feeding abilities of southern Appalachian populations are often atypical of the surrounding area, as is the case with *Populus tremuloides* (Fig. 20) (Scriber, 1983).

In addition to preliminary evidence from morphometric discriminant analyses of "spring form" adult wing traits, we know that the Wisconsin populations of *P. g. canadensis* exhibit approximately 60% neonate survival on mountain magnolia, *Magnolia acuminata* L. (a favorite food of *P. g. glaucus* throughout the Appalachian Mountains; see Scriber *et al.*, 1982; Scriber, 1986b). This ability of *P. g. canadensis* appears in direct contrast to that on tulip tree and sweetbay (Table V) and may provide some evidence for the parapatric origin hypothesis of the proto-*canadensis* with proto-*glaucus* in Eastern North America. A feeding study on *Magnolia acuminata* of *P. g. canadensis* populations from across Canada into Alaska might provide useful information for distinguishing among the four hypotheses of "canadensis" origin (see Fig. 18).

We have conducted hybridization/developmental compatability studies with the various *Papilio* taxa described in this chapter to help assess the phylogenetic relationships (see Oliver, 1972, 1979; Ae, 1979, for discussions of the genetic differentiation index). While such relationships in the *Papilio* generally coincide with those of classical taxonomy (Ae, 1979), a number of problems with hand-pairing studies exist (Scriber and Lederhouse, unpublished). For example, (1) all of the matings of appropriate

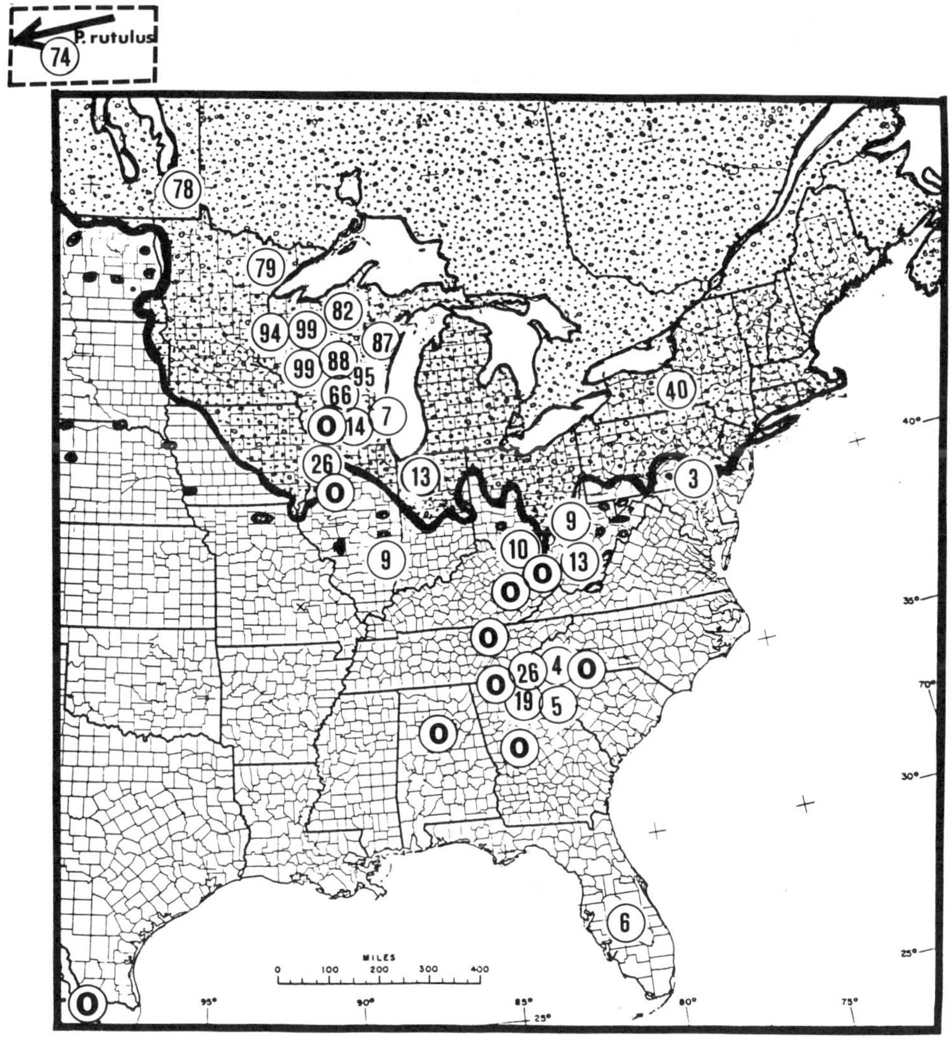

Fig. 20. First-instar larval survival of various *Papilio glaucus* and Mexican *P. alexiares* populations in relation to the geographic range of quaking aspen (cf. Fig. 9b).

durations (usually more than 30 minutes will suffice for *Papilio*) will result in successful transmission of a male spermatophore, (2) all spermatophores transferred are not of equal size or "quality" (e.g., differential nutrient investment), (3) while some eggs are fertile, the larvae are often too weak to eclose successfully, and (4) we know that fertility from single matings declines significantly with time after copulation, regardless of the genetic developmental compatability of hybrids (Lederhouse and Scriber, 1987b). Our hybridization studies (primary, backcrosses, and F_2's) have nevertheless been extremely useful in exploring the genetic bases of other life history traits (e.g., foodplant detoxication, diapause control, dark form color polymorphism, and numerous morphometric characters). Resolving the several biological interpretations of phylogenetics, ecological, and behavioral relationships among taxa will certainly be aided by the contributions from this hybridization program (see Fig. 19). With these breeding/rearing capabilities we will continue attempts to discern the relative importance of genetically based differences in host utilization abilities of various geographic populations and recognized taxa (see Scriber, 1986a; Scriber *et al.*, 1987b).

Experimental assessment of limitations of the rates and directions of evolution in foodplant use will depend on the degree to which adult behavioral (oviposition) host preferences will result in a physiologically suitable foodplant (Wiklund, 1975; Bush and Diehl, 1982; Feeny *et al.*, 1983; Futuyma, 1983). The various roles of foodplants in the allopatric/sympatric speciation of the North American tree-feeding *Papilio* may be understood or better elucidated only after the degree of "linkage disequilibrium" is known for these behavioral/biochemical capabilities. A combination of theoretical (e.g. Rausher, 1985) and empirical (e.g., Singer, 1983; Grossmueller and Lederhouse, 1987, Hagen and Scriber, 1988) approaches will be useful in such endeavors.

IV. Conclusions

Foodplant specialization is the rule rather than the exception in the Papilionidae. Only 23 species (less than 5%) of the species of Papilionidae of the world feed on more than two foodplant families. The North American tree-feeders are the most polyphagous of all swallowtail butterflies and as such represent a difficult yet valuable group to study. Evaluation of various hypotheses for the origins of polyphagy in this group has been complicated by the lack of chemical common denominators between plant taxa currently utilized and the shifting biogeographical ranges of these plant species.

We have elsewhere elucidated some of the most logical biochemical limitations in the differential toxicities of Salicaceae, Magnoliaceae, and Lauraceae in the *P. glaucus* and *P. troilus* complex (Scriber, 1986a; Lindroth *et al.*, 1987a,b, 1988). Negative genetic correlations, while suggestive, do not necessarily imply reciprocal evolutionary relationships (trade-offs) however. Also, while these differences are known to have a phytochemical basis (Lindroth *et al.*, 1986b), it seems unlikely that these solitary foliage-feeding larvae have had a major physiological impact on their host trees, and even less likely that they have shaped either the phytochemistry or the genetic control of this chemistry.

A new hypothesis is proposed herein that suggests that isolation of the proto *canadensis* in the Beringial refuge during the entire 36,000 years of Pleistocene glaciation is feasible, based on pollen/fossil records. Quaking aspen and/or balsam poplar of the Salicaceae could have supported these proto-*canadensis* type *Papilio*. The evolution of photoperiodically obligate diapause (see Rockey *et al.*, 1987b) as well as the inability to utilize foodplants of the Magnoliaceae and Lauraceae would be a logical development during this time frame. Alternatively, isolation in parapatry/sympatry is generally felt to be possible (see Remington, 1951; Brower, 1959a,b).

A testable foodplant phylogeny is proposed, based on current neonate survival patterns. The genetic basis of these differential detoxication abilities appears to be relatively straightforward (Scriber, 1986a, and unpublished); however, additional hybrid and backcross studies will be needed between various taxa to determine the precise number of loci involved. Nothing is currently known about the behavioral or phytochemical mechanisms of host selection in these two groups of butterflies (the *P. troilus* and the *P. glaucus* complex). It sometimes appears that *Papilio glaucus* oviposition stimulation is elicited by plants toxic to larvae, however (see Brower, 1958a,b; Scriber, 1975; Berenbaum, 1981; Feeny *et al.*, 1983). It remains unknown whether the physiologically efficient genotypes may have preceded or followed the evolution of host specificity in locally specialized populations of *P. glaucus*, however (Scriber, 1986b). We propose to test these alternatives by comparisons of "true" phylogenies based on numerous morphometric and allozymic character analyses, with subsequent comparisons with several adaptive traits as currently differentiated and distributed among populations.

Acknowledgments

This research was supported by the Michigan State University College of Natural Science and the Agricultural Experiment Station (project 8051) and in part by the National Science Foundation (grants DEB 7921749, BSR 8306060, and BSR 8503464), U.S. Department of Agriculture grants 85-CRCR1-1598 and 87-CRCR-1-2851, and the Graduate School and College of Agricultural and Life Sciences (Hatch 5134) of the University of Wisconsin. I would particularly like to thank the following people for valuable discussion and/or their assistance in field collections for this study: Phillip Ackery, Matt Ayres, May Berenbaum, William Bergman, Sue Borkin, Don Brockway, Murray Blum, Lincoln Brower, Keith Brown, John Bryant, Guy Bush, Michael Collins, Hugh Dingle, Robert Dowell, Phil DeVries, Gene Drecktrah, Mark Evans, Les Ferge, Doug

Futuyma, Bruce Giebink, Larry Gilbert, Fred Gould, Walt Gould, Eric Grabstein, Robert Hagen, John Hainze, Ellen Heininger, William Houtz, Phil Kingsley, Robert Krebs, John Lawton, Robert Lederhouse, Rick Lindroth, Greg Lintereur, Heidi Luebke, Syafrida Manuwoto, Lee Miller, Jacqueline Miller, Bill Mattson, Jim Maudsley, Mo Neilsen, Ric Peigler, Stephen Peterson, Ted Pike, Austin Platt, Peter Price, John Rawlins, David Robacker, Sarah Rockey, Howard Romack, David Ritland, MaryCarol Rossiter, Lauren Schroeder, Art Shapiro, Oakley Shields, Mike Singer, Frank Slansky, Jr., Kevin Spencer, Jim Sternberg, Maurice Tauber, John Thompson, Gilbert Waldbauer, Bill Warfield, Wayne Wehling, Tom Whitham, David West, Christer Wiklund, and Allen Young. In addition, I wish to thank John Bryant, Peggy Kuropat, Hans Kubler, Polly O'Hare, the Archbold Biological Station Staff, Ed Hasselkus and the University of Wisconsin Arboretum Committee for the generous use of their facilities and trees. We were fortunate to have a number of energetic (and durable) students working on the mass-rearing operations during the last two summers, including, Yvonne Allen Penny Barker, Patrick Christie, Judy Johnson, Kathy Hale, Maret Pajutee, Jeanne Pomraning, Christina Plzak, Ellen Schuh, Jane Sibenhorn, Shawn Sippl, Jeff Thorne, David Ware, and Vicki Viegut.

References

Ackery, P. R. (1975). A guide to the genera and species of parnassiinae (Lepidoptera: Papilionidae). *Bull. Br. Mus. Nat. Hist. Entomol.* **31,** 73–98.

Ae, S. A. (1979). The phylogeny of some *Papilio* species based on interspecific hybridization data. *Syst. Entomol.* **4,** 1–16.

Axelrod, D. I. (1958). Evolution of the Madro-Tertiary geoflora. *Bot. Rev.* **24,** 433–509.

Barnes, B. V. (1975). Phenotypic variation of trembling aspen in western North America. *For. Sci.* **21,** 319–328.

Benson, W. W. (1978). Resource partitioning in passion vine butterflies. *Evolution* **32,** 493–518.

Benson, W. W., Brown, K. S., and Gilbert, L. E. (1975). Coevolution of plants and herbivores: Passion flower butterflies. *Evolution* **29,** 659–680.

Berenbaum, M. (1981). An oviposition "mistake" by *Papilio glaucus* Papilionidae. *J. Lepid. Soc.* **35,** 75.

Berenbaum, M. (1983). Coumarins and caterpillars: A case for coevolution. *Evolution* **37,** 163–179.

Berenbaum, M., and Feeny, P. (1981). Toxicity of angular furonocoumarins to swallowtail butterflies: Escalation in a coevolutionary arms race? *Science* **212,** 927–929.

Berlocher, S. H. (1984). Insect molecular systematics. *Annu. Rev. Entomol.* **29,** 403–433.

Berlocher, S. H., and Bush, G L. (1982). An electrophoretic analysis of Rhagoletis (Diptera: Tephritidae) phylogeny. *Syst. Zool.* **31,** 136–155.

Beutelspacher, C. R. (1975). Una especie nueva de *Papilio* L. (Papilionidae). *Rev. Soc. Mex. Lepid.* **1,** 3–6.

Beutelspacher, C. R., and Howe, W. H. (1984). "Mariposas de Mexico." La Prensa Medica Mexicana, Mexico.

Blau, P. A., Feeny, P. P., Contardo, L., and Robson, D. (1978). Allylglucosinolate and herbivorous caterpillars: A contrast in toxicity and tolerance. *Science* **200,** 1296–1298.

Braun, E. L. (1955). The phytogeography of unglaciated Eastern United States and its interpretation. *Bot. Rev.* **21,** 297–375.

Brower, L. P. (1958a). Larval foodplant specificity in butterflies of the *Papilio glaucus* group. *Lepid. News* **12,** 103–114.

Brower, L. P. (1958b). Bird predation and foodplant specificity in closely related procryptic insects. *Am. Nat.* **92,** 183–187.

Brower, L. P. (1959a). Speciation in butterflies of the *P. glaucus* group. I. Morphological relationships and hybridization. *Evolution* **13,** 40–63.

Brower, L. P. (1959b). Speciation in butterflies of the *Papilio glaucus* group. II. Ecological relationships and interspecific sexual behavior. *Evolution* **13,** 212–228.

Brown, J. H., and Gibson, A. C., eds. (1983). Reconstructing biogeographic histories. *In* "Biogeography," pp. 248–277. Mosby, St. Louis.

Brown, K. S., Damman, A. J., and Feeny, P. P. (1980). Troidine swallowtails (Lepidoptera: Papilionidae) in southeastern Brazil: Natural history and foodplant relationships. *J. Res. Lepid.* **19,** 199–226.

Brues, C. T. (1920). The selection of food-plants by insects, with special reference to lepidopterous larvae. *Am. Nat.* **54,** 313–332.

Bush, G. L., and Diehl, S. R. (1982). Host shifts, genetic models of sympatric speciation and the origin of parasitic insect species. *Proc. Int. Symp. Insect-Plant Relat., 5th, Pudoc, Wageningen* pp. 297–305.

Chemsak, J. A. (1963). Taxonomy and bionomics of the genus *Tetraopes. Univ. Calif. Publ. Entomol.* **30,** 1–90.

Cheverud, J. M., Rutledge, J. J., and Atchley, W. R. (1983). Quantitative genetics of development: Genetic correlations among age-specific trait values and the evolution of ontogeny. *Evolution* **37,** 895–905.

Clark, A. H., and Clark, L. F. (1951). The butterflies of Virginia. *Smithsonian Misc. Collect.* **116,** 1–239.

Clarke, C. A., and Sheppard, P. M. (1955). The breeding in captivity of the hybrid *Papilio rutulus* female × *P. glaucus* male. *Lepid. News* **9,** 46–48.

Clarke, C. A., and Sheppard, P. M. (1962). The genetics of the mimetic butterfly, *Papilio glaucus. Ecology* **43,** 159–167.

Coley, P., Bryant, J. P., and Chapin, S. (1985). Resource availability and plant antiherbivore defense. *Science* **230,** 895–899.

Colinvaux, P. (1981). Historical ecology in Beringia: The south land bridge coast at St. Paul Island. *Quat. Res.* **16,** 18–36.

Collins, M. M. (1984). Genetics and ecology of a hybrid zone in *Hyalophora* (Lepidoptera: Saturniidae). *Univ. Calif. Publ. Entomol.* **104,** 1–93.

Couper, W. (1874). A dissertation on northern butterflies. *Can. Entomol.* **6,** 91–96.

Curtis, J. T. (1959). "The Vegetation of Wisconsin." Univ. of Wisconsin Press, Madison.

Danks, H. V. (1981). The composition, distribution and ecology of arctic insects, with some speculations on the evolution of arctic communities. *Proc. Int. Congr. Syst. Evol. Biol., 2nd* pp. 21–23.

Dethier, V. G. (1941). Chemical factors determining the choice of foodplants by *Papilio* larvae. *Am. Nat.* **75,** 61–73.

Dethier, V. G. (1954). Evolution of feeding preferences in phytophagous insects. *Evolution* **8,** 33–54.

DeVries, P. J. (1986). Host plant records and natural history notes on Costa Rican butterflies (Papilionidae, Pieridae, and Nymphalidae). *J. Res. Lepid.* **24,** 290–333.

DeVries, P. J. (1987). "The Butterflies of Costa Rica and Their Natural History." Princeton Univ. Press, Princeton, New Jersey.

Diehl, S. R., and Bush, G. L. (1984). An evolutionary and applied perspective of insect biotypes. *Annu. Rev. Entomol.* **29,** 471–504.

Dirr, M. A. (1983). "Manual of Woody Landscape Plants: Their Identification, Ornamental Characteristics, Culture, Propagation and Uses," 3rd Ed. Stipes Publ., Chicago.

Dixon, W. J., and Brown, M. B. (1979). "BMDP-79, Biomedical Computer Program, P-series." Univ. of Calif. Press, Berkeley.

D'Urban, W. S. M. (1857). On the order Lepidoptera, with the description of 2 species of Canadian butterflies. *Can. Nat. Geol.* **2,** 223–236.

D'Urban, W. S. M. (1860). Natural history of the river range, and the surrounding counties of Argentenil and Ottawa. *Can. Nat. Geol.* **5,** 87, 243.

Durden, C. J., and Rose, H. (1978). Butterflies from the Middle-Eocene: The earliest fossil occurrence of fossil Papilionidae (Lepidoptera). *Pearce-Sellards Ser. Tex. Meml. Mus.* **29,** 1–25.

Edwards, W. H. (1884). "Butterflies of North America." Houghton Mifflin, Boston.

Edwards, W. H. (1885a). On willow as food-plant of *Papilio turnus*. *Can. Entomol.* **17,** 232.

Edwards, W. H. (1885b). Experiments with larvae as to foodplants. *Can. Entomol.* **17,** 112–113.

Edwards, W. H. (1886). On willow as food-plant of *P. turnus*. *Can. Entomol.* **18,** 139–140.

Ehrlich, P. R., and Raven, P. H. (1964). Butterflies and plants: A study in coevolution. *Evolution* **18,** 586–608.

Endler, J. A. (1977). "Geographic Variation, Speciation, and Clines." Princeton Univ. Press, Princeton, New Jersey.

Endler, J. A. (1982). Problems in distinguishing historical from ecological facts in biogeography. *Am. Zool.* **22,** 441–452.

Farris, J. S. (1981). Distance data in phylogenetic analysis. In Funk, V. A. and D. R. Brooks (eds.), Advances in Cladistics, vol. 1. *Proc. Meet. Willi Hennig Soc., 1st, N.Y. Bot. Garden* pp. 3–23.

Feeny, P. P. (1975). Biochemical coevolution between plants and their insect herbivores. *In* "Coevolution of Animals and Plants" (L. E. Gilbert and P. R. Ravens, eds.), pp. 3–19. University of Texas Press, Austin.

Feeny, P., Rosenberry, L., and Carter, M. (1983). Chemical aspects of oviposition behavior in butterflies. *In* "Herbivorous Insects: Host-Seeking Behavior and Mechanisms" pp. 27–76. Academic Press, New York.

Felsenstein, J. (1982). Numerical methods for inferring evolutionary trees. *Q. Rev. Biol.* **57,** 379–404.

Felsenstein, J. (1984). Distance methods for inferring phylogenies: A justification. *Evolution* **38,** 16–24.

Felsenstein, J. (1985). Phylogenies and the comparative method. *Am. Nat.* **125,** 1–15.

Forbes, W. T. M. (1932). How old are the Lepidoptera? *Am. Nat.* **98,** 452–460.

Forbes, W. T. M. (1951). Footnote on *Papilio. Lepid. News* **5,** 16.

Forbes, W. T. M. (1958). Caterpillars as botanists. *Proc. Int. Congr. Entomol., 10th* **1,** 313–317.

Forbes, W. T. M. (1960). Lepidoptera of New York and neighboring states. Part IV. *Cornell Univ. Agric. Exp. Sta. Mem.* **371,** 1–188.

Fox, L. R. (1981). Defense and dynamics in plant-herbivore systems. *Am. Zool.* **21,** 853–864.

Fox, L. R., and Morrow, R. A. (1981). Specialization: Species property or local phenomenon? *Science* **211,** 887–893.

Frances, A. D., and Elvira, J. M. (1978). Guia ilustrada de las mariposas Mexicanas. Parte I. Familia Papilionidae. Special Publ. Sociedad Mexicana de Lepdiopterologia, A-C.

Futuyma, D. J. (1983). Evolutionary interactions among herbivorous insects and plants. *In* "Coevolution" (D. J. Futuyma and M. Slatkin, eds.), pp. 207–231. Sinauer, Sunderland, Massachusetts.

Futuyma, D. J., and Slatkin, M. (1983). Introduction. *In* "Coevolution" (D. J. Futuyma and M. Slatkin, eds.), pp. 1–13. Sinauer, Sunderland, Massachusetts.

Futuyma, D. J., and Peterson, S. C. (1985). Genetic variation in the use of resources by insects. *Annu. Rev. Entomol.* **30,** 217–238.

Futuyma, D. J., and Wasserman, S. S. (1981). Foodplant specialization and feeding efficiency in the tent caterpillars *Malacosoma disstria* and *M. americanum*. *Entomol. Exp. Appl.* **30,** 106–110.

Gilbert, L. E. (1979). Development of theory in the analysis of insect-plant interactions. *In* "Analysis of Ecological Systems" (D. J. Horn, G. R. Stairs, and R. D. Mitchell, eds.), pp. 117–154. Ohio State Univ. Press, Columbus.

Gilbert, L. E., and Raven, P. R. (1975). "Coevolution of Animals and Plants." Univ. of Texas Press, Austin.

Gleason, H. A., and Cronquist, A. (1963). "Manual of Vascular Plants of Northeastern United States and Adjacent Canada." Willard Grant Press, Boston, Massachusetts.

Gosse, P. H. (1840). "The Canadian Naturalist." Van Voorst, London, Ontario.

Gould, F. (1979). Rapid host range evolution in a population of the phytophagous mite *Tetranychus urticae* Koch. *Evolution* **33,** 791–802.

Gould, F. (1983). Genetics of plant-herbivore systems. *In* "Variable Plants and

Herbivores in Natural and Managed Systems" (R. F. Denno and M. McClure, eds.), pp. 599–654. Academic Press, New York.

Grabstein, E. M., and Scriber, J. M. (1982a). The relationship between restriction of host plant consumption, and post-ingestive utilization of biomass and nitrogen in *Hyalophora cecropia*. *Entomol. Exp. Appl.* **31,** 202–210.

Grabstein, E. M., and Scriber, J. M. (1982b). Host-plant utilization by *Hyalophora cecropia* as affected by prior feeding experience. *Entomol. Exp. Appl.* **32,** 262–268.

Graham, A. (1964). Origin and evolution of the biota of southeastern North America: Evidence from the fossil plant record. *Evolution* **18,** 571–585.

Grossmueller, D. W., and Lederhouse, R. C. (1985). Oviposition site selection: An aid to rapid growth and development in the tiger swallowtail butterfly, *Papilio glaucus*. *Oecologia* **66,** 68–73.

Guttman, S. I., Wood, T. K., and Karlin, A. A. (1981). Genetic differentiation along host plant lines in the sympatric *Echenopa binotata* Say complex (Homoptera: Membracidae). *Evolution* **35,** 205–217.

Hafernik, J. (1982). Phenetics and ecology of hybridization in buckeye butterflies (Lepidoptera: Nymphalidae). *Univ. Calif. Publ. Entomol.* **96,** 1–109.

Hagen, R. H. (1986). The Evolution of Host Plant Use by the Tiger Swallowtail Butterfly, *Papilio glaucus*. Ph.D. thesis, Cornell Univ., Ithaca, New York.

Hagen, R. H., and Lederhouse, R. C. (1985). Polymodal emergence of the tiger swallowtail, *Papilio glaucus* (Lepidoptera: Papilionidae): Source of a false second generation in New York state. *Ecol. Entomol.* **10,** 19–28.

Hagen, R. C., and Scriber, J. M. (1988). Genetic analyses of sex-linked loci in the tiger swallowtail butterfly. *J. Heredity* (submitted).

Hainze, J. H., and Scriber, J. M. (1985). The influence of environmental factors on diapause and voltinism in *Papilio glaucus* subspecies. *Bull. Ecol. Soc. Am.* **66,** 185.

Hancock, D. L. (1983). Classification of the Papilionidae (Lepidoptera): A phylogenetic approach. *Smithersia* **2,** 1–48.

Hare, J. D., and Kennedy, G. G. (1986). Genetic variation in plant-insect associations survival of *Leptinotarsa decemlineata* populations on *Solanum carolinense*. *Evolution* **40,** 1031–1043.

Harrison, R. G. (1983). Barriers to gene exchange between closely related cricket species. I. Laboratory hybridization studies. *Evolution* **37,** 245–251.

Hoffman, R. S. (1981). Different voles for different holes: Environmental restrictions on refugial survival of mammals. *Int. Congr. Syst. Evol. Biol., 2nd* pp. 25–45.

Hopkins, D. M., Smith, P. A., and Matthews, J. V. (1981). Dated wood from Alaska and the Yukon: Implications for forest refugia in Beringia. *Quat. Res.* **15,** 217–249.

Hsiao, T. H. (1982). Geographic variation and host plant adaptation of the Colorado potato beetle. pp. 315–324. *Proc. Int. Symp. Insect-Plant Relat., 5th, Pudoc, Wageningen.*

Igarashi, S. (1984). The classification of the Papilionidae mainly based on the morphology of their immature stages. *Tyo to Ga* **34,** 41–95.

Jackson, J. R., and Pounds, J. A. (1979). Comments on assessing the dedifferentiating effect of gene flow. *Syst. Zool.* **28,** 78–85.

Jaenike, J., and Grimaldi, D. (1983). Genetic variation for host preference within and among populations of *Drosophila tripunctata*. *Evolution* **37,** 1023–1033.

Janzen, D. H. (1980). When is it coevolution? *Evolution* **34,** 611–612.

Jermy, T. (1976). Insect-host-plant relationship: Co-evolution or sequential evolution? *In* "The Host-plant in Relation to Insect Behavior and Reproduction," (T. Jermy, ed.), pp. 109–113. Plenum, New York.

Jermy, T. (1984). Evolution of insect/host plant relationships. *Am. Nat.* **124,** 609–630.

Kimball, C. P. (1965). "The Lepidoptera of Florida: An Annotated Checklist." State of Florida Dept. of Agric. Pub., Gainesville.

Klots, A. B. (1951). "A Field Guide to the Butterflies of North America East of the Great Plains." Houghton Mifflin, Boston.

Lederhouse, R. C., and Scriber, J. M. (1987a). Increased relative frequency of dark morph females in the tiger swallowtail, *Papilio glaucus* (Lepidoptera: Papilionidae) in south central Florida. *Am. Midl. Nat.* **118,** 211–213.

Lederhouse, R. C., and Scriber, J. M. (1987b). The ecological significance of post-mating decline in egg viability in the tiger swallowtail, *Papilio glaucus* L. (Papilionidae). *J. Lepid. Soc.* **41,** 83–93.

Lederhouse, R. C., Finke, M. D., and Scriber, J. M. (1982). The contributions of larval growth and pupal duration in the black swallowtail butterfly, *Papilio polyxenes*. *Oecologia* **53,** 296–300.

Levin, M. P., and Angleberger, M. A. (1972). Observations on foodplant records for *Papilio glaucus* (Papilionidae). *J. Lepid. Soc.* **26,** 177–180.

Lindroth, R. L., and Pajutee, M. S. (1987). Chemical analysis of phenolic glycosides: Art, facts, and artifacts. *Oecologia* **74,** 144–148.

Lindroth, R. L., Scriber, J. M., and Hsia, M. T. S. (1986a). Effects of catechol, salicin, and isoniazid on performance of two subspecies of tiger swallowtails. *Am. Midl. Nat.* (in press).

Lindroth, R. L., Scriber, J. M., and Hsia, M. T. S. (1986b). Differential responses of tiger swallowtail subspecies to secondary metabolites from tulip tree and quaking aspen leaves. *Oecologia* **70,** 13–19.

Lindroth, R. L., Scriber, J. M., and Hsia, M. T. S. (1988). Chemical ecology of the tiger swallowtail: Mediation of host use by phenolic glycosides. *Ecology* (in press).

Lindroth, R. L., Hsia, M. T. S., and Scriber, J. M. (1987a). Characterization of phenolic glycosides from quaking aspen (*Populus tremuloides*). *Biochem. Syst. Ecol.* **15,** 677–680.

Lindroth, R. L., Hsia, M. T. S., and Scriber, J. M. (1987b). Seasonal patterns in the phytochemistry of three *Populus* species. *Biochem. Syst. Ecol.* **15,** 681–687.

Luebke, H. (1985). Differentiation of Hybrids in a Wisconsin Blend Zone: Morphometric Analysis with *Papilio glaucus glaucus* and *P. g. canadensis*. MS thesis, University of Wisconsin, Madison.

Luebke, H., Scriber, J. M., and Yandell, B. S. (1988). Use of multivariate discriminant analysis of male wing morphometrics to delineate a hybrid zone for *Papilio glaucus glaucus* and *P. g. canadensis*. *Am. Midl. Nat.* (in press).

McDonnell, L. J., Gartside, D. F., and Littlejohn, M. J. (1978). Analysis of a narrow hybrid zone between two species of Pseudophryne (Anura: Leptodactylidae) in southeastern Australia. *Evolution* **32,** 602–612.

McGugan, B. M. (1958). "Forest Lepidoptera of Canada. Vol. 1. Papilionidae to Arctiidae." Forest Biology Division: Canad. Dept. Agric. Publ. No. **1034,** 1–76.

Makielski, S. K. (1972). Polymorphism in *Papilio glaucus* L. (Papilionidae): Maintenance of the female ancestral form. *J. Lepid. Soc.* **26,** 109–111.

Mattson, W. J., and Scriber, J. M. (1987). Feeding ecology of insect folivores of woody plants: Nitrogen, water, fiber, and mineral considerations. *In* "The Nutrition Ecology of Insects, Mites, and Spiders" (F. Slansky and J. G. Rodriguez, eds.), pp. 105–146. Wiley, New York.

Miller, J. S. (1986). Phylogenetic Systematics and Chemical Constraints on Host Plant Associations in the Papilionidae (Lepidoptera: Papilionidae). Ph.D. thesis, Cornell Univ., Ithaca, New York.

Miller, J. S., and Feeny, P. (1983). Effects of benzylisoquinoline alkaloids on the larvae of polyphagous Lepidoptera. *Oecologia* **58,** 332–339.

Mitter, C., and Brooks, D. R. (1983). Phylogenetic aspects of coevolution. *In* "Coevolution" (D. J. Futuyma and M. Slatkin, eds.), pp. 65–98. Sinauer, Sunderland, Massachusetts.

Mitter, C., and Futuyma, D. J. (1983). An evolutionary-genetic view of host-plant utilization by insects. *In* "Variable Plants and Herbivores in Natural and Managed Systems" (R. F. Denno and M. S. McClure, eds.), pp. 427–459. Academic Press, New York.

Morgan, A., Morgan, A. V., and Elias, S. A. (1985). Holocene insects and paleoecology of the An Sable River, Michigan. *Ecology* **66,** 1817–1828.

Munroe, E. (1961). The generic classification of the Papilionidae. *Can. Entomol. Suppl.* **17,** 1–51.

Munroe, E., and Ehrlich, P. R. (1960). Harmonization of concepts of higher classification of the Papilionidae. *J. Lepid. Soc.* **14,** 169–175.

Murray, D. F. (1981). The role of arctic refugia in the evolution of the arctic vascularflora—A Beringian perspective. *In Proc. Int. Congr. Syst. Evol. Biol., 2nd* pp. 11–20.

Okano, K. (1983). The revision of classification on the genera of Papilionidae in the world. (Preliminary report) Part I: With description of new genus. *Tokurana* (*Acta Rhaphocerol.*) **5,** 1–75.

Oliver, C. G. (1972). Genetic and phenotypic differentiation and geographic distance in four species of Lepidoptera. *Evolution* **26,** 221–241.

Oliver, C. G. (1979). Genetic differentiation and hybrid viability within and between some Lepidoptera species. *Am. Nat.* **114,** 169–175.

Oliver, C. G. (1983). Disturbance of eclosion sequence in hybrid Lepidoptera. *Can. Entomol.* **115,** 1445–1452.

Opler, P. A., and Krizek, G. O. (1984). "Butterflies East of the Great Plains." Johns Hopkins Univ. Press, Baltimore.

Peterson, G. M., Webb, T., Kutzbach, J. E., vander Hammen, T., Wijmstra, T. A., and Street, F. A. (1979). The continental record of environmental conditions at 18,000 yr V.P.: An initial evaluation. *Quat. Res.* **12,** 47–82.

Pielou, E. C. (1979). "Biogeography." Wiley, New York.

Platnick, N. I., and Nelson, G. (1978). A method of analysis for historical biogeography. *Syst. Zool.* **27,** 1–16.

Platt, A. P., Harrison, S. J., and Williams, T. F. (1984). Absence of differential mate selection in the North American tiger swallowtail, *Papilio glaucus. In* "The Biology of Butterflies" (R. I. Vane-Wright and P. R. Ackery, eds.), pp. 245–250. Academic Press, London.

Preston, R. J. (1976). "North American Trees." Iowa State Univ. Press, Ames.

Rausher, M. D. (1981). Host plant selection by *Battus philenor* butterflies: The roles of predation, nutrition, and plant chemistry. *Ecol. Monogr.* **51,** 1–20.

Rausher, M. D. (1982). Population differentiation in *Euphydryas editha* butterflies: Larval adaptation to different hosts. *Evolution* **36,** 581–590.

Rausher, M. D. (1983). Ecology of host-selection behavior in phytophagous insects. *In* "Variable Plants and Herbivores in Natural and Managed Systems" (R. F. Denno and M. S. McClure, eds.), pp. 223–257. Academic Press, New York.

Rausher, M. D. (1984). Tradeoffs in performance on different hosts: Evidence from within- and between-site variation in the beetle, *Deloyala guttata. Evolution* **38,** 582–595.

Rausher, M. D. (1985). Variability for host preference in insect populations: Mechanistic and evolutionary models. *J. Insect Physiol.* **31,** 873–889.

Remington, C. L. (1951). Geographic subspeciation in the Lepidoptera. I. Introduction. A general outline of subspeciation. *Lepid. News* **5,** 17–20.

Remington, C. L. (1968). Suture-zones of hybrid interaction between recently joined biotas. *Evol. Biol.* **2,** 321–428.

Reyment, R. A., Blackick, R. E., and Campbell, N. A. (1984). "Multivariate Morphometrics." Academic Press, New York.

Richard, D., and Guedes, M. (1983). The Papilionidae (Lepidoptera): Coevolution with the angiosperms. *Phyton* **23,** 117–126.

Riska, B., Atchely, W. R., and Rutledge, J. J. (1984). A genetic analysis of targeted growth in mice. *Genetics* **107,** 79–101.

Ritland, D. B., and Scriber, J. M. (1985). Larval developmental rates of three putative subspecies of tiger swallowtail butterflies, *Papilio glaucus,* and their hybrids in relation to temperature. *Oecologia* **65,** 185–193.

Rockey, S. J., Hainze, J. H., and Scriber, J. M. (1987a). Evidence of sex-linked diapause response in *Papilio glaucus* subspecies and their hybrids. *Physiol. Ecol.* **12,** 181–184.

Rockey, S. J., Hainze, J. H., and Scriber, J. M. (1987b). Diapause in three subspecies of the Eastern tiger swallowtail, *Papilio glaucus* (Lepidoptera: Papilionidae): Evidence of a latitudinal and obligatory diapause. *Am. Midl. Nat.* **118,** 162–168.

Rothschild, W., and Jordan, K. (1906). A revision of the American papilios. *Novitates Zool.* **13,** 411–744.

Saunders, W. (1874). On some of our common insects II the tiger swallowtail, *P. turnus. Can. Entomol.* **6,** 2–5.

Saunders, W. (1883). Miscellaneous. *Can. Entomol.* **17,** 204.

Schoonhoven, L. M., and Meerman, J. (1978). Metabolic cost of changes in diet and neutralization of allelochemics. *Entomol. Exp. Appl.* **24,** 689–693.

Schroeder, L. A. (1976). Energy, matter and nitrogen utilization by the larvae of the monarch butterfly *Danaus plexippus*. *Oikos* **27,** 254–264.

Schroeder, L. A. (1977). Energy, matter and nitrogen utilization by larvae of the milkweed tigermoth, *Euchaetias egle*. *Oikos* **28,** 27–31.

Scott, J. A. (1986). "The Butterflies of North America. A Natural History and Field Guide." Stanford Univ. Press, Stanford, California.

Scott, J. A., and Shepard, J. H. (1976). Simple and computerized discriminant functions for difficult identifications: A rapid nonparametric method. *Pan Pacif. Entomol.* **52,** 23–28.

Scriber, J. M. (1972). Confirmation of a disputed foodplant of *Papilio glaucus* (Papilionidae). *J. Lepid. Soc.* **26,** 235–236.

Scriber, J. M. (1973). Latitudinal gradients in larval feeding specialization of the world Papilionidae (Lepidoptera). *Psyche* **80,** 355–373.

Scriber, J. M. (1975). Comparative Nutritional Ecology of Herbivorous Insects: Generalized and Specialized Feeding Strategies in the Papilionidae and Saturniidae (Lepidoptera). Ph.D. thesis, Cornell Univ., Ithaca, New York.

Scriber, J. M. (1977). Limiting effects of low leaf-water content on the nitrogen utilization, energy budget, and larval growth of *Hyalophora cecropia* (Lepidoptera: Saturniidae). *Oecologia* **28,** 269–287.

Scriber, J. M. (1979a). The effects of sequentially switching foodplants upon the biomass and nitrogen utilization by polyphagous and stenophagous *Papilio* larvae. *Entomol. Exp. Appl.* **25,** 203–215.

Scriber, J. M. (1979b). Post-ingestive utilization of plant biomass and nitrogen by Lepidoptera: Legume feeding by the southern armyworm. *J. N.Y. Entomol. Soc.* **87,** 141–153.

Scriber, J. M. (1981). Sequential diets, metabolic costs, and growth of *Spodoptera eridania* (Lepidoptera: Noctuidae) feeding upon dill, lima bean, and cabbage. *Oecologia* **51,** 175–180.

Scriber, J. M. (1982a). The behavior and nutritional physiology of southern armyworm larvae as a function of plant species consumed in earlier instars. *Entomol. Exp. Appl.* **31,** 359–369.

Scriber, J. M. (1982b). Foodplants and speciation in the *Papilio glaucus* group. *Proc. Int. Symp. Insect-Plant Relat., 5th, Pudoc, Wageningen* pp. 307–314.

Scriber, J. M. (1983). The evolution of feeding specialization, physiological efficiency, and host races in selected Papilionidae and Saturniidae. *In* "Variable Plants and Herbivores in Natural and Managed Systems." (R. F. Denno and M. S. McClure, eds.), pp. 373–412. Academic Press, New York.

Scriber, J. M. (1984a). Host plant suitability. *In* "Chemical Ecology of Insects" (W. J. Bell and R. T. Carde, eds.), pp. 159–202. Sinauer, Sunderland, Massachusetts.

Scriber, J. M. (1984b). Larval foodplant utilization by the world Papilionidae (Lep.): Latitudinal gradients reappraised. *Tokurana* (*Acta Rhopalocerol.*) **6/7,** 1–50.

Scriber, J. M. (1984c). Nitrogen nutrition of plants and insect invasion. Ch. 34. *In* "Nitrogen in Crop Production" (R. D. Hauck, ed.), pp. 441–460. Amer. Soc. Agron. C.S.S.A. and S.S.S.A. Madison, Wisconsin.

Scriber, J. M. (1985). The ecological and genetic factors determining geographic limits to the dark morph polymorphism in *Papilio glaucus*. *Bull. Ecol. Soc. Am.* **66,** 267.

Scriber, J. M. (1986a). Allelochemicals and alimentary ecology: Heterosis in a hybrid zone? *In* "Molecular Mechanisms in Insect Plant Interactions" (L. B. Brattsten and S. Ahmad, eds.), Ch. 2, pp. 43–71. Plenum, New York.

Scriber, J. M. (1986b). Origins of the regional feeding abilities in the tiger swallowtail butterfly: Ecological monophagy and the *Papilio glaucus australis* subspecies in Florida. *Oecologia* **71,** 94–103.

Scriber, J. M. (1986c). Population genetics and foodplant use among the North American tree-feeding Papilionidae. *Proc. Int. Symp. Insect/Plant Relat., 6th,* pp. 221–230. W. Junk, Dordrect, Netherlands.

Scriber, J. M., and Evans, M. H. (1986a). An exceptional case of paternal transmission of the dark form female trait in the tiger swallowtail butterfly, *Papilio glaucus* (Lepidoptera: Papilionidae). *J. Res. Lepid.* **25,** 110–120.

Scriber, J. M., and Evans, M. H. (1986c). Bilateral gynandromorphs, and sexual/color mosaics in the tiger swallowtail butterfly, *Papilio glaucus* (Papilionidae: Lepidoptera). *J. Res. Lepid.* (in press).

Scriber, J. M., and Feeny, P. P. (1979). The growth of herbivorous caterpillars in relation to degree of feeding specialization and to growth form of foodplant. *Ecology* **60,** 829–850.

Scriber, J. M., and Hainze, J. H. (1987). Geographic invasion and abundance as facilitated by differential host plant utilization abilities. *In* "Insect Outbreaks" (P. Barbosa and J. C. Schultz, eds.), pp. 433–468. Academic Press, New York.

Scriber, J. M., and Lederhouse, R. C. (1983). Temperature as a factor in the development and feeding ecology of tiger swallowtail caterpillars. *Oikos* **40,** 95–102.

Scriber, J. M., and Lederhouse, R. C. (1987). Hand-pairing of *Papilio glaucus* and *Papilio pilumnus* and hybrid survival on various host plants. *J. Lepid. Soc.* (in press).

Scriber, J. M., and Slansky, F., Jr. (1981). The nutritional ecology of immature insects. *Annu. Rev. Entomol.* **26,** 183–211.

Scriber, J. M., Lederhouse, R. C., and Contardo, L. (1975). Spicebush, *Lindera benzoin,* a little known foodplant of *Papilio glaucus* (Papilionidae). *J. Lepid. Soc.* **29,** 10–14.

Scriber, J. M., Lintereur, G. L., and Evans, M. H. (1982). Foodplant suitabilities and a new oviposition record for *Papilio glaucus canadensis* (Lepidoptera: Papilionidae) in northern Wisconsin and Michigan. *Great Lakes Entomol.* **15,** 39–46.

Scriber, J. M., Hsia, M. T., Lindroth, R., and Sunarjo, P. (1987). Allelochemicals as determinants of insect damage across the North American continent: Biotypes and biogeography. *Proc. ACS, Washington, D.C.* No. 330, pp. 439–448.

Scriber, J. M., Evans, M. H., and Ritland, D. B. (1987a). Hybridization as a causal mechanism of mixed color broods and unusual color morphs in the eastern tiger swallowtail, *Papilio glaucus*. *In* "Evolutionary Genetics of Invertebrate Behavior" (M. Huettel, ed.), pp. 119–134. Univ. Florida Press, Gainesville.

Scriber, J. M., Evans, M. H., and Lederhouse, R. C. (1988a). Hybridization of the Mexican Tiger Swallowtail, *Papilio alexiares garcia* with other *Papilio glaucus* group species and the survival of pure and hybrid neonate larvae on potential host plants. *J. Research Lepid.* (submitted).

Scudder, S. H. (1889). "The Butterflies of the eastern United States and Canada," Vol. II. Published by the author, Cambridge, Massachusetts. (see also *Psyche* **5,** 274–278).

Service, P. M., and Lenski, R. E. (1982). Aphid genotypes, plant phenotypes, and genetic diversity: A demographic analysis of experimental data. *Evolution* **36,** 1276–1282.

Shapiro, A. M. (1963). The butterflies of the Morris Arboretum. *Morris Arbor. Bull.* **14,** 32–36.

Shapiro, A. M. (1974). Butterflies and skippers of New York state. *Cornell Univ. Agric. Exp. Sta. Search* **4,** 1–60.

Shields, O., and Dvorak, S. K. (1984). Butterfly distribution and continental drift between the Americas, the Caribbean and Africa. *Jul. Nat. Hist.* **13,** 221–250.

Shields, O., Emmel, J. F., and Breedlove, D. E. (1970). Butterfly larval foodplant records and a procedure for reporting foodplants. *J. Res. Lepid.* **8,** 21–36.

Singer, M. C. (1983). Determinants of multiple host use by a phytophagous insect population. *Evolution* **37,** 389–403.

Slansky, F., and Scriber, J. M. (1985). Food consumption and utilization. Ch. 3. *In* "Comprehensive Insect Physiology, Biochemistry and Pharmacology" (G. A. Kerkut and L. I. Gilbert, eds.), Vol. 4, pp. 87–163. Pergamon, Oxford.

Smiley, J. (1978). Plant chemistry and the evolution of host specificity: New evidence from *Heliconius* and *Passiflora*. *Science* **201,** 745–747.

Sperling, F. A. H. (1986). Evolution of the *Papilio machaon* Species Group in Western Canada. M.S. thesis, Univ. of Edmonton, Alberta.

Strong, D. R., Lawton, J. H., and Sir Southwood, R. (1984). "Insects on Plants; Community Patterns and Mechanisms." Harvard Univ. Press, Cambridge, Massachusetts.

Tabashnik, B. (1983). Host range evolution: The shift from native legume hosts to alfalfa by the butterfly *Colias philodice eriphyle*. *Evolution* **37,** 150–162.

Thompson, J. N. (1982). "Interaction and Coevolution." Wiley, New York.

Thompson, J. M. (1986). Patterns in coevolution. *In* "Coevolution and Systematics" (A. R. Stone and D. I. Hawksworth, eds.), pp. 119–143. Clarendon Press, Oxford.

Thorpe, R. S. (1984). Primary and secondary transition zones in speciation and population differentiation: a phylogenetic analysis of range expansion. *Evolution* **38,** 233–243.

Tietz, H. M. (1972). "Index to the Described Life Histories, Early States and Hosts of the Macrolepidoptera of the Continental United States and Canada." Allyn, Sarasota, Florida.

Tyler, H. (1975). "The Swallowtail Butterflies of North America." Naturegraph, Healdsburg, California.

Via, S. (1984). The quantitative genetics of polyphagy in an insect herbivore. II.

Genetic correlations in larval performance within and among host plants. *Evolution* **38,** 896–905.
Via, S., and Lande, R. (1985). Genotype-environment interaction and the evolution of phenotypic plasticity. *Evolution* **39,** 505–522.
Watts, W. A. (1979). Late quaternary vegetation of central Appalachia and the New Jersey coastal plain. *Ecol. Monogr.* **49,** 427–469.
Watts, W. A. (1980). The late quaternary vegetation history of the southeastern United States. *Annu. Rev. Ecol. Syst.* **11,** 387–409.
Wiklund, C. (1975). The evolutionary relationship between oviposition preferences and larval host range in *Papilio machaon* L. *Oecologia* **18,** 185–197.
Wiley, E. O. (1981). "Phylogenetics: The Theory and Practice of Phylogenetic Systematics." Wiley, New York.
Wolfe, J. A. (1978). A paleobotanical interpretation of tertiary climates in the northern hemisphere. *Am. Sci.* **66,** 694–703.
Wood, T. K. (1980). Divergence in the *Enchenopa binotata* Say complex (Homoptera: Membracidae) effected by host plant adaptation. *Evolution* **34,** 147–160.
Worth, C. B., Platt, A. P., and Williams, T. F. (1982). Differential growth and utilization of three foodplants by first instar larvae of *Citheronia regalis* (Saturniidae). *J. Lepid. Soc.* **36,** 76–82.
Yamada, Y. (1962). Genotype × environment interaction and genetic correlation of the same trait under different environments. *Jpn. J. Genet.* **37,** 498–509.

CHAPTER 9

Comparative Mechanisms of Host Selection by Insects Attacking Pine Trees and Crucifers

J. A. A. RENWICK
Boyce Thompson Institute, Ithaca, New York

THE RELATIONSHIP BETWEEN INSECTS and their host plants is often considered to be a result of continuous adaptation and counteradaptation throughout evolutionary time. This dynamic relationship has sometimes been referred to as an arms race, where the attackers and defenders are continuously producing new weapons and developing new strategies in their fight for survival. From a practical point of view, we are interested in finding ways to alter the balance of power in such confrontations to favor the plants that we need, or to favor the insects that destroy unwanted plants. It is useful, therefore, to consider the likely outcome of natural battles involving widely differing participants.

When we compare two plant–insect systems in which the host plants are very large and very small, many differences are quite obvious. The relationships between bark beetles attacking pine trees and caterpillars feeding on crucifers provide us with two extreme cases. The greatest difference from an evolutionary perspective is in the generation time of the host plants. Many pine trees require 25 years for reproduction, whereas most crucifers are annuals. Some bark beetles, such as the southern pine beetle, *Dendroctonus frontalis* Zimm., can complete seven generations in a year, whereas the imported cabbageworm, *Pieris rapae* L., is restricted to two or three generations per year. Thus, despite their size, pine trees are likely to be at a distinct disadvantage in an arms race with bark beetles, while crucifers appear to have a much better chance of adopting new defenses against their insect attackers. Closer examination of these two plant–insect systems might reveal whether or not this deduction is reasonable.

I. Pine Tree–Bark Beetle Relationships

Pine trees are protected against insects and fungi by a system of resin canals in the cortex and secondary xylem of the stem as well as in the leaf mesophyll. When a tree is wounded, resin is exuded to repel potential invaders, physically and/or chemically.

A. Adaptation to Physical Defense

The physical defense alone is often enough to abort an attack by bark beetles. The beetles are essentially flushed out by the flow of resin, and appear to have little chance of success when they encounter a very vigorous tree. Both flow capacity and resin pressure are involved. Oleoresin exudation pressure is a property that can be measured by hydrostatic gauges inserted into the tree xylem and can actually provide an estimate of susceptibility of individual trees to certain bark beetles (Vité and Wood, 1961). The success of a bark beetle depends on overcoming the physical barrier of resin flow. After initial penetration of the outer bark, the beetles must excavate a pattern of galleries in the phloem. Eggs are laid along the sides of these galleries, and the survival of emerging larvae depends on the absence of significant quantities of resin. Most bark beetles have developed two evolutionary means of ensuring that this is the case. First, they attack target trees *en masse* to reduce the overall resin pressure, and second, they introduce a blue stain fungus that plays a role in killing the trees, probably by interrupting water conductance.

Different bark beetle species differ considerably in their tolerance for resin. *Ips* species are relatively intolerant and have developed the ability to find trees that have been weakened in some way. Pines that have been recently felled, windblown, stressed, or attacked by other beetles are usually chosen for colonization by *Ips*. Such a colonization is usually characterized by frass production during the early stages of attack, which indicates that feeding and gallery construction are successful. However, beetles in the genus *Dendroctonus* are quite tolerant of resin and have developed an aggressive strategy of mass attack to overpower their host trees. The early stage of these attacks is characterized by the appearance of pitch tubes, which are the result of persistent penetration of the cortical tissue by invading beetles and subsequent crystallization of the exuding resin. Individual beetles can spend several days during this stage fighting the flow of resin to become established in the phloem. One group of species, the turpentine beetles (e.g., *D. valens* LeConte and *D. terebrans* Olivier), appears to be particularly well adapted to resin. These species attack pine trees close to the base of the trunk, where they must cope with high resin pressures for long periods of time. Morphological adaptation of *Dendroctonus* species includes the use of an air space beneath the elytra to carry their own air supply while "swimming" in resin. The beetles later appear to use the gallery to scrape off the resin coating when the flow of resin ceases (P. R. Hughes, personal communication).

Adaptation of bark beetles to the physical resistance provided by resin has been accompanied by utilization of resin components by the beetles in their orientation to host trees. Many species are attracted to the volatile terpenes present in pine resin. The turpentine beetles appear to find their hosts primarily by means of terpenes, whereas other species of *Dendroctonus* rely on a combination of pheromones and monoterpenes to concentrate their attack on specific trees (Renwick and Vité, 1970). Most *Ips* species avoid high concentrations of terpenes and rely primarily on their aggregation pheromones for locating susceptible hosts. However, resin components may enhance the attraction of some *Ips* species to sources of pheromone (Billings, 1985).

Trees are often attacked by a complex of bark beetle species in a sequence that reflects species tolerance for resin. The southern pine beetle, *D. frontalis*, attacks the mid to lower bole of apparently healthy trees. Apparently at the same time, black turpentine beetles, *D. terebrans*, often invade the base of the trunk. After initial weakening of the tree, the engraver beetle, *Ips calligraphus* Germ., colonizes the midsection of the bole. As the physiological condition of the tree further deteriorates, the less aggressive *I. grandicollis* Eichh. colonizes the upper bole. Meanwhile, the top of the tree is attacked by *I. avulsus* Eichh., which survives well in the branches, where resin flow is minimal.

B. Adaptation to Chemical Defense

Bark beetles have obviously been quite successful in adapting to the physical resistance of pine resin. But many components of resin are also toxic to insects, including bark beetles. Contact with, or exposure to, monoterpenes can be fatal to most bark beetles in the laboratory (Smith, 1965). Under field conditions, during the first stages of attack, bark beetles are exposed to high concentrations of terpene vapors for varying degrees of time. How do these beetles cope with such potentially lethal toxins? The answer to this question came from our search for the aggregation pheromones responsible for the mass attack of standing trees. The pheromones were found to be present in hindguts of attacking beetles, and the first compounds to be identified from *D. frontalis* and *D. brevicomis* LeConte were *trans*-verbenol from females and verbenone from males (Renwick, 1967). Both of these compounds are oxidation products of the ubiquitous terpene α-pinene. *trans*-Verbenol appears to play a role in the aggregation pheromone systems of several bark beetles and is the major pheromone involved in aggregation of the mountain pine beetle, *D. ponderosae* Hopk. (Pitman and Vité, 1969). Other compounds that could be considered oxidation products of terpenes were identified from frass of *I. paraconfusus* Lanier (formerly *confusus*) and shown to be pheromones for this species (Silverstein *et al.*, 1966). *cis*-Verbenol can be derived from α-pinene, ipsdienol from myrcene, and ipsenol appears to be a reduction product of ipsdienol. These terpene alcohols are now known to be key chemicals in the aggregation pheromone systems of a large number of bark beetles, especially within the genus *Ips* (Vité and Francke, 1985).

Although the relationship of these pheromones to terpene hydrocarbons present in pine resin is quite apparent, the first definitive proof that bark beetles could perform the necessary oxidation came from experiments performed by Hughes (1973, 1974). Recently emerged adult beetles were exposed to terpene hydrocarbon vapors for a period of 12 h, and the oxidation products were then detected in the beetles' hindguts. Exposure of emergent *D. ponderosae* females to α-pinene results in the production of copious amounts of *trans*-verbenol, which is in fact the aggregation pheromone of this species. Both sexes of other *Dendroctonus* species are capable of detoxifying α-pinene and other terpenes in a similar manner (Renwick *et al.*, 1976b). Some beetles have oxidation products of terpenes in their hindguts as soon as they emerge from their pupal chambers in the bark. These appear to be derived from resin components encountered by larvae and pupae (Hughes, 1975). But when they begin to attack a new host, new compounds appear as soon as the mature adults encounter fresh resin. Emergent males of *D. frontalis* already have large quantities of verbenone and small amounts of myrtenal and myrtenol in their hindguts. However, when exposed to α-pinene vapor, they produce *cis*- and *trans*-verbenol, and β-pinene is oxidized to the corresponding pinocarvone and *trans*-pinocarveol (Renwick *et al.*, 1973). Other sex-specific oxidations by *Dendroctonus* have also been demonstrated. Exposure of *D. brevicomis* males and females to myrcene results in the production of myrcenol by both sexes (Fig. 1). But only the males produce ipsdienol, which is an important pheromone of many *Ips* species (Renwick *et al.*, 1976b).

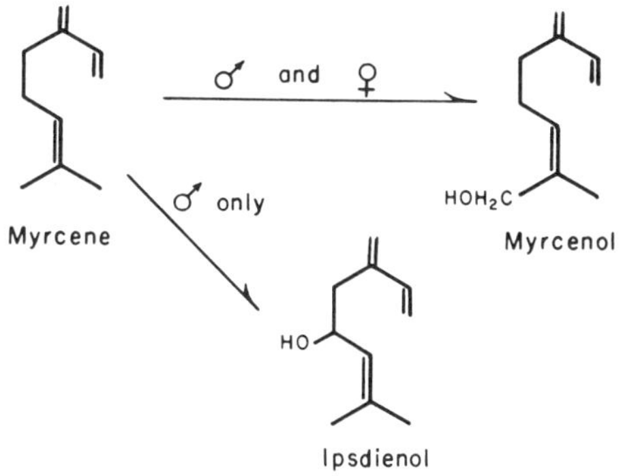

Fig. 1. Oxidation of myrcene by males and females of *Dendroctonus brevicomis*.

Dendroctonus bark beetles clearly have developed the ability to detoxify terpene components of pine resin by oxidizing these hydrocarbons to the corresponding alcohols and ketones. Some of these products have been utilized as pheromones that serve to recruit the large number of beetles needed to overcome the physical resistance of a tree. On the other hand, *Ips* species cannot colonize a tree unless the resin flow is minimal and the invading male beetles are capable of feeding in the phloem. These beetles have therefore developed a safety mechanism to ensure that these conditions are met. The process of detoxification of terpenes by oxidation to alcohols is carefully controlled so that the complete pheromone complex is not released until feeding is successfully initiated. The pheromones ipsdienol and ipsenol are produced by oxidation of myrcene. Yet newly emerged males are not capable of performing this oxidation until their guts are filled. Gut stretching apparently removes the neural inhibition of juvenile hormone release, which is necessary for activation of brain hormone, which in turn seems to be responsible for switching on the oxidation mechanism (Hughes and Renwick, 1977). Some species have also developed highly specialized oxidation systems to deal with different terpenes. Stereoselectivity has been found in *Ips paraconfusus* in its hydroxylation of α-pinene. When presented with pure optical isomers of α-pinene, males of *I. paraconfusus* produce *trans*-verbenol from the (+)-enantiomer and *cis*-verbenol from the (−)-enantiomer (Renwick *et al.*, 1976a).

The oxidation of some terpenes by microorganisms is well known in the fermentation industry. The possibility of microflora involvement within the hindguts of bark beetles has been suggested (Brand *et al.*, 1975), and this has led to a widespread misconception that gut microorganisms are entirely responsible for the necessary terpene conversion to pheromone. However, the amount of bacteria in the hindgut of a beetle does not appear to be sufficient for the immediate conversion of terpenes to large

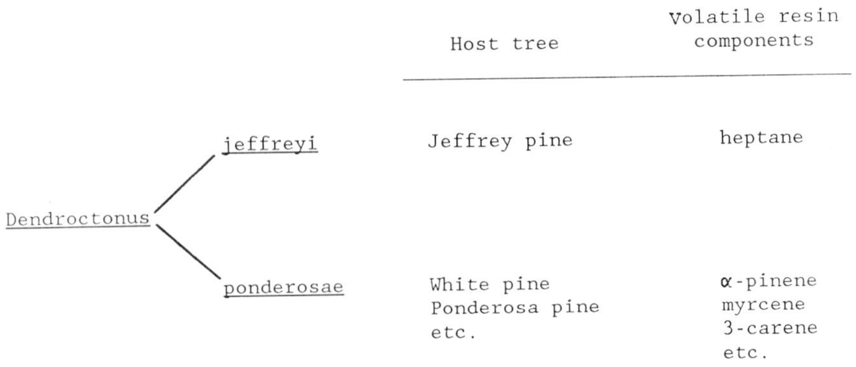

Fig. 2. Postulated divergence of *D. ponderosae* and *D. jeffreyi*.

quantities of the corresponding alcohols. Also, no sex-specific microorganisms have been demonstrated to explain the different mechanisms of males and females, and hormonal control of microbial activity is extremely unlikely. The fact that terpene oxidation can be accomplished in the absence of microorganisms has been clearly demonstrated by the production of pheromones by axenically reared *D. ponderosae* and *I. paraconfusus* beetles (Conn *et al.*, 1984).

The ability of bark beetles to convert hydrocarbons to alcohols appears to be a general phenomenon. Even nonterpenes such as cyclohexene are oxidized to the corresponding allylic alcohols (Renwick and Hughes, 1975). Thus beetles are well equipped to detoxify a wide array of possible terpenes and other hydrocarbons. But what if the precursor of a pheromone is missing from a tree? The adaptation of bark beetles to such a situation can be seen in the case of the Jeffrey pine beetle, *D. jeffreyi* Hopk. This beetle is closely related to and was formerly synonymous with the mountain pine beetle, *D. ponderosae*. However, these beetles have diverged and become adapted to quite different host trees (Fig. 2). Jeffrey pine resin contains very little α-pinene, which is necessary for production of the mountain pine beetle pheromone. Instead, the most abundant volatile constituent of this resin is heptane. Hindguts of attacking Jeffrey pine beetles contain large quantities of 1-heptanol, and this compound appears to be the aggregation pheromone of the species (Renwick and Pitman, 1979).

We might conclude that the evolutionary battle between pine trees and bark beetles appears to be quite one-sided. Despite an extensive arsenal of terpenes which the trees might manipulate to combat their invaders, the bark beetles have a remarkable ability to deal with most of these chemicals. The dependence of bark beetles on specific precursors for their pheromones might be expected to make them vulnerable to changes in resin composition. But their general oxidation mechanism provides them with an array of products that might be brought into use as pheromones.

II. Crucifer–Insect Relationships

Members of the family Cruciferae appear to rely more on chemical than physical defense to protect themselves from invading organisms. The primary chemical defense is generally considered to be provided by glucosinolates (mustard oil glucosides). These compounds give rise to volatile hydrolysis products (mustard oils) by the action of myrosinases when plant tissues are damaged (Feeny, 1977). Both glucosinolates and their breakdown products may have antibiotic properties and inhibit the growth of insects (Virtanen, 1965; Lichtenstein *et al.*, 1964). The best known of these glycosides is sinigrin, which is prominent in mustard, and which releases allylisothiocyanate upon hydrolysis. Sinigrin is quite toxic to larvae of the black swallowtail butterfly, which normally feed on umbellifers; however, inhibition of feeding by a generalist insect such as the southern armyworm, *Spodoptera eridania* Cramer, requires higher concentrations (Blau *et al.*, 1978).

Despite the presence of glucosinolates, crucifers are attacked by a wide array of insects. Many generalist insects appear to be quite tolerant of low concentrations, and feeding by crucifer specialists may actually be stimulated by the glucosinolates. Several of these specialist insects are also attracted to their host plants by the volatile mustard oils. These include the cabbage fly, *Delia radicum* (L.), the vegetable weevil, *Listroderes obliquus* (Klug), the flea beetles, *Phyllotreta cruciferae* (Goeze) and *P. striolata* (Fabr.), the diamondback moth, *Plutella xylostella* (L.), and larvae of the imported cabbageworm, *Pieris rapae* (Schoonhoven, 1972).

Because of its economic importance, *P. rapae* is probably one of the most extensively studied of the crucifer specialists. The host range of this insect is remarkably well correlated with the presence of glucosinolates and includes plants within the Capparaceae, Resedaceae, and Tropaeolaceae (Verschaffelt, 1911). However, despite this correlation, the chemical factors involved in host selection by *P. rapae* butterflies are not clear. Also, not all crucifers are acceptable to *P. rapae*. The presence of secondary defense chemicals other than glucosinolates is thought to be responsible for avoidance of these plants, but little is known about the mechanisms involved (Feeny, 1977).

Our recent research has focused on the host selection behavior of *P. rapae* and the chemical factors that determine whether a plant will be accepted or rejected. Since the early-instar larvae of this insect are relatively immobile, the choice of initial food plant is the responsibility of the ovipositing female butterfly. Oviposition involves orientation toward the potential host, tarsal contact, and acceptance based on the presence of the right chemical stimuli. The major force in the orientation phase appears to be visual. When offered choices between different colored cards, butterflies show a distinct preference for landing on those having a maximum reflectance at 550 nm. The spectrum obtained from the preferred colored card closely resembles that of a cabbage leaf. Based on these observations, a bioassay for the chemical oviposition stimulant has been developed using green index cards that can be painted with test extracts of plants. When the stimulant is present, gravid females readily lay eggs on these cards. For bioassays, caged butterflies are offered a choice of a test card treated with extract, or control cards painted with solvent alone. In the absence of stimulant, no eggs are laid, and when extracts are active, all the eggs are laid on the test card (Renwick and Radke, 1983).

Although the orientation of *P. rapae* butterflies to their host plants appears to be primarily based on color, the effect of volatiles has also been suggested. In order to test this notion, bioassays were conducted using cards coated with stimulatory cabbage extract placed over cups containing homogenized cabbage tissue. Holes were punched in the cards to allow the volatiles to escape. Control cards were placed over cups containing water. The results indicated no preference of butterflies for the cards with volatiles. In fact, there appeared to be a slight deterrent effect produced by the volatiles (Renwick and Radke, 1983). Thus we have concluded that visual and contact stimuli are the major factors regulating orientation to and acceptance of host plants by this insect.

The role of sinigrin in host acceptance by some crucifer-feeding insects has been clearly indicated. The large cabbage butterfly, *P. brassicae* L., oviposits in response to sinigrin taken up by bean plants (Ma and Schoonhoven, 1973). The strong correlation between the host range of *P. rapae* and the presence of glucosinolates might suggest that glucosinolates also stimulate oviposition by this species. This suggestion has led to a false assumption by many researchers in the field. Experiments in our laboratory have shown that sinigrin does not elicit the precise ovipositional response that is obtained with extracts of host plants (Renwick and Radke, 1983). Furthermore, fractionation of host extracts has led to the isolation of active fractions that do not contain glucosinolates. However, the effect of sinigrin on the behavior of these butterflies is interesting. If a group of females is provided with a choice of cabbage extract or a blank card, all the eggs are laid on the extract card. When the same butterflies are offered a card coated with sinigrin on the next day, they generally withhold their eggs. When provided with cabbage extract on the third day, they oviposit again with perfect discrimination between stimulant and control. Occasionally this pattern is broken and sinigrin will stimulate oviposition, but some eggs are also laid on the blank card. On those occasions, when cabbage extract is introduced on the third day, the butterflies no longer have the ability to discriminate. Eggs are consistently laid on blank cards as well as the cards treated with cabbage extract. These results would suggest that some butterflies are stimulated to some extent by sinigrin, but this stimulation leads to confusion.

The sporadic effect of sinigrin on groups of *P. rapae* butterflies indicates that individual butterflies could differ in their response to this compound. When 19 individual butterflies were tested for their reaction to sinigrin as described above (Table I), 47% of these ignored the sinigrin. Another 32% laid some eggs on sinigrin cards and some eggs on control cards. When offered a choice of sinigrin and cabbage extract, these distinctly preferred the cabbage extract. The remaining 21% responded strongly to sinigrin, and also laid a large number of eggs on control cards. When presented with a choice, these individuals showed only a slight preference for cabbage extract over sinigrin. So it appears that there is considerable variation in the behavioral response of individual butterflies to a representative glucosinolate, and those individuals that are stimulated lose their ability to discriminate in their choice of oviposition sites.

Efforts to identify the actual oviposition stimulants of *P. rapae* in cabbage have resulted in the development of a procedure to isolate the active materials. The final step involves high-pressure liquid chromatography (HPLC), and more than one active fraction has been isolated. When water extracts of other host plants are taken through the same isolation scheme, the most active material is found in different fractions of the different plants. Chromatograms have been divided into three major fractions: A, B, and C. The most stimulatory fraction from cabbage is C, whereas the B fraction from mustard and A from nasturtium, respectively, show the greatest activity. It appears likely, therefore, that any one of several compounds present in host plants may stimulate oviposition by *P. rapae*.

Many crucifers are avoided by the usual crucifer-adapted insects, including *P. rapae*. A secondary line of defense chemicals is believed to be responsible for discouraging oviposition and/or feeding. Some of these

Table I. Oviposition Response of Individual *Pieris rapae* Butterflies to a Choice of Cards Treated with Sinigrin Solution or Water Alone (Controls)

Response	Number of Females
Ignore sinigrin (withhold eggs)	9
Oviposit on sinigrin and control. Some discrimination.	6
Strong response to sinigrin. No discrimination between sinigrin and control (confusion).	4

plants are known to contain cardenolides, alkaloids and cucurbitacins (Feeny, 1977). These compounds have been shown to deter some specialist insects (Nielsen, 1978), but are not toxic to *P. rapae* (Usher and Feeny, 1983). *Pieris rapae* butterflies readily land on nonhost plants, but will not oviposit. The search for chemicals that affect the choice of host plants by these butterflies has led to the discovery of oviposition deterrents in host as well as nonhost plants (Renwick and Radke, 1985). A simple bioassay for deterrents offers butterflies a choice of two cabbage plants. One is sprayed with extract or fractions, and the other with solvent alone. When ether and water extracts of cabbage were compared, the nonpolar extract was deterrent, but the water extract was not. This is not surprising, since the water extract is known to contain the stimulant for oviposition. Nonpolar extracts of other host plants are also deterrent, and water extracts have little or no effect. However, water extracts of nonhost plants are quite deterrent to ovipositing cabbage butterflies, indicating the presence of polar deterrents that may be lacking in host plants. Among those unacceptable plants are a few crucifers including shepherd's purse, *Capsella bursa-pastoris* L., and wormseed mustard, *Erysimum cheiranthoides* L.

The question arises whether unacceptable crucifers are avoided because they lack the necessary chemical stimulants for oviposition, or because the presence of a deterrent is responsible for discouraging the gravid butterflies. Recent comparative studies on *Capsella* and *Erysimum* have provided an answer to this question. Extracts of *Erysimum* foliage were fractionated by partitioning between water and butanol. The deterrent was extracted into the butanol, and the remaining water fraction was stimulatory in oviposition bioassays. When *Capsella* tissue was extracted and subjected to the same separation procedure, the butanol-soluble material was deterrent, but the water fraction was not stimulatory. We can conclude, therefore, that *Erysimum* contains both stimulants and deterrents, and the deterrent effect is sufficient to block the butterfly response to stimulant. Although *Capsella* also contains deterrent material, avoidance of this plant can be explained as well by the lack of stimulatory compounds.

The presence of stimulants and deterrents in the same plants means that different ratios of the positive and negative signals are possible. Differences may occur due to genetic variation or as a result of different environmental conditions. We now need to know how the balance of stimulants and deterrents affects the acceptance or rejection of potential hosts by the butterflies. In any case, it appears possible that plants may avoid attack by *P. rapae* by producing more deterrent or by producing less stimulant.

The variability in individual insect responses to chemicals and the ability of butterflies to use different plant constituents as oviposition cues would suggest that *P. rapae* could easily adapt to shifts in plant chemistry. However, crucifers also have the potential to escape from this insect by discarding attractive compounds or by introducing new defensive chemicals. Thus a dynamic relationship exists where no clear winner emerges from the arms race.

III. Conclusions

Plant generation time may indeed play a major role in the evolution of plant–insect relationships. Bark beetles have capitalized on their advantage over pine trees by adapting to both physical and chemical barriers. This adaptation involves the development of morphological as well as physiological characteristics to deal with the hazards of resin. The cabbage butterfly, however, has been faced with a relatively wide diversity in crucifer chemistry, so the extent of its adaptation has been limited.

The most significant difference between these two plant–insect systems is that while *Pieris rapae* has developed the ability to cope with the major defensive chemicals of crucifers, bark beetles have further developed the ability to utilize the defensive chemicals of pine trees in their colonization process. The production of pheromones is critical for the success of the beetles, and different mechanisms of attack have been adopted to use the pheromones for different purposes. The aggressive *Dendroctonus* species must attack healthy hosts *en masse* to overcome the physical barrier of resin flow. In this case, the pheromone serves primarily as a recruitment signal. The "secondary" *Ips* species must quickly locate declining host material before it is no longer suitable for brood development. Here the pheromone functions to pinpoint the location of such breeding sites. In contrast, cabbage butterflies have no requirement for aggregation and actually depend on distribution of their eggs on different plants for maximum survival of their progeny. Although specific plant chemicals are obviously used as cues for host recognition by *P. rapae*, utilization of these compounds as part of a communication system has not occurred.

The possible impact of plant generation time on the relationship between pine trees and attacking beetles can be visualized when we consider countermeasures that the trees might resort to in the arms race. One simple change in chemistry would be to alter the terpene composition of the resin to deprive the beetles of their pheromone precursors. But the bark beetles are capable of oxidizing an array of monoterpenes that might confront them and would have ample time to adapt to the use of new terpene oxidation products as pheromones. The production of aggregation pheromones from terpenes by bark beetles might represent the ultimate in adaptation and exploitation. On the other hand, the cabbage butterfly presumably has been confronted with changing crucifer chemistry, which seems to be effective in preventing attack by this insect for a significant period of evolutionary time.

One might argue that pine trees are surviving remarkably well, despite their apparent disadvantage in the arms race. But in fact, many pine stands have been destroyed by bark beetles, especially in Central America and Mexico, where a great diversity of tree and beetle species provides a relatively natural environment for evolutionary forces to be manifested. Probably no pine species are immune to attack by bark beetles. However, in crucifers, the incorporation of secondary defense compounds appears to be quite effective in discouraging attack by cabbage butterflies and other crucifer "specialists." The level of chemical sophistication used in the host selection process by the two groups of insects appears to support the hypothesis that differences in generation time between insect and plant can profoundly affect the degree of chemical evolution in the attackers and defenders.

Acknowledgments

I thank Celia D. Radke for her role in the experimental investigations on *Pieris rapae* and Patrick R. Hughes for comments on the manuscript.

References

Billings, R. F. (1985). Southern pine bark beetles and associated insects. Effects of rapidly-released host volatiles on response to aggregation pheromones. *Z. Angew. Entomol.* **99,** 483–491.

Blau, P. A., Feeny, P., Contardo, L., and Robson, D. S. (1978). Allylglucosinolate and herbivorous caterpillars: A contrast in toxicity and tolerance. *Science* **200,** 1296–1298.

Brand, J. M., Bracker, J. W., Markovetz, A. J., Wood, D. L., and Browne, L. E. (1975). Production of verbenol pheromone by a bacterium isolated from bark beetles. *Nature (London)* **254,** 136–137.

Conn, J. E., Borden, J. H., Hunt, D. W. A., Holman, J., Whitney, H. S., Spanier, O. J., Pierce, H. D., Jr., and Oehl-Schlager, A. C. (1984). Pheromone production by axenically reared *Dendroctonus ponderosae* and *Ips paraconfuses* (Coleoptera: Scolytidae). *J. Chem. Ecol.* **10,** 281–290.

Feeny, P. (1977). Defensive ecology of the Cruciferae. *Ann. Missouri Bot. Gard.* **64,** 221–234.

Hughes, P. R. (1973). *Dendroctonus:* Production of pheromones and related compounds in response to host monoterpenes. *Z. Angew. Entomol.* **73,** 294–312.

Hughes, P. R. (1974). Myrcene: A precursor of pheromones in *Ips* beetles. *J. Insect Physiol.* **20,** 1271–1275.

Hughes, P. R. (1975). Pheromones of *Dendroctonus:* Origin of α-pinene oxidation products present in emergent adults. *J. Insect. Physiol.* **21,** 687–691.

Hughes, P. R., and Renwick, J. A. A. (1977). Neural and hormonal control of pheromone biosynthesis in the bark beetle, *Ips paraconfuses. Physiol. Entomol.* **2,** 117–123.

Lichtenstein, E. P., Morgan, D. G., and Muller, C. H. (1964). Naturally occurring insecticides in cruciferous crops. *J. Agric. Food Chem.* **12,** 158–161.

Ma, W. C., and Schoonhoven, L. M. (1973). Tarsal contact chemosensory hairs of the large white butterfly *Pieris brassicae* and their possible role in oviposition behavior. *Entomol. Exp. Appl.* **16,** 343–357.

Nielsen, J. K. (1978). Host plant discrimination within Cruciferae: Feeding responses of four leaf beetles (Coleoptera: Chrysomelidae) to glucosides, cucurbitacins, and cardenolides. *Entomol. Exp. Appl.* **24,** 41–54.

Pitman, G. B., and Vité, J. P. (1969). Aggregation behavior of *Dendroctonus ponderosae* (Coleoptera: Scolytidae) in response to chemical messengers. *Can. Entomol.* **101,** 143–149.

Renwick, J. A. A. (1967). Identification of two oxygenated terpenes from the bark beetles *Dendroctonus frontalis* and *Dendroctonus brevicomis. Contrib. Boyce Thompson Inst.* **23,** 355–360.

Renwick, J. A. A., and Hughes, P. R. (1975). Oxidation of unsaturated cyclic hydrocarbons by *Dendroctonus frontalis. Insect Biochem.* **5,** 459–463.

Renwick, J. A. A., and Pitman, G. B. (1979). An attractant isolated from female Jeffrey pine beetles, *Dendroctonus jeffreyi. Environ. Entomol.* **8,** 40–41.

Renwick, J. A. A., and Radke, C. D. (1983). Chemical recognition of host plants for oviposition by the cabbage butterfly, *Pieris rapae* (Lepidoptera: Pieridae). *Environ. Entomol.* **12,** 446–450.

Renwick, J. A. A., and Radke, C. D. (1985). Constituents of host- and non-host plants deterring oviposition by the cabbage butterfly, *Pieris rapae. Entomol. Exp. Appl.* **39,** 21–26.

Renwick, J. A. A., and Vité, J. P. (1970). Systems of chemical communication in *Dendroctonus. Contrib. Boyce Thompson Inst.* **24,** 283–292.

Renwick, J. A. A., Hughes, P. R., and Ty, T. D. (1973). Oxidation products of pinene in the bark beetle, *Dendroctonus frontalis. J. Insect Physiol.* **19,** 1735–1740.

Renwick, J. A. A., Hughes, P. R., and Krull, I. S. (1976a). Selective production of *cis*- and *trans*-verbenol from (−)- and (+)-α-pinene by a bark beetle. *Science* **191,** 199–201.

Renwick, J. A. A., Hughes, P. R., Pitman, G. B., and Vité, J. P. (1976b). Oxidation products of terpenes identified from *Dendroctonus* and *Ips* bark beetles. *J. Insect Physiol.* **22,** 725–727.

Schoonhoven, L. M. (1972). Secondary plant substances and insects. *Recent Adv. Phytochem.* **5,** 197–224.

Silverstein, R. M., Rodin, J. O., and Wood, D. L. (1966). Sex attractants in frass produced by male *Ips confusus* in ponderosa pine. *Science* **154,** 509–510.

Smith, R. H. (1965). Effect of monoterpene vapors on the western pine beetle. *J. Econ. Entomol.* **58,** 509–510.

Usher, B. F., and Feeny, P. (1983). Atypical secondary compounds in the family

Cruciferae: Tests for toxicity to *Pieris rapae,* an adapted crucifer-feeding insect. *Entomol. Exp. Appl.* **34,** 257–265.

Verschaffelt, E. (1911). The cause determining the selection of food in some herbivorous insects. *Proc. K. Ned. Akad. Wet.* **13,** 536–542.

Virtanen, A. I. (1965). Studies on organic sulphur compounds and other labile substances in plants. *Phytochemistry* **4,** 207–228.

Vité, J. P., and Francke, W. (1985). Waldschutz gegen Borkenkäfer: Vom Fangbaum zur Falle. *Chemie in unserer Zeit* **19,** 11–21.

Vité, J. P., and Wood, D. L. (1961). A study on the applicability of the measurement of oleoresin exudation pressure in determining susceptibility of second growth Ponderosa pine to bark beetle infestation. *Contrib. Boyce Thompson Inst.* **21,** 67–78.

CHAPTER 10

Variation in the Terpene Chemistry of Douglas-Fir and Its Relationship to Western Spruce Budworm Success

REX G. CATES
Department of Botany
Chemical Ecology Laboratory
Brigham Young University, Provo, Utah

RICHARD A. REDAK
Department of Entomology
Colorado State University, Fort Collins, Colorado

I. Introduction

Initially, most theories of plant–herbivore interactions dealt with the evolution and production of secondary metabolites among growth forms, among sympatric plant species, and between tissues such as young versus mature photosynthetic tissues. Feeny (1976), Rhoades and Cates (1976), and Cates and Rhoades (1977), among others (McKey, 1979; Janzen, 1973), proposed that the type of defense as well as the amount of energy allocated to defense were primarily a function of the predictability or apparency of a plant and/or tissue to herbivores. The degree of predictability or apparency of a resource was suggested to be the result of the evolved life history characteristics of plants due to plant–plant interactions along with plant interactions with the abiotic environment. Additionally, Bryant *et al.* (1983) and Coley *et al.* (1985) proposed that the evolutionary response of plants to herbivory was determined largely by the influence of resource availability on the carbon/nutrient balance of plants.

However, none of the above provided a rigorous discussion of the role of chemical variation in the mediation of plant–herbivore interactions. This was partially due to the fact that ecologists saw variation as a problem in experimental design and tried to minimize its effect in averages. In the early 1980s, several studies showed that variation in plant defensive chemistry, nutrition, and morphology was a most important influence on herbivores (Denno and McClure, 1983). Variation in the production of secondary metabolites has been correlated with reduced herbivore success within and among plant populations (Crankshaw and Langenheim, 1981; Cates *et al.*, 1983a; Louda and Rodman, 1983; Raffa and Berryman, 1983; Whitham, 1983, Redak and Cates, 1984; Cates and Redak, 1986), with tissue age and season (Chew and Rodman, 1979; Horner, 1984; McClure and Hare, 1984; Schultz *et al.*, 1982; Cates *et al.*, 1983a; Rhoades, 1979; Waring *et al.*, 1985), and with abiotic and biotic factors (Haukioja and Niemela, 1979; Wratten *et al.*, 1984; Rhoades, 1979). In addition, subspecific variation in the production of secondary metabolites is well documented (Cates and Rhoades, 1977; Welch and McArthur, 1981). Currently, variation is suggested to decrease the predictability of a plant and/

or tissue to herbivory, increase interactions among herbivores for suitable resources, and make herbivores more predictable to natural enemies (Whitham, 1983).

The theoretical importance of chemical variation in the mediation of coevolutionary interactions among herbivores and their host plants has been well discussed, but few data are available indicating how extensive chemical variation may be in a plant species. The objective of this chapter is to elucidate some of the variation that occurs in terpenoids in the current year's needles of 300 Douglas-fir trees from nine populations throughout western United States. Much of the data dealing with chemical variation in Douglas-fir come from a 5-year study of a single population in New Mexico. Specifically, for a single population, we will address the level of variation that occurs in individual terpenoids, total terpene content, and nitrogen and water content of the current-year's foliage (1) among trees, (2) through the growing season, and (3) among years. In addition, we will demonstrate that chemical variation among trees in a population growing on different aspects exists in response to water stress. Finally, we will show that geographical variation in the current year's foliage exists among populations in both qualitative and quantitative terpenoid production. The role of chemical variation as a potential regulatory influence on the western spruce budworm will be explored, as well as the importance of chemical variation in the silvicultural management of herbivores.

II. Materials and Methods

A. Study Sites

The main research site was located at Barley Canyon, Santa Fe National Forest, approximately 2.4 km northeast of Fenton Lake, New Mexico. The average elevation is 2420 m. Dominant tree species include *Pseudotsuga menziesii* var. *glauca* (Beisn.) Franco, *Abies concolor* (Gord. and Glend.) Lindl., *Populus tremuloides* Michx., *Pinus ponderosa* Dougl., and *Picea pungens* Englem. Further details of the site are found in Redak and Cates (1984) and Cates and Redak (1986). Details of the sites selected for the geographical variation study are found below.

Of the 200 trees initially screened at Barley Canyon in May 1980, 105 were selected for the studies reported here (Redak and Cates, 1984). Of the 105 trees, 30 were trenched in May 1980 for studies on the effects of moisture stress on secondary metabolites and western spruce budworm biomass production. Trenching involved digging a 1-m-deep trench around each of the trees at canopy dripline. Six-mil black plastic was then laid from and surrounding the bole of each tree to the trench, down into the trench, and up the outer edge of the trench. The trench was backfilled and black plastic was secured to the bole. In this way, water and snow was shed from the roots increasing water stress. Every trenched tree was paired with an untrenched control tree. Foliage quality of the trenched trees was compared to that of nontrenched trees. In addition, budworm larvae were reared in nylon screen bags on both the trenched and untrenched trees as outlined in Cates *et al.* (1983a). The remaining 75 trees (including the controls for the trenched trees) were used for studies dealing with seasonal variation in terpenes, variation in terpenes among trees within a population, developmental variation in terpenes, and the effects of this variation (as well as that of water and nitrogen) on western spruce budworm survival and dry weight biomass production.

The budworm used as the bioassay of the effect of stress and variation in foliage quality was collected near Taos, New Mexico, Carson National Forest. Details of collecting and rearing budworm in bags on sample trees are in Redak and Cates (1984).

Approximately 15 g of current year's foliage was collected from each tree during the period when larvae were in the fifth and sixth instars. All foliage was collected from the same north-facing, midcrown quarter of the trees where the experimental insects were placed. In 1980, tissue was put in labeled Ziploc bags (for tannin and nitrogen analyses) or indium tubing (terpene analysis), placed on ice, and taken to the laboratory and frozen in a Revco ultra-cold freezer (Redak and Cates, 1984). In 1981–1983, tissue was collected in 2-oz bottles, frozen in liquid nitrogen in the field, and stored at −75°C in the ultra-cold. No difference in terpene chemistry was found between tissue collected within the bag containing budworm as compared to tissue collected just outside the bag on the same branch; consequently, tissue for chemical analyses described above was collected outside the budworm-rearing bags.

B. Chemical Methods

1. Nitrogen Analysis

Approximately 5 g of the current year's foliage was weighed to the nearest 1.0 mg, dried at 60°C for 72 h, and used to determine total, alcoholic-soluble, and insoluble foliar nitrogen content using micro-Kjeldahl techniques. Total nitrogen content, the ratio of soluble to insoluble nitrogen, and tissue dry weight and water content were used in statistical analyses.

2. Monoterpene Analysis

In 1980, terpenes were analyzed according to Redak and Cates (1984). In 1981–1983, terpenes were analyzed according to Cates and Redak

(1986). Both methods yielded 95% or greater of the total amount of terpenoids in the tissue. Analyses were performed on a Perkin-Elmer Sigma II-B gas chromatograph equipped with a Hewlett-Packard recorder/integrator. In 1980, a 1% V-930 column was used, while in 1981–1983 a 25 m × 0.25 mm ID fused silica column coated with OV-101 was used. For 1980, identifications were made using known standards. Terpenes were identified in the 1981–1983 samples using cochromatography of the components with authentic standards on two different columns, and by comparison of mass spectra generated by a Finnigan gas chromatograph/mass spectrograph (GC/MS) equipped with a SE-54 column with those from the National Bureau of Standards. Quantitation of the terpenes analyzed in the 1981–1983 samples was by internal standard (fenchone; Aldrich Chemical Co., Inc.).

C. Statistical Methods

Stepwise discriminant analysis was used to describe the variation in terpenes (1) within a single population of Douglas-fir through the growing season; (2) due to aspect and water stress within a single population of Douglas-fir trees at one point in time; (3) within a single population among growing seasons; and (4) between geographically distinct populations of Douglas-fir within the same growing season. Additionally, discriminant analysis was used to determine how adult female and male budworm dry weight were related to differences in foliage chemistry of trees growing on different aspects and under different levels of water stress.

Statistical analyses followed BMDP Statistical Software (Dixon, 1983). The discriminant analysis technique used here generates a linear combination of the original variables (the discriminant function score) such that the variation among the a priori groups is maximized relative to that within the groups. There may be as many as $b - 1$ discriminant functions generated, where b is the number of groups. The first function is constructed so as to explain the largest fraction of the between-group variance, the second function the next largest fraction, and so on until 100% of the between-group variance is explained. For all analyses, more than 90% of the between-group variation was explained by the first two discriminant functions (see Figs. 1–5). A particular discriminant function is of the form $D = a_1x_1 + a_2x_2 + \cdots + a_nx_n$, where D is the discriminant function score (plotted on Figs. 1–5, for discriminant functions 1 and 2 in each case), x_n are the original variables, and a_n are the coefficients for the original variables. Large coefficients (when standardized) indicate that a particular variable is relatively important in distinguishing between groups. When discriminant function scores are plotted, positive coefficients drive the discriminant function score to the positive end of that score's axes. Negative coefficients drive the score to the negative end of the axes. Tests of statistical significance among groups were made using the variance and covariance matrices in a manner analagous to MANOVA techniques. For a more detailed explanation of the techniques used refer to Neff and Marcus (1980).

1. Within-Season Variation

About 1 g of the current year's foliage was collected each of five times in 1980 (June 18, 26, and 30, and July 3 and 8) at the midcrown, north-facing level from the nontrenched sample trees, and each set of foliage samples was analyzed for the variables listed below (Redak and Cates, 1984). Stepwise discriminant analysis was used to determine how foliage quality changed, if at all, throughout the growing season. For this analysis, the foliage quality was described by 22 chemical and nutritional variables. These were α-pinene, camphene, β-pinene, myrcene, limonene, carene/α-terpinene, terpinolene, bornyl acetate, citronellyl acetate, geranyl acetate, and unidentified compounds 2, 3, 4, 5, 8, 9, and 10; total terpene content, terpene evenness, total nitrogen; the ratio of soluble to insoluble nitrogen; and percent water content of the foliage.

2. Variation within a Population at One Point in Time Due to Aspect and Trenching

In 1980, each tree was assigned to one of four categories depending on location of the tree at the site (growing on a south-facing slope or in the valley) and whether the tree was trenched or not. Consequently, the four categories were south-facing and trenched, south-facing and nontrenched, valley and trenched, valley and nontrenched. Stepwise discriminant analysis was then used in 1981 to determine how the groups of trees differed with respect to terpene content. The foliage quality was described by 13 chemical variables. These were thujene, α-pinene, camphene, β-pinene, myrcene, carene, limonene, terpinolene, linalool, bornyl acetate, sesquiterpenes 3 and 4, and terpene evenness as measured by Simpson's diversity index (Redak and Cates, 1984). A different method for terpenoid analysis, and elimination of most unidentified compounds, resulted in fewer variables used in the discriminant analysis. In no case are comparisons made where methodologies differ.

3. Variation in Terpene Chemistry of the Same Population among Growing Seasons

Seventy-five trees were sampled at the same phenological stage (75% budbreak) for three successive growing seasons (1981–1983). Stepwise discriminant analysis was used to determine year-to-year variation in foliage quality based on 14 terpene variables. The variables were thujene, α-pinene, camphene, β-pinene, myrcene, limonene, carene, linalool, terpinolene, bornyl acetate, cadinene, sesquiterpenes 3 and 4, and terpene evenness (as measured by Simpson's diversity index).

4. Variation in Resistance–Susceptibility Characteristics of Trees between Years as Assayed by Budworm Biomass Production

Details of the methods used for budworm collection, rearing, collection of tissue, and chemical analyses are outlined in Cates and Redak (1986). Given below is a description of methods used to determine if trees varied from year to year in their resistance–susceptibility factors. During 1981–1983, budworm were reared to pupation on the 75 nontrenched sample trees. Third-instar larvae were placed on the north-facing, midcrown area of each tree about June 20–24, and they pupated between July 10 and 14 each year. Pupae were removed from each tree and taken to a controlled environmental chamber at the University of New Mexico, and adults were permitted to emerge. Adults were killed with ethyl acetate, sexed, and labeled, and fresh weight was determined. They were then dried for 72 h at 60°C and weighed. Male and female dry weights were used as the dependent variables in the multiple correlation analysis to determine the effect of current year's foliage nitrogen, water content, and terpene chemistry on biomass production.

Five grams of the current year's foliage was collected from the same area of each tree when the fifth-instar larvae were feeding. In total, 30 chemical and morphological variables of trees were used as independent variables in the multiple stepwise correlation analysis (Cates and Redak, 1986; for complete listing of variables; Harris, 1975).

5. Geographic Variation in the Production of Terpenes among Douglas-Fir Populations

A general location for each is as follows: Montana site = 3.5 km north of Boulder, Montana on left fork of Whitetail Creek road; Mt. Sawtelle site = 15 km northeast of Mac's Inn, Idaho; McCall site = located on Red Ridge about 20 km west of McCall, Idaho; Jack's Creek site = 15 km west of Sisters, Oregon; Santium Pass site, Oregon = 21 km west of Sisters, Oregon; Sequim site = 8 km east of Sequim, Washington; Issaquah site = 20 km east of Issaquah, Washington. Five grams of the current year's foliage was collected from the north-facing, midcrown area of each of 10–20 trees per population during late June and early July 1983. Foliage from each tree was put in separate 2-oz bottles, labeled, and placed in a liquid nitrogen tank in the field. Terpenes were analyzed according to the method used for all foliage collected from 1981 to 1983 (Cates and Redak, 1986). Twenty terpenoid variables were used: thujene, α-pinene, camphene, sabinene, β-pinene myrcene, carene, limonene, ocimene, γ-terpinene, terpinolene, linalool, terpinen-4-ol, citronellol, bornyl acetate, citronellyl acetate, geranyl acetate, cadinene, α-humulene, and an unidentified sesquiterpene (sesq. 2). Stepwise discriminant analysis was used to determine if populations differed in terpene chemistry.

III. Results

Results are presented beginning with the within-season variation in the types and quantities of terpenes in the current year's foliage. This is followed by variation in the response of sample trees to trenching and the budworm's response to trenching. The data dealing with variation in foliage quality among seasons within the same Douglas-fir population are presented next along with the budworm's response to this variation. Finally, geographical variation in terpenoids in current year's foliage among populations is discussed.

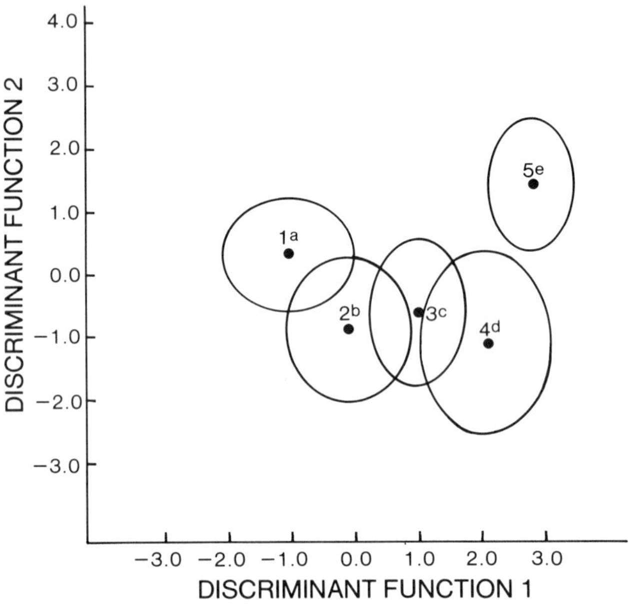

Fig. 1. Variation in the production of terpenes in the current year's foliage of Douglas-fir through a season (1980) within a population ($p < 0.001$ for the overall model). 1, Collection of July 8; 2, July 3; 3, June 30; 4, June 26; and 5, June 18. Numbers followed by different letters are significantly different at $p < 0.001$. Centroids are shown with a one-standard deviation ellipse.

A. Within-Season Variation

Discriminant analysis of the terpene chemistry in the current year's foliage of 30 trees indicated significant variation among all five sampling dates (Fig. 1). Of the seven variables important in the two discriminant functions, four (or 57%) of the variables were related to terpenes (Table I). Discriminant function 1, which accounted for 76% of the discriminant variation, shows that early in the growing season, the current year's foliage is characterized by high water content, total nitrogen, terpinolene, unidentified terpene 4, and terpene evenness (an expression, when compared to the population average, of the degree of evenness in the quantitative distribution of terpenes as opposed to a skewed distribution). As the foliage matured (July 3 and 8 collections), the ratio of soluble to insoluble nitrogen increased, and the total concentration of terpenes increased significantly. Discriminant function 2, which accounted for an additional 19% of the discriminant variation, follows the same general pattern. However, the only variable associated with the later collection dates is the total terpene content of the current year's foliage. Since there is a nega-

Table I. Discriminant Analysis Model Showing the Within-Season Variation in Terpenes in the Current Year's Foliage of Douglas-Fir

Variable	Standardized Coefficient[a]
Discriminant function 1	
Percent water content	0.807
Total nitrogen	0.707
Terpinolene	0.406
Terpene evenness	0.349
Unknown terpene 4	0.302
Soluble/insoluble nitrogen	−0.178
Total terpenes	−0.737
Discriminant function 2	
Terpinolene	1.074
Total nitrogen	0.459
Unknown terpene 4	0.145
Terpene evenness	0.101
Soluble/insoluble nitrogen	0.093
Percent water content	0.086
Total terpenes	−0.310

Note: $F = 12.14$; df = 28,549; $p < 0.001$.

[a] Positive coefficients drive the discriminant functions in a positive direction along their respective axes in Fig. 1. Negative coefficients drive the discriminant functions in a negative direction along the respective axes in Fig. 1.

tive coefficient for total terpene content for both functions (Table I), the score is driven toward the negative end of both axes (Fig. 1).

B. Within-Population Variation in Terpene Chemistry Due to Aspect and Trenching

Discriminant analysis of the terpene chemistry of the current year's foliage of four groups of trees sampled in 1981 (1 year after trenching) indicated that the only significant difference was the comparison of the trenched trees on the south-facing slope (group 1) versus all other categories (groups 2, 3, and 4) (Fig. 2). For discriminant function 1, which accounted for 58% of the discriminant variation, the trees growing in the valley that were trenched, the trees in the valley that were not trenched, and the trees growing on the south-facing slope that were not trenched were characterized by high quantities of terpinolene, whereas the trenched trees on the south-facing slope were higher in linalool and unidentified sesquiterpene 4, and had a more even distribution of terpenes as

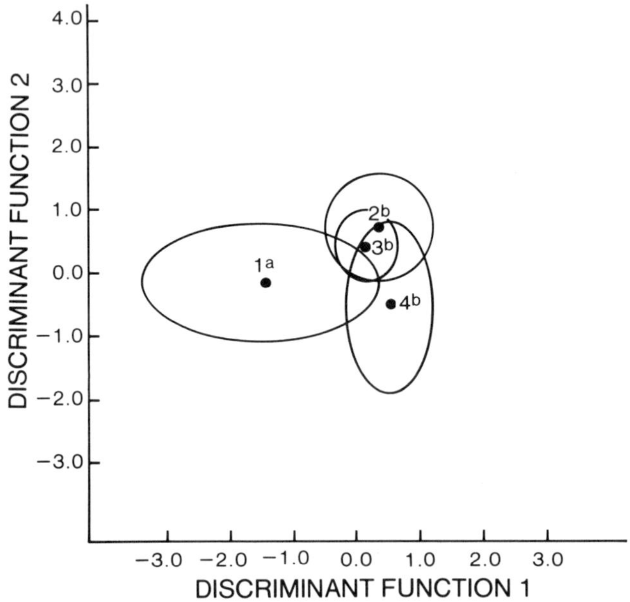

Fig. 2. Variation in the production of terpenes in the current year's foliage of Douglas-fir due to aspect and trenching. 1, Trenched trees growing on south-facing slope; 2, nontrenched trees growing on the south-facing slope; 3, nontrenched trees growing in the valley; 4, trenched trees growing in the valley. Numbers followed by different letters are significantly different ($p <$ 0.002). Centroids are shown with a one-standard deviation ellipse. (After Cates and Redak, 1988).

Table II. Discriminant Analysis Model Showing the within-Population Variation in Terpene Chemistry Due to Aspect and Trenching at Canopy Dripline

Variable	Standardized Coefficient[a]
Discriminant function 1	
Terpinolene	1.622
Linalool	−0.619
Terpene evenness	−0.678
Sesquiterpene 4	−1.621
Discriminant function 2	
Sesquiterpene 4	2.912
Terpene evenness	−0.665
Linalool	−1.231
Terpinolene	−3.253

Notes: $F = 2.509$; df = 12,127; $p < 0.002$. (After Cates and Redak, 1988).
[a] See Table I, footnote 1.

compared to the trenched trees on the south-facing slope (Table II). For discriminant function 2, the same four variables accounted for an additional 34% of the discriminant variation. However, in this case sesquiterpene 4 was higher in the two nontrenched groups as compared to the

Table III. Discriminant Analysis Model Showing the within-Population Variation in Budworm Dry Weight Production When Reared on Douglas-Fir Trees Varying in Aspect and Trenching at the Barley Canyon Site

Variable	Coefficient[a,b]
Discriminant function 1	
Male average dry weight	0.351
Female average dry weight	0.146
Constant	−6.842
Discriminant function 2	
Male average dry weight	−0.677
Female average dry weight	0.254
Constant	1.390

Notes: $F = 5.653$; $p < 0.001$. (After Cates and Redak, 1988).
[a] See Table I, footnote 1.
[b] Unstandardized coefficients, since all variable units and ranges are the same.

trenched groups, while linalool and terpene evenness were higher in the two groups that were trenched as compared to the nontrenched groups.

Table III and Fig. 3 show the budworm response to the variation among the groups of trees. The dry weights of budworm reared on trees that were growing on the south-facing slope were significantly less than those of budworm reared on the other three groups of trees. Discriminant function 1 accounted for 96% of the discriminant variation in budworm weight and discriminant function 2 accounted for an additional 4%. No other comparisons in budworm dry weight production among the groups of trees were significantly different.

C. Variation in the Same Population among Seasons in Terpene Chemistry

In an effort to determine if terpene chemistry varied among years, the current year's foliage of the same 75 trees was sampled in 1981, 1982, and

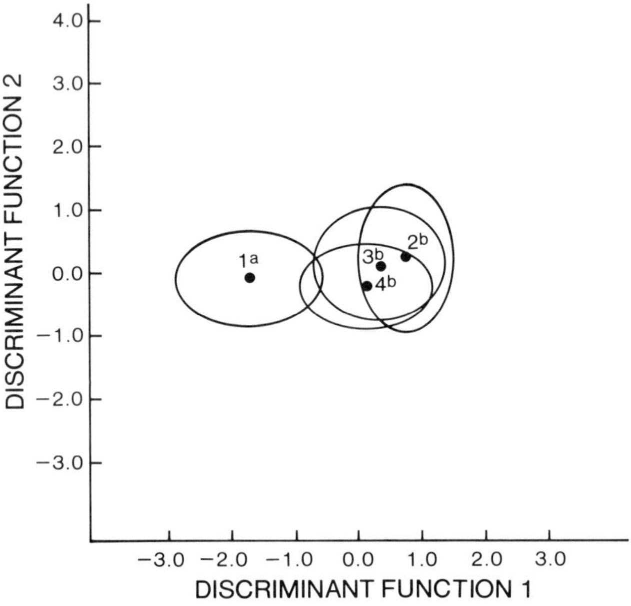

Fig. 3. Response in dry weight production of the western spruce budworm reared on Douglas-fir trees varying in aspect and trenching at the Barley Canyon site. Categories followed by different letters are significantly different ($p < 0.001$). 1, Budworm reared on trenched trees growing on the south-facing slope; 2, budworm reared on nontrenched trees growing in the valley; 3, budworm reared on trenched trees growing in the valley; 4, budworm reared on nontrenched trees growing in the south-facing slope. Centroids are shown with a one-standard deviation ellipse. (After Cates and Redak, 1988).

1983, and the terpene chemistry was determined. Discriminant analysis showed that all comparisons among years were significantly different (Fig. 4). Discriminant function 1, which accounted for 79% of the discriminant variation, indicates that the trees sampled in 1982 were significantly higher in limonene, sesquiterpene 4, terpinolene, and linalool, and had a more even distribution of terpenes in the foliage, while trees sampled in 1983 were higher in thujene and β-pinene (Table IV). The trees sampled in 1981 were intermediate in these terpene constituents in the current year's foliage. The same terpenes were involved in discriminant function 2, which explained 20% of the discriminant variation, but sesquiterpene 4, linalool, terpinolene, and thujene were highest in the trees sampled in 1981 while limonene, β-pinene, and terpene evenness were the predominant factors for the trees sampled in 1982 and 1983.

D. Budworm Response to Variation in Terpene Chemistry

Based on the data from Fig. 4 and Table IV, the hypothesis was generated that there was no difference in terpene resistance variables to budworm between years. As described in Section II, budworms were reared

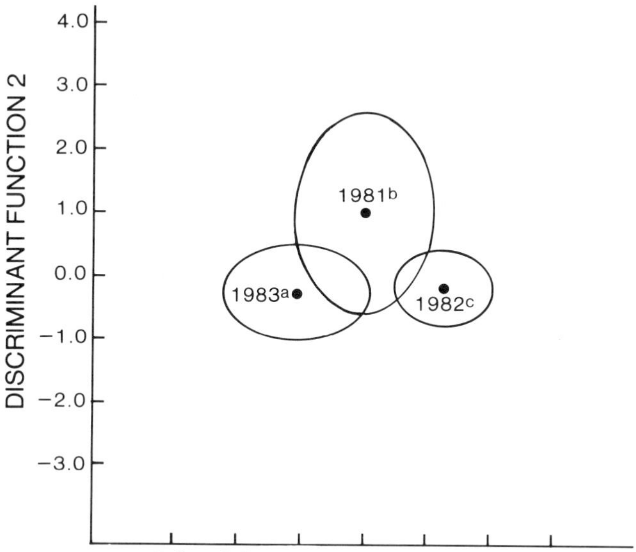

Fig. 4. Discriminant analysis showing the among-year (1981–1983) variation in terpene content of the current year's foliage of Douglas-fir at the Barley Canyon site. Years followed by different letters are significantly different ($p < 0.001$). Centroids are shown with a one-standard deviation ellipse.

Table IV. Discriminant Analysis Model Showing the among-Year Variation in Terpenes in the Current Year's Foliage of Douglas-Fir

Variable	Standardized Coefficient[a]
Discriminant function 1	
Limonene	3.081
Sesquiterpene 4	1.125
Terpene evenness	0.515
Terpinolene	0.399
Linalool	0.103
β-Pinene	−0.030
Thujene	−3.955
Discriminant function 2	
Sesquiterpene 4	2.719
Linalool	0.691
Terpinolene	0.255
Thujene	0.128
Terpene evenness	−0.039
β-Pinene	−1.269
Limonene	−1.686

Note: $F = 20.793$; df = 14,498; $p < 0.001$.
[a] See Table I, footnote 1.

on the trees and the trees were sampled for morphological tree variables and for foliage quality variables during the fifth and sixth instar. For the 1981 data, multiple correlation analysis using female dry weight as the dependent variable indicated that female biomass increased when terpinolene was in high concentration in the foliage and when the quantitative distribution of terpenes was more even in the foliage (terpene evenness) than when skewed toward certain terpenes or a group of terpenes (Table V). Female dry weight production, however, was inversely related to β-pinene concentration in 1981. For the 1982 data, female biomass production was positively related to terpene evenness and to high concentrations of myrcene and camphene in the foliage. Alternatively, female dry weight production was inversely related to high concentrations of bornyl acetate and sesquiterpene 3. Dry weight production of the male budworm reared on the trees showed the same pattern in that different terpene variables were inversely related to male biomass production between years (data not shown). A detailed discussion of male budworm response to within-year and between-year variation in terpene chemistry is found in Section IV and in Cates and Redak (1986).

Table V. Correlation Models Depicting the Relationship between Female Dry Weight Production and Morphological Tree Variables and Foliage Quality of Douglas-Fir

1981 Variable Entered	P/I	*p*	1982 Variable Entered	P/I	*p*
β-Pinene	I	0.0005	1982 Internode length	P	0.001
5-Year growth increment	P	0.002	Terpene evenness	P	0.016
Terpene evenness	P	0.005	Bornyl acetate	I	0.021
1981 Internode length	I	0.022	Myrcene	P	0.024
			Sesquiterpene 3	I	0.052
Terpinolene	P	0.045	Camphene	P	0.062
			Crown diameter	P	0.082

Notes: Modified after Cates and Redak (1986). P, Positive relationship between female dry weight and independent variables; I, inverse relationship between female dry weight and independent variables.

E. Geographic Variation in the Production of Terpenes among Douglas-Fir Populations

Several studies have shown that mature current year's foliage collected in September and October is different qualitatively and quantitatively among populations (von Rudloff and Rehfeldt, 1980). However, to our knowledge an examination to determine if the foliage is different in terpene chemistry among populations at the time that budworms feed on the tissue (June, early July) is not available. Examinations of the current year's foliage collected from trees during late June and early July from seven populations were significantly different qualitatively and/or quantitatively in their terpene production except for the Jack's Creek–Santium Pass populations, which are within 6 km of one another (Fig. 5, Table VI). Populations from western Washington (Issaquah, population 7; Sequim, population 6) and from Oregon (Jack's Creek and Santium Pass, populations 5 and 4, respectively) were characterized by high concentrations of sabinene, ocimene, limonene, geranyl acetate, and cadinene as compared to lower amounts of these compounds in the other populations. Foliage from the Montana population (population 1) was higher in bornyl acetate, terpinolene, myrcene, and terpinene as compared to the other populations. Foliage from the Mt. Sawtelle, Idaho, population (population 2) tended to be intermediate in the concentration of these nine terpenes, while foliage of the population from McCall, Idaho (population 3), was higher in bornyl acetate, terpinolene, myrcene, and terpinene, and lower in sabinene, ocimene, limonene, geranyl acetate, and cadinene.

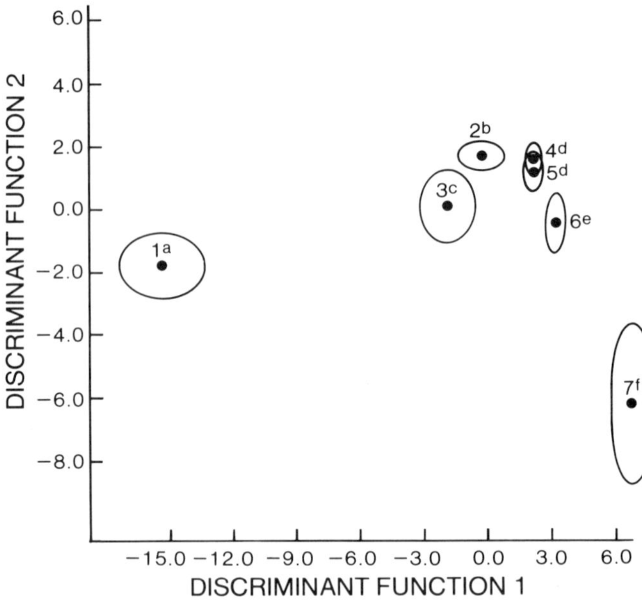

Fig. 5. Discriminant analysis showing the geographical variation in terpene content of the current year's foliage of Douglas-fir. Numbers followed by different letters are significantly different ($p < 0.001$). 1, Boulder, Montana; 2, Mt. Sawtelle, Idaho; 3, McCall, Idaho; 4, Santium Pass, Oregon; 5, Jack's Creek, Oregon; 6, Sequim, Washington; 7, Issaquah, Washington. Populations are arranged from east (Boulder, Montana) to west. Centroids are shown with a one-standard deviation ellipse.

IV. Discussion

A. Within-Season Variation

As the current year's foliage matures, significant changes occur, particularly a decrease in water and nitrogen content of the foliage and an increase in total terpene content (Table I, Fig. 1). In some trees a significant five- to six-fold increase in total terpenes occurred between June 13 and July 29, and the increase continued until at least September 20 (Gambliel and Cates, 1988). However, interactions between individual terpenes and budworm success also have been noted. For example, terpinolene is high in the younger foliage (Table I), and when budworms are reared in the field on current year's foliage, terpinolene is positively associated with female dry weight production (Table V) (Redak and Cates, 1984; Cates and Redak, 1986). Evidently, female budworm are well adapted to terpinolene, and its high concentration in the current year's

Table VI. Discriminant Analysis Model Showing the Geographic Variation in Terpenes in the Current Year's Foliage of Douglas-Fir

Variable	Standardized Coefficient[a]
Discriminant function 1	
Sabinene	5.561
Ocimene	0.686
Limonene	0.586
Geranyl acetate	0.515
Cadinene	0.328
Terpinene	−0.031
Myrcene	−0.217
Terpinolene	−0.268
Bornyl acetate	−0.433
Discriminant function 2	
Terpinolene	0.642
Cadinene	0.280
Myrcene	0.053
Limonene	0.026
Bornyl acetate	−0.073
Sabinene	−0.190
Terpinene	−0.608
Ocimene	−0.679
Geranyl acetate	−0.944

Note: $F = 15.81$; df = 54,260; $p < 0.001$.
[a] See Table I, footnote 2.

foliage may function as an attractant to female budworm. However, the inverse is true with terpinolene and male budworm dry weight production (Cates and Redak, 1986, data not shown). Male budworm appear to be different from female budworm in that terpenes do adversely influence male budworm but these are not always the same terpenes that adversely affect female budworm. It should be noted that these studies are correlative in design, and the difference between sexes needs to be resolved with synthetic diet studies.

Alternatively, bornyl acetate, which has been shown using agar diets to be an effective toxin (Cates *et al.*, 1987), increases from about 0.25 mg/g fresh weight on June 18 to about 1 mg/g on July 8 in the current year's foliage. Similar patterns occur for most of the terpenes including the sesquiterpenes, which begin low but are high in the September foliage (Gambliel and Cates, 1988). As for many other herbivores that feed on young tissues (Rhoades and Cates, 1976; Feeny, 1976), the window of optimal feeding for the budworm is defined by abiotic and genetic factors that control budbreak (hence, food availability) versus increasing inhibitory levels of secondary metabolites that may serve as defenses, and decreasing nutrient content. Variation in this window due to variation in the phenology of budbreak among sympatric trees, as well as among years due to cold temperatures, is known to have severe adverse effects on budworm (Cates *et al.*, 1983a; Wulf and Cates, 1987).

B. Within-Population Variation in Terpene Chemistry Due to Aspect and Trenching

Not all trees that were trenched responded in the same way to the treatment. Trenched trees on the south-facing slope were primarily segregated by higher contents of sesquiterpene 4 and linalool, and by having a more even distribution of terpenes in the foliage (see Table II). Terpene evenness is interesting because in every case where female budworm were reared on trees in New Mexico, terpene evenness always was associated positively with increased female dry weight regardless of whether trees were trenched or not trenched. Similar studies conducted in Montana showed a significant positive correlation between terpene evenness and greater densities of budworm as well as higher levels of defoliation (Cates *et al.*, 1983b). The conclusion we draw is that terpene evenness is an excellent predictor of tree susceptibility. Again, male budworm (data not shown) do not conform to the pattern observed for females in that when males were reared on nontrenched trees, terpene evenness and percent water content were inversely related to male biomass production (Cates and Redak, 1986).

That the chemical pattern observed in the trenched trees growing in the valley is significantly different from that of trenched trees growing on the south-facing slope is likely due to the trees in the valley being supplied water by capillary action from underground sources in sufficient amount to negate the effect of trenching. Evidence to support this is found in water potential data (not shown) and in the data dealing with the response of budworm to trenching (Fig. 3). Another possibility is that different ecotypes are present in the valley versus those on the south-facing slope. However, the chemistry in the foliage of the two groups of nontrenched trees is not significantly different (Fig. 2).

Interestingly, budworm dry weight production tracked the terpene chemistry very closely in that budworm reared on the trenched trees growing on the south-facing slope were significantly different in their biomass production when compared to the budworm reared on the other three sets of trees (Table III, Fig. 3). Budworm dry weight production was not significantly different when reared on the trenched trees in the valley as opposed to being reared on the nontrenched trees growing in the valley or on the nontrenched trees growing on the south-facing slope. Often leaf-chewing insect success is suggested to be greater on stressed plants

(White, 1969, 1974; Cates *et al.*, 1983a). This was not the case in this study, because budworm dry weight production was *significantly lower* on the trenched trees growing on the south-facing slope as compared to budworm weights from larvae reared on the other three groups of trees (Cates and Redak, unpublished). A closer examination of the level and duration of stress may be the cause of this pattern. Furthermore, the data reported here on the budworm's response to changes in foliage quality, when combined with another study in which trees were under greater water stress (Cates *et al.*, 1983a), strongly suggest that the amelioration of variation in a host-plant's defensive system may be an important factor in herbivore outbreaks.

C. Variation in the Same Population among Seasons in Terpene Chemistry and Budworm Response to This Variation

To our knowledge, among-year variation of the nature described here in terpene chemistry of the same tree (Fig. 4) has not been reported before in the foliage of herbaceous plants or woody perennials. An obvious interest was to determine whether trees resistant in 1981 were also relatively resistant in 1982 to the budworm. Therefore, using the multiple regression equations predicting female budworm dry weight (Table V), trees were assigned ranked scores as to their relative resistance or susceptibility. For both the 1981 and 1982 data sets, trees for which the multiple regression equations predicted low female budworm dry weights were defined as being relatively resistant and assigned lower rankings. Susceptible trees for which the regression equation predicted high female budworm dry weights were assigned high rankings. When the ranked scores for sample trees were compared between years, a positive and significant relationship was found (Spearman's $r = 0.518$, $p = 0.0034$). Thus, relatively resistant trees in 1981 remained relatively resistant in 1982 in spite of the vicissitudes in the abiotic factors between these two years. However, the terpenes determining resistance were different each year.

Three alternatives, or combinations thereof, may explain this correlation. First, a possibility exists that during the sampling of the trees in each of the 3 years, we selected buds of different sizes. As the tissue develops from budbreak to maturity, there is a significant change in the quantities of individual terpenes and in total terpenes through the season (Table I; also see Gambliel and Cates, 1988). Buds, or young foliage shoots, that are in the 100-mg fresh-weight range have about a fourth the quantity of terpenes as do buds in the 300-mg range. We deem this possibility of systematic error unlikely since a deliberate effort was made to select buds within and among trees that were similar in size and age. Second, budworm were increasing in the population of Douglas-fir, and resulting increased defoliation of the trees may have resulted in changes in the terpene chemistry by shifting the allocation of energy to different compounds in the terpene metabolic pathway possibly through changes in the carbon/nutrient balance. However, we note that there was not a significant difference in the total quantity of terpene produced each year in the foliage of trees, but just a quantitative shift in the amounts among individual terpenes among years. Field and laboratory studies are being initiated to test this hypothesis.

Third, long-lived hosts may have been selected to rapidly alter or vary quantitatively and qualitatively their chemical defenses (Cates and Redak, 1986). Variation in the production of secondary metabolites among years may introduce an element of "surprise" and increase the difficulty of insects adaptating to host defenses. Consequently, it may not be unusual to find different compounds deterring budworm in different years. It is interesting to note that we observed this pattern only when a natural budworm population was beginning to increase in numbers. One of the planned laboratory experiments is to vary terpene quantity and quality in agar diets to determine if budworm are adversely affected by such variation.

D. Douglas-Fir Population Structure and Herbivory

In the previous section, discussion centered on the maintenance of relative resistance in Douglas-fir trees among seasons from the same population. In addition, a smaller subset of trees that were susceptible to the budworm in 1981, remained susceptible in 1982, and appeared to maintain their same relative level of susceptibility in 1983 (Cates and Redak, 1986; Cates, unpublished data).

These data suggest that this population of Douglas-fir is composed primarily of resistant trees but also included in the population is a smaller number of susceptible trees. Susceptible trees do not appear, at least as a group, to be able to completely switch to the *modus operandi* of resistant trees, although it is possible that some individuals within the susceptible group may change their terpene chemistry over time resulting in increased resistance (Cates, unpublished). Populations of Douglas-fir that appear to have a structure of primarily resistant trees infiltrated with a smaller number of susceptible trees also have been observed at the Whitetail Deer Creek site south of Boulder, Montana (Cates *et al.*, 1983a) and at two research sites near Island Park, Idaho (Targhee National Forest) (Cates, unpublished).

The evolution of such a population structure may occur through an interaction between selection pressures dealing with the physical environment and those related to the frequency and intensity of herbivore attack. For example, in Douglas-fir populations that experience infrequent attacks or sparse budworm densities, trees with low resistance to budworm would be favored since presumably most of their energy would be shunted from defense and toward adaptations that increase their ability to acquire resources needed for growth and reproduction (Rhoades and Cates, 1976; Bryant *et al.*, 1983). However, in a population of trees frequently attacked by budworm, selection might favor the population structure to swing toward resistance with susceptible trees surviving only in microsites that are most favorable for growth. We suggest the hypothesis that susceptible trees may play a key role in the interaction between the level of evolved host tree resistance and budworm resistance to the tree's defensive system. The presence of susceptible trees may dilute the evolution of a highly resistant gene pool in the budworm population. This would operate best if random mating occurred within and among populations of a panmictic herbivore species. To our knowledge this hypothesis has not been tested. In testing this hypothesis in the field, considerable attention would need to be given to the populations of Douglas-fir and budworm used in the investigation. Site and stand characteristics, such as tree species composition, tree and stand vigor, stocking density, crown classes, stand structure, and tree and stand maturity greatly influence the outcome of the interaction between the host tree species and the herbivore (Wulf and Cates, 1987).

E. Geographic Variation in the Production of Terpenes among Douglas-Fir Populations

Data presented in this paper confirm that the geographic variation observed in mature, current year foliage (von Rudloff and Rehfeldt, 1980) is present in the developing current year foliage upon which budworm feed. Similarly, some of the populations differ in qualitative production of terpenes. For example, the Montana and Mt. Sawtelle and McCall, Idaho, populations were very high in bornyl acetate, which has been demonstrated to be an effective toxin against the budworm (Cates *et al.*, 1983a,b, 1986). However, none was recorded in the foliage of the other populations. These three populations also were much higher in thujene, limonene, and camphene in comparison to the foliage of the other populations. Differences existed among the three eastern populations (Montana, Mt. Sawtelle, and McCall, Idaho) in that the Mt. Sawtelle population was relatively higher in the sesquiterpene cadinene (as were the four western populations) but the Montana and McCall, Idaho, populations were very low (in some cases none was recorded in the gas chromatographic analysis). Alternatively, sabinene was consistently an important component in the foliage of the four western populations, whereas sabinene was unimportant in the three eastern populations. Certainly, if various individual terpenes or terpene combinations function as defenses in the foliage of

Douglas-fir populations, there appears to be sufficient qualitative and quantitative variation among populations to select for host races in the budworm as well as other oligophagous herbivores feeding on Douglas-fir.

Preliminary experiments suggesting host-race development have shown that budworm taken from an Ashton, Idaho population and reared on trees at an Oquirrh Mountain, Utah site experience a significant ($p < 0.05$) decrease in larval survival and pupal weight when compared to Ashton budworm reared on their natural Ashton trees. Similarly, budworm from the Oquirrh Mountain, Utah site when reared on Ashton trees also showed a significant decrease in larval survival and pupal weight when compared to control budworm reared on their natural Oquirrh Mountain, Utah trees (Cates, unpublished). The role of much of the variation that has been described for Douglas-fir in this paper will become much more apparent when studies take into account both insect and host plant population dynamics.

F. Induction of Secondary Metabolites in Douglas-Fir

Variation due to long-term induction of secondary metabolites in leaves has been noted for several deciduous species (e.g. Haukioja and Niemela, 1979; Denno and McClure, 1983; Whitham, 1983; Tuomi *et al.*, 1984; Haukioja *et al.*, 1985) but this same phenomenon appears to be less common in the needles and leaves of evergreens. For example, none of the data that we have accumulated from field studies dealing with terpene chemistry in Douglas fir needles suggests induction in terpenes to the same extent as that observed in the above reports for deciduous species (Cates *et al.*, 1983a; Redak and Cates, 1984; Cates and Redak, 1986). In addition, a laboratory study was designed to assess the potential of long-term induction in Douglas-fir needles following mechanical injury. Although the data from this study are in a preliminary stage of statistical analysis, induction at the level described above for deciduous species was not observed in Douglas-fir (Sheldahl and Cates, unpublished).

Our preliminary data, along with observations from the literature, suggest that the lack of long-term induction of secondary metabolites in Douglas-fir needles, and possibly for most evergreen conifers and angiosperms (J. Bryant, personal communication), may be related to needles functioning as a storage organ to a greater capacity when compared to leaves of deciduous species. Carbohydrates and nutrients appear to be stored to a larger extent in the needles of pines, Douglas-fir, and the leaves of evergreen species growing on sites less favorable for growth than on sites or under conditions more favorable for growth (Webb and Karchesy, 1977; Kramer and Kozlowski, 1979, Ch. 7,9; Gosz, 1981; Horner *et al.*, 1988). Consequently, for evergreen species removal of photosynthetic tissue by herbivores may reduce the carbohydrate and nutrient levels required for the induced production of secondary metabolites. Whether or not the new tissues produced to replace severely damaged leaves or needles will be more nutritious may depend on several factors including the availability of reserves from other leaves, twigs, or branches that could be used in the production of new leaves and associated defenses, time of defoliation, and the number of leaf-chewing generations of the herbivore within a growing season of the plant. Studies designed to investigate replacement of photosynthetic tissue due to herbivore or pathogen damage, the potential for induction in the new tissues, and the relationship between new tissue nutrition and susceptibility to natural enemies are needed.

G. Host-Tree Resistance–Susceptibility and Budworm Population Dynamics

In every field study conducted to date, the quantity of an individual terpenoid, the acetate fraction, and/or terpene evenness (the evenness in the quantitative distribution of terpenes in the foliage among the sampled trees) were the most important variables in the correlation models (Cates *et al.*, 1983a; Redak and Cates, 1984; Cates and Redak, 1986). Using synthetic diets, bornyl acetate, which was inversely related to female and male budworm success in two of three field studies cited above, significantly reduced larval growth rate and survival and increased adult mortality when administered in concentrations normally found in the foliage of over 300 Douglas-fir trees (Cates *et al.*, 1987). Tannin capacity never entered any of the models for either female or male budworm, suggesting that these compounds are not in high enough concentration to adversely influence budworm, or that the budworm has ameliorated the complexing capacity of the condensed tannins in the current year's foliage of Douglas-fir (Martin and Martin, 1982). In these same studies, primary nutrition, as estimated by total, soluble, insoluble, and the ratio of soluble to insoluble nitrogen was a minor contributor to the significance of the models; frequently it did not occur in the models. In only one study (Redak and Cates, 1984) was nitrogen an important component, and in that study it was inversely related to female dry weight production. We suggested that in this case total foliar nitrogen appeared to be an indicator of tree vigor and was not functioning as a toxic component (Redak and Cates, 1984).

The tree morphological variables measured were shown not to reliably predict budworm success, tree resistance, or tree susceptibility. Most of the physical variables measured were related to age and bole characteristics such as 5-year growth increments, diameter at breast height (DBH), sapwood/basal area ratios, and bole radius. None of these, nor crown ratio or crown diameter variables, reliably predicts the susceptibility of a tree to the budworm, or budworm preference (Cates *et al.*, 1983a; Redak and Cates, 1984; Cates and Redak, 1986). The most consistent morphological variable entering the correlation models was 5-year growth increment, but after appearing as an important parameter in the 1980 and 1981 female models, it did not appear in the 1982 female model. Studies are planned to determine the effects of thinning on the level of susceptibility or resistance of trees to the budworm. Perhaps these, in conjunction with appropriate measurements of physical parameters of foliage and crown dynamics, will provide a clue as to what morphological or physical tree variables may predict accurately budworm success and/or tree resistance at low to moderate budworm population levels. It is important to note that in the Montana study where the budworm was allowed to interface with foliage phenology naturally, trees that were characterized by late budbreak were less susceptible to budworm, a result predicted by theory (Rhoades and Cates, 1976; Cates and Rhoades, 1977).

The idea that qualitative or toxic defenses may primarily offer protection against generalized or nonadapted herbivores (e.g., Rhoades and Cates, 1976; Krischik and Denno, 1983) is not supported by the above results. In every case, increased titers of terpenes (low-molecular-weight, potentially toxic secondary metabolites) in the current year's foliage of Douglas-fir were always inversely associated with budworm density, and dry weight production. Agar diet studies have shown that bornyl acetate is extremely toxic to the well-adapted, monophagous–oligophagous western spruce budworm (Cates *et al.*, 1983a, 1987). Alternatively, these data do support the contention that herbivore success is adversely affected when the pattern in terpene chemistry production significantly deviates from the population average (Cates and Rhoades, 1977).

V. Summary

This chapter has delineated variation in chemistry that has been observed in the current year's foliage of Douglas-fir. A summary of the types of variation observed in terpenoids, and in nitrogen and water content includes:

1. Variation through the growing season in individual terpenoids, total terpenoid content, terpene evenness, and total nitrogen and water content.
2. Variation due to aspect and degree of water stress in individual terpenoids and terpene evenness among trees within a population.
3. Variation among seasons in terpenoids of the same trees that were sampled in 1981, 1982, and 1983.
4. Variation in the same trees between seasons in potential and actual terpenoid resistance factors.
5. Geographical variation in terpenoids among populations.

The variation observed within a growing season, among individual trees within a population, and between years has been shown to influence significantly budworm dry weight production and, hence, egg mass production (Redak and Cates, 1984). Studies of the population dynamics of both the budworm and its host plant will undoubtedly reveal in greater detail the major role that variation plays as a plant defensive mechanism as well as a strong selection pressure in the formation of herbivore host races.

These data, as well as those in the literature (Schultz *et al.*, 1982; Arrhenius and Langenheim, 1983; Cates *et al.*, 1983a,b; Denno and McClure, 1983; Whitham, 1983; Gambliel *et al.*, 1985; Cates and Redak, 1986; among many others), indicate that variation in the qualitative and quantitative production of secondary metabolites is a most important factor in the mediation of coevolutionary processes at the plant–herbivore interface. This same variation plays a major role in mediating the plant–herbivore–predator interface (Price *et al.*, 1980; Smiley *et al.*, 1985). Appropriate hypotheses must be tested so that the role of variation in plant–herbivore systems can be assessed and intertwined with current knowledge dealing with patterns in the production of secondary metabolites among growth forms and due to variation in carbon/nutrient balance considerations (Bryant *et al.*, 1983). Finally, we will bear the fruits of this research at another level when we are able to incorporate the information dealing with variation in defensive mechanisms in natural forest systems and into other integrated pest-management schemes.

Acknowledgments

We wish to thank T. McMurray, J. Horner, K. Petersen, H. Alexander, C. Henderson, H. Gambliel, M. Tandysh, C. Nixon, and Steve Andrews for their advice and help. In addition, we thank the numerous other graduate and undergraduate students (and including everyone above) for the many hours of sorting budworm, collecting tissue, analyzing samples, and the interesting experiences we had in obtaining the predawn water potentials. We thank James R. Gosz for his advice and help as well. For the research support provided by the National Science Foundation grants DEB 7619950, DEB 7922067, Research Allocations Committee at the University of New Mexico, and the Professional Development Funds at Brigham Young University, we are grateful. Work leading to this publication also was funded by the Canada/United States Spruce Budworms program, and Accelerated Research, Development, and Application program, sponsored by the U.S. Department of Agriculture Forest Service.

References

Arrhenius, S., and Langenheim, J. (1983). Inhibitory effects of *Hymenaea* and *Copaifera* leaf resins on the leaf fungus, *Pestalotia subcuticularis*. *Biol. Syst. Ecol.* **11,** 361–366.

Bryant, J., Chapin, F., and Klein, D. (1983). Carbon/nutrient balance of boreal plants in relation to vertebrate herbivory. *Oikos* **40,** 357–368.

Cates, R. G., and Redak, R. (1986). Between-year population variation in resistance of Douglas-fir to the western spruce budworm. *Am. Chem. Soc. Symp. Ser.* (296), 106–115.

Cates, R. G., and Redak, R. (1988). Response of budworm to water stress induced changes in foliage quality. Submitted.

Cates, R. G., and Rhoades, D. F. (1977). Patterns in the production of antiherbivore chemical defenses in plant communities. *Biochem. Syst. Ecol.* **5,** 185–193.

Cates, R. G., Redak, R., and Henderson, C. (1983a). Patterns in defensive natural product chemistry: Douglas-fir and western spruce budworm interaction. *Am. Chem. Soc. Symp. Ser.* (208), pp. 3–19.

Cates, R. G., Redak, R., and Henderson, C. (1983b). Natural product defensive chemistry of Douglas-fir, western spruce budworm success, and forest management practices. *Z. Angew. Entomol.* **96,** 173–182.

Cates, R. G., Henderson, C., and Redak, R. (1987). Responses of the western spruce budworm to varying levels of nitrogen and terpenes. *Oecologia* **73,** 312–316.

Chew, F. S., and Rodman, J. E. (1979). Plant resources for chemical defense. *In* "Herbivores: Their Interaction with Secondary Metabolites" (G. Rosenthal and D. Janzen, eds.), pp. 271–307. Academic Press, New York.

Coley, P., Bryant, J., and Chapin, F. (1985). Resource availability and plant antiherbivore defense. *Science* **230,** 895–899.

Crankshaw, D., and Langenheim, J. (1981). Variation in terpenes and phenolics through leaf development in *Hymenaea* and its possible significance to herbivory. *Biochem. Syst. Ecol.* **9,** 115–124.

Denno, R., and McClure, M. (1983). "Variable Plants and Herbivores in Natural and Managed Systems." Academic Press, New York.

Dixon, W. J., ed. (1983). "BMDP Statistical Software." Univ. of California Press, Berkeley.

Feeny, P. (1976). Plant apparency and chemical defenses. *Recent Adv. Phytochem.* **10,** 1–40.

Gambliel, H., and Cates, R. G. (1988). Terpene changes due to maturation and canopy level in Douglas-fir (*Pseudotsuga menziesii*) flush needle oil. *Can. J. Bot.* (submitted).

Gambliel, H., Cates, R. G., Caffey-Moquin, M., and Paine, T. D. (1985). Variation in the chemistry of Loblolly pine in relation to infection by the blue-stain fungus. *Proc. Integrated Pest Manage. Res. Symp. South. For. Exp. Sta., New Orleans* pp. 177–184. Louisiana.

Gosz, J. (1981). Nutrient cycling in coniferous ecosystems. *In* "Nitrogen Cycling in Terrestrial Ecosystems Processes, Ecosystem Strategies, and Management Impacts." Ecological Bulletins, Vol. 33. (F. Clark and T. Rosswall, eds.), pp. 405–426. Stockholm, Norway.

Harris, R. (1975). "A Primer of Multivariate Statistics." Academic Press, New York.

Haukioja, E., and Niemela, P. (1979). Birch leaves as a resource for herbivores: Seasonal occurrence of increased resistance in foliage after mechanical damage of adjacent leaves. *Oecologia* **39,** 151–159.

Haukioja, E., Suomela, J., and Neuvonen, S. (1985). Long-term inducible resistance in birch foliage: triggering cues and efficacy on a defoliator. *Oecologia* **65,** 363–369.

Horner, J. (1984). Phenological and stress-induced changes in phenolic and tannin composition of Douglas-fir foliage. M.S. thesis, University of New Mexico, Albuquerque.

Horner, J., Gosz, J., and Cates, R. G. (1988). Carbon-based plant secondary metabolites and terrestrial ecosystem function. *Am. Nat.* In Press.

Janzen, D. (1973). Community structure of secondary compounds in plants. *Pure Appl. Chem.* **34,** 529–538.

Kramer, P., and Kozlowski, T. (1979). "Physiology of Woody Plants." Academic Press, New York.

Krischik, V., and Denno, R. (1983). Individual, population, and geographic patterns in plant defense. *In* "Variable Plants and Herbivores in Natural and Managed Systems" (R. Denno and M. McClure, eds.), pp. 463–512. Academic Press, New York.

Louda, S., and Rodman, J. (1983). Concentration of glucosinolates in relation to habitat and insect herbivory for the native crucifer *Cardamine cordifolia*. *Biochem. Syst. Ecol.* **11,** 199–207.

McClure, M., and Hare, J. (1984). Foliar terpenoids in *Tsuga* species and the fecundity of scale insects. *Oecologia* **63,** 185–193.

McKey, D. (1979). The distribution of secondary compounds within plants. *In* "Herbivores: Their Interaction with Secondary Plant Metabolites" (G. Rosenthal and D. Janzen, eds.), pp. 56–133. Academic Press, New York.

Martin, J., and Martin, M. (1982). Tannin assays in ecological studies: Lack of correlation between phenolics, proanthocyanidins and protein-precipitating constituents in mature foliage of six oak species. *Oecologia* **54,** 205–211.

Neff, N., and Marcus, L. (1980). A Survey of Multivariate Methods for Systematics. A Workshop in Numerical Methods in Systematic Mammalogy. *Am. Soc. Mammalog. Annu. Meet. 9 June.*

Price, P., Bouton, C., Gross, P., McPhreon, B., Thompson, J., and Weis, A. (1980). Interactions among three trophic levels: Influence of plants on interactions between insect herbivores and natural enemies. *Annu. Rev. Ecol. Syst.* **11,** 41–266.

Raffa, K., and Berryman, A. (1983). The role of host plant resistance in the colonization behavior and ecology of bark beetles (Coleoptera: Scolytidae). *Ecol. Monogr.* **53,** 27–49.

Redak, R., and Cates, R. G. (1984). Douglas-fir (*Pseudotsuga menziesii*)—spruce budworm (*Choristoneura occidentalis*) interactions: The effect of nutrition, chemical defenses, tissue phenology, and tree physical parameters on budworm success. *Oecologia* **62,** 61–67.

Rhoades, D. F. (1979). Evolution of plant chemical defenses against herbivores. *In* "Herbivores: Their Interaction with Secondary Plant Metabolites" (G. Rosenthal and D. Janzen, eds.), pp. 3–54. Academic Press, New York.

Rhoades, D. F., and Cates, R. G. (1976). Toward a general theory of plant antiherbivore chemistry. *Rec. Adv. Phytochem.* **10,** 168–213. Plenum, New York.

Schultz, J., Nothnagle, P., and Bladwin, I. (1982). Seasonal and individual variation in leaf quality of two northern hardwood tree species. *Am. J. Bot.* **69,** 753–759.

Smiley, J., Horn, J., and Rank, N. (1985). Ecological effects of salicin at three trophic levels: New problems from old adaptations. *Science* **229,** 649–651.

Tuomi, J., Niemela, P., Haukioja, E., Siren, S., and Neuvonen, S. (1984). Nutrient stress: An explanation for plant anti-herbivore responses to defoliation. *Oecologia* **61,** 208–210.

von Rudloff, E., and Rehfeldt, G. (1980). Chemosystematic studies in the genus *Pseudotsuga*. IV. Inheritance and geographical variation in the leaf oil terpenes of Douglas-fir from the Pacific Northwest. *Can. J. Bot.* **58,** 546–556.

Waring, R. H., McDonald, A. J. S., Larsson, S., Ericsson, T., Wiren, A., Arwidsson, E., Ericsson, A., and Lohammar, T. (1985). Differences in chemical composition of plants grown at constant relative growth rates with stable mineral nutrition. *Oecologia* **65,** 157–160.

Webb, W. (1980). Starch content of conifers defoliated by the Douglas-fir tussock moth. *Canad. J. For. Res.* **10,** 535–540.

Webb, W., and Karchesy, J. (1977). Starch content of Douglas-fir defoliated by the tussock moth. *Canad. J. For. Res.* **7,** 186–188.

Welch, B., and McArthur, C. (1981). Variation of monoterpenoid content among subspecies and accessions of *Artemisia tridentata* grown in a uniform garden. *J. Range Manage.* **34,** 380–384.

White, T. C. R. (1969). An index to measure weather-induced stress of trees associated with outbreaks of psyllids in Australia. *Ecology* **50,** 905–909.

White, T. C. R. (1974). A hypothesis to explain outbreaks of looper caterpillars with special reference to populations of *Selidosema suavis* in a plantation of *Pinus radiata* in New Zealand. *Oecologia* **16,** 279–301.

Whitham, T. (1983). Host manipulation of parasites: Within-plant variation as a defense against rapidly evolving pests. *In* "Variable Plants and Herbivores in Natural and Managed Systems" (R. Denno and M. McClure, eds.), pp. 15–41. Academic Press, New York.

Wratten, S., Edwards, P., and Dunn, I. (1984). Wound-induced changes in the palatability of *Betula pubescens* and *B. penula*. *Oecologia* **61,** 372–375.

Wulf, W., and Cates, R. G. (1987). Site and stand characteristics. *In* "Western Spruce Budworm" (R. Brooked, R. Campbell, J. Colbert, R. Mitchell, and R. Stark, eds.). USDA Forest Service, Washington, D.C.

CHAPTER 11

The Induced Defense Hypothesis: Does It Apply to the Population Dynamics of Insects?

JUDITH H. MYERS
The Ecology Group and
Departments of Plant Science and Zoology
University of British Columbia
Vancouver, Canada

I. Introduction

One of the most popular ideas in evolutionary ecology in the last 10 years is that secondary plant compounds are defensive adaptations against herbivores. This "arms race" view of the world fits into the broader perspective of coevolution, the mutual and repetitive adjustment of two closely related species. But it is difficult to demonstrate reciprocal adaptation between two species in evolutionary time, and more rapid responses have not been found (Strong *et al.*, 1984). Evidence of plants evolving new chemical defenses in response to herbivores in natural systems is scarce, although the feeding and oviposition choices of insect herbivores are often dictated by the chemical composition of plants (Futuyma, 1983). Either the time scale of these evolutionary changes is too long for observation, or the proposed coevolutionary "arms race" is not a common process.

Induced chemical plant defense, a direct phenotypic response by plants to damage from insects, appears to be a much more testable phenomenon. Because a genetic shift is not a prerequisite for the change in plant quality, induced defenses do not involve the reciprocal genetic change necessary for coevolution. However, if plants respond to herbivory by increasing levels of compounds detrimental to the herbivores, a specific adaptation is implied. This then could be interpreted as a coevolutionary change. Increases in plant phenolic compounds and decreases in nitrogen content of leaves following foliar damage or defoliation have been observed (Rhoades, 1983a,b; Schultz and Baldwin, 1982; Tuomi *et al.*, 1984). These chemical changes are assumed to be "defensive" responses, but while their impact on the population dynamics of insect herbivores is frequently suggested, it has never been investigated to my knowledge.

The "induced defense" hypothesis has been expressed in a variety of ways (Rhoades, 1985). In this chapter I define the induced defense hypothesis to state that *the feeding of herbivorous insects stimulates a biochemical change in plants that protects the plants against herbivory by reducing the growth, survival, and/or fecundity of insects, and therefore causing a decline in the insect population.* In this situation the insect is adapted to oviposit and/or feed on the plant, but the plant, through an active process of chemical change, becomes less suitable as food following insect attack. I define defense to be a reaction of the plant that takes place at low levels of herbivory and reduces herbivore damage. I do not consider a passive change such as deterioration of nutrient quality following a major reduction in photosynthetic activity to be a defense. A dead tree is not a well defended tree, even though insect survival on a tree in this condition is zero. Physiological changes by the plants that do not translate into a reduction of the insect population are not a defense by this definition.

In this chapter I consider three questions about induced chemical defenses in an attempt to evaluate the relevance of this hypothesis to insect population dynamics. First, I consider whether chemical changes reported to occur in damaged plants are detrimental to insect herbivores and whether they are specific responses to herbivory. Next, from my own studies of red alder (*Alnus rubra* Bong.) and western tent caterpillar (*Malacosoma californicum pluviale* Dyar), I evaluate laboratory bioassays of foodplant quality and compare those to field assays of foodplant quality. Since much of the evidence on the induced defense hypothesis comes from laboratory bioassays, the comparison of field and laboratory results is necessary to determine the relevancc of this type of information. Finally, I consider whether speculations about the impact of plant "defenses" on insect population dynamics are consistent with parameters observed in fluctuating populations of the western tent caterpillar, and those reported of Douglas-fir tussock moth (*Orgyia pseudotsugata* McDonnough) and larch budmoth (*Zeiraphera diniana* Hubner).

A. Are Increased Concentrations of Phenolic Compounds Specific Defensive Responses to Insects?

Concentrations of particular or total phenolics have been observed to increase in damaged leaves (Niemela *et al.*, 1979; Schultz and Baldwin, 1982), and larvae fed damaged leaves sometimes show reduced growth and smaller final size (Haukioja and Hankimaki, 1985; Wallner and Walton, 1979). These observations led to the interpretation of phenolic compounds as being induced plant defenses. However, this interpretation is based on two assumptions: (1) that phenolics are detrimental to the insects adapted to feed on the plant in question, and (2) that phenolics are specific adaptations for defense against insects. These two assumptions may not be valid.

Phenolic compounds such as tannins, lignins, resins, and phenolic acids are frequently called "defense chemicals." Sometimes these compounds are called defenses without evidence of any insects being deterred by them (Mihaliak and Lincoln, 1985). In many cases the only evidence that a particular substance is detrimental to insect growth or survival comes from adding the chemical to artificial medium (e.g., Feeny, 1970; Reese *et al.*, 1982). However, as Campbell and Duffy (1979) show, these results should be interpreted with caution. Allelochemicals may have no effect on insects when in the natural situation in plants. Other constituents of the food medium can influence the reaction of an insect to an allelochemical (Elliger *et al.*, 1976). Klocke and Chan (1982) observed a decrease in growth of *Heliothis zea* (Boddie) fed artificial medium with increased levels of condensed tannins. They attributed this to reduced feeding by caterpillars on food with higher tannin levels. However, caterpillars were not deterred from feeding on leaf discs to which very high concentrations of tannins were applied. This may indicate a difference in the influence of tannins in medium and in association with leaves.

Insects do not always respond to different levels of phenols in their host plants. Lawson *et al.* (1982) found that variation in the tannin concentration of leaves had no effect on the consumption, utilization, or growth of larval oakworms, *Anisota senetoria* J. E. Sm. Bernays (1978) showed that growth and survival of *Schistocerca gregaria* Forskal was influenced by additional condensed or hydrolyzable tannins. The tree locust *Anacridium melanorhodon* actually had improved growth and survival when tannic acid was added to the diet (Bernays *et al.*, 1983). Fox and Macauley (1977) found no difference in the nutritional physiology of *Paropsis atomaria* fed *Eucalyptus* leaves with different levels of condensed tannins. Martin and Martin (1984) have shown that surfactants in insect guts can reduce the protein-binding property of tannins and therefore eliminate their impact on the digestion of insects. Zucker (1983) suggests that condensed tannins are more likely defenses against microbes and pathogens rather than insects. Leaf miners in damaged leaves that had higher condensed tannin levels had reduced mortality from fungus and bacterial infection but higher rates of parasitization (Faeth, 1986).

The relative growth rate of the California oak moth *Phryganidia californica* Packard was the same when fed new and mature leaves of *Quercus lobata* Neé even though the new leaves were significantly higher in total polyphenols. On the other hand, mean growth rate, fecundity, and pupal mass were higher when oak moth caterpillars were fed on *Q. lobata* than on *Q. agrifolia* Neé, although mature leaves of both species had similar levels of total polyphenols (Puttick, 1986). Other characteristics such as leaf toughness and water content appeared to be the important variables. The classic study of Feeny (1970) showed a stronger effect of leaf toughness than tannin levels on winter moth growth and survival.

To summarize, certainly not all insects respond in a negative way to phenols or particular tannins. Biochemical change in leaves following damage or with maturity is frequently evaluated by levels of total phenols. Higher levels are interpreted to indicate increased plant defense (e.g., Schultz and Baldwin, 1982). However, without an evaluation of the impact of total or particular phenols on the insects, any extrapolation to plant defense theory is unwarranted. Insects adapted to plants high in phenol content such as oaks and *Eucalyptus* may be totally uninfluenced by changing levels of these compounds.

The association between leaf damage and increased levels of phenols is partly responsible for the interpretation that phenols must be defensive chemicals. But are these changes in leaf biochemistry specific adaptations to insect attack? In general, foliar damage and stressful conditions for plants and trees seem to initiate similar biochemical changes. When leaves are damaged, respiration increases, ethylene is produced, and concentrations of phenylalanine ammonia lyase (PAL) increase (Yang and Pratt, 1978). The increased levels of PAL cause a shift in the shikimic acid pathway toward the production of phenolic acids and away from the production of phenylalanine. These biochemical changes in stressed plants can be interpreted as a reallocation of resources from the production of compounds containing nitrogen, (phenylalanine and tyrosine) to nonnitrogenous compounds (cinnamic acid and other phenolic acids). Increases in phenolic acids occur when plants are nutrient- or water-stressed (del Moral, 1972).

Phenolic compounds are highest in concentration when trees are stressed, such as after defoliation, and concentrations decline when trees are fertilized (Mihaliak and Lincoln, 1985; Tuomi *et al.*, 1984). Therefore, if phenolic compounds are defensive, stressed trees should be better defended against insect attack. However, Mattson and Addy (1975), Rhoades (1983b), and White (1984) propose just the opposite: that environmental stress triggers insect outbreaks. Little data are available to resolve this contradiction. Haukioja (1980) observed that birch trees living in a harsh environment on the edge of their distribution showed the strongest "defensive" response (increase in phenols immediately following damage), but moths reached outbreak densities at these sites.

According to my observations on tent caterpillars and fall web worms, isolated trees have the highest densities of caterpillars (Moore *et al.*, 1988). These trees receive high levels of solar radiation, which, according to Schultz (1983), increases total phenol content of leaves. This situation also suggests that phenols are in some cases not effective at defending trees against insect attack. These observations suggest that it is unlikely that variations in the foliar concentrations of phenols dictate the population dynamics of insect herbivores.

Finally, it is very difficult to evaluate the impact of changes in phenolic compounds on insects in a natural situation because leaf damage and/or repeated defoliation is associated with a suite of chemical modifications: reduced water, reduced nitrogen, and/or increased fiber content and leaf toughness (Baltensweiler *et al.*, 1977; Feeny, 1970; Schultz and Baldwin, 1982; Valentine *et al.*, 1983). Any and all of these changes could be detrimental to insect growth, and are the types of responses expected if the photosynthetic activity of plants is reduced (Bryant *et al.*, this volume), as with resource depletion after repeated defoliation. Therefore, these changes are more likely to be passive responses than active defenses (Williams and Myers, 1984).

In conclusion, the increase in concentration of phenolic compounds observed in damaged leaves or following repeated defoliation is a general response to stress, wounding, or reduced nutrient availability. Similar patterns of response are observed in plants subjected to physical damage or stimulated by bacteria, fungi, or virus (Cramer *et al.*, 1985; Sequeira, 1983). It cannot be assumed that these changes have an impact on the reproductive attributes or population dynamics of insect defoliators. The occurrence of chemical change following foliar damage is not evidence for a defensive adaptation against insects.

B. Are Laboratory Bioassays Indicative of Field Foodplant Quality?

To test the "induced defense" hypothesis the quality of foliage from plants attacked by insects must be compared to that of undamaged plants. Food quality has been assayed either by chemical composition or by the growth, survival, and ultimate size of insects reared on foliage from the two types of plants. Some of the problems with trying to make these comparisons have been outlined by Neuvonen and Haukioja (1985). The major problem is that to control for microclimatic differences, the insects must be reared in the same environment, which almost always means feeding them cut vegetation in the laboratory. Before conclusions can be drawn from laboratory bioassays of foodplant quality, it must be established that the results of insect growth experiments using cut foliage are comparable to the field situation. I have investigated this for a population of western tent caterpillars.

The natural history of the western tent caterpillars has been described previously (Myers, 1981; Wellington, 1960). Female tent caterpillars lay all of their eggs in a single mass in late June to early July. In mid-April larvae hatch and they form a silken tent. Caterpillars from a single egg mass remain together, and feed gregariously off the tent. Because of this gregarious feeding the caterpillars rapidly devour total leaves once they reach the second or third instar. Western tent caterpillars feed on a variety of tree species such as alder, apple, birch, cherry, and wild rose.

II. Methods

The methods used in the caterpillar growth experiments have been outlined by Myers and Williams (1984). Early-instar caterpillars were divided into groups and maintained in plastic containers. Caterpillars or egg masses were collected from the field each year for the experiments. Foliage was collected from red alder trees (*A. rubra*) every several days, from an area with prolonged and intense insect herbivory (Cambie St. in Myers and Williams, 1984) and from another area where trees had not been previously attacked by tent caterpillars (University of British Columbia campus). A different control area was used each year, but the area with intense herbivory, where the tent caterpillar population had been high since 1980, was the same in each of the four years of the study. Food quality was assayed as the proportion of caterpillars surviving to the fifth instar, proportion of caterpillars surviving from the fifth instar to pupate, and the size of female pupae. In the field population, early larval survival was estimated by measuring the tents produced by the caterpillars in the fourth instar. In the area of high caterpillar density, tents occurred naturally. Alder trees used as controls were isolated groups of young trees that had never supported a natural population of tent caterpillars. Egg masses were transferred from the high-density population (Cambie St.) to these control areas in February or March (Myers and Williams, 1984). The tents of these caterpillars were measured in June. The index of tent size was the length of the tent multiplied by the width. This measurement is correlated to the number of caterpillars forming the tent (1984, $r = 0.70$, slope = 1.1, $N = 17$; 1985, $r = 0.94$, slope = 1.0, $N = 11$) and therefore gives an estimate of early larval survival in the field. Pupae were collected from the field sites by searching tents and curled leaves on the alder trees. Pupae were weighed, and only the results for female pupae are presented since they are most relevant to the rate of increase of the population.

III. Results

The weights of female pupae from caterpillars reared in the laboratory on cut foliage are shown in Fig. 1. In 1982, pupal weights were similar for caterpillars reared on foliage from an area with intense herbivory and control foliage. From 1983 through 1985, female pupal weights declined progressively on both types of foliage: control and with intense herbivory.

The lower weights of pupae reared on foliage from the area of intensive herbivory can be explained by lower nitrogen content (percent dry weight of total nitrogen) of leaves of trees from that site (Myers and Williams, 1987). In 1982, the nitrogen content of foliage from the area of intense herbivory and the control area was 2.23%. In subsequent years, the foliage from control areas had higher nitrogen content, and in 1984 and 1985 the difference was approximately 1% (Myers and Williams, 1987). The higher nitrogen content of control foliage can explain the heavier weight of pupae of larvae fed leaves from control trees.

The most interesting observation is that the weight of pupae from the area of intense herbivory declined over the years on both types of foliage. This seems to indicate a deterioration in the caterpillar quality with progressive years of high density. The decline in pupal weight from 1983 to 1985 cannot be explained by reduced nitrogen content, since the nitrogen content of foliage of trees in the area of high caterpillar density varied only between 2.23% in 1982 and 2.35% in 1985 (Myers and Williams, 1987).

Field-collected pupae from the area of intense herbivory weighed less than those from low-density control areas. This also reflects the reduced nitrogen content of foliage at the high density site. However, weights of pupae collected from the field sites did not decline from 1983 to 1985. No deterioration was detected in the field populations to confirm the findings of the laboratory studies.

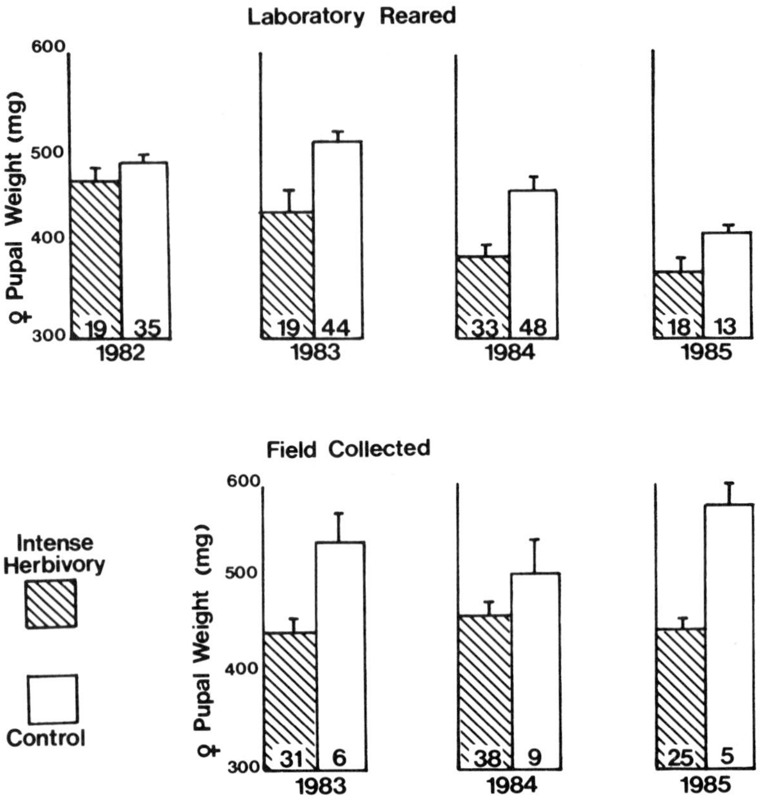

Fig. 1. Mean weight (±1 SE) of female pupae of western tent caterpillars reared in the laboratory on foliage from red alder trees exposed to intense herbivory or control trees, and pupae collected from comparable field populations. Sample size is given on the bars. For laboratory-reared pupae the differences in weights are significant for 1983, 1984, and 1985. Weights of field-collected pupae were significant in 1983 and 1985.

Female pupal weight is frequently used in bioassays of food quality because it is related to egg production. Since tent caterpillars lay all of their eggs in one batch, the mean number of eggs per batch from field populations is a measure of fecundity. These data (Table I) show an increase in fecundity of moths from 1982 to 1984, when the laboratory studies showed a decline in pupal weight of females. There is no indication of a reduction of fecundity over 3 years of high density.

Early survival of caterpillars fed cut foliage was significantly lower in 1982 and 1985 than in intervening years, but survival did not differ with food type (Table II). However, in 1982 and 1985, early larval survival, indicated in the field by the size of the tents, was poorer in the area of high

Table I. Mean Number of Eggs per Mass for a Western Tent Caterpillar Population

Year	Number of Eggs/Mass [mean (SE) *N*]
1981	196 (7) 58
1982	146 (5) 29
1983	165 (9) 26
1984	198 (5) 41

Notes: Population remained at high density and caused intense herbivore damage to red alder trees each year. Differences between subsequent years are significant ($p \leq 0.05$, t-test).

tent caterpillar density than in the control areas. In 1983 and 1984 tent size of the high density population was large. This measure does not indicate a trend for reduced quality in either foliage or the caterpillars. To summarize, laboratory rearings indicate no influence of food source on early larval survival. Survival of caterpillars in the laboratory was reduced for both groups in 1982 and 1985. Field measurements showed reduced survival of early instar caterpillars in the area of high caterpillar density only in 1982 and 1985.

In the laboratory, survival from the fifth instar to the pupal stage was quite constant from year to year for caterpillars fed foliage from trees exposed to intense herbivory (Table III). In 1983, caterpillars fed foliage from the control area survived significantly better and in 1985 survival of control caterpillars was significantly worse than caterpillars fed foliage from the area with prolonged defoliation. Therefore, no consistent trend

Table II. Larval Survival to Late Instar and Tent Size of the Western Tent Caterpillar

	1982	1983	1984	1985
Larval survival (%, *N*)				
Intense herbivory	70, 115	96, 153	90, 163	71, 300
Control	71, 120	90, 145	93, 160	78, 300
Tent size (cm^2) [mean (SE) *N*]				
Intense herbivory	53 (5) 50[a]	103 (11) 34	104 (6) 76	63 (3) 229[a]
Control	73 (9) 22	94 (9) 21	113 (16) 17	99 (7) 49

Notes: Larvae reared in laboratory from cut foliage from trees exposed to intense herbivory, and little or no herbivory (control). Tent size of groups of caterpillars occurring naturally in the area (intense herbivory) or introduced to the area (Control).
[a] Tent size significantly larger in control area ($p \leq 0.05$, t-test).

Table III. Proportion of Western Tent Caterpillars Surviving from the Fifth Instar to Pupation

	Larval Survival (%, *N*)			
Herbivory	1982	1983	1984	1985
Intense	52, 115	42, 153[a]	48, 163	44, 234[b]
Control	58, 120	59, 145	46, 327	1, 213

Note: Caterpillars reared in the laboratory on cut foliage from trees exposed to intense herbivory and control trees with little or no herbivory.

[a] Survival significantly better on control foliage.

[b] Survival significantly better on foliage from the area of intense herbivory.

indicating a defensive reaction by the trees was apparent in the survival from late instar larva to pupa.

To conclude, this analysis does not support our earlier interpretation of food-quality deterioration in red alders exposed to a high density of tent caterpillars for a number of years (Myers and Williams, 1984). Pupal weight differences between defoliation and control areas can be explained by lower nitrogen content of leaves from the defoliated site. Laboratory assays indicate deterioration in caterpillar quality over 3 years of high density, but pupal weight and fecundity of field populations do not support this suggestion.

IV. Discussion

Bioassays of foodplant quality based on cut foliage may be very misleading. Of particular interest are the measurements of the size of the female pupae. Why should the weight of pupae decline from 1983 to 1985 in the laboratory assays while it remains constant in the field? The egg mass counts indicate that the reproducing female moths are actually increasing in size over this same time period. One explanation is that the laboratory environment stresses the caterpillars and therefore exposes their deteriorating condition. Laboratory-reared pupae tend to weigh less than field-collected pupae (Fig. 1). The quality of leaves may deteriorate from being cut, or under laboratory conditions caterpillars may be forced to be less selective about the leaves they eat. In the field caterpillars eat large portions of an individual leaf during one feeding bout. In containers, they may be forced to eat previously damaged and deteriorating leaves. Even enclosing caterpillars on foliage of field trees (Haukioja and Neuvonen, 1985) can give misleading results if caterpillars are not able to choose freely the leaves on which they feed.

Possibly other aspects of the field conditions, such as temperature, rainfall, or tree age, improved conditions for larval growth and compensated for any declining quality. Another explanation is that under field conditions there is a weight threshold for survival. Laboratory conditions might favor the pupation of caterpillars that do not weigh enough to undergo successful pupation in the field. However, the weights of pupae from larvae collected in the field shortly before pupation were not less than field-collected pupae, and in some cases were significantly heavier (mean weight of field-collected male pupae = 286 mg, SE = 7.2; mean weight of field-collected larvae that pupated in the laboratory = 321 mg, SE = 5.7; t = 3.78). Therefore, this does not explain the less massive pupae resulting from the laboratory rearings.

The difference in pupal weights that we originally interpreted to indicate deterioration in food quality in the area of high density (Myers and Williams, 1984) now appears to be explained by the lower nitrogen content of leaves of trees in that area (Myers and Williams, 1986).

The fecundity of moths monitored in the area of high density increased from 1982 to 1984. This does not indicate any deterioration in food or insect quality in the field population with successive years of defoliation. Pupal weight remained constant over this time, and early larval survival measured by size of the tents increased and then declined in 1985. These measures of insect "fitness" do not show a pattern of deterioration.

The inconsistency of results from field observations and laboratory bioassays of food plant quality makes general conclusions difficult. Interpretations from laboratory bioassays should be viewed cautiously.

A. Plant Defenses and Population Dynamics of Western Tent Caterpillars

Studies showing a change in the biochemistry of plants following defoliation or damage either implicitly or explicitly predict an influence of the chemical change on the population density of insects feeding on the plants. Rhoades (1983b, 1985) has more explicitly described how changes in plant quality could influence the population dynamics of insects through environmental stress causing a decline in plant defenses and an increase in growth rates, survival, and reproductive rates, leading to a population eruption. Rhoades (1985) predicts several alternative interactions following the initial outbreak. One of these is continued suppression of any plant defenses through stress from defoliation. This hypothesis is not easily tested since it seems impossible to determine whether plant defenses are suppressed by insect attack or just do not exist. However, if plant defenses do not exist or are suppressed, the outbreak phase should last several years, during which insect fecundity, survival, and growth should be high. This hypothesis varies from the earlier hypothesis

(Rhoades, 1983b), in which the eruption phase was predicted to be short and terminated by the inducement of plant defenses.

Two testable predictions arise from these hypotheses: (1) that insect fecundity should be high during the population increase, and (2) the decline of the population, stimulated at least in part by food quality deterioration, should be associated with the length of time insects have attacked the plants and stimulated the induced plant defenses (Rhoades, 1983b), or suppressed plant defenses (Rhoades, 1985). Much of the original speculation on induced defenses arises from work on the western tent caterpillar (Rhoades, 1983a,b). Therefore, it is appropriate to test the predictions of these general hypotheses using observations on the population parameters of tent caterpillars.

On Vancouver Island and the lower mainland of British Columbia, tent caterpillars fluctuate with a periodicity of 6–10 years. Populations in this area were monitored by Wellington (1960, 1964) during the late 1950s, and since 1975 I have studied the population dynamics of several populations. In 1977, I observed a high population of tent caterpillars on Westham Island. Since 1978, I have counted the number of fourth-instar tents on crab apple trees growing along a slough, have estimated the fecundity of the moths by counting the number of eggs in masses, and have estimated early larval survival by measuring the size of the tents produced by the late instar caterpillars. The population increased in 1983 with no change in fecundity (Fig. 2). The largest egg masses were produced in 1977, the year of population decline. Early larval survival indicated by tent size shows no trend consistent with the prediction that insect populations increase with a sudden improvement in food quality allowing better larval survival (Fig. 3). Tent size was large during the period of low numbers, the year before the increase, and was as high in the third year of the outbreak as it was in the first year. Neither fecundity nor early larval survival indicates

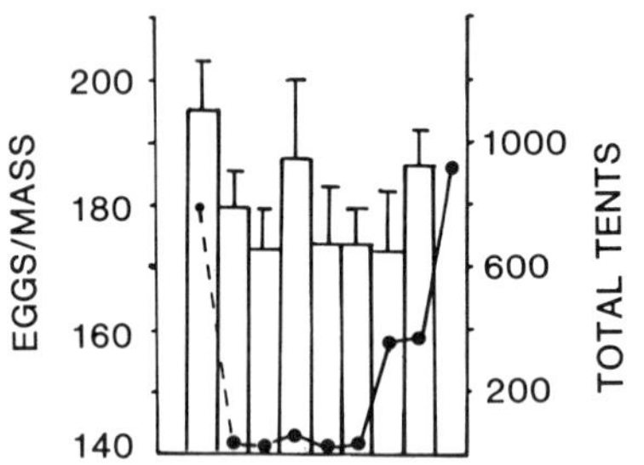

Fig. 2. Mean number of eggs per mass (± 1 SE) for the Westham Island western tent caterpillar population. Observed changes in density (total number of tents) are indicated by the solid line. Density in 1977 was not quantified, but was observed to be high.

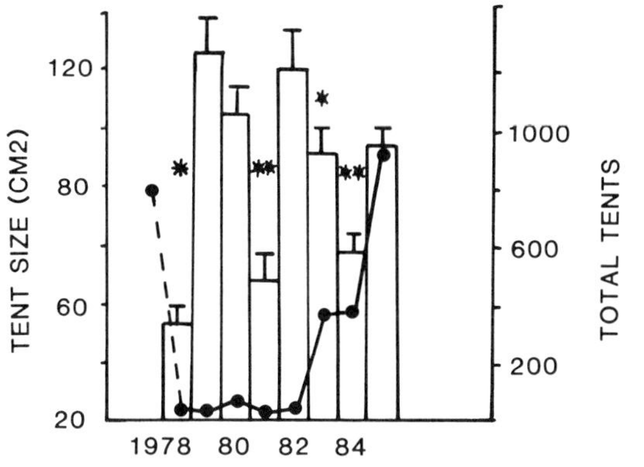

Fig. 3. Mean tent size (±1 SE) for the Westham Island western tent caterpillar population. Years between which tent size changed significantly are indicated by an asterisk. Observed changes in density (total number of tents) are indicated by the solid line. Density in 1977 was not quantified, but was observed to be high.

that the outbreak of the insects is associated with a sudden improvement in food quality followed by a defensive response by the tree causing food quality deterioration.

The second prediction of the inducible defense hypothesis is that declining foodplant quality should be indicated by reduced survival and fecundity as the herbivore damage to trees increases. The decline of the insect population should be associated with the length of time insects have attacked the plants and stimulated the induced plant defenses.

Wellington (1960) monitored the decline of western tent caterpillars on Vancouver Island from peak density in 1956 through low numbers in 1959. The population density was measured as the number of tents observed along a consistent route, and the observed number of tents declined from 74,000 in 1956, to 3,164 in 1957, and eventually declined to 250 in 1959. The fecundity of the moths, measured as the number of eggs per mass, did not decrease until the third year of the decline. Early larval survival was also the lowest in the third year of the decline. Therefore, rather than fecundity and early larval survival being reduced when herbivore density was high, these parameters that could be indicators of food quality did not diminish until late in the population decline.

Wellington (1960) also observed tent caterpillars in several areas that had different histories of infestation. These areas were scored as having no tent caterpillars, having low density, or moderate to high density in 1955. With the exception of one area in which the decline was delayed by 1 year, the caterpillar populations declined synchronously regardless of their density in 1955 (Fig. 4).

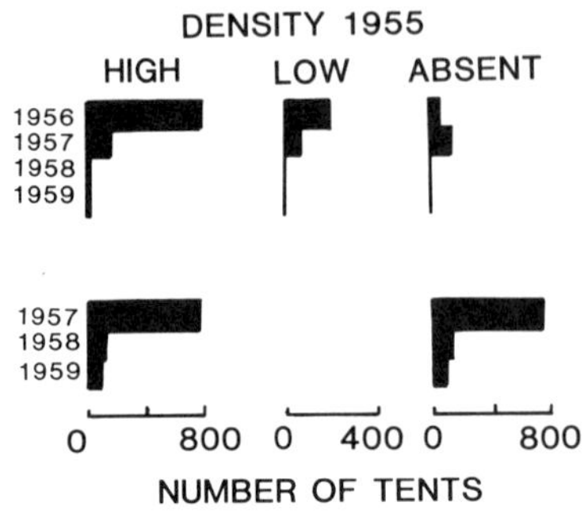

Fig. 4. Change in density in five subpopulations of western tent caterpillar on Vancouver Island, British Columbia, with different histories indicated by their density in 1955. (Data from Wellington, 1960).

The history of insect attack on trees had little influence on the dynamics of the insect population. The seeds of destruction were carried with the caterpillars, rather than being dependent on the history of insect attack on the food plants. These data do not support Rhoades's (1983b) hypothesis, but do concur with experiments in which protecting foliage from insect attack did not prevent the decline of a tent caterpillar population (Myers, 1981).

In this search for the attributes that account for the decline in tent caterpillar populations, fecundity and early larval survival have been emphasized because they are the characteristics most likely to reflect food plant quality. But they do not explain the decline. In a more detailed study of the decline of a tent caterpillar population on Mandarte Island (Myers, 1981), I observed that the initial reduction in the population occurred between the late larval instar stage and the oviposition of eggs by the next moth generation approximately a month later. The initial decline occurred when fecundity and early larval survival were still high, and continued with little reduction in the fecundity but with a subsequent decline in early larval survival indicated by the size of the tent (Fig. 5). Again, the measurements most frequently used to evaluate food quality, fecundity and early larval survival, do not explain the initial decline of western tent caterpillar populations.

B. The Induced Defense Hypothesis

The induced defense hypothesis has great appeal because it seems to provide the density-dependent mechanism necessary to explain the population dynamics of herbivorous insects. However, this attractive hypothe-

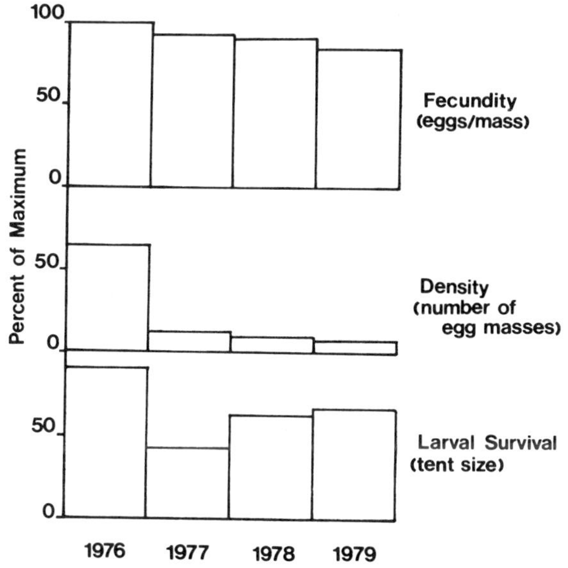

Fig. 5. Percent of maximum fecundity and tent size (early larval survival) during the decline in the western tent caterpillar population on Mandarte Island. The number of egg masses is that produced by the generation of moths flying in the summer indicated. The number of eggs per mass is that measured at the beginning of the summer indicated, and represents the fecundity of moths in the previous summer.

sis suffers when one takes a detailed look at the attributes of the insect populations that we are trying to understand. The simple approach of looking for chemical changes in foliage has been misleading because a suite of biochemical changes occur in damaged plant foliage. Because tannins and phenolics are widely viewed as plant defenses, increases in these have been assumed to indicate adaptations to herbivory by the plants. However, if increased levels of phenolics are associated with reduced levels of nitrogen, interpreting the effect of the biochemical modifications of the food plants becomes difficult.

The next level of complexity, looking for foliage quality changes through bioassay, has been faulted by problems with experimental design (Fowler and Lawton, 1985). Many of the experiments investigating the responses of trees to defoliation have been done with artificial defoliations. This does not necessarily mimic the attack of insects (Valentine *et al.*, 1983; Haukioja and Neuvonen, 1985). For example, the response of oak trees to having two-thirds of each leaf removed was different from that of birch trees having two of every three leaves removed (Valentine *et al.*, 1983). The timing and pattern of defoliation may greatly influence the observed responses of trees, and therefore the most realistic results come from allowing the insects to carry out the herbivory.

The studies of the western tent caterpillar reported here show that bioassays of food quality measured with cut foliage do not necessarily mimic the characteristics of insects in the field populations. Pupal sizes of laboratory-reared caterpillars were smaller than field-collected pupae, indicating that the laboratory conditions may have been stressful for them. The progressive reduction in pupal weight of laboratory-reared pupae from 1982 to 1985 was not evident in field-collected pupae and may be an artifact of the laboratory situation. Whether this criticism applies to studies of other insect species will only be determined by studies designed to carefully consider the characteristics of the field populations they are ultimately trying to explain.

To evaluate a general hypothesis for insect population fluctuations, it is necessary to identify the patterns of fecundity, survival, and population density as well as the synchrony of fluctuations that must be explained. For the western tent caterpillar the characteristics of the fluctuating populations observed in the field do not fit the predictions of the induced plant defense hypothesis. The eruption is not preceeded by a sudden increase in fecundity, nor is the beginning of the decline associated with a reduction in fecundity. The decline begins following a summer when early larval survival is good and is not associated with the history of herbivory of trees in a particular area.

What does explain the population decline of western tent caterpillars? Wellington (1964) attributed the decline to a combination of reduced quality of the caterpillars induced through maternal effects and mediated through reduced survival of sluggish colonies, differential dispersal of active moths, and deterioration of the weather during the time of caterpillar activity. However, the population he studied from 1956 to 1959 showed almost no change in the frequency of active larval types, and the major decline in the population occurred between 1956 and 1957 without reduced early larval survival in 1956. Populations declined at the same time even though the frequency of elongate tents, an indicator of active colonies, varied from 26% to 61% for populations monitored in different areas. It is unlikely that dispersal caused the decline, since populations declined in synchrony over the whole general area (Wellington, 1960).

In an analysis of the years of population decline recorded since 1932 we can find no association between the beginning of the decline and increased rainfall, lower than average temperatures, or reduced sunshine during the time of caterpillar activity, April to June (Myers, unpublished data). Western tent caterpillars caused extensive defoliation in southwestern British Columbia in 1986 even though the rainfall in May was the second highest in 28 years. The amount of sunshine in April 1986 was 30% less than average, and yet the caterpillar survived and maintained sufficient densities through June to cause total defoliation of alder trees in many areas. The poor weather in the spring of 1986 was not sufficient to prevent good survival and dense populations of late instar tent caterpillars. Therefore, neither poor weather during the time of larval development nor changes in caterpillar activity as defined by Wellington (1960) explain the cyclic declines of tent caterpillar populations.

Viral disease may explain the synchronous declines of tent caterpillars. Deaths from viral disease are always noted in declines of western tent caterpillars recorded by the Canadian Forest Insect and Disease Survey. The initial decline occurs between the late larval stage and the egg stage of the next generation, a stage during which deaths from virus are observed in the laboratory. Wellington's observations indicate that the decline is dependent on the history of the insects, not on the history of the foliage. Virus can be expected to be carried with the insects (Hostetter and Bell, 1985). In the first year of the decline, viral disease may be expressed most strongly during the crucial pupation period, causing the decline before the next oviposition period. In subsequent years, disease may be expressed earlier in the larval stages as indicated by poorer survival. Nonlethal disease levels may reduce the vigor of surviving moths, causing the reduction in fecundity and increased frequency of sluggish caterpillar colonies observed late in the population decline.

Characteristics of fluctuations of other forest-defoliating insects are similar to those of the western tent caterpillar and also do not fit the predictions of the induced defense hypothesis. Mason *et al.* (1977) observed a synchronous decline in the fecundity of tussock moth (*O. pseudotsugata*), regardless of the defoliation history of the trees in the area inhabited. Like the western tent caterpillar, a maternal effect on fecundity was expressed in moths reared in the laboratory. Moths from an area of heavy defoliation produced smaller offspring in the laboratory, although in the field populations the fecundity of all moths was lower and similar, regardless of the herbivore pressure on the trees in the previous year. As with the tent caterpillars, this suggests a widespread change in the quality of the insects.

The decline phase of tussock moth populations begins following a summer of high caterpillar density and defoliation. High mortality between late-instar caterpillars and the emergence of moths results in a sharp reduction of egg mass numbers (Mason, 1974). Therefore, as in the situation with tent caterpillars, early larval survival is good in the summer prior to population crash.

In studies of the larch bud moth (*Z. diniana*), protection from defoliation of two areas of more than 1000 hectares did not prevent the collapse of the insect population (Baltensweiler *et al.*, 1977). Therefore, the increased fiber content and reduced nutritional value of the larch foliage after defoliation were not necessary for the population decline. Subpopulations monitored in three different outbreaks differed by an order of magnitude in density but declined in synchrony (Fischlin, 1982). This suggests that populations can decline after different amounts of impact on the foliage of trees.

V. Conclusions

Plants are subjected to a wide array of potentially damaging situations. Any damage to leaf tissue is likely to make it more susceptible to invasion by bacteria, fungi, or virus. Limitations in nutrients such as nitrogen, or in solar radiation or water can reduce the photosynthetic activity of the plants or some of the leaves on the plants. This can cause modifications in the activity of various biochemical pathways and result in changing concentrations of both primary and secondary plant chemicals. Plant defenses to disease involve relatively nonspecific responses to pathogens or their products (Sequeira, 1983).

Given the generality of chemical modification following damage or defoliation, one must exercise caution interpreting the phenomena as specific adaptations to particular damaging agents. Rather than representing a tightly coevolved system, the defense system of plants represents a general adaptation to many "elicitors." Insects, through their damage and removal of photosynthetically active tissue, may elicit this general response in plants.

But do these chemical changes have an impact on their insect defoliators? It is very important to carry out biologically relevant tests of food plant quality to evaluate this question. Using artificial media with unnatural protein sources or using insects not adapted to feed on the plant in question does not provide an appropriate test. It is likely that insects adapted to plants with a certain array of secondary chemicals will not be influenced by moderate changes in concentration of these chemicals.

Finally, the characteristics of population fluctuations of at least some insect defoliators do not match the predictions of the induced defense hypothesis for two major reasons. First, the history of attack of plants in a particular area is not closely related to the initiation of the decline. A consistent relationship between the intensity of herbivore damage to plants and population trends of insects should exist if the induced defense hypothesis is valid. Second, fecundity, a parameter that should be influenced by food quality, does not always increase during outbreak of insect populations or decrease in the early stages of decline. Either insect size and fecundity are not good indicators of food quality or food quality is not changing as predicted by the induced defense hypothesis.

In the future, more emphasis should be placed on identifying and explaining the details of insect population fluctuations such as synchrony among populations, changes in fecundity and mortality and heterogeneity in population trends, levels of defoliation, and foodplant species. A more holistic view of the selection pressures acting on plants will lead to a more realistic analysis of the responses of plants to damage.

Acknowledgments

This research was supported by Natural Science and Engineering Research Council (N.S.E.R.C.) of Canada, through an Operating Grant. I would like to thank Linda Edwards, Irene Wilkin, and Elsie Krebs for help with laboratory and field work. I would particularly like to acknowledge Kathy Williams for her help with this research and many discussions of topics presented in this paper.

References

Baltensweiler, W., Beny, G., Bovey, P., and Delucci, V. (1977). Dynamics of larch bud moth populations. *Annu. Rev. Entomol.* **22,** 79–100.

Bernays, E. A. (1978). Tannins: An alternative viewpoint. *Entomol. Exp. Appl.* **24,** 244–253.

Bernays, E. A. (1981). Plant tannins and insect herbivores: An appraisal. *Ecol. Entomol.* **6,** 353–360.

Bernays, E. A., Chamberlain, D. J., and Woodhead, S. (1983). Phenols as nutrients for a phytophagous insect *Anacridium melanorhodon*. *J. Insect Physiol.* **29,** 535–539.

Campbell, B. C., and Duffy, S. S. (1979). Tomatine and parasitic wasps: Potential incompatability of plant antibiosis with biological control. *Science* **205,** 700–702.

Cramer, C. L., Ryder, T. B., Bell, J. N., and Lamb, C. J. (1985). Rapid switching of plant gene expression induced by fungal elicitor. *Science* **227,** 1240–1243.

del Moral, R. (1972). On the variability of chlorogenic acid concentration. *Oecologia* **9,** 289–300.

Elliger, C. A., Zinkel, D. F., Chan, B. G., and Waiss, A. C. (1976). Diterpene acids as larval growth inhibitors. *Experientia* **32,** 1364–1366.

Faeth, S. H. (1986). Indirect interactions between temporally separated herbivores mediated by the host plant. *Ecology* **67,** 479–494.

Feeny, P. P. (1970). Seasonal changes in oak leaf tannins and nutrients as a cause of spring feeding by winter moth caterpillars. *Ecology* **51,** 565–581.

Fischlin, A. (1982). Analyse eines Wald-Insekten-Systems: Der subalpine Larchen-Arvenwald und der graue Larchenwickler *Zeiraphera diniana* Gn. (Lep., Tortricidae). Diss ETH Nr. 6977. ADAG Administration and Druck AG, Zurich.

Fowler, S. V., and Lawton, J. H. (1985). Rapidly induced defenses and talking trees: The devil's advocate position. *Am. Nat.* **126,** 181–195.

Fox, L. R., and Macauley, B. J. (1977). Insects grazing on *Eucalyptus* in response to variation in leaf tannins and nitrogen. *Oecologia* **29,** 145–162.

Futuyma, D. J. (1983). Evolutionary interactions among herbivorous insects and plants. *In* "Coevolution" (D. J. Futuyma and M. Slatkin, eds.), Ch. 10, pp. 207–231. Sinauer, Sunderland, Massachusetts.

Haukioja, E. (1980). On the role of plant defenses in the fluctuations of herbivore populations. *Oikos* **35,** 202–213.

Haukioja, E., and Hanhimaki, S. (1985). Rapid wound-induced resistance in white birch (*Betula pubescens*) foliage to the geometrid *Epirrita autumnata:* A comparison of trees and moths within and outside the outbreak range of the moth. *Oecologia* **65,** 223–228.

Haukioja, E., and Neuvonen, S. (1985). Induced long-term resistance of birch foliage against defoliators: Defensive or incidental? *Ecology* **66,** 1303–1306.

Hostetter, D. L., and Bell, M. R. (1985). Natural dispersal of baculoviruses in the environment. *In* "Viral Insecticides for Biological Control" (K. Maramorosch and K. Sherman, eds.), pp. 249–284. Academic Press, New York.

Klocke, J. A., and Chan, B. G. (1982). Effects of cotton condensed tannin on feeding and digestion in the cotton pest *Heliothis zea. J. Insect Physiol.* **28,** 911–915.

Lawson, D. L., Merrit, R. W., Klug, M. J., and Martin, J. S. (1982). The utilization of late season foliage by the orange striped oakworm, *Anisota senatoria. Entomol. Exp. Appl.* **32,** 242–248.

Martin, M. M., and Martin, J. S. (1984). Surfactants: Their role in preventing the precipitation of proteins by tannins in insect guts. *Oecologia* **61,** 342–345.

Mason, R. R. (1974). Population change in an outbreak of the Douglas-fir tussock moth, *Orgyia pseudotsugata* (Lepidoptera: Lymantriidae) in central Arizona. *Can. Entomol.* **106,** 1171–1174.

Mason, R. R., Beckwith, R. C., and Paul, H. G. (1977). Fecundity reduction during collapse of a Douglas-fir tussock moth outbreak in Northeast Oregon. *Environ. Entomol.* **6,** 623–626.

Mattson, W. J., and Addy, N. D. (1975). Phytophagous insects as regulators of forest primary production. *Science* **190,** 515–522.

Mihaliak, C. A., and Lincoln, D. E. (1985). Growth pattern and carbon allocation to volatile leaf terpenes under nitrogen-limiting conditions in *Heterotheca subaxillaris* (Asteraceae). *Oecologia* **66,** 423–426.

Moore, L. V., Myers, J. H., and Eng, R. (1988). Western tent caterpillars prefer the sunny side of the tree, but why? *Oikos* **51,** 321–326.

Myers, J. H. (1981). Interactions between western tent caterpillars and wild rose: A test of some general plant herbivore hypotheses. *J. Anim. Ecol.* **50,** 11–25.

Myers, J. H., and Williams, K. S. (1984). Does tent caterpillar attack reduce the food quality of red alder foliage? *Oecologia* **62,** 74–79.

Myers, J. H., and Williams, K. S. (1987). Lack of short or long term inducible defenses in the red alder-western tent caterpillar system. *Oikos* **48,** 73–78.

Neuvonen, S., and Haukioja, E. (1985). How to study induced plant resistance. *Oecologia* **66,** 456–457.

Niemela, P., Aro, A. M., and Haukioja, E. (1979). Birch leaves as a resource for herbivores. Damage-induced increase in leaf phenols with trypsin-inhibiting effects. *Rep. Kevo Subarctic Res. Sta.* **15,** 37–40.

Puttick, G. M. (1986). Utilization of evergreen and deciduous oaks by the Californian oak moth *Phryganidia californica. Oecologia* **68,** 589–594.

Reese, J. C., Chan, B. G., and Waiss, A. C., Jr. (1982). Effects of cotton condensed tannin, maysin (corn) and pinitol (soybeans) on *Heliothis zea* growth and development. *J. Chem. Ecol.* **8,** 1429–1436.

Rhoades, D. F. (1983a). Responses of alder and willow to attack by tent caterpillars and webworms: Evidence of pheromonal sensitivity to willows. *Am. Chem. Soc. Symp. Ser.* **208,** 55–68.

Rhoades, D. F. (1983b). Herbivore population dynamics and plant chemistry. *In* "Variable Plants and Herbivores in Natural and Managed Systems" (R. F. Denno and M. S. McClure, eds.), pp. 155–222. Academic Press, New York.

Rhoades, D. F. (1985). Offensive-defensive interactions between herbivores and plants: Their relevance in herbivore population dynamics and ecological theory. *Am. Nat.* **125,** 205–238.

Schultz, J. C. (1983). Habitat selection and foraging tactics of caterpillars in heterogeneous trees. *In* "Variable Plants and Herbivores in Natural and Managed Systems" (R. F. Denno and M. S. McClure, eds.), pp. 61–89. Academic Press, New York.

Schultz, J. C., and Baldwin, I. T. (1982). Oak leaf quality declines in response to defoliation by gypsy moth larvae. *Science* **217,** 149–151.

Sequeira, L. (1983). Mechanisms of induced resistance in plants. *Annu. Rev. Microbiol.* **37,** 51–80.

Strong, D. R., Lawton, J. H., and Southwood, R. (1984). "Insects on Plants." Harvard Univ. Press, Cambridge, Massachusetts.

Tuomi, J., Niemela, P., Haukioja, E., Siren, S., and Neuvonen, S. (1984). Nutrient stress: An explanation for anti-herbivore responses to defoliation. *Oecologia* **61,** 208–210.

Valentine, H. T., Wallner, W. E., and Wargo, P. M. (1983). Nutritional changes in host foliage during and after defoliation, and their relation to the weight of gypsy moth pupae. *Oecologia* **57,** 298–302.

Wallner, W. E., and Walton, G. S. (1979). Host defoliation: A possible determinant of gypsy moth population quality. *Ann. Entomol. Soc. Am.* **72,** 62–67.

Wellington, W. G. (1960). Qualitative changes in natural populations during changes in abundance. *Can. J. Zool.* **38,** 290–314.

Wellington, W. G. (1964). Qualitative changes in populations in unstable environments. *Can. Entomol.* **96,** 436–451.

White, T. C. R. (1984). The abundance of invertebrate herbivores in relation to the availability of nitrogen in stressed food plants. *Oecologia* **63,** 90–105.

Williams, K. S., and Myers, J. H. (1984). Previous attack of red alder may improve food quality for fall webworm larvae. *Oecologia* **62,** 74–79.

Yang, S. F., and Pratt, H. K. (1978). The physiology of ethylene in wounded plant tissues. *In* "Biochemistry of Wounded Plant Tissue" (G. Kahl, ed.), pp. 595–622. De Gruyter, New York.

Zucker, W. V. (1983). Tannins: Does structure determine function? An ecological perspective. *Am. Nat.* **121,** 335–365.

CHAPTER 12

Environmental Constraint Of Constitutive and Long-Term Inducible Defenses in Woody Plants

JOHN P. BRYANT
Institute of Arctic Biology
University of Alaska
Fairbanks, Alaska

JUHA TUOMI AND
PEKKA NIEMALA
Department of Biology
University of Turku
SF-20500 Turku
Finland

I. Introduction

The title of this volume clearly indicates that coevolution is considered an important cause of chemically mediated interactions between organisms. For example, coevolutionary interactions between herbivores and their plant prey have been invoked to explain the types and quantities of resources allocated by plants to chemical defense against herbivores (Feeny, 1976; Rhoades and Cates, 1976; Rhoades, 1979). However, both circumstantial and direct experimental evidence suggests that the allocation of resources by plants to potentially defensive phytochemicals is strongly influenced by soil fertility and shade (e.g., Butler and Bailey, 1973; Janzen, 1974; McKey *et al.*, 1978; Chew and Rodman, 1979; Waterman *et al.*, 1984; Waring *et al.*, 1985; Mihaliak and Lincoln, 1985; Bryant and Chapin, 1986). Coevolutionary models of plant defensive evolution cannot account for such environmental constraint of plant defensive chemistry.

In response to this insufficiency, alternative models have been developed that attempt to explain much of plant defensive chemistry as the consequence of plant physiological responses to resource availability. Physiological adaptations by plants to insufficient mineral nutrition and light have been suggested to be major constraints upon the evolution of woody plant constitutive defenses (Bryant *et al.*, 1983; Coley, 1983; Coley *et al.*, 1985). Similarly, phenotypic responses by woody plants to mineral nutrient availability and shade (plant carbon/nutrient balance) have been invoked to explain how resource availability affects the phenotypic expression of such defenses (Bryant *et al.*, 1983; Waterman *et al.*, 1984). More recently, Tuomi *et al.* (1984) have provided a carbon/nutrient balance model that explains long-term inducible (*sensu* Haukioja, 1980) defenses in leaves of severely defoliated woody plants as a result of nutrient stress. Although each of these models provides insight into how the physical environment may influence a single aspect of woody plant defensive chemistry, there is no comprehensive theory that attempts to explain how plant–environment interactions influence the allocation of resources by woody plants to both constitutive and inducible antiherbivore defenses. In this chapter we initiate such a synthesis. Specifically, we consider how plant physiological responses to nutrient stress influence the evolution and phenotypic expression of constitutive and long-term inducible defenses in woody plants. Although we recognize the importance of short-term inducible defenses (sensu Haukioja, 1980) in woody plant–insect herbivore interactions, we do not consider them in this chapter.

II. Environmental Constraint of Constitutive Defenses

A. Evolutionary Responses: Effects of Inherently Slow Growth

Herbivores of very different ecosystems feed more heavily on tissues of inherently fast-growing species associated with productive environments (e.g., fertile soils) than on tissues of inherently slow-growing species that persist in unproductive environments (e.g., infertile soils) (Table I). This relationship between the availability of resources in the environment, plant inherent growth rate, and resistance to herbivore attack has been predicted to be one of causality: adaptation to low resource availability, inherently slow growth (Grime, 1977; Chapin, 1980a), has been predicted to favor the evolution of constitutive antiherbivore defenses by plants (Bryant *et al.*, 1983; Coley, 1983; Coley *et al.*, 1985).

Inherently slow growth is the consequence of a suite of interdependent traits that limit the capacity of plants to acquire resources required for growth (mineral nutrients, carbon) (Grime, 1977). Inherently slow-growing plants have comparatively low nutrient absorption capacities (Chapin, 1980a; Chapin *et al.*, 1986) and photosynthetic rates (Mooney, 1972). Thus such plants grow slowly even in the most productive environments (Grime, 1977; Chapin, 1980a). For example, in boreal forests and arctic shrub tundra rapidly growing deciduous species such as willow (*Salix*), poplar (*Populus*), and birch (*Betula*) have high nutrient absorption capacities (Chapin, 1980b; Chapin *et al.*, 1983b, 1986) and photosynthetic rates (Small, 1972; Chapin, 1980b) in comparison to more slowly growing evergreen species such as spruce (*Picea*) and *Ledum* (Table II). There are many other examples where inherently slowly growing plant species associated with resource-limited environments have been found to have a limited capacity to acquire resources (e.g., Mooney, 1972; Grime, 1977, 1979; Chapin, 1980a; Sarmiento *et al.*, 1985).

A limited capacity to acquire resources and an inherently slow growth rate limit a plant's ability to compensate for herbivory through growth (Bryant *et al.*, 1983; Coley *et al.*, 1985). For example, studies of the effect of winter browsing on the growth of boreal woody plants have consistently demonstrated that inherently slow-growing species are less capable of compensating for herbivory by growth than are more rapidly growing species (e.g., Aldous, 1952; Fox and Bryant, 1984; Bryant and Chapin, 1986) (Fig. 1, Table II). Similar results are being obtained in studies of plant–herbivore interactions in South African savannas (Bryant, unpublished; O'Regan, unpublished) and Brazilian caatinga (Kirmse, 1984; Gobina, unpublished). Moreover, a given rate of herbivory (grams of tissue eaten per day) represents a larger fraction of the net production of a slow grower than of a fast grower (Coley *et al.*, 1985). Such physiological constraints over the capacity of woody plants to replace tissue eaten by herbivores have been suggested as a major cause of the evolution of comparatively effective antiherbivore defenses by inherently slow-growing woody plants (Bryant *et al.*, 1983; Coley *et al.*, 1985). In short, trees and shrubs that cannot easily compensate for herbivory by growth must deter herbivory by defense.

Tree and shrub species that have evolved in productive environments have been selected for rapid growth in order to outcompete their neighbors

Table I. Preferences for Mature Foliage by Herbivores of Several Major Ecosystems

Ecosystem	High-Resource Fast Grower		Low-Resource Slow Grower	Deciduous versus Evergreen	Source
Shrub tundra	(*Salix–Betula*)	versus	(*Ledum–Empetrum–Cassiope*)		
Generalist insects	9		1	1	MacLean and Jensen (1984)
Collard lemming	8		2	1	Batzli and Jung (1980)
Brown lemming	10		0	1	Batzli and Jung (1980)
Tundra vole	9		Trace	1	Batzli and Jung (1980)
Ground squirrel	9		1	2	Batzli and Jung (1980)
Muskoxen	10		0	1	Robus (1981)
Caribou	10		0	1	Kuropat (1984)
Reindeer	9		1	1	Trudell and White (1981)
Willow ptarmigan	10		0	1	Williams *et al.* (1984)
Rock ptarmigan	10		0	1	Weeden (1969)
White-tailed ptarmigan	10		0	1	Weeden (1969)
Boreal forest	(*Salix–Populus–Betula*)	versus	(*Ledum–Picea*)		
Snowshoe hare	9		Trace	1	Bryant (unpublished)
Mountain hare	8		2	2	Pulliainen (1973)
Moose	10		0	1	Miquelle (1982)

Temperate forest	(Ephemeral herb)	versus	(Evergreen herb)		
Slug (*Arion subfuscus*)	9		Trace	1	Rathcke (1985)
Slug (*Arion fasciatus*)	8		2	2	Rathcke (1985)
Slug (*Derocerus reticulatum*)	8		2	2	Rathcke (1985)
California oak woodland	(Deciduous oak)	versus	(Evergreen oak)		
Leaf-mining insects	8		2	2	Opler (1974)
Tropical rainforest	(Pioneer species)	versus	(Persistent species)		
Insects	8		2	2	Coley *et al.* (1985)
Black colobus monkey	8		2	2	McKey *et al.* (1981)
Red colobus monkey	3		7	3	Struhsaker (1978)
Subtopical savanna	(Fertile soil)	versus	(Infertile soil)		
Kudu	9		1	2	Cooper and Owen-Smith (1985)
Impala	9		1	2	Cooper and Owen-Smith (1985)
Boer goat	7		3	2	Cooper and Owen-Smith (1985)
Caatinga	(Fire-adapted)	versus	(Shade-tolerant)		
Goat	High preference		Low preference	3	Pfister (1983)
Sheep	High preference		Low preference	3	Pfister (1983)
Temperate marine intertidal	(*Laminariales*)	versus	(Fucales)		
Snail (*Tegula funebralis*)	High preference		Low preference	NA	Steinberg (1985)

Notes: Preference indices comparing high-resource-adapted fast-growing plants and low-resource-adapted slowly-growing plants were computed from results of cafeteria feeding trials and field measurements of preference (use/availability ratios) and range from 0 (never eaten) to 10 (always preferred). Relative preference for deciduous and evergreen plants was ranked from 1 (deciduous always preferred) to 5 (evergreen always preferred). Plant groups compared are indicated for each ecosystem. Detailed results and methods are in cited references. Taxonomy follows cited references.

Table II. Growth, Herbivore Preference for and Defensive Traits of Mature Foliage of High- and Low-Resource-Adapted Shrubs Common in Alaskan Arctic Tundra and Boreal Forest

	High-Resource *Salix pulchra*	Low-Resource *Ledum palustre*	Reference
Growth			
Growth form	Deciduous	Evergreen	Viereck and Little (1972)
Leaf relative growth rate ($mg \cdot g^{-1} \cdot day^{-1}$)	29	6	Chapin (1980b)
Photosynthetic capacity ($mg\ CO_2 \cdot h^{-1} \cdot g^{-1}$)	40 ± 3	14 ± 1	Chapin (1980b)
Phosphate absorption rate ($\mu mol \cdot h^{-1} \cdot g^{-1}$)	2.6 ± 0.1	0.8 ± 0.1	Chapin (1980b)
Maximum leaf nitrogen (percent dry mass)	4.3 ± 0.1	3.0 ± 0.1	Chapin (1980b)
Below-ground biomass (percentage of total)	70 ± 4	43 ± 6	Chapin (1980b)
Spring decrease in percent total nitrogen–carbon (TNC) in below-ground stems	23	18	Chapin (1980b)
Spring decrease in percent N in below-ground stems	33	25	Chapin (1980b)
Effect of defoliation on shoot weight (percent of control)[a]	48 ± 3	6 ± 1	Archer and Tiezen (1980)
Preference index			
Generalist insects			
Polia sp.	10	0	MacLean and Jensen (1984)
Apantesis sp.	9	1	MacLean and Jensen (1984)
Parasemia parthenost	10	0	MacLean and Jensen (1984)
Gynaephora rossi	8	2	MacLean and Jensen (1984)

Mammals			
Collared lemming (*Dicrostonyx torquatus*)	10	0	Batzli and Jung (1980)
Tundra vole (*Microtus oeconomus*)	10	0	Batzli and Jung (1980)
Ground squirrel (*Spermophilus parryi*)	8	2	Batzli and Sobaski (1980)
Snowshoe hare (*Lepus americanus*)	10	0	Bryant (unpublished)
Muskoxen (*Ovibos moschatus*)	10	0	Robus (1981)
Caribou (*Rangifer tarandus*)	10	0	Kuropat (1984)
Moose (*Alces alces*)	10	0	Milke (1969)
Willow ptarmigan	10	0	Williams *et al.* (1980)
Mean for all herbivores considered	9.6 ± 0.2	0.4 ± 0.2	
Defensive traits			
Leaf methanol-soluble phenolics (percent dry mass)[b]	4.2 ± 0.2	11.3 ± 0.3	Kuropat (1984)
Leaf resins (percent dry mass)	0.0	10.0 ± 0.4	Kuropat (1984)
Leaf lignin (percent dry mass)	5.0	11.0	Kuropat (1984)
In vitro digestibility in caribou rumen fluid[c]	61.0	37.4	Kuropat (1984)
In vivo digestibility in caribou rumen[d]	58.3	41.8	Kuropat (1984)
Effect of ethanol extract on microtine efficiency of growth (growth/intake)			
Collard lemming (*Dicrostonyx torquatus*)	0.142 ± 0.013	0.016 ± 0.046	Jung and Batzli (1981)
Brown lemming (*Lemmus sibiricus*)	0.036 ± 0.008	−0.201 ± 0.041	Jung and Batzli (1981)
Tundra vole (*Microtus oeconomus*)	0.075 ± 0.004	−0.036 ± 0.065	Jung and Batzli (1981)

Notes: Preference indices were computed from results of cafeteria feeding trials and field measurements of preference (use/availability ratios) and range from 0 (never eaten) to 10 (always preferred). Detailed methods and results are in cited references. Taxonomy follows cited references. Mean ± SE presented.

[a] Shoot weight measured 1 year after 100% defoliation in mid growing season.

[b] Folin–Denis method using tannic acid standard.

[c] Two-stage technique (Tilley and Terry, 1963), 48-h dry-matter disappearance (DMD).

[d] Nylon bag technique (Pehrson *et al.*, 1980), 48-h dry-matter disappearance (DMD).

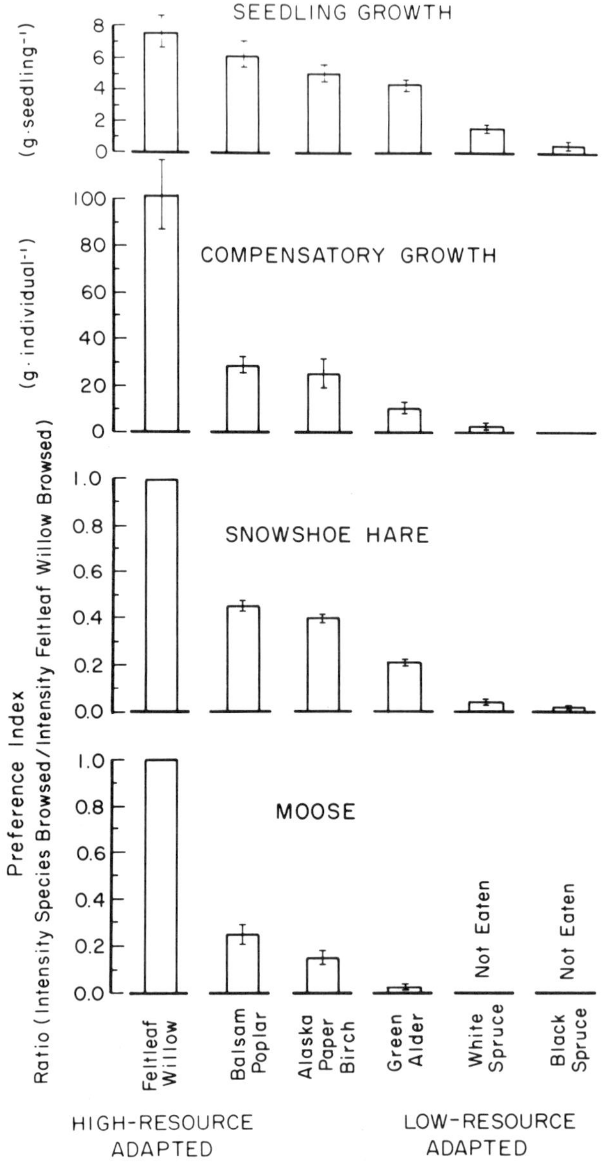

Fig. 1. Seedling growth and compensatory growth following simulated winter browsing, and palatability of some Alaskan woody plants to snowshoe hare and moose in winter. Seedling growth calculated from Walker (1985) and Bryant *et al.* (1983); compensatory growth calculated from Bryant and Chapin (1986); palatability indices calculated from Wolff and Zasada (1979) and Bryant (unpublished). Detailed results and methods in cited references. Mean and SE presented.

and thus dominate the available light and mineral nutrient resources (Grime, 1977). Thus in inherently rapidly growing woody plants selection has favored the allocation of resources to growth at the expense of chemical defense (Bryant *et al.*, 1983; Coley *et al.*, 1985). Furthermore, in many ecosystems high-resource environments are the result of disturbances that destroy the aboveground parts of plants, for example, wildfire (Kozlowski and Algren, 1974; Chapin and Van Cleve, 1981). Trees and shrubs have adapted to such disturbances by evolving below-ground storage organs that contain large nutrient and carbohydrate reserves that can be used for vegetative regeneration after disturbance (Kramer and Kozlowski, 1979). These reserves can be used for regrowth following herbivory as well (Garrison, 1972; Bryant and Kuropat, 1980; Bryant *et al.*, 1983). Thus adaptation to physical destruction of above ground parts is preadaptation for the capacity to compensate for herbivory by growth. Such preadaptation further reduces the intensity of selection for antiherbivore defenses in rapidly growing woody plants.

Although limited, the available experimental evidence supports this hypothesis. In boreal forests and arctic shrub tundra, inherently slow-growing woody species that have a limited capacity for compensatory growth are also less palatable to a variety of herbivores, in comparison to more rapidly growing species that have a greater capacity for compensatory growth (Bryant *et al.*, 1983) (Fig. 1, Table II). Similar results are being obtained in studies of woody plant–herbivore interactions in African savannas (Cooper and Owen-Smith, 1985; Owen-Smith and Cooper, 1987; Bryant, unpublished; O'Regan, unpublished) and Brazilian caatinga (Pfister, 1983; Kirmse, 1984; Gobina, unpublished).

B. Evolutionary Responses: Effects of Being Evergreen

The evergreen growth form is frequently associated with unproductive environments and low palatability to herbivores (e.g., Janzen, 1974; Grime, 1977; Chapin, 1980a; Bryant *et al.*, 1983; Coley, 1983; Coley *et al.*, 1985; Sarmiento *et al.*, 1985). Slow leaf turnover is advantageous in a low-nutrient environment, because each time a leaf is shed it carries with it approximately half the nitrogen and phosphorus contained in it at mid growing season (Chapin, 1980a; Chapin and Kedrowski, 1983a). These nutrients are not easily replaced by plants growing in infertile soils (Janzen, 1974). Similarly, in deep shade where the potential for energy (carbon) acquisition is low, carbon loss can be minimized by a slow leaf turnover rate (Coley *et al.*, 1985). However, greater leaf longevity incurs certain costs. If an herbivore eats an evergreen leaf, an expensive long-term investment in capital resources (nutrients or carbon) and machinery (the leaf) is lost (Mooney and Gulmon, 1982). The obvious evolutionary response of evergreens to herbivory has been increased effectiveness of antiherbivore defenses (e.g., Janzen, 1974; Bryant *et al.*, 1983; Coley *et al.*, 1985). A long-lived leaf is also more likely to encounter unfavorable physical conditions before being shed than is a short-lived leaf. The high fiber content and thick cuticle of many evergreen leaves are ultimately evolutionary responses to unfavorable conditions such as drought (Levitt, 1972; Sarmiento *et al.*, 1985). These traits are, however, "defensive" in that they lower the palatability and digestibility of foliage (e.g., Scriber and Slansky, 1981; Van Soest, 1982; Robbins, 1983).

Unlike deciduous species, evergreen species retain a comparatively large fraction of the reserves used for early-season growth in leaves rather than in stems and below-ground parts (e.g., Keller and Kozlowski, 1966; Garrison, 1972; Chapin, 1980a; Shaver, 1983). This pattern of dormant season reserve storage seriously limits the capacity of evergreens to compensate for herbivory through growth (Garrison, 1972; Chapin, 1980b) (Fig. 1, Table II) and favors selection for defenses that deter herbivory (Bryant *et al.*, 1983).

C. Phenotypic Responses: Effects of Plant Carbon/Nutrient Balance

The phenotypic expression of plant chemical defense against herbivores is influenced by soil fertility and shade. For example, fertilization with growth-limiting nutrients or shading often results in increased browsing of woody plants by mammals (e.g., Heiberg and White, 1951; Behrend, 1973; Loyttyniemi, 1981; Sullivan and Sullivan, 1982; Rousi, 1983; Waterman *et al.*, 1984; Bryant and Chapin, 1986; Bryant *et al.*, 1987). Although the data are more limited, there is some evidence that fertilization and/or shade can result in increased consumption of woody plant leaves by insects (e.g., Schmidt and Fellin, 1983; Larsson *et al.*, 1986). Furthermore, both better mineral nutrition of woody plants (Oh *et al.*, 1970) and shade (Waterman *et al.*, 1984; Hanley *et al.*, 1987; Robbins *et al.*, 1987) have been correlated with increased digestibility of woody plant tissues by mammals.

Insufficient mineral nutrition limits plant growth more severely than it does photosynthesis (Chapin, 1980a), so that soluble carbohydrate accumulates in tissues of nutrient-stressed woody plants (e.g., Kramer and Kozlowski, 1979; Chapin, 1980a; Waring *et al.*, 1985; Bryant and Chapin, 1986). Accumulation of soluble carbohydrate in excess of growth demands in plant tissues favors synthesis of nonnitrogenous secondary metabolites (phenolics and terpenes) (Bryant *et al.*, 1983). Thus it is not surprising that nutrient-stressed plants have high concentrations of such metabolites in their tissues in comparison to those found in tissues of conspecifics whose growth is not nutrient-limited (e.g., Wilde *et al.*, 1948; del Moral, 1972; Wong, 1973; Chew and Rodman, 1979; Gershenzon,

1984; Waring *et al.*, 1985; Mihaliak and Lincoln, 1985; Bryant and Chapin, 1986). Conversely, insufficient light limits photosynthesis, thereby directly limiting synthesis of soluble carbohydrate and its derivatives, carbon-based secondary plant metabolites. Thus concentrations of carbon-based secondary metabolites are often lower in tissues of shaded plants than in tissues of conspecifics whose growth is not light-limited (e.g., Waterman *et al.*, 1984; Waring *et al.*, 1985; Bryant and Chapin, 1986; Bryant *et al.*, 1987; Larsson *et al.*, 1986). Such reduced concentrations of carbon-based secondary plant metabolites are presumed to be a major cause of the increased palatability and digestibility of fertilized or shaded woody plants (Bryant *et al.*, 1983, 1987; Waterman *et al.*, 1984; Hanley *et al.*, 1987; Robbins *et al.*, 1987).

III. Environmental Constraint of Long-Term Inducible Defenses

Defoliation of some woody plants reduces the food value of leaves for defoliating insects for 1 or more years following defoliation (Table III). This response by woody plants to defoliation has been called *long-term defensive induction* (LTI) (Haukioja, 1980) and has been implicated as an important control over defoliating insect population dynamics: the reduced food value of leaves of induced plants is presumed to be a major cause of the decline phase and subsequent time lag in defoliator population growth required to generate periodic defoliator outbreaks (e.g., Haukioja and Hakala, 1975; Baltensweiler *et al.*, 1977; Fischlin and Baltensweiler, 1979; Baltensweiler, 1985; Rhoades, 1979; Haukioja, 1980; Haukioja *et al.*, 1983).

Because LTI is a potentially important control over forest defoliator population outbreaks, several hypotheses attempt to explain its mechanism. Haukioja and his co-workers (e.g., Haukioja and Hakala, 1975; Haukioja, 1980; Haukioja *et al.*, 1983; Haukioja and Neuvonen, 1985) and Rhoades (1979) have proposed that LTI is an evolved defense response by the plant targeted directly at the defoliator. On the other hand, Myers and Williams have proposed that LTI is a passive response to defoliation that has little or no effect on the population dynamics of insect defoliators (Myers and Williams, 1984; Williams and Myers, 1984; Myers, this volume). Alternatively, Tuomi *et al.* (1984) have suggested that LTI is the result of a defoliation-caused nutrient stress that reduces the nutrient concentrations of leaves of induced plants and favors synthesis of nonnitrogenous secondary metabolites that further reduce the food value of such leaves for insect herbivores.

Any hypothesis that correctly describes the mechanism of LTI must be capable of accounting for a suite of traits (Table III) that characterize

Table III. Characteristics of Forest Defoliating Insect Outbreaks Where Reduced Food Value of Reflushed Foliage to the Defoliating Insect is Well Documented

Characteristic	Defoliator			
	Zeiraphera diniana	*Epirrita atumnata*	*Choristoneura conflictana*	*Rheumaptera hastata*
Defoliation event				
Defoliation phenology[a]	E	E	E	E-M
Duration of severe defoliation (years)	1–2	1–2	1–3	1–3
Host				
Species[b]	*L. decidua*	*B. pubescens*	*P. tremuloides*	*B. papyrifera*
Growth form	Deciduous	Deciduous	Deciduous	Deciduous
Changes in reflushed foliage				
Growth rate	Decrease	?	Decrease	Decrease
Nitrogen concentration	Decrease	Decrease	Decrease	Decrease
Phenolic concentration	?	Increase	Increase	Increase
Fiber concentration	Increase	?	?	?
Food value	Decrease	Decrease	Decrease	Decrease
References[c]	1, 2, 3	4, 5, 6, 7	8, 9	9, 10

Note: Taxonomy follows cited references.

[a] E, Early growing season; M, middle growing season.

[b] *L.*, *Larix* (larch); *B.*, *Betula* (birch); *P.*, *Populus* (quaking aspen).

[c] (1) Benz, 1974; (2) Baltensweiler *et al.*, 1977; (3) Fischlin and Baltensweiler, 1979; (4) Haukioja and Niemala, 1976; (5) Haukioja, 1980; (6) Haukioja *et al.*, 1983; (7) Kallio and Lehtonen, 1973; (8) Beckwith, 1968, 1970; (9) Werner, 1988a,b; (10) Werner, 1977.

defoliation events resulting in LTI. First, LTI appears to be a characteristic of deciduous woody plants rather than more slowly growing evergreens. Second, defoliation events associated with LTI are severe: frequently induced plants have been completely or almost completely defoliated for several years in sequence (e.g., Werner 1977, 1979, 1988a,b; Myers and Williams, 1984; Williams and Myers, 1984; Myers, this volume). Third, results of Werner (1979; 1988a), Tuomi *et al.* (1984), and Myers and Williams (Myers and Williams, 1984; Williams and Myers, 1984; Myers, this volume) indicate that such defoliation results in a progressive, rather than a sudden, decline in leaf food value that is directly proportional to the severity of defoliation and is cumulative over defoliation events. And finally, defoliation reduces the nitrogen concentration of leaves and their size and results in an increase in their nonnitrogenous secondary metabolite and/or fiber concentrations (e.g., Baltensweiler *et al.*, 1977; Miquelle, 1983; Tuomi *et al.*, 1984; Baltensweiler, 1985; Werner, 1988a,b).

A. Evolutionary Responses: Effects of Reserve Storage in Woody Plants

Allocation of nitrogen from reserves stored in stems and below-ground parts to new leaf growth depletes the nitrogen reserves of all woody plants (Kramer and Kozlowski, 1979). If severe defoliation, such as that characteristic of LTI, occurs before a deciduous species translocates a significant fraction of the nitrogen contained in its leaves at mid growing season to sites of dormant season reserve storage, then most of the nitrogen allocated from such reserves to new leaf growth will be lost. Furthermore, if this loss cannot be easily replaced by uptake from the soil, then severe defoliation will result in a depletion of the nitrogen reserves of such a deciduous species as well as evergreens. On the other hand, defoliation may have very different effects on the soluble carbohydrate reserves of disturbance-adapted deciduous species as compared to slow-growing evergreens. Disturbance-adapted deciduous species have large below-ground carbohydrate reserves in comparison to evergreens (Garrison, 1972), where a greater proportion of the plant's reserve carbohydrate is stored in leaves (Bryant *et al.*, 1983). Thus severe defoliation will deplete the carbohydrate reserves of evergreen species more severely than it does those of disturbance-adapted deciduous species (defoliation of an evergreen results in loss of major carbohydrate reserves stored in leaves, whereas defoliation of a disturbance-adapted deciduous species does not). For example, consider the potential effects of defoliation by an outbreak population of *Rheumaptera hastata* L., the spear-marked black moth, on the nutrient and carbohydrate reserves of Alaska paper birch (*Betula resinifera* Br., Dugle, 1966). Alaska paper birch leaves attain their maximum mass and pool sizes of new growth nitrogen by late June, at which time 90–93% of their new growth nitrogen is in leaves (Chapin and Kedrowski, 1983a). Subsequently, there is a gradual loss of nitrogen from the leaf (presumably due to translocation to storage organs and recurrently flushing leaves). However, by the time *R. hastata* has completed pupation, mid August (Werner, 1977), only 25% of the nitrogen contained in Alaska paper birch's leaves at mid growing season has been removed by translocation (Chapin and Kedrowski, 1983a). Clearly, severe defoliation repeated over several years, as occurs in *R. hastata* outbreaks (Werner, 1977, 1979; 1988a,b), could significantly deplete Alaska paper birch's nitrogen reserves unless these reserves are rapidly replaced by uptake from the soil. However, 3 years of 100% artificial defoliation timed to simulate *R. hastata* defoliation did not significantly reduce below-ground carbohydrate reserves of Alaska paper birch (Werner, in review a). Moreover, leaves produced by previously severely defoliated and induced birches have higher photosynthetic rates than do those of undefoliated birches (Prudhomme, 1982). Such an increase in photosynthesis can partially offset any carbohydrate loss resulting from defoliation. Conversely, severe defoliation of boreal evergreen conifers results in an increase in the nitrogen concentration of their needles (Piene, 1980; Gezelius *et al.*, 1981; Ericsson *et al.*, 1985) and a decrease in needle starch (Webb and Karchesy, 1977; Ericsson *et al.*, 1980, 1985; Webb, 1980). Such results indicate that defoliation of evergreens, unlike defoliation of deciduous species, depletes carbohydrate reserves more severely than nitrogen reserves.

Tuomi *et al.* (1984) have predicted that such a defoliation-caused reduction in the nitrogen reserves of deciduous species results in a nutrient stress that reduces the concentration of nitrogen in leaves of induced plants. Furthermore, Tuomi *et al.* (1984) have predicted that such a reduction of foliar nitrogen results in nutrient limitation of leaf growth and increased synthesis of carbon-based secondary metabolites in leaves of induced plants. The low nitrogen concentration and the high nonnitrogenous secondary metabolite concentration of leaves of induced plants is presumed by the Tuomi model to be the cause of LTI. On the other hand, the increase in leaf nitrogen (Piene, 1980; Gezelius *et al.*, 1980; Ericsson *et al.*, 1985) and the decrease in leaf starch (Webb and Karchesy, 1977; Ericsson *et al.*, 1980, 1985; Webb, 1980) caused by severe insect defoliation of evergreen conifers result in a carbohydrate status that is unfavorable for the synthesis of nonnitrogenous secondary plant metabolites (Bryant *et al.*, 1983). Furthermore, such changes in leaf chemistry indicate that severe defoliation results in an increase rather than a decrease in the quality of evergreen leaves as food for insect herbivores. In short, consideration of the effects of defoliation on woody plant nitrogen and carbohydrate reserves indicates that LTI will be more characteristic of comparatively rapidly growing disturbance-adapted deciduous species than more slowly growing low-resource-adapted evergreens. Although data are scarce, the Niemala *et al.* (1984) demonstration that defoliation of Scots pine (*Pinus sylvestris* L.) by the sawfly (*Neodiprion sertifer*) decreases resistance of Scots pine to further *N. sertifer* attack is consistent with this suggestion.

B. Fine-Rootlet Mortality: Implications for LTI

As mentioned above, results of Tuomi *et al.* (1984), Werner (1979; 1988a,b), and Myers and Williams (Myers and Williams, 1984; Williams and Myers, 1984; Myers, this volume) all indicate that the food value of leaves of induced plants declines progressively rather than suddenly and in direct proportion to the severity of defoliation. We propose that defoliation-caused fine-rootlet mortality in concert with a defoliation-caused nutrient stress such as that hypothesized by Tuomi *et al.* (1984) can account for these observations.

Defoliation causes fine-rootlet mortality in woody plants. Moreover, this mortality is directly proportional to the severity of defoliation (Kozlowski, 1971b). For example, Redmond (1959) found that less than 15% of the fine rootlets of healthy *Abies balsamifera* L. were dead, whereas in trees where spruce budworm had eaten 70% of the new leaves, 30% of the fine rootlets were dead. When 100% of the foliage was consumed, fine-rootlet mortality exceeded 75%. Such mortality greatly limits the capacity of trees to absorb mineral nutrients (Kozlowski, 1971b; Kramer and Kozlowski, 1979). Furthermore, if (1) over several years of an insect outbreak a significant fraction of a deciduous tree's leaves and the nitrogen contained in them are eaten by insects and if (2) this defoliation simultaneously results in significant fine-rootlet mortality that limits nitrogen uptake from the soil, then (3) it must also progressively deplete the plant's nitrogen reserves. Such a depletion of plant nitrogen reserves can explain the progressive decline in the concentration of leaf nitrogen that is associated with sequential severe defoliation of aspen (Werner, 1988a) and birch (Tuomi *et al.*, 1984; Werner, 1988a). Moreover, such a progressive decline in leaf nitrogen will result in nutrient limitation of leaf growth and thus favor increased synthesis of nonnitrogenous secondary metabolites. This sequence of events can account for the progressive increase in the concentrations of carbon-based secondary metabolites reported for induced aspen and birch (Tuomi *et al.*, 1984; Werner, 1988a). These changes in leaf chemistry can explain a progressive decline in leaf food value for insect herbivores if either (1) a low nitrogen concentration or (2) a high concentration of nonnitrogenous secondary metabolites reduces the food value of leaves for insect herbivores.

In the case of nitrogen-fixing plants such as red alder (*Alnus rubra*

Bong.) (Kramer and Kozlowski, 1979), it is even more likely that decay of leaf food value for insects will occur progressively and only after repeated very severe defoliations, as reported by Myers and Williams (Myers and Williams, 1984; Williams and Myers, 1984; Myers, this volume). Nitrogen-fixing deciduous woody plants have large nitrogen reserves and high tissue nitrogen concentrations in comparison to other deciduous woody plants (Kramer and Kozlowski, 1979). Such large reserves of nitrogen may buffer the nutrient stress presumed to be the cause of LTI by the Tuomi *et al.* (1984) model. Furthermore, nitrogen fixing nodules are strong "sinks" for photosynthate because nitrogen-fixing bacteria require photosynthate as an energy source (Kramer and Kozlowski, 1979). The strength of this "sink" will further decrease the difference in the nitrogen and carbohydrate reserves predicted to be the cause of LTI by the Tuomi model. Thus the Myers and Williams observation that repeated severe defoliation is required to initiate LTI in red alder may, in part, reflect the initial high nitrogen status of red alder and the energy demands of nitrogen fixation.

IV. Summary and Synthesis

Physiological adaptation by woody plants to resource limitation has resulted in a limited capacity to compensate for herbivory by growth. This limitation has been predicted to be a major cause of the evolution of highly effective constitutive antiherbivore defenses by inherently slow-growing species that persist in unproductive environments (Bryant and Kuropat, 1980; Bryant *et al.*, 1983; Coley, 1983; Coley *et al.*, 1985). In this chapter we suggest further that one of the physiological adaptations that limits compensatory growth by slow-growing evergreens, storage of major carbohydrate reserves in comparatively long-lived leaves, has also limited the evolution of long-term inducible defenses in evergreens. Thus we predict that slow-growing evergreens such as boreal evergreen conifers have either weak or no long-term inducible defenses. Furthermore, we expect that in many cases severe and repeated defoliation will result in an increase in the quality of their foliage as food for herbivores rather than a decrease in quality such as that associated with long-term induction.

Plant carbon/nutrient balance constrains the phenotypic expression of constitutive antiherbivore defense by woody plants (e.g., Chew and Rodman, 1979; Bryant *et al.*, 1983; Waterman *et al.*, 1984). Of particular importance to this chapter is the comparatively low nitrogen concentration and high concentrations of nonnitrogenous secondary plant metabolites frequently observed in leaves of plants whose growth is nitrogen-limited (e.g., Wong, 1973; Chew and Rodman, 1979; Waring *et al.*, 1985; Mihaliak and Lincoln, 1985). In this chapter we have provided a model that can account for the progressive decline in leaf food value associated with long-term inducible defenses (e.g., Werner, 1979; 1988a,b; Myers, this volume; Myers and Williams, 1984; Williams and Myers, 1984). This model predicts that the cause of LTI is a nitrogen stress in leaves of induced plants that results from defoliation-caused fine-rootlet mortality in concert with loss of nitrogen contained in leaves eaten by herbivores. If this model correctly describes the mechanism of LTI, then increasing the nitrogen reserves of woody plants by nitrogen fertilization well before defoliation will result in a reduced induction response in comparison to unfertilized plants. We further predict that such an increase in the nitrogen status of nitrogen-fixing woody plants explains why *A. rubra* only exhibits an induction response after several severe and sequential defoliations (Myers and Williams, 1984; Williams and Myers, 1984; Myers, this volume).

In conclusion, we suggest that plant physiological responses to the availability of resources in the environment strongly constrain the evolution and phenotypic expression of constitutive and long-term inducible defenses of woody plants. The importance of such constraints must be determined before coevolutionary interactions between plants and herbivores can be presumed to be the primary cause of allocation of resource by plants to potentially defensive phytochemicals.

Acknowledgments

This work was funded by National Science Foundation grant BSR-8416461 and the Academy of Finland. The ideas presented are the result of joint research supported by the United States–Finland Cooperative Science Agreement supported by the U.S. National Science Foundation and the Academy of Finland. We thank E. Bernays, F. S. Chapin, III, P. Coley, T. T. Kozlowski, R. Lindroth, K. Raffa, J. M. Scriber, and P. Waterman for reviewing various drafts of this paper.

References

Aldous, S. E. (1952). Deer browse clipping study in the Lake States Region. *J. Wildl. Manage.* **16,** 401–409.

Archer, S., and Tiezen, L. L. (1980). Growth and physiological responses of tundra plants to defoliation. *Arct. Alp. Res.* **12,** 531–552.

Baltensweiler, W. (1985). On the mechanisms of the outbreaks of the larch bud moth (*Zeiraphera diniana* Gn., *Lepidoptera, Tortricidae*) and its impact on the subalpine larch-Cembran pine ecosystem. *Proc. Int. Workshop IUFRO Project Group P1.07-00 Ecol. Subalp. Zones, 3rd, Sept. 3–5, 1984, Riederalp, Switzerland.*

Baltensweiler, W., Benz, G., Bovey, P., and Delucchi, V. (1977). Dynamics of larch bud moth populations. *Annu. Rev. Entomol.* **22,** 79–100.

Batzli, G. O., and Jung, H. G. (1980). Nutritional ecology of microtine rodents: Resource utilization near Atkasook, Alaska. *Arct. Alp. Res.* **12,** 483–499.

Batzli, G. O., and Sobaski, S. T. (1980). Distribution, abundance, and foraging patterns of ground squirrels near Atkasook, Alaska. *Arct. Alp. Res.* **12,** 501–510.

Beckwith, R. C. (1968). The large aspen tortrix, *Choristoneura condlictana* (Wlkr.), in interior Alaska. USDA For. Serv. Res. Note PNW-81.

Beckwith, R. C. (1970). Influence of host on larval survival and adult fecundity of *Choristoneura conflictana* (*Lepidoptera: Toricidae*). *Can. Entomol.* **102,** 1474–1480.

Behrend, D. F. (1973). Wildlife management–forest fertilization relations. *In* Forest Fertilization, pp. 108–110. U.S. For. Serv. Gen. Tech. Rep. NE-3.

Bryant, J. P., and Chapin, F. S., III. (1986). Browsing-woody plant interactions during boreal forest plant succession. *In* "Forest Ecosystems in the Alaskan Taiga" (K. Van Cleve, ed.), pp. 213–225. Springer-Verlag, New York.

Bryant, J. P., and Kuropat, P. J. (1980). Selection of winter forage by subarctic browsing vertebrates: The role of plant chemistry. *Annu. Rev. Ecol. Syst.* **11,** 261–285.

Bryant, J. P., Chapin, F. S., III, and Klein, D. R. (1983). Carbon/nutrient balance of boreal plants in relation to vertebrate herbivory. *Oikos* **40,** 357–368.

Bryant, J. P., Chapin, F. S., III, Clausen, T. P., and Reichardt, P. R. (1987). Effects of resource availability upon the woody plant-mammal interaction. *Proc. Annu. Wildl. Shrub Symp., 4th, August 7–9, 1985.* USDA, Forest Service, General Technical Report.

Butler, G. W., and Bailey, R. W. (1973). "Chemistry and Biochemistry of Herbage." Academic Press, New York.

Chapin, F. S., III. (1980a). The mineral nutrition of wild plants. *Annu. Rev. Ecol. Syst.* **11,** 233–260.

Chapin, F. S., III. (1980b). Nutrient allocation and responses to defoliation in tundra plants. *Arct. Alp. Res.* **12,** 553–563.

Chapin, F. S., III, and Kedrowski, R. A. (1983a). Seasonal changes in nitrogen and phosphorus fractions and autumn retranslocation in deciduous and evergreen taiga trees. *Ecology* **64,** 376–391.

Chapin, F. S., III, and Van Cleve, K. (1981). Plant nutrient absorption and retention under differing fire regimes. *In* "Fire Regime and Ecosystem Properties" (H. A. Mooney, T. M. Bonnicksen, J. L. Christensen, J. E. Lotan, and W. A. Reiners, eds.), USDA Forest Service, General Tech. Rept., Washington, D.C.

Chapin, F. S., III, Tryon, P. R., and Van Cleve, K. (1983b). Influence of phosphorus on growth and biomass distribution of Alaskan taiga tree seedlings. *Can. J. For. Res.* **13,** 1092–1098.

Chapin, F. S., III, Van Cleve, K., and Tryon, P. R. (1986). Relationship of ion absorption to growth rate in taiga trees. *Oecologia* **69,** 238–242.

Chew, F. S., and Rodman, J. B. (1979). Plant resources for chemical defense. *In* "Herbivores, Their Interaction with Secondary Plant Metabolites" (G. A. Rosenthal and D. H. Janzen, eds.), Academic Press, New York.

Coley, P. D. (1983). Herbivory and defensive characteristics of tree species in a lowland tropical forest. *Ecol. Monogr.* **53,** 209–233.

Coley, P. D., Bryant, J. P., and Chapin, F. S., III. (1985). Resource availability and plant antiherbivore defense. *Science* **230,** 895–899.

Cooper, S. M., and Owen-Smith, N. (1985). Condensed tannins deter feeding by browsing ruminants in a South African Savanna. *Oecologia* **67,** 142–146.

del Moral, R. (1972). On the variability of chlorogenic acid concentration. *Oecologia* **9,** 289–300.

Dugle, J. R. (1966). A taxonomic study of western Canadian species in the genus *Betula*. *Can. J. Bot.* **44,** 929–1007.

Ericsson, A., Larsson, S., and Tenow, O. (1980). Effects of early and late season defoliation on growth and carbohydrate dynamics in Scots pine. *J. Appl. Ecol.* **17,** 747–769.

Ericsson, A., Hellqvist, C., Langstrom, B., Larsson, S., and Tenow, O. (1985). Effects of growth of simulated and induced shoot pruning by *Tomicus piniperda* as related to carbohydrate and nitrogen dynamics in Scots pine. *J. Appl. Ecol.* **22,** 105–124.

Feeny, P. (1976). Plant apparency and chemical defense. *Rec. Adv. Phytochem.* **10,** 1–40.

Fischlin, A., and Baltensweiler, W. (1979). Systems analysis of the larch bud moth system. Part 1: The larch–larch bud moth relationship. *Bull. Soc. Entomol Suisse* **52,** 273–289.

Fox, J. F., and Bryant, J. P. (1984). Instability of the snowshoe hare and woody plant interaction. *Oecologia* **63,** 128–135.

Garrison, G. A. (1972). Carbohydrate reserves and response to use. pp. 271–286. *In* Wildland Shrubs—Their Biology and Utilization. U.S.D.A. For. Serv. Gen. Tech. Rep. In T-1. Utah State Univ., Logan, Utah.

Gershenzon, J. (1984). Changes in the levels of plant secondary metabolites under water and nutrient stress. *Recent Adv. Phytochem.* **18,** 273–320.

Gezelius, K., Ericsson, A., Hallgren, J. E., and Brunes, L. (1981). Effects of bud removal in Scots pine (*Pinus sylvestris*) seedlings. *Physiol. Plant.* **51,** 181–188.

Grime, J. P. (1977). Evidence for the existence of three primary strategies in plants and its relevance to ecological and evolutionary theory. *Am. Nat.* **111,** 1169–1194.

Grime, J. P. (1979). "Plant Strategies and Vegetation Processes." Wiley, New York.

Hanley, T. A., Cates, R. G., Van Horne, B., and McKendrick, J. D. (1987). Forest stand age-related differences in apparent nutritional quality of forage for deer in southeastern Alaska. *Proc. Annu. Wildl. Shrub Symp., 4th, August 7–9, 1985.* USDA, Forest Service, General Technical Report, pp. 9–17.

Haukioja, E. (1980). On the role of plant defenses in the fluctuations of herbivore populations. *Oikos* **35,** 202–213.

Haukioja, E., and Hakala, T. (1975). Herbivore cycles and periodic outbreaks. *Rep. Kevo Subarct. Res. Sta.* **12,** 1–9.

Haukioja, E., and Niemala, P. (1976). Does birch defend itself actively against herbivores? *Rep. Kevo Subarct. Res. Sta.* **13,** 44–47.

Haukioja, E., and Neuvonen, S. (1985). Induced long-term resistance of birch foliage against defoliators: Defensive or incidental. *Ecology* **66,** 1303–1308.

Haukioja, E., Kapianinen, K., Niemala, P., and Toumi, J. (1983). Plant availability hypothesis and other explanations of herbivore cycles: Complementary or exclusive alternatives. *Oikos* **40,** 419–432.

Heiberg, S. O., and White, D. P. (1951). Potassium deficiency of reforested pine and spruce stands in northern New York. *Soil Sci. Soc. Am. Proc.* **15,** 309–376.

Janzen, D. H. (1974). Tropical blackwater rivers, animals, and mast fruiting by the Dipterocarpaceae. *Biotropica* **6,** 69–103.

Jung, H. G., and Batzli, G. O. (1981). Nutritional ecology of microtine rodents: Effects of plant extracts in the growth of Arctic microtines. *J. Mamm.* **62,** 286–292.

Kallio, P., and Lehtonen, J. (1973). Birch forest damage caused by Oporinia autumnata (Bkh.) in 1965–1966 in Utsjoki, N. Finland. *Rep. Kevo Subarct. Res. Sta.* **10,** 55–69.

Keller, T., and Kozlowski, T. T. (1966). Food relations of woody plants. *Bot. Rev.* **22,** 293–382.

Kirmse, R. D. (1984). Effects of Clearcutting on Forage Production, Quality and Decomposition in the Caatinga Woodland of Northeastern Brazil: Implications to Goat and Sheep Nutrition. Ph.D. dissertation, Utah State University.

Kozlowski, T. T. (1971a). "Growth and Development of Trees," Vol. 1. Academic Press, New York.

Kozlowski, T. T. (1971b). "Growth and Development of Trees," Vol. 2. Academic Press, New York.

Kozlowski, T. T., and Algren, C. E. (1974). "Fire and Ecosystems." Academic Press, New York.

Kramer, P. J., and Kozlowski, T. T. (1979). "Physiology of Woody Plants." Academic Press, New York.

Kuropat, P. J. (1984). Foraging behavior of caribou on a calving ground in northwestern Alaska. M.S. thesis, Univ. Alaska, Fairbanks.

Larsson, S., Wiren, A., Ericsson, T., and Lundgren, L. (1986). Effects of ligant and nutrient stress on defensive chemistry and susceptibility to *Galerucella lineola* (Col., *Chrysomelidae*) in two *Salix* species. *Oikos* **47,** 205–210.

Levitt, J. (1972). "Responses of Plants to Environmental Stresses." Academic Press, New York.

Loyttyniemi, K. (1981). Nitrogen fertilization and nutrient contents in Scots pine in relation to the browsing preference by moose (*Alces alces*). *Folia For.* (*Helsinki*) **487,** 1–14.

MacLean, S. F., and Jensen, T. J. (1984). Food plant selection by insect herbivores: the role of plant growth form. *Oikos* **44,** 211–221.

McKey, D., Waterman, P. G., Mbi, C. N., Gartlan, J. S., and Struhsaker, T. T. (1978). Phenolic content of vegetation in two African rain forests: Ecological implications. *Science* **202,** 61–63.

McKey, D. B., Gartlan, J. S., Waterman, P. G., and Choo, G. M. (1981). Food selection by black colobus monkeys (*Colobus satanus*) in relation to plant chemistry. *Biol. J. Linn. Soc.* **16,** 115–146.

Mihaliak, C. A., and Lincoln, D. E. (1985). Growth pattern and carbon allocation to volatile leaf terpenes under nitrogen-limiting conditions in *Heterotheca subaxallaris* (*Asteraceae*). *Oecologia* **66,** 423–426.

Milke, G. C. (1969). Some moose-willow relationships in the interior of Alaska. M.S. thesis, Univ. Alaska.

Miquelle, D. G. (1983). Browse regrowth and consumption following summer defoliation by moose. *J. Wildl. Manage.* **47,** 17–24.

Mooney, H. A. (1972). The carbon balance of plants. *Annu. Rev. Ecol. Syst.* **3,** 315–346.

Mooney, H. A., and Gulmon, S. L. (1982). Constraints on leaf structure and function in reference to herbivory. *Bioscience* **32,** 198–206.

Myers, J., and Williams, K. (1984). Does tent caterpillar attack reduce the food quality of red alder foliage? *Oecologia* **62,** 74–79.

Niemala, P., Toumi, J., Mannila, R., and Ojala, P. (1984). The effect of previous damage on the quality of Scotch pine (*Pinus sylvestris*) foliage as food for diprionid sawflies. *Z. Agnew. Entomol.* **98,** 33–43.

Oh, H. L., Jones, M. B., Longhurst, W. M., and Connolloy, G. E. (1970). Deer browsing and rumen microbial fermentation of Douglas fir as affected by fertilization and growth stage. *For. Sci.* **16,** 21–37.

Opler, P. A. (1974). Biology, ecology, and host specificity of microlepidoptera associated with *Quercus agrifolia* (Fagaceae). *Univ. Calif. Publ. Entomol.* **75.**

Owen-Smith, N., and Cooper, S. M. (1987). Classifying African savanna trees and shrubs in terms of their palatability to browsing ungulates. *Interactions Proc. Annu. Wildl. Shrub Symp., 4th, August 7–9, 1985.* USDA, Forest Service, General Technical Report.

Pfister, J. A. (1983). Nutrition and Feeding Behavior of Goats and Sheep Grazing Deciduous Shrub-Woodland in Northeastern Brazil. Ph.D. dissertation, Utah State University.

Piene, H. (1980). Effects of insect defoliation on growth and foliar nutrients in young balsam fir. *For. Sci.* **26,** 665–673.

Prudhomme, T. I. (1982). The effect of defoliation history on photosynthetic rates in mountain birch. *Rep. Kevo Subarct. Res. Sta.* **18,** 5–9.

Pulliainen, E. (1972). Nutrition of mountain hare (*Lepus timidus*) in northeastern Lapland. *Ann. Zool. Fenn.* **9,** 17–22.

Rathcke, B. (1985). Slugs as generalist herbivores: Tests of three hypotheses on plant choice. *Ecology* **66,** 828–836.

Redmond, D. R. (1959). Mortality of rootlets in balsam fir defoliated by the spruce budworm. *For. Sci.* **5,** 64–69.

Rhoades, D. F. (1979). Evolution of plant defenses against herbivores. *In* "Herbivores: Their Interaction with Secondary Plant Metabolites" (G. A. Rosenthall and D. H. Janzen, eds.), pp. 3–54. Academic Press, New York.

Rhoades, D. F., and Cates, R. G. (1976). Toward a general theory of plant antiherbivore chemistry. *In* "Biochemical Interactions Between Plants and Insects" (J. W. Wallace and R. L. Mansell, eds.), pp. 168–213. Plenum, New York.

Robbins, C. T. (1983). "Wildlife Feeding and Nutrition." Academic Press, New York.

Robbins, C. T., Hanley, T. A., Haggerman, A. E., Hjeljord, O., Baker, D. L., Schwartz, C. C., and Mautz, W. W. (1987). The role of phenolics in defending plants against ruminants: Reduction in protein availability due to tannins. *Ecology* **68,** 98–107.

Robus, M. (1981). Foraging behavior of muskoxen in arctic Alaska. M.S. thesis, Univ. of Alaska, Fairbanks.

Rousi, M. (1983). Susceptibility of pine to mammalian herbivores in northern Finland. *Silva Fenn.* **17,** 301–312.

Sarmiento, G., Goldstein, G., and Meinser, F. (1985). Adaptive strategies of woody species in neotropical savannas. *Biol. Rev.* **60,** 31–355.

Schmidt, W. C., and Fellin, D. G. (1983). Effect of fertilization on western spruce budworm feeding in young western larch stands. *In* "Forest Defoliator—Host Interactions: A Comparison Between Gypsy Moth and Spruce Budworms" (R. L. Talerico and M. M. Montgomery, eds.), pp. 87–96. USDA Forest Serv. Gen. Tech. Rep. NE-85.

Scriber, J. M., and Slansky, F., Jr. (1981). The nutritional ecology of immature insects. *Annu. Rev. Entomol.* **26,** 183–211.

Shaver, G. (1983). Mineral nutrition and leaf longevity in *Ledum palustre:* The role of individual nutrients and the timing of leaf mortality. *Oecologia* **56,** 160–165.

Small, E. (1972). Photosynthetic rates in relation to nitrogen recycling as an adaptation to nutrient deficiency in peat bog plants. *Can. J. Bot.* **50,** 2227–2233.

Steinberg, P. D. (1985). Feeding preferences of *Tegula funebralis* and chemical defenses of marine brown algae. *Ecol. Monogr.* **55,** 333–349.

Struhsaker, T. T. (1968). Interrelationships of red colobus monkeys and rain-forest trees in the Kibale Forest, Uganda. *In* "The Ecology of Arboreal Foliavores" (G. G. Montgomery, ed.), pp. 397–437. Smithsonian Inst. Press, Washington, D.C.

Sullivan, T. P., and Sullivan, D. S. (1982). Influence of fertilization on feeding attacks to lodgepole pine by snowshoe hares and red squirrels. *For. Chron.* **58,** 263–266.

Tuomi, J., Niemala, P., Haukioja, E., Siren, S., and Neuvonen, S. (1984). Nutrient stress: An explanation for plant anti-herbivore response to defoliation. *Oecologia* **61,** 208–210.

Trudell, J., and White, R. G. (1981). The effect of forage structure and availability on food intake, biting rate, bite size and daily eating time of reindeer. *J. Appl. Ecol.* **18,** 63–81.

Van Soest, P. (1982). "Nutritional Ecology of the Ruminant." Durham & Downey.

Viereck, L. A., and Little, E. J., Jr. (1972). Alaska Trees and Shrubs. U.S. Dept. Agric. Handbook 410.

Waring, R. H., McDonald, A. J. S., Larsson, S., Ericsson, T., Wiren, A., Arwidsson, E., Ericsson, A., and Lohammar, T. (1985). Differences in chemical composition of plants grown at constant relative growth rates with stable mineral nutrition. *Oecologia* **65,** 157–160.

Waterman, P. G., Ross, J. A. M., and McKey, D. B. (1984). Factors affecting levels of some phenolic compounds, digestibility, and nitrogen content of the mature leaves of *Barteria fistulosa* (*Passifloraceae*). *J. Chem. Ecol.* **10,** 387–401.

Webb, W. L. (1980). Starch content of conifers defoliated by the Douglas-fir tussock moth. *Can. J. For. Res.* **10,** 535–540.

Webb, W. L., and Karchesy, J. J. (1977). Starch content of Douglas-fir defoliated by the tussock moth. *Can. J. For. Res.* **7,** 186–188.

Weeden, R. B. (1969). Foods of rock and willow ptarmigan in central Alaska with comments on interspecific competition. *Auk* **86,** 271–81.

Werner, R. A. (1977). Biology and behavior of the spear-marked black moth, *Rheumaptera hastata,* in interior Alaska. *Ann. Entomol. Soc. Am.* **70,** 328–336.

Werner, R. A. (1979). Influence of host foliage on development, survival, fecundity, and oviposition of the spear-marked black moth, *Rheumaptera hastata* (*Lepidoptera: Geometridae*). *Can. Entomol.* **111,** 317–322.

Werner, R. A. (1988a). Effect of early and late season defoliation. II. Defensive chemicals, foliar nutrients, and carbohydrate reserves. *Oecologia.* In review.

Werner, R. A. (1988b). Effect of early and late season defoliation. I. Biomass production, plant growth, and survival. *Oecologia.* In review.

Wilde, S. A., Nulbandov, O. G., and Yu, T. M. (1948). Ash, protein and organosolubles of jack-pine seedlings in relation to soil fertility. *J. For.* **46,** 829–831.

Williams, J. B., Best, D., and Warford, C. (1984). Foraging ecology of ptarmigan at Meade River, Alaska. *Wilson Bull.* **92,** 341–351.

Williams, K., and Myers, J. (1984). Previous herbivore attack of red alder may improve food quality for fall webworm larvae. *Oecologia* **63,** 166–170.

Wong, E. (1973). Plant phenolics. *In* "Chemistry and Biochemistry of Herbage" (G. W. Butler and R. W. Bailey, eds.), pp. 265–316. Academic Press, New York.

CHAPTER 13

Plant-Mediated Interactions between Seasonal Herbivores: Enough for Evolution or Coevolution?

STANLEY H. FAETH
Department of Zoology
Arizona State University
Tempe, Arizona

I. Introduction

A. Competition and Coevolution

Although coevolution was first described in the parasite–host systems of herbivores and plants (Ehrlich and Raven, 1964), similar ideas of coevolution were already inextricably woven into community theories of competing species (Roughgarden, 1983a). One of the dominant themes of competition theory is that competition causes tit-for-tat evolutionary changes in participating populations to produce such community-wide patterns as character displacement, resource partitioning, density compensation, ecological release, and constant and minimum size ratios. More recently, the notion that interspecific competition is the major force structuring communities, especially communities of phytophagous insects, has been questioned. Despite some erosion of confidence in the competition edifice, this interaction is still considered by many as one of the most important in natural communities. But even those advocates of competition as a force in structuring communities have been willing to concede that competition is an unlikely candidate for structuring phytophagous insect communities (Orians and Paine, 1983; Roughgarden, 1983b; Price, 1983; Schoener, 1983).

If these ideas sound like a familiar tune, it is because they are essentially restatements of the Hairston, Slobodkin, and Smith hypothesis of 1960, which proposed that herbivores are not resource-limited and are therefore unlikely to interact within their trophic level. Instead, Hairston *et al.* (1960) and most current workers regard herbivore populations as most strongly regulated by interactions with natural enemies (between-trophic level or *vertical interactions*) and autecological factors, which include aspects of the host plants (Lawton and Strong, 1981). The return to the earlier hypothesis was precipitated by a general lack of evidence for interspecific competition between phytophagous insect species (see Lawton and Strong, 1981; Strong *et al.*, 1984, for examples).

The most obvious assumption of coevolution is that species must interact. If within-trophic level interactions (hereafter termed *horizontal interactions*) such as competition are indeed nugacious and infrequent, then the prospects for evolutionary or coevolutionary relationships and patterns among species seem bleak. Instead, we would be left with the not very interesting, at least biologically, null hypothesis of Lawton and Strong (1981) that "species coexist independently, without effective interaction, and without evolution having made any combinations of species more compatible." On the other hand, if horizontal interactions are common and intense, then at least the potential exists for ecological and evolutionary changes in the participating species. Current opinion seems to be running strongly for the lack of importance of horizontal interactions in phytophagous insect communities.

Have we been too hasty in discarding horizontal interactions as a factor in the organization and possible coevolution of insect communities? Perhaps a caveat can be gleaned from Schoener's (1983) and Connell's (1983) reviews of evidence for interspecific competition. Both noted that studies examining the role of competition between phytophagous insect species were relatively scarce in proportion to their diversity. In addition to a paucity of studies, we may have overlooked the subtlety of interactions between phytophagous insects, and accepted failure to find our preconceived notions of interactions as the absence of interaction. In this chapter, I will suggest unique and complex ways that phytophagous insects can interact.

Documentation of interaction by no means implies that coevolution is occurring or has occurred (Janzen 1980); interaction is, however, a prerequisite for coevolution. For folivorous insects, the interactions may occur through changes in their common food plants.

B. Interactions via the Host Plant

Plants are mercurial food resources for insects. Plants change seasonally and yearly in water and nutrient content, "defensive" chemistry, and in phenology, such as when leaves are produced and when they fall. Additionally, plants vary spatially in these factors both within and between individuals (Denno and McClure, 1983). The insects (or other herbivores) themselves may alter the plant, either chemically or phenologically. Induction of chemical changes resulting from herbivory has been documented in many trees [firs (*Abies* sp.), Puritch and Nijholt (1974); birch (*Betula* sp.), Haukioja and Niemela (1976, 1979), Niemela *et al.* (1979); oaks (*Quercus* sp.), Schultz and Baldwin (1982), Faeth (1986a); alder (*Alder* sp.) and willow (*Salix* sp.), Rhoades (1983a); maple (*Acer* sp.) and poplar (*Populus* sp.), Baldwin and Schultz (1983)] as well as other plants (see Rhoades, 1983b, 1985, for reviews). Induced chemical responses may be immediate and short-lived, lasting only a few hours (Schultz and Baldwin, 1982), or may accumulate slowly and persist for 1 (Faeth, 1986a) or more years (Haukioja, 1980; Neuvonen and Haukioja, 1984). It is clear that many plants change chemically after partial or complete defoliation; this sets up the possibility for interaction with later feeders, and at first glance seems to be no different than competition for a resource where one species consumes part of the resource and reduces its suitability for the next species (Schoener, 1983). Thus one might expect early feeders to negatively affect later feeders, and if the effects of the late feeders are carried over to the next growing season, the late feeders may have a detrimental impact on the next season's early feeders. A coevolutionary circuit of natural selection may thus be established.

In reality, the consequences of previous herbivory for subsequent feeders are often puzzling. While many have assumed that herbivory should always produce negative effects on later feeders (hence the frequent use of "defensive"), Fowler and Lawton's (1985) review showed that the actual effects can range from negative to positive to none at all for later feeders. For example, Niemela *et al.* (1984) found in laboratory studies that one Dioprionid sawfly species had faster larval development and higher survivorship when fed foliage previously damaged, two species showed no effect, and a fourth species had higher mortality on damaged foliage. Myers and Williams (1984) showed that foliage quality for the western tent caterpillar was reduced only at very high levels of defoliation sustained for at least 3 years. Under some circumstances, previous damage may actually improve quality of foliage for fall webworm (Williams and Myers, 1984) and tent caterpillars (Myers and Williams, 1987). It is far less certain how these effects, be they positive or negative, are translated into changes at the population or community level (Fowler and Lawton, 1985).

Why such disparity in effects of herbivory on subsequent feeding insects? I think a large part of the problem in our understanding of this stems from an understandable eagerness to generalize in ecology. We have a tendency to view all chemical responses by the plant to herbivory as universally "defensive." We should no more expect that all chemical and phenological changes in plants in response to herbivory are detrimental ("defensive") to subsequent feeders than we should expect that a particular glycoside in a milkweed plant will be toxic to all phytophagous insects. As with any other variable plant characteristic, induced responses are likely to have very different effects depending on when and how they occur, and the life histories of the insect species involved and their evolutionary association with the plant. One might even predict that if such induced responses (both chemical and phenological) are frequent and strong enough then some late feeders should evolve to exploit these changes, and this may explain why some herbivory appears to benefit some insects.

In the following sections, several experimental studies are presented that show that the consequences for late seasonal insects feeding on damaged foliage are quite variable, depending on the extent, scale, and timing of early-season herbivory and the phenology of the insect species themselves. The evolutionary and coevolutionary implications of interactions between early- and late-season herbivores are then discussed in relation to changing plant chemistry and phenology and their indirect effects on natural enemies.

II. Experimental Studies of Interactions between Seasonal Folivores

A. Leaf-Chewer and Leaf-Miner Interactions at the Level of Individual Leaves

1. Survival of Leaf Miners on Damaged and Intact Leaves

Leaf miners on oaks are typically late feeders; that is, most species begin to feed as larvae after most herbivory by the major leaf-chewing insects has occurred (Faeth *et al.*, 1981b; Faeth, 1985a, 1986; West, 1985). As such, ovipositing adults are confronted with an array of leaf types based on inherent and induced chemical and phenological variation. Selection of leaves by ovipositing adults is critical for survival of larval offspring because most larval leaf miners are confined to feed on a single leaf where the adult has placed an egg. One therefore expects strong selection pressures on adults to choose leaves that are optimal for larval growth and survival.

Alternatively, leaf chewers are usually early-season feeders, although secondary episodes of late-season feeding may occur (Feeny, 1970; Faeth, 1985b). For example, most herbivory by leaf chewers on the temperate Emory oak (*Quercus emoryi* Torr.), which holds leaves for 1 year, occurs within 30 days after budbreak in the spring, with occasional episodes of leaf-chewing folivory occurring in the fall months (Faeth, 1985b).

One would therefore predict that leaf chewers might affect leaf miners, and possibly concurrent or later chewers. To test the effect of leaf-chewing damage on the leaf-mining guild (eight species) on Emory oak, I monitored survival and causes of mortality of leaf miners in undamaged leaves and in leaves with leaf-chewing and artificial damage (hole-punched) over two complete growing seasons (Faeth, 1986). Survival of leaf miners in damaged leaves was significantly less than miners in intact leaves (1981–1982, 27.8% versus 46.7%; 1982–1983, 23.0% versus 36.6%) because parasitism by hymenopterans was greatly increased for miners in damaged leaves (1981–1982, 44.2% versus 21.1%; 1982–1983, 56.6% versus 20.1%).

2. Changes in Larval Survival via Phytochemistry and Phenology

Examination of the chemistry and phenology of damaged and intact leaves suggests possible mechanisms by which survival and mortality are altered for miners in partially damaged leaves [all phytochemical methods are detailed in Faeth (1986)]. Condensed tannin levels were significantly higher in damaged than in intact leaves (Fig. 1). Protein levels were significantly lower in damaged leaves and water content was slightly less than in intact leaves (Faeth, 1986). Of course, the physical structure of leaves was also altered by leaf-chewing damage. Increased parasitism of miners in damaged leaves may result from parasites using these physical and chemical changes as cues to locate hosts, as has been suggested by others (Vinson, 1976, Weseloh 1981). Alternatively, increased tannins and decreased protein may have slowed larval development such that miners were exposed to parasite attack for longer periods. Apparently, both these mechanisms may contribute to increased parasitism in this leaf miner/oak system. For one species of leaf miner [*Stilbosis juvantis* (Hodges)], mines were significantly smaller on damaged leaves, which

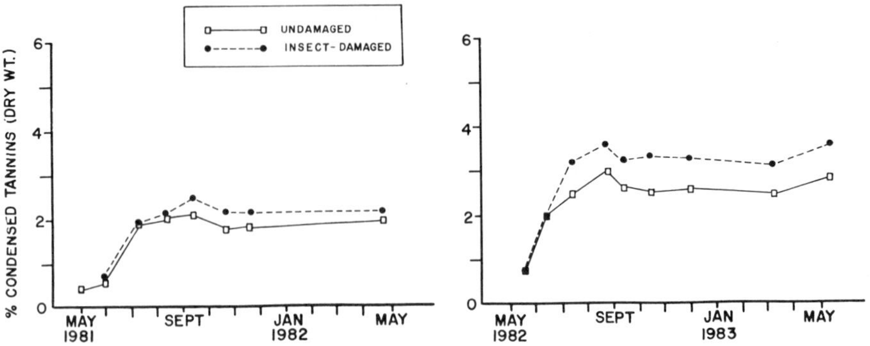

Fig. 1. Seasonal changes in condensed tannin content of intact and insect-damaged leaves of Emory oak in 1981–1982 and 1982–1983. Each point is the mean of six trees at a sample date.

suggests that development and feeding were slow (Faeth, 1985a). In another experiment, waxed, damaged leaves were attached to intact leaves that harbored *Cameraria* sp. nov. caterpillars. This experiment was designed to provide the physical cues of damage to parasites without altering chemistry or nutrients of the mined leaves. Parasitism was greater on miners in leaves with attached damaged leaves (49.3%, $n = 108$) compared with control leaves (26.8%, $n = 399$), suggesting that the physical cues of damage only can also serve to increase parasitism.

Are there only negative effects associated with partially damaged leaves? Investigation of categories of mortality other than parasitism indicate that leaf miners on partially damaged leaves could benefit from mining damaged leaves (although not enough to compensate for the negative effects of increased parasitism). Predation and death from early leaf abscission did not differ for miners on damaged and intact leaves, but death from other causes was significantly less for miners on damaged leaves. This category of mortality includes, presumably, a large fraction of deaths from bacterial and fungal attack in the early instars of leaf miners (Faeth and Bultman, 1986). I have proposed (Faeth, 1986, Faeth and Bultman, 1986) that leaf miners feeding on damaged leaves with increased levels of tannins may be protected because of the bactericidal and fungicidal properties of tannins (Swain, 1979). Taper *et al.* (1985) also found that mortality of endophagous, cynipid gallwasps caused by fungal attack was reduced when galls were formed on leaves with high levels of tannins. Conceivably, under circumstances of reduced parasite attack, mining damaged leaves could be selected for, provided that the benefits outweigh the potential risks (e.g., from reduced leaf area for mining or from increased leaf abscission).

To more closely examine the role of tannins on leaf-miner mortality, we coated leaves of *Quercus emoryi* Torr. containing the first larval instar of a leaf-mining species, *Cameraria* sp., with tannins in solution (Faeth and Bultman, 1986). The tannins were extracted from *Q. emoryi* and on average contained 15.6% gallotannins and 28.9% condensed tannins. The mined leaves had no leaf-chewing damage, and thus we could separate the effects of physical from chemical cues used by parasitoids (this is the reciprocal of the waxed, hole-punched leaf experiment). Miners in tannin-enriched leaves suffered higher rates of parasitism (57.1% versus 35.3% on control leaves), suggesting that tannins or tannin degradation products can increase attack by parasitoids. Coupled with evidence from the hole-punched, waxed-leaf experiments already discussed, it seems that both physical and chemical cues are used by parasitoids of the leaf miners. Because this species of leaf miner is attacked by many species of parasitoids in the families Eulophidae, Braconidae, Encyrtidae, and Ichneumonidae, it is not yet clear whether all or just a subset of parasitoids is attracted by cues associated with damaged leaves.

What about mortality from causes other than parasitism? Death from bacterial and fungal attack was decreased on painted leaves (40.0% versus 62.9% on control leaves), again suggesting the protective nature of increases in tannin levels. The beneficial effects of increased tannin levels, usually considered as defensive chemicals, remains an interesting question that deserves further experimentation.

3. Changes in Body Size via Phytochemistry and Phenology

Larval survival is not the only component of fitness. To determine whether leaf damage also affects body size and therefore possibly fecundity, leaves that harbored first-instar larvae of *Cameraria* sp. nov. (Davis) were hole-punched to simulate leaf-chewing damage, and pupal masses of successful miners were compared to control (undamaged) leaves with larval miners. When hole-punching leaves, only small amounts of the leaves were removed and the mines themselves were not breached. Pupal masses of miners in damaged leaves were significantly less than those in intact leaves (damaged leaves, mean mass = 1.598 ± 0.293 mg, $n = 55$; intact leaves, mean mass = 1.792 ± 0.173 mg, $n = 59$; $t = 4.29$, $p < 0.001$). Reduction of pupal mass suggests that larval development is impaired in damaged leaves and, because body size and fecundity are typically correlated in a positive fashion (Campbell, 1962; Marks, 1976; Hough and Pimentel, 1978), fecundity is also likely reduced.

The reduction in pupal mass in damaged leaves may be caused by phytochemical changes in these leaves, particularly increases in tannin levels, but could also be due to decreased feeding area or decreased water content. To separate the phytochemical effects on pupal mass from other changes in damaged leaves, we compared pupal weights of successful leaf miners on leaves painted with tannin solutions to controls to determine whether increases in tannins might affect body size and thus possibly fecundity. Pupal weights were significantly less in tannin-enriched leaves (1.50 ± 0.25 mg versus 1.84 ± 0.18 mg in control leaves). Apparently, phytochemical inductions alone can influence development and body size, and therefore are also likely to affect fecundity.

The above experiments indicate that early-season leaf chewers can indeed interact with the late-feeding leaf miners by altering individual host leaves. Changes in leaf chemistry are no doubt involved, as evidenced by experiments where only phytochemistry is altered, but a damaged leaf may represent a poorer resource to a later feeder because of a greater propensity to abscise (Faeth *et al.*, 1981a; Bultman and Faeth, 1986b), lower water content (Faeth, 1986), or simply less surface area for feeding and development. Chemical alteration of individual leaves is not exclusively detrimental, but may actually ameliorate certain mortality factors.

Other recent studies have also shown that phytophagous insects may negatively interact via changes in the host plant. Karban (1986) found that the presence of plume moths (*Platyptilia williamsii* Grinnell) adversely affected persistence of spittlebugs (*Philaenus spumarius* L.) on the common host plant *Erigeron glaucus* Ker. In this case, the plume moth removed new leaves, a necessary resource for the spittlebugs. Silva-Bohorquez (1986) reports that leaf damage by winter moths (*Opherophtera brumata* L.) reduces densities, fecundity, and longevity of aphids (*Tuberculoides annulatus* Hartig) on *Quercus robur* L. Raupp *et al.* (1986) showed that the presence of the chysomelid willow leaf beetle (*Plagiodera versicolora* Laich.) deters feeding by another herbivore (*Nyphalis antiopa* L.) on willow through glandular secretions of monoterpenes. These studies and the results presented here demonstrate that early feeders can clearly influence the presence and survival of the subsequent feeders by altering the quantity or suitability of a common plant resource and at least set the stage for an evolutionary or coevolutionary response from the participating species.

4. Evolutionary Response: Oviposition Choice

One such evolutionary response is change in oviposition behavior. Many phytophagous insects are known to discriminate within host-plant species based on leaf chemistry or age (Whitham, 1980; Raupp and Denno, 1983; Rausher, 1983). Ovipositing females may also discriminate leaves based on the presence of eggs or pheromones deposited by other insects (Prokopy *et al.,* 1978; Rausher, 1979; Williams and Gilbert, 1981; McNeil and Quiring, 1983; Bultman and Faeth, 1986). However, little work has been done on the effects of previous folivory on selection of leaves by ovipositing females. Because previous folivory may alter leaf quality and reduce leaf quantity, one might expect evolution of discriminating behavior by females since their offsprings' survival and fecundity may depend on where eggs are posited. Evolution of host selection based on previous leaf damage should be most clearly evidenced in endophagous leaf feeders, such as leaf miners and gall formers, since the feeding sites of these sedentary larvae are solely determined by the maternal adult.

To experimentally test selection of leaves based on previous folivory, colonization of the late-feeding, leaf-mining guild was monitored on 12 trees of Emory oak over two growing seasons (Faeth, 1985a, 1986). Six of these trees (controls) had endemic levels of early-season folivory by leaf chewers (~25% of leaves were damaged). On six other trees, I experimentally increased the proportion of damaged leaves to ~75% by hole-punching leaves (~25% of leaves damaged by endemic insects plus 50% hole-punched leaves). On average, hole-punching removed ~16% of the area of individual leaves.

Leaf miners occurred more frequently than expected by chance in intact than in damaged leaves on both experimental and control trees in both growing seasons. Leaf-mining adults avoid damaged leaves, or, alternatively, prefer intact leaves. The selective pressures associated with lower survivorship in damaged leaves (Faeth, 1985a, 1986) could have resulted in this discriminatory behavior.

There is yet another twist to the already complicated web of interaction between early and late folivores. Survival of leaf miners in damaged leaves on experimental trees was significantly greater than miners in damaged leaves on control trees (1982–1983, 29.3% versus 4.2%, $\chi^2 = 6.46$, $p < 0.01$), although survival in damaged leaves was always less than survival in intact leaves within respective treatments. The increase in survival on damaged leaves within experimental trees was due to a significant drop in rates of parasitism (1982–1983, 46.2% versus 78.3% on control trees, $\chi^2 = 6.31$, $p < 0.05$). The relationship between intensified parasitism and number of damaged leaves seems to be inversely density-dependent; that is, it is disadvantageous to feed within a damaged leaf, but when fewer leaves are damaged within a tree (e.g., control trees), the risk is even greater. When many leaves are damaged, parasitoids may be swamped with physical and chemical cues from damaged leaves, and become less efficient at locating leaf miners that are relatively rare. Selective pressures would change depending not only on whether a leaf is damaged but also on how many surrounding leaves are damaged. One might expect that as the number of partially damaged leaves increases, the advantages (decreased parasitism, increased leaf area, less likely abscission) of mining intact leaves may wane relative to the benefits of mining damaged leaves (decreased fungal and bacterial attack). A hypothetical relationship between the selective advantage of mining damaged or intact leaves based on the proportion of damaged leaves within a tree is shown in Fig. 2. Whether leaf-mining adults can somehow assess frequency of damaged leaves and adjust oviposition accordingly is yet to be determined.

Selective pressures leading to preference or avoidance of leaves by leaf miners are likely to be highly variable depending on the species, locality, host plant, and natural enemy complex. For example, in another locality in Arizona and with a different species of leaf miner (*Cameraria* sp. nov. Davis), I found no relationship between mining damaged leaves and changes in parasitism. First-instar larvae on hole-punched leaves of *Q. emoryi* suffered 30.0% parasitism ($n = 177$); those in intact leaves showed 26.8% parasitism ($n = 399$). However, death from premature leaf abscission was higher (25.8%) for miners in hole-punched leaves than for those in intact ones (8.4%), and, as indicated above, pupal mass was lower for miners in hole-punched leaves. In Florida, Simberloff and Stiling (1988) also report that a leaf-mining species, *Stilbosis quadricustatella* Cham.,

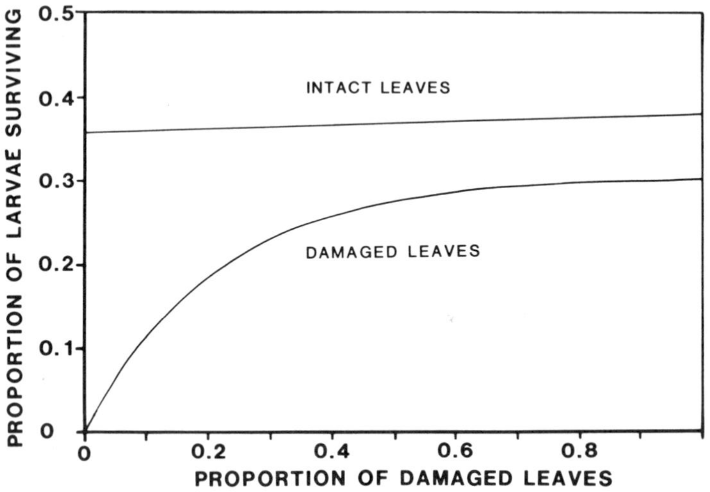

Fig. 2. Hypothetical relationship between probability of larval survival and the proportion of leaves damaged within a tree for leaf miners in damaged and intact leaves. As proportion of damaged leaves increases, the probability of survival in damaged leaves also increases because parasitoids that rely on physical and chemical cues from damaged foliage must search more and more leaves to find the relatively rare larval miners. Although probability of survival increases for larvae in damaged leaves with percentage of damaged leaves, it is still less than probability of survival for larvae in intact leaves. Parameters for the curves are derived from survival estimates in Faeth (1986). The curve for miners in damaged leaves is described by the equation: $y = a + b(p)^X$ where a is the asymptote, $(a + b)$ is the y-intercept, and $0 < p < 1$.

avoids damaged leaves for oviposition, not because of changes in parasitism but because much greater probability of early leaf abscission (and, consequently, death from desiccation) is associated with damaged leaves.

Damage by leaf chewers and associated effects on survival and possibly fecundity seem to have been of sufficient selective pressure to modify oviposition behavior in adult leaf miners. Certainly, other factors also affect oviposition behavior of leaf miners; examples are inherent variability in leaf chemistry and phenology (Faeth, 1986; Bultman and Faeth, 1986a–c; Godfray, 1986), interaction with other leaf miners (Bultman and Faeth, 1986b; Stiling *et al.*, 1987), and abiotic factors (Bultman and Faeth, 1988a,c).

B. Leaf-Chewer and Leaf-Miner Interactions when Trees Are Defoliated

Leaf chewers interact with leaf miners via altered chemistry and phenology when portions of individual leaves are consumed. But not all leaf chewers are so restrained in their gastronomical habits; some devour entire leaves. Obviously, the chewers deplete, at least temporarily, part or all of the leaf resources for later folivores, and thus could decrease fitness and population sizes of later feeders. But many plants, particularly trees, can refoliate within a growing season. For example, oak trees are sometimes partially (some leaves entirely removed) or completely (all leaves removed) defoliated by folivorous insects (e.g., gypsy moth defoliations in northeastern United States). Defoliations are usually followed by refoliation (Crawley, 1983), and the reflushed leaves may be of poorer quality for folivorous insects. For example, Schultz and Baldwin (1982) showed that replacement leaves after defoliation contained higher levels of tannins that presumably slowed growth and reduced body size of gypsy moths.

Are these refoliated leaves chemically and phenologically the same as partially damaged leaves? In other words, is all leaf damage the same in terms of its impact on fitness and population dynamics of later feeders, as is often assumed (e.g., Fowler and Lawton, 1985)? To test this hypothesis, experiments were designed where oak trees were completely defoliated (Faeth, 1987).

To test the role of complete defoliation on late-feeding insects, I manually defoliated (completely) six trees of Emory oak at different times in two growing seasons. Colonization of the late-feeding, leaf-mining and the late-feeding, leaf-chewing guilds were monitored on refoliated trees relative to control trees with mature, primary foliage. Survivorship was determined for leaf miners on refoliated and mature (control) leaves. Phytochemical differences between mature and reflushed foliage were also compared.

1. Phytochemistry

Complete defoliation alters the leaf resource so that it is quite different phytochemically from that changed through partial consumption of leaves. Refoliated leaves had significantly lower amounts of condensed tannins than did control leaves (Fig. 3), while the reverse was true for partially damaged leaves (Fig. 1). The difference in condensed tannin content between refoliated and control (mature) leaves was sustained throughout the remainder of the growing seasons. Although hydrolyzable tannins increased briefly after refoliation, they quickly returned to comparable levels of hydrolyzable tannins in mature leaves. Furthermore, the temporary increase in hydrolyzable tannins did not approach the high levels that are typically found in newly emerged, spring leaves. Protein content did not differ between refoliated and mature leaves, and water content was greater in refoliated than mature leaves. Clearly, refoliated leaves represent a much different resource to late-seasonal feeders than do partially damaged leaves, yet both are a product of the same mechanism—herbivory—and only differ by degree. In fact, one would expect refoliated leaves to be a superior resource in terms of quality; these leaves are lower in condensed tannin content and higher in water content than leaves usually present at this time. Do late-season folivores respond correspondingly to the quality of this altered resource?

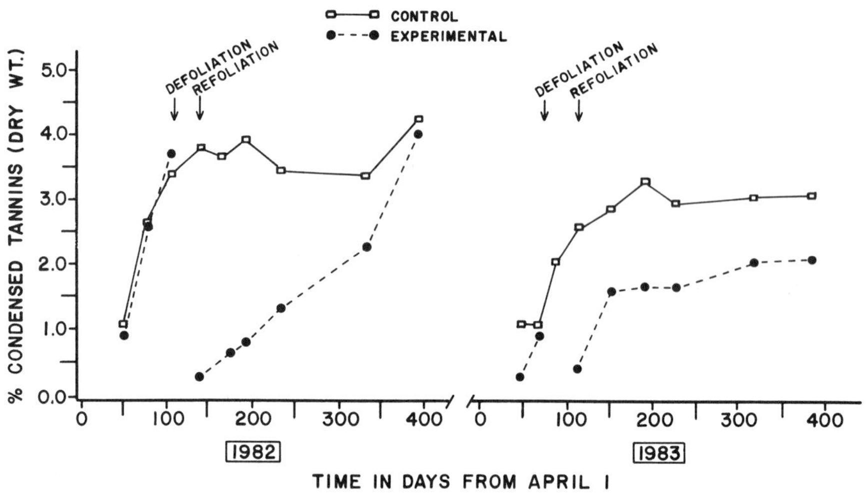

Fig. 3. Seasonal changes in condensed tannin content of Emory oak trees (experimentals) before defoliation and after refoliation at different times in two growing seasons (1982–1983; 1983–1984), relative to control trees (not defoliated). Each point is the mean of six trees at a sample date.

a. The Leaf Miners. The answer for the leaf miners depends greatly on when trees were defoliated and how long before refoliation occurred. Trees that were defoliated on July 7, 1982, refoliated 32 days (±5 days) later. Densities of leaf miners were significantly less on refoliated trees than on control trees (refoliated mean densities = 3.98 ± 2.96 mines per 1000 leaves; control = 9.83 ± 4.54 mines per 1000 leaves; $z = 2.14$, $p = 0.03$). However, the decrease in densities was caused by the absence of a dominant leaf-mining species, *Stilbosis juvantis,* that oviposits coincidentally when leaves were absent on the defoliated trees. When trees were defoliated earlier (June 8, 1983) in the next season, the lag between defoliation and refoliation was much shorter (18 ± 3 days), and densities of leaf miners were greater on refoliated trees than on control trees (refoliated mean densities = 16.53 ± 12.49; control = 9.00 ± 5.46; $z = 1.78$, $p = 0.08$). *Stilbosis juvantis,* locally extinct on refoliated trees the year before, was now a dominant species on these trees as well as on control trees because defoliated trees had refoliated before the critical oviposition window of time.

Apparently, the adult female leaf miner responds to the altered resource by preferentially selecting refoliated leaves for oviposition if they are available. I say "apparently" because I must assume that mine distribution accurately reflects egg distribution (eggs are minute and difficult to census; however, observations suggest that this assumption is valid). Is leaf selection reflected in increases in parameters of fitness? For at least larval survival, the answer is "yes." Survival of leaf miners on refoliated trees was significantly greater than on control trees in the 1983–1984 season [refoliated 40 (44.9%) survived, 49 (55.1%) dead, versus control 86 (32.5%) survived, 178 (67.5%) dead; $\chi^2 = 4.44$, $p < 0.05$). It was not possible to pinpoint exactly why more leaf miners survived in refoliated leaves, but it is likely that both chemical and phenological factors were involved. Since these leaves have lower condensed tannins and higher water content, survival could be greater because of the direct action of improved chemistry or nutrition. Indirectly, survival could be higher because development was faster, and either death by parasitism or premature leaf abscission was reduced (refoliated leaves are less prone to early abscission, anyway). Parasitism could also have been reduced because refoliated leaves are less likely to have damage by leaf chewers, which could alter chemistry and phenology and act to increase mortality.

None of the leaf-mining species in this study were specialists on new foliage, yet they still responded to a resource in a manner predicted based on survival of offspring. For leaf miners that specialize on new foliage produced late in the growing season, the response is even more pronounced. Auerbach and Simberloff (1985) found that atypical reflushing of leaves on water oak (*Q. nigra* L.) after water and fertilization stress was followed by large increases in the density of two leaf-mining species that specialize on new, secondary leaves. Again, timing of refoliation was critical; if reflushing occurred too soon and ovipositing adults were not present, then densities of these two species were unaffected. Opler (1974) also reports that the leaf miner *Neurobathra bohartiella* Opler, which feeds only on new foliage, depends on either physical damage or defoliation by the dioptid moth, *Phryganidia californica* Packard, to produce atypical flushes of new leaves after the initial spring budbreak (Puttick, 1986). In this case, one expects the presence of the leaf chewer to increase survival and population size of the specialist leaf miner.

b. The Leaf Chewers. Contrary to the leaf miners, the leaf chewers showed no response to the atypical production of refoliated leaves in either season. Despite chemical differences, leaf chewers damaged virtually the same cumulative fraction (i.e., total percent of leaves that were damaged) of refoliated and mature leaves (1982–1983, control = 31.9%, experimental = 31.1%; 1983–1984, control = 17.2%, experimental = 19.2%) and consumed the same mass per leaf (1982–1983, mean mass per leaf consumed, control = 24.6 ± 7.24 mg versus refoliated 24.0 ± 6.66 mg, $t = 0.17$, $p > 0.40$; 1983–1984, control = 21.8 ± 8.72 versus refoliated = 18.8 ± 4.36, $t = 0.75$ $p > 0.40$). In fact, seasonal patterns of herbivory on refoliated trees closely paralleled that on control trees, which suggests that late-season leaf chewers do not respond to changes in leaf quality.

This lack of response by the leaf chewers is surprising since, in terms of phytochemistry, refoliated leaves should represent a superior resource. Lack of response may be due to the fact that Emory oak is only occasionally defoliated in nature and it does not typically produce secondary flushes of new leaves late in the season (except when completely or partially defoliated) as do some other oaks. Late-season leaf chewers may not respond in ecological time simply because this resource has not been historically available, and selection has not operated for insects to exploit these leaves. Consequently, these late feeders may be generalists, as Niemela (1983) showed for insects that feed late in the season on other oak species. Species that specialize on fresh, reflushed leaves would be expected to increase in densities following refoliation. For example, Rockwood (1974) showed that leaves flushed after manual defoliation of the calabash tree (*Crescentia alata* H. B. K.) were heavily attacked by flea beetles (*Oedionychus* sp.) that specialized on such foliage. More detailed studies on the growth and survivorship of individual species of late-season leaf chewers on *Q. emoryi* are needed to ascertain the direct and indirect effects (e.g., change in efficiencies of natural enemies) of previous herbivory on specific populations.

Although in this study leaf chewers showed no response to changes in foliage quality post defoliation, evidence from other studies suggest that previous or concurrent feeding by leaf chewers does affect other leaf chewers and has resulted in evolutionary changes in feeding behavior. For example, Edwards and Wratten (1983) observed that leaf grazing by insects on individual leaves was ovcrdispersed. They surmised that wound-induced defenses surrounding damaged areas on leaves result in avoidance of partially damaged leaves by other chewers. Fowler and Lawton (1985) proposed an alternative explanation for overdispersed leaf damage; leaf chewers avoid damaged leaves because of greater risk of attraction of natural enemies to these leaves (Heinrich, 1979; Heinrich and Collins, 1983; Faeth, 1985a, 1986). Besides simply avoiding such leaves, apparently some folivorous insects have evolved behaviors to circumvent the risks, be they chemical, phenological, or from natural enemies, associated with damaged leaves. Some folivorous insects actively sever the petiole of the leaf they have just fed on, either to eliminate cues used by natural enemies (Feeny, 1976; Heinrich and Collins, 1983), or to prevent translocation of inducible chemical (Tallamy, 1986) or mechanical defenses (Dillon *et al.*, 1983). Tallamy (1986) demonstrated that the coccinellid beetle *Epilachna borealis* Fabr. avoids induced chemical defenses that are mobilized within squash (*Curcurbita pepo* L.) leaves by first trenching the area to be eaten. Although others have suggested that insects have evolved specific behaviors to counter induced "defenses" of plants, Tallamy's study is the first to clearly document that the plant does indeed induce chemicals in response to herbivory, that such chemicals reduce the fitness of the insect, and that these chemicals are evaded by a specific behavior in the insect.

The experiments described above clearly indicate that early-season folivores can have significant impacts on some late-season folivores in ecological time despite the fact that these folivore groups are separated temporally. It should be apparent that the effects of early folivores on later ones are highly complex and variable, depending on the species involved, the extent, scale, and degree of folivory, and whether the subsequent folivores are specialist or generalists. The selective pressures imposed by folivory apparently have led to evolutionary modifications in feeding and oviposition behaviors. Are there other patterns that might be explained by interactions between folivores? In the next section, I examine seasonal patterns of folivory and how these might also be explained by such horizontal interactions.

III. Evolution of Seasonal Feeding Patterns

On many trees, episodes of folivory are concentrated on new leaves early in the spring (Feeny, 1970; Coley, 1983; Faeth, 1985b). Several hypotheses have been proposed to account for seasonal patterns of folivory. One popular hypothesis was made by Feeny (1970, 1975) and extended by Rhoades and Cates (1976), who suggested that polyphenols (tannins) and toughness increase, while protein decreases, through the growing season such that leaf quality declines progressively. Most insects are thought to have evolved feeding phenologies to avoid declining leaf quality as the season proceeds. More recent analyses of phytochemistry and patterns of folivory on trees, however, do not always support this view (Coley, 1983; Karban and Ricklefs, 1984; Faeth, 1985b). Another hypothesis (Holmes *et al.*, 1979) claims that the intensity of bird predation by nesting adults and fledglings in midsummer has selected for lepidopterous insects to feed in the early spring or fall. A third hypothesis is based on weather changes (Faeth, 1985b). As summer temperatures increase, development of larval insects may decline and fecundity of adults may decrease. Feeding activity and changes in temperature and precipitation do seem to be correlated, at least for insects feeding on Emory oak in central Arizona (Faeth, 1985b). The same appears to be true for the feeding patterns of the California oak worm (*Phryganidia californica*) on coastal live oak (*Q. agrifolia* Neé); feeding by these moth larvae is unrelated to tannin concentrations but instead is explained by temperature changes (Puttick, 1986; Y. Mauffette, personal communication).

Few if any hypotheses include horizontal interactions to account for seasonal feeding patterns. If partial damage to leaves is usually harmful to later feeders, either by direct changes in chemistry, indirect changes in attack by natural enemies, or accelerated leaf abscission, then selection would favor insects that feed as early as possible. Insects that feed first on leaves would avoid any harmful effects associated with "used" leaves. Of course, the other hypotheses discussed above are not mutually exclusive of this plant-mediated interaction hypothesis, and may operate together to produce the observed feeding patterns. The possibility of plant-mediated interactions is emphasized here simply because recent trends have tended to discount the importance of horizontal interactions in either spatial or temporal patterns of folivorous insect communities.

Some insects, such as the leaf miners, are notorious late feeders (Feeny, 1970; Faeth *et al.*, 1981a; Faeth, 1986; West, 1985). Feeny (1970) suggested that leaf miners can feed late in the season on poorer quality leaves because they only mine leaf layers that are low in tannin content. However, many leaf-mining species do not discriminate leaf layers in feeding, and seasonal changes in leaf chemistry alone are not good explanations of leaf-mining activities (Faeth *et al.*, 1981b; but see Kimmerer and Potter, 1987). The late feeding of many leaf miners instead appears to be related to their endophagous feeding habits and interactions with other folivores. Because they are restricted to feed on single leaves, feeding early in the growing season could be disastrous if other insects damage leaves on which the egg or larval miner occurs. West (1985) demonstrated that leaf miners that are experimentally forced to feed early on *Quercus robur* L. suffer much higher mortality because many of these early leaves are fated to be partially damaged by leaf chewers. West surmised that leaf chewing induces defenses that cause greater leaf-miner mortality. Considering the other possible effects associated with damaged leaves (increased parasitism, reduced fecundity, slower development, and desiccation once a mine has been breached), strong selective pressures may have resulted in the late-feeding habits of sedentary leaf-mining species.

Not all leaf miners feed late in the growing season; a few species feed on new foliage early in the spring (Opler, 1974; Faeth *et al.*, 1981b). These species tend to have rapid larval developmental times compared to late-season leaf miners. For example, the leaf-mining periods of species feed-

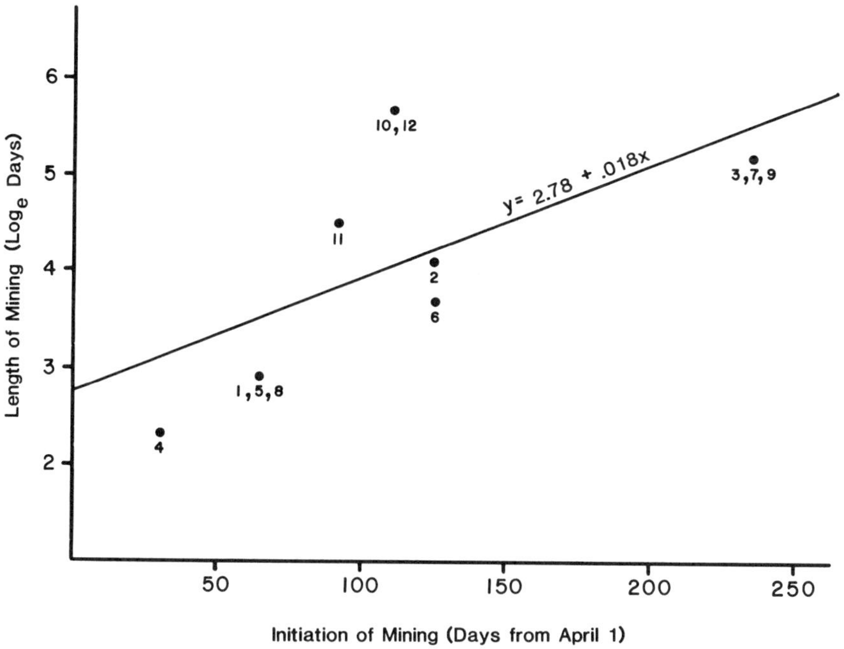

Fig. 4. Relationship of initiation of leaf mining with the length of the larval mining period for various generations of leaf-mining species occurring on Emory oak in central Arizona. Length of mining periods are natural log transformed to normalize variances. Points represent the following species or generation of species: 1, *Cameraria* sp. nov., first generation; 2, second generation; 3, third generation; 4, *Caloptilia* sp.; 5, *Tischeria* sp., first generation; 6, second generation; 7, third generation; 8, *Stigmella* sp., first generation; 9, second generation; 10, *Stilbosis juvantis;* 11, *Coptodisca* sp.; and 12, *Brachys cephalicus* Fisher.

ing on Emory oak are significantly (and positively) correlated with time of initiation of mining in the growing season (Fig. 4; correlation coefficient = 0.72, $p < 0.01$). This correlation could be evidence that selective pressures from interactions with leaf chewers has resulted in differences in larval development rates between early- and late-season leaf miners. Since new leaves are the preferred foliage of leaf chewers, leaf miners that feed on new foliage may complete development quickly to avoid the risks associated with leaf-chewing damage. Of course, it could be argued that leaf miners on new foliage develop quickly because these leaves are of superior quality. However, it should be noted that new foliage may contain higher levels of phenols and tannins than does mature foliage (Coley, 1983; Faeth, 1985b, 1986). Further experimentation with both leaf miners and leaf chewers will be necessary to determine whether or not selection pressures associated with damaged leaves have resulted in evolutionary patterns of seasonal feeding.

IV. Coevolution of Seasonal Folivores on Shared Host Plants

It seems likely that seasonal folivores do indeed interact and that these interactions have resulted in evolutionary changes, at least in the spatial and temporal aspects of ovipositing and feeding. To suggest that coevolution, as a very narrow subset of evolution (Janzen, 1980), has taken place is a very different question. Even more restrictive are the conditions necessary for the coevolution of competing species to occur (Connell, 1980; Roughgarden, 1983a). Although the evidence suggests that early feeders affect the fitness via chemical and phenological changes, the reciprocal (late feeders affecting fitness of early feeders) is much more difficult to establish. It would be necessary to document, first, that the late feeders induce chemical or phenological changes in the shared host plant, second, that these changes are somehow transferred to the next growing season when early feeders begin to feed, and third, that the plant-mediated changes affect the fitness of the next season's early feeders. Obviously, this scenario requires that the participating insects are sharing a perennial or at least biennial host plant. Although there is some evidence that the effects of folivory are carried over to the following year or years in perennial plants (e.g., Haukioja, 1980; Neuvonen and Haukioja, 1984), it is unclear how widespread this is or how it affects the next generation of early feeders.

For leaf chewers and leaf miners on Emory oak, it seems unlikely that all three conditions have been satisfied. The leaf miners are usually at very low densities and damage relatively little of a tree's leaf area. At present, there is no evidence that the leaf miners induce chemical changes that are transferred to the next growing season. Leaf mining does induce earlier leaf abscission, but this effect is confined to the current growing season. Unless leaf-miner infestations were much higher, chances for reducing survival or fecundity of leaf chewers on next season's crop of leaves seems remote. It should be noted that outbreaks of leaf miners do occasionally occur (e.g., Mazanec, 1974), but these are usually rare. Also, it appears that leaf-chewing caterpillars avoid birch leaves mined by *Eriocrania* spp. (Lepidoptera: Eriocraniidae), which suggests a possible negative interaction within a growing season (Hartley and Lawton, 1987). In most cases, however, it may be more appropriate to term the interaction between early and late feeders as amensalistic rather than competitive, since the effects seem to be overwhelmingly asymmetrical, as are many interactions between insect species (Lawton and Hassell, 1981).

Even if one assumes that leaf chewers and leaf miners, or any other folivorous guilds separated in time, do indeed affect each other's survival and fecundity by altering the shared host plant, it is improbable that stepwise and pairwise coevolution of species would result. Although the amount and timing of folivory is very important, it probably matters very little exactly which species does the damage, at least in terms of the effects of previous folivory on subsequent insects. This should be especially true of the indirect effects of folivory. For example, if parasites are attracted or a leaf falls early simply because of physical damage, then the effects on late feeders are the same regardless of whether a beetle or a caterpillar ate the leaf. For Emory oak, mechanical damage with a hole-punch evokes that same response, at least qualitatively, in changes in tannin and protein levels as does damage by chewing insects (Faeth, 1986). If pairwise coevolution were to occur, a closed loop of very specific chemical and phenological responses and recognition by early insect species, plant, late insect species, and plant again would be necessary. The probability of pairwise coevolution would drop dramatically because of the number of steps, species, and time lags involved. This scenario also requires acquired immunity to specific insects on the part of the plant, which has not yet been demonstrated (Rhoades, 1985). It seems that if any coevolution has occurred between seasonal folivores on trees, it would be diffuse coevolution (Janzen, 1980) because of the complex web of interacting species (Fox, 1981; Futuyma, 1983).

V. Summary

Folivores can affect survival, developmental times, and fecundity of subsequent folivores by altering the chemical, phenological, or physical aspects of the shared plant. These interactions may occur directly by changes in the quality or quantity of the leaf resource, or indirectly by making insects more prone to parasitism, predation, or early leaf abscission. The precise effects of folivory on later feeders is highly dependent on when and how much folivory occurs, and the life histories of the insects and plants. Certainly, the effects of folivory on later feeders are not always negative, but may increase probability of survival under certain circumstances.

As with any other plant trait, the chemical, physical, or phenological changes in plants induced by folivory are expected to cause evolutionary changes in the late-feeding insects. The oviposition behavior and feeding patterns of some late-season insects suggests that such plant-mediated changes have exerted strong selective pressures. This evidence is in contrast to prevailing notions that phytophagous insect communities are poor candidates for organization through such horizontal interactions. Although early-season folivores affect the late-season ones, at present the converse does not seem to be true, and therefore the likelihood of coevolution, especially stepwise and pairwise coevolution, seems remote.

Acknowledgments

I thank T. Bultman, K. Spencer, and three anonymous reviewers for critical reviews of this manuscript. M. Auerbach, B. Axelrod, M. Axelrod, H. Berg, C. T. Faeth, G. C. Faeth, T. G. Faeth, T. J. Faeth, A. Geischen, W. T. Tiller, and N. J. Tiller participated greatly in all aspects of this research. This research has been supported by a Faculty Grant-in-Aid from Arizona State University and National Science Foundation grants BSR 8118032 and BSR 8415616.

References

Auerbach, M. J., and Simberloff, D. (1985). Responses of leaf miners to atypical leaf production patterns. *Ecol. Entomol.* **9,** 361–367.

Baldwin, I. T., and Shultz, J. C. (1983). Rapid changes in tree leaf chemistry induced by damage: Evidence for communication between plants. *Science* **221,** 277–279.

Bultman, T. L., and Faeth, S. H. (1986a). Selective oviposition by a leaf miner in response to temporal variation in abscission. *Oecologia* **69,** 117–120.

Bultman, T. L., and Faeth, S. H. (1986b). Experimental evidence for intraspecific competition in a leaf-mining insect, *Cameraria* sp. nov. (Lepidoptera: Gracillariidae). *Ecology* **67,** 442–448.

Bultman, T. L., and Faeth, S. H. (1986c). Leaf size selection by leaf-mining insects on *Quercus emoryi* (Fagaceae). *Oikos* **46,** 311–316.

Campbell, J. M. (1962). Influence of larval environment on adult size and fecundity in the moth, *Panaxia dominula* L. *Nature* (*London*) **192,** 282.

Coley, P. D. (1983). Herbivory and defensive characteristics of tree species in a lowland tropical forest. *Ecol. Monogr.* **53,** 209–233.

Connell, J. H. (1980). Diversity and the coevolution of competitors. *Oikos* **35,** 131–138.

Connell, J. H. (1983). On the prevalence and relative importance of interspecific competition: Evidence from field experiments. *Am. Nat.* **122,** 661–696.

Crawley, M. J. (1983). "Herbivory. The Dynamics of Animal-Plant Interactions." Studies in Ecology, Vol. 10. University of California, Berkeley.

Denno, R. F., and Mclure, M. S., eds. (1983). "Variable Plants and Herbivores in Natural and Managed Systems." Academic Press, New York.

Dillon, P. H., Lowrie, S., and McKey, D. (1983). Disarming the "evil women"! Petiole constriction by a sphingid larva circumvents mechanical defenses of its host plant, *Cnidoscolus urnes* (Euphorbiaceae). *Biotropica* **15,** 112–116.

Edwards, P. J., and Wratten, S. D. (1983). Wound-induced defenses in plants and their consequences for patterns of insect grazing. *Oecologia* **59,** 88–93.

Ehrlich, P. R., and Raven, P. H. (1964). Butterflies and plants: A study in coevolution. *Evolution* **18,** 586–608.

Faeth, S. H. (1985a). Host leaf selection by leaf miners: Interactions among three trophic levels. *Ecology* **66,** 870–875.

Faeth, S. H. (1985b). Quantitative defense theory and patterns of feeding by oak insects. *Oecologia* **68,** 34–40.

Faeth, S. H. (1986). Indirect interactions between temporally-separated herbivores mediated by the host plant. *Ecology* **67,** 479–494.

Faeth, S. H. (1987). Community structure and folivorous insect outbreaks: The roles of vertical and horizontal interactions. *In* "Insect Outbreaks." (P. Barbosa and J. C. Schultz, eds.), pp. 135–171. Academic Press, New York.

Faeth, S. H., and Bultman, T. L. (1986). Interacting effects of increased tannin levels on leaf-mining insects. *Entomol. Exp. Appl.* **40,** 297–300.

Faeth, S. H., Connor, E. F., and Simberloff, D. (1981a). Early leaf abscission: A neglected source of mortality for folivores. *Am. Nat.* **117,** 409–415.

Faeth, S. H., Mopper, S., and Simberloff, D. (1981b). Abundance and diversity of leaf-mining insects on three oak species: Effects of host-plant phenology and nitrogen content of leaves. *Oikos* **37,** 238–251.

Feeny, P. (1970). Seasonal changes in oak leaf tannins and nutrients as a cause of spring feeding by winter moth caterpillars. *Ecology* **51,** 565–581.

Feeny, P. (1975). Biochemical coevolution between plants and their insect herbivores. *Symp. Int. Congr. Syst. Evol. Biol., 1st* pp. 3–19.

Feeny, P. (1976). Plant apparency and chemical defense. *Recent Adv. Phytochem.* **10,** 1–40.

Fowler, S. V., and Lawton, J. H. (1985). Rapidly induced defenses and talking trees: The devil's advocate position. *Am. Nat.* **126,** 181–195.

Fox, L. R. (1981). Defense and dynamics in plant-herbivore systems. *Am. Zool.* **21,** 853–854.

Futuyma, D. J. (1983). Evolutionary interactions among herbivorous insects and plants. *In* "Coevolution" (D. J. Futuyma and M. Slatkin, eds.), pp. 207–231. Sinauer, Sunderland, Massachusetts.

Godfray, H. C. J. (1986). Clutch size in a leaf-mining fly (*Pegomyza nigrotarsis:* Anthomyiidae). *Ecol. Entomol.* **11,** 75–81.

Hairston, N. G., Smith, F. E., and Slobodkin, L. B. (1960). Community structure, population control, and competition. *Am. Nat.* **94,** 421–425.

Hartley, S. E., and Lawton, J. H. (1987). Effects of different types of damage on the chemistry of birch foliage, and the responses of birch feeding insects. *Oecologia* **74,** 432–437.

Haukioja, E. (1980). On the role of plant defenses in the fluctuation of herbivore populations. *Oikos* **35,** 202–213.

Haukioja, E., and Niemela, P. (1976). Does birch defend itself actively against herbivores? *Rep. Kevo Subarct. Res. Sta.* **13,** 44–47.

Haukioja, E., and Niemela, P. (1979). Birch leaves as a resource for herbivores. Seasonal occurrence of increased resistance in foliage after mechanical damage of adjacent leaves. *Oecologia* **39,** 151–159.

Heinrich, B. (1979). Foraging strategies of caterpillars: Leaf damage and possible predator avoidance strategies. *Oecologia* **42,** 325–337.

Heinrich, B., and Collins, S. L. (1983). Caterpillar leaf damage and the game of hide-and-seek with birds. *Ecology* **64,** 592–602.

Holmes, R. T., Schultz, J. C., and Nothnagle, P. (1979). Bird predation on forest insects: An enclosure experiment. *Science* **206,** 462–463.

Hough, J. A., and Pimentel, D. (1978). Influence of host foliage on development, survival, and fecundity of the gypsy moth. *Environ. Entomol.* **7,** 97–102.

Janzen, D. H. (1980). When is it coevolution? *Evolution* **34,** 611–612.

Karban, R. (1986). Interspecific competition between folivorous insects on *Erigeron glaucus*. *Ecology* **67,** 1063–1072.

Karban, R., and Ricklefs, R. E. (1984). Leaf traits and species richness of lepidopteran larvae on deciduous trees in southern Ontario. *Oikos* **43,** 165–170.

Kimmerer, T., and Potter, D. (1987). Nutritional quality of specific leaf tissue and selective feeding by a specialist leaf miner. *Oecologia* **71,** 548–551.

Lawton, J. H., and Hassell, M. P. (1981). Asymmetrical competition in insects. *Nature (London)* **289,** 793–795.

Lawton, J. H., and Strong, D. R., Jr. (1981). Community patterns and competition in folivorous insects. *Am. Nat.* **118,** 317–333.

McNeil, J. N., and Quiring, D. T. (1983). Evidence of an oviposition-deterring pheromone in the alfalfa blotch leaf miner, *Agromyza frontella* (Rondani) (Diptera: Agromyzidae). *Environ. Entomol.* **12,** 990–992.

Marks, R. L. (1976). Mating behavior and fecundity of the red bollworm *Diparapis catanea* Hmps. (Lepidoptera: Noctuidae). *Bull. Entomol. Res.* **66,** 145–158.

Mazanec, Z. (1974). Influence of the jarrah leaf miner on the growth of jarrah. *Aust. For.* **37,** 32–42.

Myers, J. H., and Williams, S. K. (1984). Does tent caterpillar attack reduce the food quality of red alder foliage? *Oecologia* **62,** 74–79.

Myers, J. H., and Williams, S. K. (1987). Lack of short or long term inducible defenses in the red alder–western tent caterpillar system. *Oikos* **48,** 73–78.

Neuvonen, S., and Haukioja, E. (1984). Low nutritive quality as defense against herbivores: Induced responses in birch. *Oecologia* **63,** 71–74.

Niemela, P. (1983). Seasonal patterns in the incidense of specialism: Macrolepidopteran larvae on Finnish deciduous trees. *Ann. Zool. Fenn.* **20,** 100–202.

Niemela, P., Aro, E. M., and Haukioja, E. (1979). Birch leaves as a resource for herbivores. Damaged-induced increase in leaf phenolics with trypsin-inhibiting effects. *Rep. Kevo Subarct. Res. Sta.* **15,** 37–40.

Niemela, P., Toumi, J., Mannila, J., and Ojala, P. (1984). The effect of previous

damage on the quality of Scots pine foliage as food for Dioprionid sawflies. *Z. Angew. Entomol.* **98,** 33–43.
Opler, P. A. (1974). Biology, ecology, and host specificity of microlepidoptera associated with *Quercus agrifolia* (Fagaceae). *Univ. Calif. Publ. Entomol.* **75.**
Orians, G. H., and Paine, R. T. (1983). Convergent evolution at the community level. *In* "Coevolution" (D. J. Futuyma and M. Slatkin, eds.), pp. 431–458. Sinauer, Sunderland, Massachusetts.
Price, P. W. (1983). Hypotheses on the organization and evolution in herbivore communities. *In* "Variable Plants and Herbivores in Natural and Managed Systems" (R. F. Denno and M. S. McClure, eds.), pp. 559–596. Academic Press, New York.
Prokopy, R. J., Ziegler, J. R., and Wong, T. T. Y. (1978). Deterrence of repeated oviposition by fruit-making pheromone in *Ceratitis capitata. J. Chem. Ecol.* **4,** 55–63.
Puritch, G. S., and Nijholt, W. W. (1974). Occurrence of juvabione-related compounds in grand firs and Pacific silver fir infested by balsam wooly aphid. *Can. J. Bot.* **52,** 585–587.
Puttick, G. M. (1986). Utilization of evergreen and deciduous oak by the Californian oak moth *Phryganidia californica. Oecologia* **68,** 589–594.
Raupp, M. J., and Denno, R. F. (1983). Leaf age as a predictor of herbivore distribution and abundance. *In* "Variable Plants and Herbivores in Natural and Managed Systems" (R. F. Denno and M. S. McClure, eds.), pp. 91–124. Academic Press, New York.
Raupp, M. J., Milan, F. R., Barbosa, P., and Leonhardt, B. A. (1986). Methylcyclopentanoid monoterpenes mediate interactions among insect herbivores. *Science* **232,** 1408–1410.
Rausher, M. D. (1979). Egg recognition: Its advantage to a butterfly. *Anim. Behav.* **27,** 1034–1040.
Rausher, M. D. (1983). Ecology of host-selection behavior in phytophagous insects. *In* "Variable Plants and Herbivores in Natural and Managed Systems" (R. F. Denno and M. S. McClure, eds.), pp. 223–257. Academic Press, New York.
Rhoades, D. F. (1983a). Responses of alder and willow to attack by tent caterpillars and webworms: Evidence for pheromonal sensitivity of willows. *Am. Chem. Soc. Symp. Ser.* **208,** 55–68.
Rhoades, D. F. (1983b). Herbivore population dynamics and plant chemistry. *In* "Variable Plants and Herbivores in Natural and Managed Systems" (R. F. Denno and M. S. McClure, eds.), pp. 155–220. Academic Press, New York.
Rhoades, D. F. (1985). Offensive-defensive interactions between herbivores and plants: Their relevance in herbivore population dynamics and ecological theory. *Am. Nat.* **125,** 205–238.
Rhoades, D. F., and Cates, R. G. (1976). Toward a general theory of plant antiherbivore chemistry. *Recent Adv. Phytochem.* **10,** 168–213.
Rockwood, L. L. (1974). Seasonal changes in the susceptibility of *Crescentia alata* leaves to the flea beetle, *Oedionychus* sp. *Ecology* **55,** 142–148.
Roughgarden, J. (1983a). Coevolution between competitors. *In* "Coevolution" (D. J. Futuyma and M. Slatkin, eds.), pp. 383–403. Academic Press, New York.

Roughgarden, J. (1983b). Competition and theory in community ecology. *Am. Nat.* **122,** 583–601.

Schoener, T. W. (1983). Field experiments on interspecific competition. *Am. Nat.* **122,** 240–285.

Schultz, J. C., and Baldwin, I. T. (1982). Oak leaf quality declines in response to defoliation by gypsy moth larvae. *Science* **217,** 149–151.

Silva-Bohorquez, I. (1986). Interspecific interactions between insects on oak trees, with special reference to defoliators and the oak aphid. Ph.D. thesis, Oxford University.

Simberloff, D., and Stiling, P. (1988). Larval dispersion and survivorship in a leaf-mining moth. *Ecology* **69,** 1647–1657.

Stiling, P. D., Simberloff, D., and Anderson, L. C. (1987). Non-random distribution patterns of leaf miners on oak trees. *Oecologia* **73,** 116–119.

Strong, D. R., Jr. (1983). Natural variability and the manifold mechanisms of ecological communities. *Am. Nat.* **122,** 636–660.

Strong, D. R., Jr., Lawton, J. H., and Southwood, R. (1984). "Insects on Plants. Community Patterns and Mechanisms." Blackwell, Oxford.

Swain, T. (1979). Tannins and lignins. *In* "Herbivores. Their Interactions with Secondary Plant Metabolites" (G. A. Rosenthal and D. J. Janzen, eds.), pp. 657–682. Academic Press, New York.

Tallamy, D. W. (1985). Squash beetle feeding behavior: An adaptation against induced cucurbit defenses. *Ecology* **66,** 1574–1579.

Taper, M. L., Zimmerman, E. M., and Case, T. J. (1985). Sources of mortality for a cynipid gall-wasp (*Dryocomus dubiosus* Hymenoptera: Cynipidae): The importance of the tannin/fungus interaction. *Oecologia* **68,** 437–445.

Vinson, S. B. (1976). Host selection by insect parasitoids. *Annu. Rev. Entomol.* **21,** 109–138.

Weseloh, R. M. (1981). Host location by parasitoids. *In* "Semiochemicals. Their Role in Pest Control" (D. A. Nordlund, R. L. Jones, and W. J. Lewis, eds.), pp. 79–95. Wiley, New York.

West, C. (1985). Factors underlying the late seasonal appearance of the lepidopterous leaf-mining guild on oak. *Ecol. Entomol.* **10,** 111–120.

Whitham, T. G. (1980). The theory of habitat selection examined and extended using *Pemphigus* aphids. *Am. Nat.* **115,** 449–466.

Williams, K. S., and Gilbert, L. E. (1981). Insects as selective agents on host plant morphology: Egg mimicry reduces egg laying by butterflies. *Science* **212,** 467–469.

Williams, K. S., and Myers, J. H. (1984). Previous herbivore attack of red alder may improve food quality for fall webworms. *Oecologia* **63,** 166–170.

CHAPTER 14

Adaptations of Mammalian Herbivores to Plant Chemical Defenses

RICHARD L. LINDROTH
Department of Entomology
University of Wisconsin
Madison, Wisconsin

I. Introduction

Secondary plant compounds strongly influence the nutritional ecology of herbivores, affecting their individual performance (e.g., Berenbaum and Feeny, 1981; Lincoln *et al.*, 1982; Berenbaum, 1983; Reichardt *et al.*,

1984; Manuwoto *et al.*, 1985; Lindroth *et al.*, 1986a), population dynamics (e.g., Haukioja, 1980; Bryant, 1981a), and community organization (e.g., Rodman and Chew, 1980; Batzli and Jung, 1980; Berenbaum, 1981). However, most work in this area has been conducted with insects; until recently studies with mammalian herbivores have simply reported correlations between secondary plant compounds and food habits (Cooper-Driver *et al.*, 1977; Bryant and Kuropat, 1980; Schwartz *et al.*, 1980; McKey *et al.*, 1981; Calvert, 1985). Moreover, it was insect ecologists who introduced the first theories of plant–animal coevolution (Ehrlich and Raven, 1964; Gilbert and Raven, 1975). Consequently, understanding of chemical mediation of plant–mammal interactions has lagged about a decade behind that of plant–insect interactions.

Because of the similarity of primary metabolic processes in all organisms, recent research with mammals has shown that many of the effects of secondary compounds are the same as those exhibited toward insects. But mammals are not simply "large furry insects." They differ from insects in many respects, and these differences have implications for how chemicals have mediated evolution in each group of herbivores. First among a variety of factors promoting differences in responses to secondary plant compounds is that of nutrient requirements. Insects, for example, require sterols in their diet whereas mammals do not. Second, the large body size and low conversion efficiencies (due to the maintenance of homeothermy) of mammals require them to consume large quantities of food, and this in turn forces them to be generalist feeders. Specialist herbivores are rare among mammals, occur only when their food source is very abundant (e.g., koalas and *Eucalyptus* species), and even then are not monophagous. Third, detoxication systems differ both quantitatively and qualitatively between the two groups of herbivores. Polysubstrate monooxygenase activity (per unit protein) is often 10 times higher in small mammals than in insects (L. Brattsten, personal communications). Mammals rely on the glucuronidation pathway for detoxication of many plant compounds, and this pathway is absent in insects. Fourth, mammals have longer lifespans and generation times, and are typically active over more seasons of the year than insects, which pass through adverse periods in a dormant stage. Substantial seasonal and annual variation in the quality and quantity of available food (e.g., Lindroth *et al.*, 1986b) requires mammals to maintain considerable plasticity in their food habits. Finally, the capability of mammals for sophisticated learning (in contrast to innate behaviors) is greater than that of insects, and this difference may profoundly influence their respective foraging behaviors. Albeit an incomplete list, these differences are sufficient to emphasize that the evolutionary and ecological responses of mammals to secondary plant compounds will not parallel exactly the better-known responses of insects.

The purpose of this chapter is first to address several aspects of plant—

mammal interactions that have received relatively little empirical study, that is, behavioral, physiological, and biochemical adaptations of mammals to secondary plant compounds. Second, I will discuss the question of whether plant chemicals have mediated coevolution between plants and mammalian herbivores. I will summarize factors contributing to the complexity of the nutritional ecology of mammalian herbivores, discuss various adaptations of mammals that most likely evolved in response to phytochemicals, and consider the constraints on and cases supporting chemically mediated coevolution. Some of the ideas reviewed here were first put forth in a seminal paper by Freeland and Janzen (1974), and I will evaluate a few of their hypotheses in light of more recent research. This chapter is not a comprehensive review, but will draw from the literature at large and will use information from domestic mammals when analogous information for wild mammals is unavailable. I will emphasize studies with microtine rodents because more is known about their responses to secondary plant compounds than is known for any other group of mammals, with the possible exception of snowshoe hares.

II. Factors Influencing Nutritional Status

A herbivore's nutritional status determines its growth, survival, and reproductive performance, and these in turn affect ecologically important characteristics such as competitive interactions, population dynamics, and habitat selection. Nutritional status itself is the product of a complex set of internal and external factors and their interactions (Fig. 1).

Plant chemical defenses may affect the nutritional status of an animal directly by altering its food selection and consumption patterns, decreasing its digestion or conversion efficiencies, or causing various degrees of toxicosis. Alternatively, they may affect an animal indirectly by changing its gut microflora community or altering the efficacy of its biochemical detoxication systems, which in turn influence the aforementioned responses. Moreover, a herbivore's response to a particular phytochemical is influenced by characteristics such as genetic makeup, age, sex, and its previous experience with the same or different phytochemicals. Clearly, the reasons for an animal's selection of food and its performance on that food are highly complex. We should not be surprised when attempts to explain the foraging behavior of generalist herbivores on the basis of one or a few forage characteristics prove unsuccessful.

An important corollary to the principle that nutritional status is the result of complex interactions among both plant and herbivore characteristics is that food quality must be regarded as the product of an interaction between a herbivore and plant, and is best measured by assessing a herbivore's performance (growth, survival, or reproduction) when fed a par-

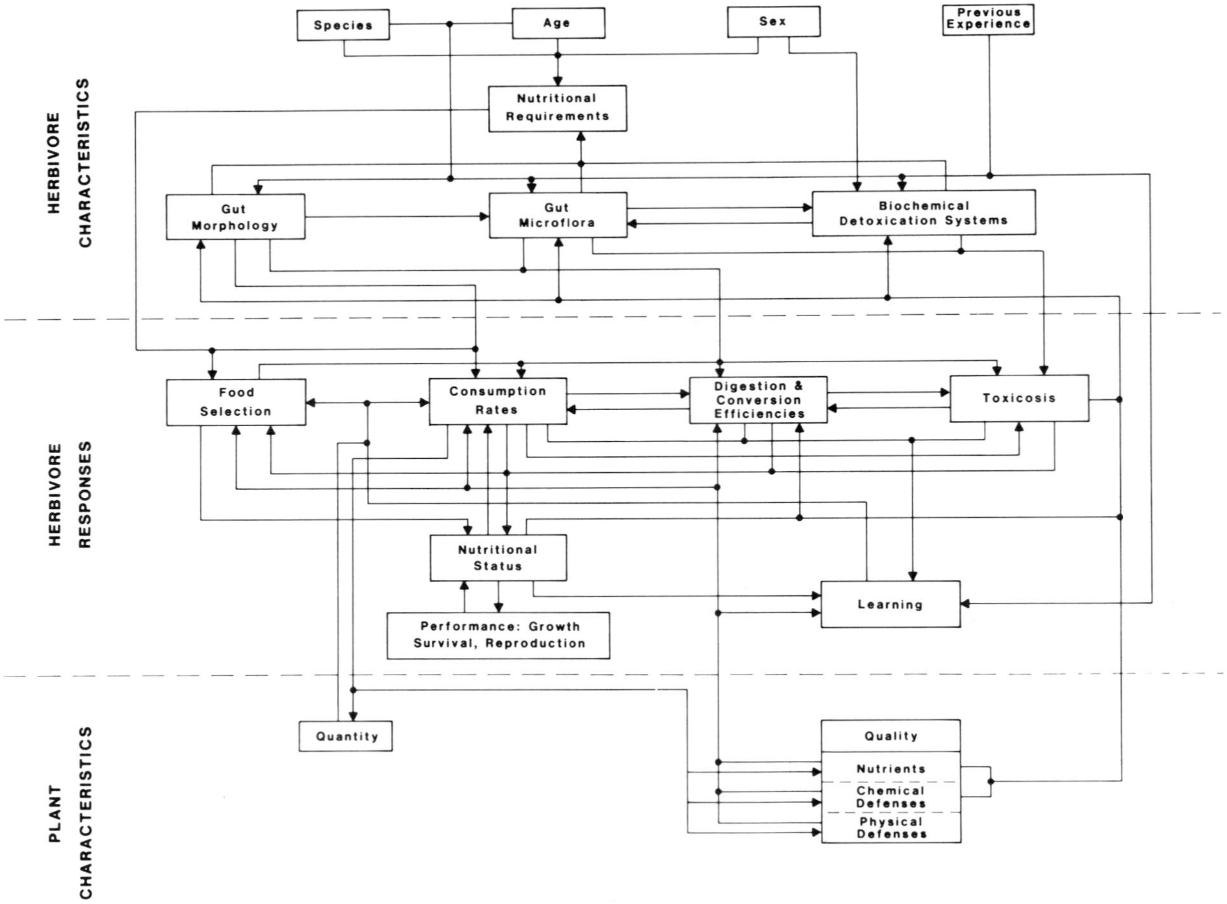

Fig. 1. Relationships among biotic factors influencing the nutritional status of mammalian herbivores. (From Lindroth, 1988. Used by permission).

ticular plant. Thus a priori classification of foods as toxic or nontoxic simply on the basis of their secondary compounds (not an uncommon practice in the ecological literature) has little value. An elementary principle of toxicology, first articulated by Paracelsus (1493–1541), is that all substances are poisonous at the proper dose. But not all secondary compounds are toxic to all herbivores at their naturally occurring concentrations. In addition to dose, the relative toxicity of a phytochemical is influenced by the herbivore characteristics illustrated in Fig. 1 as well as by an animal's previous and concomitant consumption of other secondary compounds and nutrients. Thus as the nutrient content of a food increases, its toxicity (measured in terms of effects on animal performance) may decline even if its levels of secondary compounds remain constant or increase slightly (Janzen, 1978). Animals will not necessarily prefer to feed on plants with low amounts of secondary compounds if they can feed on plants with higher concentrations whose potentially negative effects are offset by high concentrations of nutrients. For example, Palo *et al.* (1983) found that food intake and performance (maintenance of body weight) of mountain hares (*Lepus timidus* L.) were better on birch twigs of intermediate size (1.5–3.0 mm diameter) than on twigs of smaller or larger size; this diet afforded the best balance between concentrations of phenolic glycosides (highest in small twigs) and twig digestibility (lowest in large twigs).

III. Behavioral Adaptations to Plant Chemical Defenses

A. Specific Hungers and Innate Aversions

If secondary plant compounds indeed function as chemical defenses against mammals, what types of behavioral adaptations may have evolved to reduce an animal's exposure or susceptibilty to them? One possibility involves the theory of "specific hungers" (Richter, 1943), which purports that animals innately recognize not only specific states of nutrient deficiency but also the presence of required nutrients in their food, and feed accordingly. Evidence supporting this model comes from the well-defined, innate preferences of rats and many herbivores (e.g., rabbits, porcupines, deer, elk) for sodium (Denton, 1967; Zahorik and Houpt, 1977; Roze, 1985). Similarly, avoidance of certain secondary compounds by mammals may be an innate behavior. Bate-Smith (1972) suggests that bitterness, a property most notably characteristic of alkaloids and cyanogenic glycosides, is universally repellent to higher animals. Such genetically linked preferences have certainly evolved in response to the chemical characteristics of plants, but we know too little at this time to ascertain whether taste preferences and plant chemistries are coevolved traits.

In recent years, however, researchers have become increasingly reluctant to accept the theory of innate preferences and aversions as a comprehensive model for feeding behavior. Not only is it difficult to perceive of mammals having special sensory mechanisms for the detection of every conceivable nutrient deficiency or plant toxin, but innate behaviors cannot explain the development of aversions to novel compounds, such as that of rats for synthetic poisons (Zahorik and Houpt, 1977). It is possible that mammals have innate preferences (and accompanying physiology) for certain scarce but essential nutrients (e.g., moose and sodium; Botkin *et al.*, 1973) and innate aversions to a limited number of secondary compounds that generally exhibit broad-spectrum toxicity against generalist herbivores (e.g., alkaloids). But preferences for or aversions to most plant constituents require another explanation.

B. Learned Aversions and Preferences

A learning-based model for foraging behavior provides such an alternative, and is referred to in the literature by several names, including "learned aversions," "long-delay learning," and "single-trial learning." This model requires that a mammal be able to distinguish among the plants it eats, that it be able to recognize positive and negative postingestional consequences, and that it be able to associate such consequences with food cues (Zahorik and Houpt, 1977).

The learning model explains quite well the feeding behavior of laboratory rats. Associations between food cues and illness or recovery from illness can be formed in a single trial in spite of long delays between events, can be remembered for weeks or months, and can alter subsequent food preferences (Rozin, 1967, 1969; Rozin and Kalat, 1971; Garcia *et al.*, 1974; Zahorik, 1977). The associative process resembles classical conditioning in that food cues act as conditioned stimuli (CS) and gastrointestinal feedback acts as the unconditioned stimulus (UCS), but differs in that the taste CS predominates over other stimuli (e.g., visual, tactile cues) and in that long CS–UCS intervals are easily spanned (Palmerino *et al.*, 1980). This system appears to be unique to feeding behavior and is clearly adaptive to animals whose food resources vary greatly in nutritional value. Thus food items may be classified into four categories based on the amount and consequences of an animal's previous experience with the food (Zahorik, 1977) (Fig. 2), and novel foods may shift into one category or another on the basis of a single trial.

Extrapolation of the learning model to generalist mammalian herbivores is problematical in that it is difficult to define a "meal" for animals (especially large grazing mammals) that consume a mixture of species,

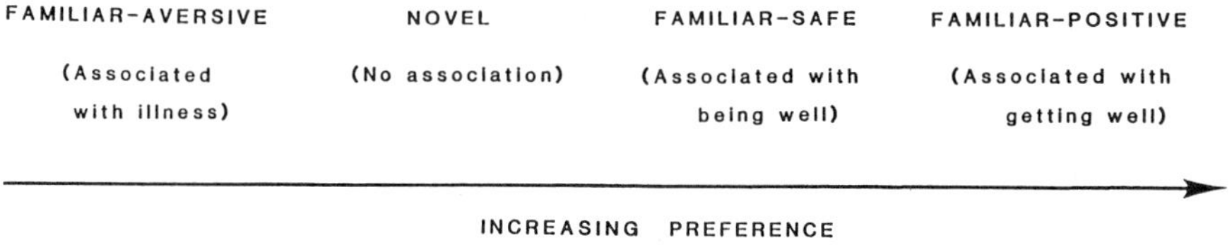

Fig. 2. Classification of food items according to the amount and results of an animal's previous experience. (Adapted from Zahorik, 1977).

often during long periods of feeding. Westoby (1974) suggests that mammals may overcome this problem by selecting foods according to a sampling procedure. For instance, mammals could feed mainly on one food during a certain time period, or could feed on a mixture of foods containing only one novel food during a time period. He concludes that herbivores probably exhibit imperfect diet optimization in which the aim to obtain the best nutrient balance within a fixed amount of food is limited by the requirement for regular sampling of foods and fallabilities in the learning process.

Learning processes in addition to "trial learning" may further enable mammals to select optimal diets. Galef and colleagues (Galef, 1977) found that milk of lactating laboratory rats contains flavor cues reflecting the diets of mothers such that young could identify safe foods during weaning. They also showed that both the presence of adults at a feeding site and residual olfactory cues influence the pups' choice of location at weaning, and thus their choice of food at weaning. The role of social interactions in the development of feeding behavior of young Stephen's woodrats (*Neotoma stephensi* Goldman) was considered by Vaughan and Czaplewski (1985). These woodrats specialize on juniper foliage, which is rich in tannins and terpenoids. Relative to other species of woodrats, Stephen's woodrats are weaned late, grow slowly, and exhibit a long, close association with their mothers. Mothers and young were often observed "sitting nose to nose and eating the same food item" (Vaughan and Czaplewski, 1985). The authors propose that the ability of young to select foliage from specific trees with low levels of defensive compounds is learned during the long preweaning period.

If mammalian herbivores do have the capacity for sophisticated feeding behaviors, then they may also avoid damage from secondary compounds by capitalizing on favorable reactions among ingested compounds. Mammals may learn to mix consumption of certain secondary compounds with other secondary compounds, nutrients, or inert substances so as to reduce or compensate for detrimental effects. For example, laboratory mice provided with a choice between diets containing saponins or tannins ate combinations of the foods that eliminated the deleterious effects (growth suppression) associated with consumption of either food alone (Freeland *et al.*, 1985). Gut surfactants in animals inhibit the precipitation of protein by tannins (Martin and Martin 1984; Mole and Waterman 1985); mammals may selectively consume plant surfactants such as saponins to accomplish the same-purpose. Goldstein and Spencer (1985) found that tannins inhibit the action of β-glucosidases and thus reduce the formation of cyanide. Herbivores feeding on cyanogenic plants may also consume plants containing tannins in order to reduce production of hydrogen cyanide. Many herbivorous mammals ingest small amounts of soil for reasons mostly unknown, but that the habit may reduce the toxicity of ingested plant compounds is suggested by the fact that a variety of clays adsorb alkaloids (Johns, 1986).

Unfortunately, very little research has been directed toward ascertaining the validity of the learning model for the feeding behavior of wild mammalian herbivores. Preliminary evidence supporting the model comes from studies on diet preferences of small mammals. Partridge (1981) found that of four species tested, all exhibited preferences for familiar foods over novel foods of similar quality, and Partridge and Maclean (1981) showed that bank voles [*Clethrionomys glareolus* (Schreber)] learned to associate nutritional effects of foods with cues such as taste or color, and altered their feeding preferences accordingly. In another study (Vickery, 1984), three species of rodents [*Clethrionomys gapperi* (Vigors), *Napaeozapus insignis* (Miller), and *Peromyscus maniculatus* (Wagner)], sampled food items even when they were not part of the optimal diet (defined as the highest caloric gain per unit handling time). Clearly, the role of learned aversions in the feeding behavior of mammalian herbivores deserves further investigation.

Research on the function of food cues in feeding behavior may prove especially interesting. If the learning model is true for mammalian herbivores, then animals should associate the value of a particular plant or plant part with cues such as taste and smell. For example, phenolic glycosides appear to influence feeding preferences of the mountain hare, which can discriminate among food items on the basis of olfactory stimuli (Tahvanainen *et al.*, 1985). Plant cues that influence feeding preferences may not necessarily be associated with the active chemical defense itself, especially if other plant compounds can be more readily and as reliably detected. Camazine (1985) found that an innocuous volatile compound (octen-3-ol) in mushrooms can act as an olfactory warning signal to opossums (*Didelphis virginiana* Kerr) with respect to delayed illness caused by the mushroom toxin muscimol. One reason (among many) why plants contain a variety of secondary metabolites is that some may serve as toxins while others provide a warning taste or odor. A logical, albeit speculative, extension of aposematic tastes or odors in plants is the possibility of Batesian or Mullerian mimicry among plants.

IV. Physiological Adaptations to Plant Chemical Defenses

Numerous factors influence the physiological fate of an ingested plant chemical, including characteristics of both the compound itself and the consuming organism (Table I). Internal (i.e., nonbehavioral) adaptations of animals to toxic compounds may be either physiological (reduced uptake or transport such that less of a toxin reaches its target site) or biochemical (enzymatic transformation of the toxin) (Brooks, 1976; Hollingworth, 1976; Berenbaum, 1985). These adaptations have developed over evolutionary time but in many cases particular adaptations remain quite flexible, producing phenotypic responses to food constituents over periods of hours to weeks.

A. Changes in Gastrointestinal Physiology

Secondary plant compounds penetrate body membranes and enter the bloodstream of mammals by the same processes that are used for the absorption of water and food nutrients, and may do so along the length of the gastrointestinal tract. Most toxins cross gastrointestinal membranes by passive diffusion and thus absorption rates are strongly affected by the lipid solubility of compounds. Lipophilicity of ionized compounds is low, and they may form ionic interactions with membrane lipids and proteins, reducing transport even further (Guthrie, 1980). Different components of the gastrointestinal tract have highly variable pH values (Table II), so the membrane permeability of a compound may change markedly as it passes through an animal. Absorption of alkaloids (ionized at low pH), for exam-

Table I. Factors Influencing the Physiological Fate of Ingested Phytochemicals

Absorption	Distribution	Elimination
Ingestion rate	Absorption rate	Absorption rate
Molecular size	Molecular size	Molecular size
Solubility	pK_a	Tissue binding
pK_a	Lipophilicity	Metabolism by tissue enzymes
Lipophilicity	Protein binding in blood	pK_a and lipophilicity in reabsorption from kidney
Interactions with other ingested compounds	Presence of barriers, active transport systems	Enterohepatic circulation
Metabolism by microbes or gut enzymes	Substrate specificity	
Rentention time	Enterohepatic circulation	

Table II. pH (mean ± SE, n = 6) of the Gastrointestinal Tract of Prairie Voles (*Microtus ochrogaster*)

Site	pH
Mouth	8.5 ± 0.1
Stomach:	
Esophageal sac	6.4 ± 0.1
Fundic region	0.9 ± 0.1
Pyloric region	1.4 ± 0.1
Duodenal loop	6.7 ± 0.1
Small intestine	7.6 ± 0.2
Cecum	8.0 ± 0.1
Large intestine	8.0 ± 0.1

Note: pH measured with narrow-range pH paper; voles fed rabbit pellets (Purina Rabbit Chow, no. 5321).

ple, is much more rapid in neutral to basic intestines than in the acid stomach.

The mammalian gastrointestinal tract has specialized active transport systems for some nutrients and electrolytes, including a carrier system for the absorption of glucose and galactose, three systems for the absorption of amino acids, a system for the absorption of pyrimidines, and separate systems for the absorption of calcium, sodium, and iron (Klaassen, 1980). Absorption of toxins via special transport systems occurs only rarely (e.g., 5-fluorouracil is carried by the pyrimidine transport system), but these systems are important in the elimination of toxins after absorption has occurred (Guthrie, 1980). This is especially true for compounds whose lipophilicity has been reduced after metabolism by detoxication enzymes.

Absorption of toxic compounds is also influenced by food retention times and by consumption of substances that alter the permeabillty of gastrointestinal membranes. Decreased intestinal motility tends to increase absorption, probably because of the high absorptive capacity of the proximal portion of the small intestine. Ethylenediamine tetraacetic acid (EDTA) increases the permeability of the gastrointestinal membrane to a variety of compounds by chelating and removing calcium (Klaassen, 1980).

To what extent mammalian herbivores utilize changes in gastrointestinal physiology to limit their uptake of potentially toxic secondary plant compounds is not known at this time, but the possibilities are intriguing. That such alterations may occur is suggested by the fact that processes for the acquisition of nutrients change in mammals as their nutritional requirements or the nutrient contents of their food change. In animals experiencing an increased requirement for all nutrients (e.g., during gestation or lactation, or as a consequence of the evolutionary development of endothermy) an increase in absorptive area through mucosal growth and increased gut length elevates absorption of nutrients (Karasov and Diamond, 1983). Activity of specific nutrient transport mechanisms (e.g., for sugar and amino acids) may be induced or suppressed as dietary availability of nutrients changes. Transport rates vary with intestinal position in proportion to the nutrient load arriving at each position, as well as with daily and seasonal changes in the nutrient mixture ingested (Karasov and Diamond, 1983). Mammalian herbivores may very likely rely on analogous physiological alterations to reduce uptake of plant toxins. In fact, the physiological changes exhibited by herbivores in response to dietary changes probably reflect a compromise between conflicting demands to maximize nutrient uptake while minimizing toxin uptake.

Preliminary work has shown that mammalian herbivores may rely on physiological responses to reduce the activity of plant compounds that disrupt digestive processes. Feeding of sorghum with high levels of tannins to rats induced production of proline-rich salivary proteins (PRPs), which have an unusually high affinity for tannins, and the onset of PRP synthesis coincided with enhanced weight gains (Mehansho *et al.*, 1983). Similarly, mule deer [*Odocoileus hemionus* (Rafinesque)] have adapted to tanniniferous forage through production of salivary proteins that bind tannins in a highly specific manner (Robbins *et al.*, 1987a). Mole and Waterman (1985) showed that glycocholic acid, a bile salt, acts as a surfactant to inhibit the *in vitro* precipitation of protein by tannins (but not protein–tannin complexing per se), and suggest that such compounds may afford protection to tannin-consuming mammals.

B. Changes in Metabolic Rates

The major factor influencing basal metabolic rates and daily energy expenditure in mammals is body size (Kleiber, 1961; McNab, 1980, 1986), but even for a given body size considerable variation in basal metabolic rate (BMR) exists. When the effect of body size is factored out, much of this residual variation correlates with food habits (McNab, 1983).

Certain food habits of mammals are correlated with low BMRs regardless of the animal's taxonomic affiliation. McNab (1983) argues that mammals with low BMRs consume foods that are so deficient as to prevent them from having higher BMRs, which in turn are associated with short gestation periods, rapid postnatal growth, and high fecundity. Food characteristics that may have led to an evolutionary decrease in BMRs include seasonal periodicity (BMRs adjusted down to the maximal level possible during the poorest season), low nutrient content or poor digestibility

(BMRs adjusted down to the highest level that can be maintained given the mechanical and energetic constraints on consumption rates), and high levels of chemical defenses (BMRs adjusted down to the highest level that can be maintained given the costs of detoxication processes) (McNab, 1983, 1986). McNab's proposals are based on a large compilation of data on food habits and metabolic rates for a variety of mammals. For example, folivores, which use food generally characterized by a high content of poorly digestible fiber and high levels of secondary compounds, have lower metabolic rates than do grazers of a similar size, which consume food that is less fibrous (herbs) or has lower levels of secondary compounds (grasses).

Daily and seasonal changes in temperature exert a multiple impact on the interactions of herbivores with secondary plant compounds. Not only do food consumption rates vary to meet energetic requirements for homeothermy, but detoxication processes, under metabolic control, are influenced by changes in ambient temperature. Variations in temperature influence the toxicity of compounds in two basic ways—either an increase in toxicity at both high and low temperatures or an increase in toxicity with an increase in temperature. For instance, toxicity of the alkaloid caffeine to mice increases at high and low temperatures, but the toxicity of *d*-amphetamine simply increases with temperature (Hodgson, 1980). Thus the problem of variation in quality or quantity of food faced by herbivores in seasonal climates may be confounded by simultaneous changes in their ability to process the secondary compounds in their food plants.

V. Biochemical Adaptations to Plant Chemical Defenses

The best-known adaptation of mammals to plant chemical defenses has involved the evolution of biochemical detoxication systems, either through mutualistic associations with gut microflora or through the development of specific enzyme systems.

A. Detoxication by Gut Microflora

Nearly all herbivorous mammals have gastrointestinal tracts with one or more organs expanded to house a microbial population that aids in the processing of plant foods (McBee, 1971). Fermentative forestomachs are found not only in ruminants but also in a variety of other mammals, including colobid monkeys, kangaroos, hippopotamuses, rock hyraxes, and many rodents. Hindgut fermentation chambers such as enlarged ceca and large intestines occur, for example, in lagomorphs (rabbits and pikas), perissodactyls (horses, rhinoceroses, and other odd-toed ungulates), howler monkeys, elephants, koalas, and rodents (McBee, 1971; Vaughan, 1972; Robbins, 1983; Hume, 1984). The role of the gut microflora in digestive processes is probably least well understood in herbivorous rodents, although their gastrointestinal anatomy and associated pH values (e.g., Table II), are appropriate for bacterial activity. Cellulolytic bacteria have been isolated from the stomachs of the vole *Microtus montebelli* (Milne-Edwards) (Kudo *et al.*, 1979) and the golden hamster [*Mesocricetus auratus* (Waterhouse)] (Hoover *et al.*, 1969).

Nutritional benefits accruing to mammalian herbivores from mutualistic relationships with microorganisms have long been known to animal scientists, and include the utilization of cellulose and nonprotein nitrogen and the synthesis of essential fatty acids and water-soluble vitamins. More recently researchers have realized that microorganisms also play important roles in the detoxication of many secondary plant compounds. Thus the gut microflora may be considered a herbivore's "first line of active defense" (Reid, 1973); indeed, some authors (Janzen, 1979; Morris and Rogers, 1982) attribute more importance to this function than to the simple acquisition of additional nutrients.

Reactions of gut microbes with phytochemicals occur under anaerobic conditions and belong to three major classes of reactions, including hydrolyses, reductions, and degradations such as dealkylation, deamination, decarboxylation, dehalogenation, and ring fission (Allison, 1978; Scheline, 1978). Ruminal microbes in sheep hydrolyze the hepatotoxic pyrrolizidine alkaloids heliotrione and lasiocarpine to nontoxic, hydroxylated methylpyrrolizidines (Lanigan and Smith, 1970). Unsaturated hydroxycinnamic acids (e.g., caffeic, coumaric, ferulic, and sinapic acids) undergo reduction by bacteria to phenylpropionic acids (Scheline, 1978). Examples of heterocyclic ring fission by gut microflora include the degradation of coumarin, its derivative umbelliferone, and many flavonoids (Scheline, 1978).

Gut microbes have a great capacity to accommodate not only a variety of secondary plant compounds, but also temporal variation in the amounts and composition of mixtures of compounds. Adaptations are accomplished via changes in the activity (i.e., alteration of the number or metabolic capacity of particular bacteria) or composition of the microbial community. Metabolism of the flavonoid robinin and its aglycone kaempferol to *p*-hydroxyphenylacetic acid by rats is elevated when the rats are maintained on a diet containing the compounds (Griffiths and Smith, 1971). The rumen microflora of deer with previous exposure to Douglas fir is less inhibited by monoterpenes from Douglas fir than is the microflora of deer with no previous exposure (Oh et al., 1967), and prior access of sheep to pyrrolizidine alkaloids increases their rate of microbial detoxication (Lanigan and Smith, 1970). In contrast to the reactions of tissue enzymes, detoxication reactions of the gut microflora may change qualitatively as well as quantitatively. Although little research has been conducted in this area, limited evidence shows that entirely new metabolic pathways may be developed by microorganisms; this potential would certainly benefit generalist herbivores. The best-known example of this phenomenon is metabolism of cyclamate, a synthetic sweetening agent. When cyclamate was administered orally to guinea pigs, rabbits, and rats with no previous exposure to the compound, it was excreted entirely unchanged in the urine and feces. When maintained for several months on a diet containing cyclamate, all three species developed the ability to convert it into cyclohexylamine (Renwick and Williams, 1972), a reaction accomplished by the gut microflora (Drasar *et al.*, 1972).

Gut microbes cannot process all secondary compounds, however, and in some cases their reactions may adversely affect herbivores. Microbial activity is inhibited by some alkaloids (e.g., perloline; Bush *et al.*, 1970), essential oils (e.g., monoterpene alcohols and esters; Oh *et al.*, 1967), and tannins (Tagari *et al.*, 1965), and consequently microbial production of nutrients is reduced. The degradative, reductive, and especially hydrolytic reactions of gut microorganisms may increase the biological activity of some plant compounds. Enterobacterial glycosidases hydrolyze cyanogenic glycosides to products that release hydrogen cyanide (Coop and Blakley, 1949), convert the glucoside cycasin to its hepatotoxic and carcinogenic aglycone (Spatz *et al.*, 1967), and are most likely responsible for the production of thiocyanates from glucosinolates (Scheline, 1978).

Gut microfloras are also largely responsible for the enterohepatic circulation of compounds, especially those conjugated with glucuronic acid. Many plant compounds are metabolized in mammals by conjugation with glucuronic acid and excretion via the bile (molecular weight greater than 400, in rats) or urine (molecular weight less than 400) (Smith and Williams, 1966). Conjugated plant compounds excreted in the bile may be hydrolyzed by microbial β-glucuronidases, and the aglycones may be absorbed again. Thus, enterohepatic circulation prolongs the persistance of potentially toxic compounds in mammals.

We know exceedingly little about the role of microorganisms in the detoxication of plant compounds in wild mammalian herbivores, although adaptations important to the survival of livestock would be equally if not more important to the survival of their wild relatives, for whom the selective force of plant allelochemicals is probably greater. Several lines of evidence indicate that microorganisms are important to the performance of mammalian herbivores. Snowshoe hares can be maintained on a diet of mature twigs of Alaska paper birch if the diet is introduced gradually (Bryant, 1981b), and growth rates of meadow voles and prairie voles fed diets containing *Lespedeza* phenolics or the alkaloid boschniakine tended to increase after 5 days on the diet (Lindroth *et al.*, 1986a), results consistent with adaptation of the gut microflora to a new diet. Game managers know that when moving ruminants from one area to another it is often important to inoculate them with ruminal fluid or fresh dung from local animals in order to prevent severe digestive problems (Pienaar, 1968). Individual porcupines [*Erethizon dorsatum* (L.)] in the Catskill Mountains limit their winter feeding to one or two of the 10 tree species used by the population as a whole, and Roze (1985) proposes that such specialization enhances the ability of an individual's microorganisms to process a suboptimal (winter) diet. Finally, and most convincingly, microbes from the esophageal sac of the vole *Microtus arvalis* Pallas degrade mimosine, a nonprotein amino acid well known for its toxicity to a variety of mammals (Kudo *et al.*, 1986).

B. Detoxication by Tissue Enzymes

Secondary plant compounds absorbed from the gut are usually metabolized by specific enzyme systems located most notably in the liver but also in many other body tissues, especially the kidneys, intestines, and lungs. These enzyme systems are associated with the cytosol, mitochondria, and endoplasmic reticulum of cells. [Unfortunately, the term "microsomal enzymes" is commonly and inaccurately substituted for "tissue enzymes." The term should refer only to systems (e.g., polysubstrate monooxygenases and glucuronyltransferases) associated with the endoplasmic reticulum, which forms microsomes upon homogenization.]

Tissue enzymes metabolize foreign compounds via a large number of pathways, which are typically classified as oxidations, reductions, hydrolyses, and conjugations. The reactions are not detoxication reactions per se (occasionally they may even activate substances, as is the case with pyrrolizidine alkaloids), but produce compounds more hydrophilic, hence more easily excreted, than the parent compounds. In contrast to reactions of the gut microflora, the most important reactions occurring in body tissues are oxidative and conjugative (Millburn, 1978). Detoxication of specific secondary plant compounds by mammals has been reviewed in detail (Scheline, 1978), so will not be addressed here. Let it suffice to say that nearly all assimilable phytochemicals may be metabolized by one or several metabolic pathways in mammals, but of course the efficacy of such reactions is influenced by a number of factors (Fig. 1) and varies greatly.

Large differences in metabolic and physiological capabilities often occur among closely related species, especially in the metabolism of foreign compounds (Smith *et al.*, 1984). In addition to genetically based differences, an animal's response to a phytochemical is influenced by its age, sex, reproductive condition, nutritional status, and previous and simultaneous consumption of the same and other plant compounds. Polysubstrate monooxygenase (PSMO) activity is very low in newborn mammals (Brattsten, 1979), male rats metabolize many foreign compounds more rapidly than females (Dauterman, 1980), PSMO and glucuronyltransferase activities in rats decrease during pregnancy (Dauterman, 1980), and the activity of enzymatic detoxication systems in general is (usually) lowered by deficiencies in dietary protein, carbohydrates, lipids, vitamins, and minerals (Boyd and Campbell, 1983). Moreover, activity of particular enzyme systems may be induced or inhibited by the presence of other foreign compounds (Brattsten, 1979). For example, measurement of uronic acid excretion indicated that glucuronyltransferase activity was induced in prairie voles fed the flavonol quercetin or tannic acid (Lindroth and Batzli, 1983).

Although the toxicological literature on natural products and domestic animals (especially laboratory rats) and on synthetic products and wild vertebrates is abundant, very little information is available on the toxicology of plant products and wild mammalian herbivores. Future research in this area is essential for improving our understanding of the role of detoxication systems in the nutritional ecology of mammals.

Clearly, the great variability in forage quality faced by long-lived (compared to invertebrates) generalist mammalian herbivores has led to the evolution of multiple, complex, and complementary behavioral, physiological, and biochemical mechanisms for averting or reducing the deleterious effects of plant defenses. Mammalian herbivores can be expected to exhibit considerable plasticity in their response to plant chemical defenses, and this plasticity holds implications for the coevolution of plant–mammal interactions.

VI. Responses of Microtine Rodents to Secondary Compounds

The responses of mammalian herbivores to particular secondary compounds and to interactions among dietary constituents have received relatively little attention, but have been most thoroughly documented with microtine rodents. As with insects, secondary compounds may function as feeding deterrents, toxins, or digestibility reducers to these mammals (Table III). In fact, evidence that some phenolic compounds inhibit digestion of protein is much more compelling for mammals (Lindroth and Batzli, 1984a; Lindroth *et al.*, 1986a; Robbins *et al.*, 1987b) than it is for insects (Martin and Martin, 1982). Plant compounds may also influence reproduction in microtines. Reproductive activity in the vole *Microtus montanus* (Peale) is synchronized with the plant growing season, being triggered in spring by the appearance of 6-methoxybenzoxazolinone in

Table III. Effects of Secondary Plant Compounds on Microtine Rodents

Compound	Species	Mode of Action	Reference
Alkaloids			
Boschniakine	Prairie vole [*Microtus ochrogaster* (Wagner)]	Toxin	Lindroth *et al.* (1986a)
Quinine	Meadow vole [*M. pennsylvanicus* (Ord)]	Deterrent	Kendall and Leath (1976)
Indole alkaloids	Meadow vole	Deterrents, toxins	Kendall and Sherwood (1975); Goelz *et al.* (1980)
Beta-nitropropionic acid	Meadow vole	Deterrent, toxin	Shenk *et al.* (1976)
Phenolics			
Quercetin	Prairie vole	Toxin	Lindroth and Batzli (1984a)
Tannic acid	Prairie vole	Toxin, digestibility reducer	Lindroth and Batzli (1984a)
Quebracho	Prairie vole	Deterrent, digestibility reducer	Lindroth and Batzli (1984a)
Lespedeza tannins	Prairie and meadow voles	Deterrent, digestibility reducer	Lindroth *et al.* (1986a)
Saponins	Meadow vole	Deterrent	Kendall and Leath (1976)
Sesquiterpene lactone helenin	Redback vole [*Clethrionomys rutilus* (Pallas)]	Deterrent	Picman *et al.* (1982)

food plants, and ceasing in autumn coincident with increased concentrations of cinnamic acids (Berger *et al.*, 1977, 1981; Sanders *et al.*, 1981).

Not surprisingly, responses of voles vary according to both the concentration of secondary plant metabolites and the concentrations of other dietary constituents. In laboratory trials, prairie voles fed diets with 8% protein exhibited decreasing growth and decreasing ability to digest protein with increasing tannic acid, but when fed diets with 20% protein were unaffected by the same levels of tannic acid (Lindroth and Batzli, 1984a). Similarly, quercetin reduced vole growth (via a toxic effect) at low protein, but not at high protein, levels. Moreover, the mode of action of a phytochemical may vary with its concentration. *Lespedeza cuneata* tannins at a concentration equivalent to a natural diet of 80% *Lespedeza* deterred feeding by prairie voles to the point of starvation. But at a concentration equivalent to a diet of 50% *Lespedeza*, food consumption increased 19%, protein digestibility decreased 57%, and growth did not differ from that of control animals (Lindroth *et al.*, 1986a). Finally, plant chemicals may have multiple modes of action at one concentration; at low protein levels tannic acid decreased protein digestibility and was toxic to prairie voles (Lindroth and Batzli, 1984a).

Field studies have shown that the diets of meadow voles in grassland habitats in east-central Illinois change seasonally in conjunction with the quality and quantity of available forage. Voles in a bluegrass oldfield ate increasing amounts of dicotyledons and decreasing amounts of monocotyledons as seasons progressed from winter to autumn; voles in a tallgrass prairie ate more dicotyledons than monocotyledons year-round (Lindroth and Batzli, 1984b). Individual voles, as well as the population as a whole, are generalist feeders. For voles in the bluegrass field during summer, 46% of the plant species consumed by the population (and contributing >2% of the average diet) were present in the stomachs of individual voles when trapped, and for voles in the prairie, 68% of the species consumed were present in individual voles.

Food preferences of voles in these habitats are influenced by both plant nutrients and secondary compounds (Table IV). Of the three major food species of meadow voles in the bluegrass area, *Poa* was avoided relative to its availability; it had the lowest levels of chemical defenses of the three species, but also the lowest level of protein. [Meadow voles are closely related to prairie voles, which require between 5% and 8% dietary protein for normal growth (Lindroth and Batzli, 1984a).] *Taraxacum* was a preferred food plant; it had substantial quantities of phenolics but also almost double the protein concentration of *Poa*. *Trifolium* was strongly preferred; it had lower phenolic levels and nearly double the protein level of *Taraxacum*. Preferences of meadow voles in the prairie showed a similar pattern. The grass *Andropogon* was avoided probably because of its small amount of protein, even though it had lower levels of secondary com-

Table IV. Preference Indices and Nutritional Quality of Major Constituents of Meadow Vole Diets during Summer

Habitat and Species	Preference Index[a]	Crude Protein (% dry weight)[b]	Protein-Binding Phenolics ($E^{1\%}_{1\,cm}$)[b]	Alkaloids ($E^{1\%}_{1\,cm}$)[b]
Bluegrass				
Poa pratensis Linnaeus (5)[c]	0.33	6.3 ± 0.2	6.0 ± 0.2	—
Taraxacum officinale Weber (5)	7.04	11.0 ± 0.7	44.9 ± 1.7	—
Trifolium pratense Linnaeus (6)	8.82	20.8 ± 0.7	20.5 ± 0.9	—
Prairie				
Andropogon gerardii Vitman (5)	0.52	3.8 ± 0.3	21.5 ± 0.9	—
Lespedeza cuneata (Dumont de Courset) (5)	1.15	11.4 ± 0.4	85.2 ± 1.8	—
Penstemon digitalis (5)	0.99	5.7 ± 0.4	46.9 ± 3.4	0.006 ± 0.001

Note: Data from Lindroth (1984) and Lindroth and Batzli (1984b).
[a] Preference index = (proportion in stomach)/(proportion in habitat).
[b] Mean ± SE.
[c] Sample size in parentheses.

pounds than did the other common food items. *Lespedeza* and *Penstemon* were neither significantly preferred or avoided. *Lespedeza* had very high levels of phenolics and three times as much protein as *Andropogon*. *Penstemon* had protein and phenolic levels intermediate between the other two species, and also contained alkaloids. Several results of this study (Lindroth and Batzli, 1984b) deserve emphasis. First, foods commonly eaten by mammalian herbivores are not necessarily preferred—a measure of food availability is essential for determinations of "preference." Second, contrary to the feeding behavior proposed by Freeland and Janzen (1974), mammalian herbivores may often not "prefer to feed on foods that contain only minor amounts of toxic plant secondary compounds." Differing levels of nutrients in forage confound the selection process, as Janzen (1978) later suggested.

That meadow voles eat substantial quantities of *Lespedeza* and *Penstemon* in the field is interesting in light of the fact that voles find each species unpalatable (relative to rabbit pellets) in laboratory trials, and when restricted to either species will eat very little and die (Lindroth *et al.*, 1986a). These results support the contention of Freeland and Janzen (1974) that the amount of any one plant that a herbivore consumes is limited by the capacity of the gut microflora and tissue enzymes to detoxify it. Because mammalian herbivores must be feeding generalists in order to meet their food requirements, their detoxication systems must remain flexible enough to handle a plethora of secondary plant compounds. Ecological dogma asserts that "a jack of all trades is master of none," and if this applies to detoxication systems then it is clear that the detoxication systems of a generalist feeder may easily be overloaded if restricted to one or a few food items. Thus one consequence of the evolution of chemical defenses in plants has been to enforce the generalist feeding pattern of most mammalian herbivores.

VII. Are Plant–Mammal Systems Coevolved?

The term coevolution was first used by Ehrlich and Raven (1964) to describe evolutionary relationships between plants and the insects that feed on them. These authors offered no definition, and since then numerous ecologists have attributed a variety of meanings to the term (Futuyma and Slatkin, 1983). In the most restrictive sense coevolution implies closely integrated, reciprocal, and simultaneous evolution between two interacting populations. At the other extreme of the continuum is the notion that coevolution is simply the unilateral adaptation of populations to aspects of their biotic environment. In this discussion of plant–mammal coevolution, neither extreme is useful. The first, by definition, rules out such a possibility. Almost all mammalian herbivores feed on a wide variety of plants, and the major chemical diversification of angiosperms long preceded the diversification of modern mammals, so the criteria of specificity (one-to-one evolution in paired populations) and simultaneity are not met. The second is so general that coevolution is no different than evolution in response to biotic factors, and the utility of the term is lost. The many adaptations of mammalian herbivores to plant chemical defenses previously reviewed are certainly evolved responses, but may not be coevolved—that is, they are necessary but not sufficient conditions to confirm chemically mediated coevolution. For this discussion I will adopt the definition advanced by Futuyma and Slatkin (1983), that coevolution has occurred when each of two or more interacting species exhibit adaptive responses to genetic changes in the others. This definition maintains the requirement of reciprocal genetic change, but allows for the possibility of "diffuse coevolution" (Janzen, 1980; Fox, 1981), in which groups of populations evolve in response to each other.

The primary constraint on the development of coevolved plant–mammal complexes has been the absence of directional selection sufficiently intense to produce them. Mammals must exert a selective pressure strong enough to promote significant changes in gene frequencies in one to many of the plant populations used as food, and these populations must be important enough to the fitness of the herbivores to promote a reciprocal genetic change. But generalist herbivores feed on many species that in turn are fed on by many different herbivores, and the selective forces involved are so diffuse that coevolution has probably rarely occurred.

What factors characterize the plant–mammal systems in which evidence for coevolution is most likely to be found? First, the phylogenetic lineages of the plants and mammals should have long histories of close association. Parallel cladogenesis (*sensu* Futuyma and Slatkin, 1983) would provide even stronger evidence of coevolution. Second, the impact of each plant or animal group must exert (or have exerted) a strong selective force upon the other. Consequently, the plants and animals involved should exhibit morphological, physiological, biochemical, and/or behavioral adaptations toward each other. This is most likely to have occurred in systems with relatively few participants (i.e., those with mammals that tend to specialize on certain plant genera or families). Third, if the geographical ranges of the participants are not congruent, then these adaptations should be most pronounced in the zones of sympatry. Finally, evidence of coevolution may be provided by the structure of ecological communities in which the plants and mammals interact strongly.

Such a combination of phylogenetic, chemical, and biological data is not available for any plant–mammal system. But for several systems enough information is known to make coevolution an attractive hypothesis. The best documentation of coevolution between plants and mammals comes from research on grasslands and large, grazing mammals (Mack and Thompson, 1981; McNaughton, 1983, 1984, 1985). The plants and mammals have long phylogenetic histories of close association, exert strong selective pressures on each other, and exhibit a variety of adaptations toward each other. But because plants in these systems rely primarily on physical defenses (sharp awns, high silica content), growth form, and compensatory growth rather than chemicals for defense (McNaughton, 1984), coevolution of grasses and grazers will not be considered here.

Plant chemicals may have mediated coevolution between leguminous species and their marsupial herbivores in Australia. Approximately 34 species of the genera *Gastrolobium* and *Oxylobium* in western Australia produce appreciable quantities (e.g., 0.25% fresh weight) of fluoroacetate, a potent toxin to most animals. The geographical range of the rat kangaroo *Bettongia penicillata* Gray includes the *Gastrolobium* and *Oxylobium* species, and laboratory studies show it is highly resistant to fluoroacetate (Mead *et al.*, 1985). *Bettongia leseur* Quoy Gaimard, whose range has not included fluoroacetate-producing vegetation for at least 7000 years, is moderately tolerant. The most primitive living rat kangaroo is *B. gaimardi* Desmaurest; its past distribution was limited to southeast Australia, and it is not resistant to fluoroacetate. Mead *et al.* (1985) suggest that the *Bettongia* genus originated in eastern Australia and radiated westward, with subsequent development of resistance to fluoroacetate in the western species. Comparison of fluoroacetate resistance values in several populations of the southern bush rat (*Rattus fuscipes* Waterhouse) showed that all were tolerant, but those whose ranges included *Gastrolobium* and *Oxylobium* were more tolerant than those whose ranges did not. This system is especially interesting because the identity of the chemical mediator is known, the mammals exhibit biochemical adaptations to it, and the phylogenetic and geographical patterns fit a hypothesis of coevolved relationships. But the data are short of conclusive. We do not know, for instance, the intensity of the selective force exerted by these and related marsupial herbivores on *Gastrolobium* and *Oxylobium* species; that is, the production of fluoroacetate in these plants may not be related to mammalian herbivory.

Chemical coevolution may also explain the interactions of snowshoe hares (*Lepus americanus* Erxleben) and deciduous trees in the boreal forests of North America. Winter browsing by hares exerts a strong selective pressure on juvenile trees, and an inadequate supply of quality food dramatically affects hare populations (Bryant, 1981a; Bryant *et al.*, 1983, 1985; Reichardt *et al.*, 1984). Twigs of juvenile deciduous trees are covered by terpene and phenolic resins (e.g., papyriferic acid in Alaska paper birch, *Betula resinifera* Britton), which most likely evolved as deterrents against winter feeding by mammals. Hares browse selectively among species, and among parts within individual trees, for food low in resin content. Preliminary studies also indicate that hares can detoxify some plant phenolics via the glucuronidation pathway (Bryant *et al.*, 1985). Coevolution appears plausible in this system because relatively few participants are involved (selection pressures are not highly diffuse), and the chemical defenses of the trees appear to be directed against mammalian herbivores. What is missing, however, is substantial information linking specific adaptations of hares to the chemical defenses of their winter food plants. Indeed, the fact that hare populations crash after extensive, browsing-induced reversion of deciduous trees to resistant juvenile forms suggests that effective counteradaptations do not exist.

Finally, the structure of microtine rodent communities on the coastal plain of northern Alaska may have been shaped by the coevolution of the animals and their food plants. The brown lemming (*Lemmus sibiricus* Kerr) and collared lemming [*Dicrostonyx torquatus* (Pallas)] are the dominant members of the community near Atkasook, Alaska, and exhibit remarkably different food habits and habitat preferences (Batzli and Jung, 1980). Brown lemmings inhabit wet areas where grasses and sedges predominate and eat mainly monocotyledons ($\geq$80% of diet during summer), especially sedges. In years of high population densities grazing by brown lemmings can reduce the total yield of grasses and sedges by 50–90% (Schultz, 1964). Collared lemmings have a longer evolutionary history than brown lemmings in tundra environments (Hoffman and Taber, 1968), inhabit drier areas where forbs and willows predominate, and consume mostly dicotyledons ($\geq$76% of diet during summer), especially willows. Food habits of the two species cannot be explained simply on the basis of habitat preferences because in some areas the lemmings overlap broadly and maintain their food preferences.

What does explain these differences in food habits is plant chemistry. In laboratory feeding trials brown lemmings found willow (*Salix pulchra* Chamisso) unpalatable and tended to grow poorly when fed rabbit pellets coated with an ethanol extract of willow leaves (Batzli and Jung, 1980; Jung and Batzli 1981). Collared lemmings found the sedge *Carex aquatilis* Wahlenberg palatable but grew very poorly when fed rabbit pellets coated with sedge extract. Apparently brown lemmings and collared lemmings have evolved detoxication systems that allow them to specialize on certain classes of foods and that prevent them from tolerating the secondary compounds in each other's food plants. Moreover, in accordance with McNab's predictions, relative basal metabolic rates of brown lemmings are 25% higher than those of collared lemmings (McNab, 1986). Thus, the evolutionary impact of plant chemistry in this system has led to the alteration of basal metabolic rates and the development of specialized detoxication systems, which in turn influence food habits, habitat choice, and community structure. Unfortunately, the chemicals responsible for the reciprocal feeding ability are not known. Neither is it clear whether the defense systems of the plants evolved in response to herbivory by rodents or to other factors.

VIII. Conclusions

Information compiled from studies on laboratory animals, domestic livestock, and wild mammalian herbivores shows that mammals have evolved complex behavioral, physiological, and biochemical systems in response to the allelochemicals in their food plants. As of yet, however, little research has straightforwardly addressed the behavioral and chemical mechanisms of interactions between wild mammalian herbivores and their food plants. We need studies to answer questions such as: What are the roles of learned preferences and aversions in food selection by wild herbivores? Do mammalian herbivores rely on changes in their gastrointestinal physiology to reduce uptake of potential toxins, and how are these functions compromised with the requirement for uptake of nutrients? What are the relationships among metabolic rates, food habits, and detoxication processes in mammals? What are the functions of gut microbes and tissue enzymes in detoxication of specific plant compounds? How are these processes affected by internal and external factors, and how do they vary among and within species?

The answer to the question of whether chemicals have mediated coevolution between plants and mammals is at best an unsatisfactory "maybe." Plants and mammals certainly exhibit chemically oriented adaptations vis-à-vis herbivory, and in certain cases these may be indicative of coevolution. But we know too little about any single system to relate these adaptations to a repetitive, reciprocal evolutionary process. In many systems mammals may simply have evolved adaptations to plant defenses that themselves evolved in response to herbivory by insects. Coevolution without the criterion of reciprocity is indistinguishable from evolution and hence a useless concept. And because of the diffuse nature of the selective forces operating between generalist herbivores and their food plants, reciprocity has probably rarely occurred. Thus I believe that chemical coevolution has had, at best, little impact on the development of interactions between plants and their mammalian herbivores.

This is not to say that the possibility should not receive serious consideration by ecologists. Researchers should focus on systems in which the plants and mammals (1) have long histories of close association, (2) exert (or did so in their evolutionary histories) strong selective forces on each other, and (3) exhibit chemically oriented adaptations and counteradaptations. Ideally, these would be systems in which differential foodplant use by closely related species, or by geographically distinct populations within a species, could be attributed to the progressive evolution of specific adaptations. Only as these data come to light will we be in a position to better assess the importance of chemicals to plant–mammal evolution and the consequences of evolved interactions for population and community dynamics.

Acknowledgments

I thank George Batzli, May Berenbaum, and Lena Brattsten for discussing some of the original ideas for this paper. Timothy Johns, Brian McNab, Simon Mole, Thomas Palo, and Uldis Roze graciously provided unpublished data or manuscripts and several anonymous reviewers provided helpful suggestions. I gratefully acknowledge support from the National Science Foundation, grant BSR-8503464, during preparation of the manuscript.

References

Allison, M. J. (1978). The role of ruminal microbes in the metabolism of toxic constituents from plants. *In* "Effects of Poisonous Plants on Livestock" (R. F. Keeler, K. R. Van Kampen, and L. F. James, eds.), pp. 101–118. Academic Press, New York.

Bate-Smith, E. C. (1972). Attractants and repellents in higher animals. *Proc. Phytochem. Soc.* **8,** 45–56.

Batzli, G. O., and Jung, H. G. (1980). Nutritional ecology of microtine rodents: Resource utilization near Atkasook, Alaska. *Arc. Alp. Res.* **12,** 483–499.

Berenbaum, M. (1981). Patterns of furanocoumarin distribution and insect herbivory in the Umbelliferae: Plant chemistry and community structure. *Ecology* **62,** 1254–1266.

Berenbaum, M. (1983). Effects of tannins on growth and digestion in two species of Papilionids. *Entomol. Exp. Appl.* **34,** 245–250.

Berenbaum, M. (1985). Brementown revisited: Interactions among allelochemicals in plants. *Recent Adv. Phytochem.* **19,** 139–169.

Berenbaum, M., and Feeny, P. (1981). Toxicity of angular furanocoumarins to swallowtail butterflies: Escalation in a coevolutionary arms race? *Science* **212,** 927–929.

Berger, P. J., Sanders, E. H., Gardner, P. D., and Negus, N. C. (1977). Phenolic plant compounds functioning as reproductive inhibitors in *Microtus montanus*. *Science* **195,** 575–577.

Berger, P. J., Negus, N. C., Sanders, E. H., and Gardner, P. D. (1981). Chemical triggering of reproduction in *Microtus montanus*. *Science* **214,** 69–70.

Botkin, D. B., Jordan, P. A., Dominski, A., Lowendorf, H., and Hutchinson, G. E. (1973). Sodium dynamics in a northern forest ecosystem. *Proc. Natl. Acad. Sci. U.S.A.* **70,** 2745–2748.

Boyd, J. N., and Campbell, T. C. (1983). Impact of nutrition on detoxication. *In* "Biological Basis of Detoxication" (J. Caldwell and W. B. Jakoby, eds.), pp. 287–306. Academic Press, New York.

Brattsten, L. B. (1979). Biochemical defense mechanisms in herbivores against plant allelochemicals. *In* "Herbivores. Their Interaction with Secondary Plant Metabolites" (G. A. Rosenthal and D. H. Janzen, eds.), pp. 199–270. Academic Press, New York.

Brooks, G. T. (1976). Penetration and distribution of insecticides. *In* "Insecticide Biochemistry and Physiology" (C. F. Wilkinson, ed.), pp. 3–58. Plenum, New York.

Bryant, J. P. (1981a). Phytochemical deterrence of snowshoe hare browsing by adventitious shoots of four Alaskan trees. *Science* **213,** 889–890.

Bryant, J. P. (1981b). The regulation of snowshoe hare feeding behavior during winter by plant antiherbivore chemistry. *Proc. World Lagomorph Conf.* pp. 720–731.

Bryant, J. P., and Kuropat, P. J. (1980). Selection of winter forage by subarctic browsing vertebrates: The role of plant chemistry. *Annu. Rev. Ecol. Syst.* **11,** 261–285.

Bryant, J. P., Chapin, F. S., III, and Klein, D. R. (1983). Carbon/nutrient balance of boreal plants in relation to vertebrate herbivory. *Oikos* **40,** 357–368.

Bryant, J. P., Wieland, G. D., Clausen, T., and Kuropat, P. (1985). Interactions of snowshoe hare and feltleaf willow in Alaska. *Ecology* **66,** 1564–1573.

Bush, L. P., Streeter, C., and Buckner, R. C. (1970). Perloline inhibition of *in vitro* ruminal cellulose digestion. *Crop Sci.* **10,** 108–109.

Calvert, J. J. (1985). Food selection by western gorillas (*C. g. gorilla*) in relation to food chemistry. *Oecologia* **65,** 236–246.

Camazine, S. (1985). Olfactory aposematism. Association of food toxicity with naturally occurring odor. *J. Chem. Ecol.* **11,** 1289–1295.

Coop, I. E., and Blakley, R. L. (1949). The metabolism and toxicity of cyanides and cyanogenetic glucosides in sheep. I. Activity in the rumen. *N. Z. J. Sci. Technol.* **30A,** 277–291.

Cooper-Driver, G., Finch, S., and Swain, T. (1977). Seasonal variation in secondary plant compounds in relation to the palatability of *Pteridum aquilinum*. *Biochem. Syst. Ecol.* **5,** 177–183.

Dauterman, W. C. (1980). Physiological factors affecting metabolism of xenobiotics. *In* "Introduction to Biochemical Toxicology" (E. Hodgson and F. E. Guthrie, eds.), pp. 133–142. Elsevier, New York.

Denton, D. A. (1967). Salt appetite. *Handb. Physiol.* **1,** 433–459.

Drasar, B. S., Renwick, A. G., and Williams, R. T. (1972). The role of the gut flora in the metabolism of cyclamate. *Biochem. J.* **129,** 881–890.

Ehrlich, P. R., and Raven, P. H. (1964). Butterflies and plants: A study in coevolution. *Evolution* **18,** 586–608.

Fox, L. R. (1981). Defense and dynamics in plant-herbivore systems. *Am. Zool.* **21,** 853–864.

Freeland, W. J., and Janzen, D. H. (1974). Strategies in herbivory by mammals: The role of secondary plant compounds. *Am. Nat.* **108,** 269–289.

Freeland, W. J., Calcott, P. H., and Anderson, L. R. (1985). Tannins and saponin: Interaction in herbivore diets. *Biochem. Syst. Ecol.* **13,** 189–193.

Futuyma, D. J., and Slatkin, M. (1983). Introduction. *In* "Coevolution" (D. J. Futuyama and M. Slatkin, eds.), pp. 1–13. Sinauer, Sunderland, Massachusetts.

Galef, B. G., Jr. (1977). Mechanisms for the social transmission of acquired food preferences from adult to weaning rats. *In* "Learning Mechanisms in Food Selection" (L. M. Barker, M. R. Best, and M. Domjan, eds.), pp. 123–148. Baylor Univ. Press, Waco, Texas.

Garcia, J., Hankins, W. G., and Rusiniak, K. W. (1974). Behavioral regulation of the milieu interne in man and rat. *Science* **185,** 824–831.

Gilbert, L. E., and Raven, P. H. (1975). "Coevolution of Animals and Plants." Univ. of Texas Press, Austin.

Goelz, M. F. B., Rothenbacher, H., Wiggins, J. P., Kendall, W. A., and Hershberger, T. V. (1980). Some hematological and histopathological effects of the alkaloids gramine and hordenine on meadow voles (*Microtus pennsylvanicus*). *Toxicology* **18,** 125–131.

Goldstein, W. S., and Spencer, K. C. (1985). Inhibition of cyanogenesis by tannins. *J. Chem. Ecol.* **11,** 847–858.

Griffiths, L. A., and Smith, G. E. (1972). Metabolism of apigenin and related compounds in the rat. Metabolite formation *in vivo* and by the intestinal microflora *in vitro*. *Biochem. J.* **128,** 901–911.

Guthrie, F. E. (1980). Absorption and distribution. *In* "Introduction to Biochemical Toxicology" (E. Hodgson and F. E. Guthrie, eds.), pp. 10–39. Elsevier, New York.

Haukioja, E. (1980). On the role of plant defenses in the fluctuation of herbivore populations. *Oikos* **35,** 202–213.

Hodgson, E. (1980). Chemical and environmental factors affecting metabolism of xenobiotics. *In* "Introduction to Biochemical Toxicology" (E. Hodgson and F. E. Guthrie, eds.), pp. 143–161. Elsevier, New York.

Hoffman, R. S., and Taber, R. D. (1968). Origin and history of Holarctic tundra ecosystems with special reference to their vertebrate faunas. *In* "Arctic and Alpine Environments." (H. E. Wright and W. H. Osburn, eds.), pp. 143–170. Indiana University, Bloomington, Indiana.

Hollingworth, R. M. (1976). The biochemical and physiological basis of selective toxicity. *In* "Insecticide Biochemistry and Physiology" (C. F. Wilkinson, ed.), pp. 431–506. Plenum, New York.

Hoover, W. H., Mannings, C. L., and Sheerin, H. E. (1969). Observations on digestion in the golden hamster. *J. Anim. Sci.* **28,** 349–352.

Hume, I. D. (1984). Microbial fermentation in herbivorous marsupials. *BioScience* **34,** 435–440.

Janzen, D. H. (1978). Complications in interpreting the chemical defenses of trees against tropical arboreal plant-eating vertebrates. *In* "The Ecology of Arboreal Folivores" (G. G. Montgomery, ed.), pp. 73–84. Smithsonian Institution Press, Washington, D.C.

Janzen, D. H. (1979). New horizons in the biology of plant defenses. *In* "Herbivores. Their Interaction with Secondary Plant Metabolites" (G. A. Rosenthal and D. H. Janzen, eds.), pp. 331–350. Academic Press, New York.

Janzen, D. H. (1980). When is it coevolution? *Evolution* **34,** 611–612.

Johns, T. (1986). The detoxification function of geophagy and the domestication of the potato. *J. Chem. Ecol.* **12,** 635–646.

Jung, H. G., and Batzli, G. O. (1981). Nutritional ecology of microtine rodents: Effects of plant extracts on the growth of arctic microtines. *J. Mamm.* **62,** 286–292.

Karasov, W. H., and Diamond, J. M. (1983). Adaptive regulation of sugar and amino acid transport by vertebrate intestine. *Am. J. Physiol.* **245,** G443–G462.

Kendall, W. A., and Leath, K. T. (1976). Effect of saponins on palatability of alfalfa to meadow voles. *Agron. J.* **68,** 473–476.

Kendall, W. A., and Sherwood, R. T. (1975). Palatability of leaves of tall fescue and reed canarygrass and of some of their alkaloids to meadow voles. *Agron. J.* **67,** 667–671.

Klaassen, C. D. (1980). Absorption, distribution, and excretion of toxicants. *In* "Casarett and Doull's Toxicology. The Basic Science of Poisons" (J. Doull, C. D. Klaassen, and M. O. Amdur, eds.), pp. 28–55. Macmillan, New York.

Kleiber, M. (1961). "The Fire of Life." Wiley, New York.

Kudo, H., Oki, Y., and Minato, H. (1979). *Microtus* species as laboratory animals. I. Bacterial flora of the esophageal sac of *Microtus montebelli* fed different rations and its relationship to the cellulolytic bacteria. *Nippon Vet. Zootech. Coll. Res. Rep.* **28,** 13–19.

Kudo, H., Cheng, K.-J., Majak, W., Hall, J. W., Arai, T., Oki, Y., and Costerton, J. W. (1986). *In vitro* degradation of mimosine by microorganisms from the esophageal sac of voles (*Microtus arvalis*). *Can. J. Anim. Sci.* **66,** 547–551.

Lanigan, G. W., and Smith, L. W. (1970). Metabolism of pyrrolizidine alkaloids in the ovine rumen. I. Formation of 7α-hydroxy-1α-methyl-8α-pyrrolizidine from heliotrine and lasiocarpine. *Aust. J. Agric. Res.* **21,** 493–500.

Lincoln, D. E., Newton, T. S., Ehrlich, P. R., and Williams, K. S. (1982). Coevolution of the checkerspot butterfly *Euphydras chalcedona* and its leaf resin. *Oecologia* **52,** 216–223.

Lindroth, R. L. (1984). The role of plant chemical defenses in the nutritional ecology of voles. Ph.D. dissertation, Univ. of Illinois, Urbana-Champaign.

Lindroth, R. L. (1988). Mammalian herbivore–plant interactions. *In* "Plant-Animal Interactions." (W. G. Abrahamson, ed.). McGraw-Hill, New York. In press.

Lindroth, R. L., and Batzli, G. O. (1983). Detoxication of some naturally occurring phenolics by prairie voles: A rapid assay of glucuronidation metabolism. *Biochem. Syst. Ecol.* **11,** 405–409.

Lindroth, R. L., and Batzli, G. O. (1984a). Plant phenolics as chemical defenses: Effects of natural phenolics on survival and growth of prairie voles (*Microtus ochrogaster*). *J. Chem. Ecol.* **10,** 229–244.

Lindroth, R. L., and Batzli, G. O. (1984b). Food habits of the meadow vole (*Microtus pennsylvanicus*) in bluegrass and prairie habitats. *J. Mamm.* **65,** 600–606.

Lindroth, R. L., Batzli, G. O., and Avildsen, S. I. (1986a). *Lespedeza* phenolics and *Penstemon* alkaloids: Effects on digestion efficiencies and growth of voles. *J. Chem. Ecol.* **12,** 713–728.

Lindroth, R. L., Batzli, G. O., and Seigler, D. S. (1986b). Patterns in the phytochemistry of three prairie plants. *Biochem. Syst. Ecol.* **14,** 597–602.

McBee, R. H. (1971). Significance of intestinal microflora in herbivory. *Annu. Rev. Ecol. Syst.* **2,** 165–176.

Mack, R. N., and Thompson, J. N. (1982). Evolution in steppe with few large, hooved mammals. *Am. Nat.* **119,** 757–773.

McKey, D. B., Gartlan, J. S., Waterman, P. G., and Choo, G. M. (1981). Food selection by black colobus monkeys (*Colobus satanas*) in relation to plant chemistry. *Biol. J. Linn. Soc.* **16,** 115–146.

McNab, B. K. (1980). Food habits, energetics, and the population biology of mammals. *Am. Nat.* **116,** 106–124.

McNab, B. K. (1983). Ecological and behavioral consequences of adaptation to various food resources. *In* "Advances in the Study of Mammalian Behavior" (J. F. Eisenberg and D. G. Kleiman, eds.), pp. 664–697. Spec. Publ. 7, Am. Soc. Mamm.

McNab, B. K. (1986). The influence of food habits on the energetics of eutherian mammals. *Ecol. Monogr.* **56,** 1–19.

McNaughton, S. J. (1983). Physiological and ecological implications of herbivory. *Encycl. Plant Physiol. New Ser.* **12C,** 657–677.

McNaughton, S. J. (1984). Grazing lawns: Animals in herds, plant form, and coevolution. *Am. Nat.* **124,** 863–886.

McNaughton, S. J. (1985). Ecology of a grazing ecosystem: The Serengeti. *Ecol. Monogr.* **55,** 259–294.

Manuwoto, S., Scriber, J. M., Hsia, M. T., and Sunarjo, P. (1985). Antibiosis/antixenosis in tulip tree and quaking aspen leaves against the polyphagous southern armyworm, *Spodoptera eridania*. *Oecologia* **67,** 1–7.

Martin, J. S., and Martin, M. M. (1982). Tannin assays in ecological studies: Lack of correlation between phenolics, proanthocyanidins and protein-precipitating constituents in mature foliage of six oak species. *Oecologia* **54,** 205–211.

Martin, M. M., and Martin, J. S. (1984). Surfactants: Their role in preventing the precipitation of proteins by tannins in insect guts. *Oecologia* **61,** 342–345.

Mead, R. J., Oliver, A. J., King, D. R., and Hubach, P. H. (1985). The coevolutionary role of fluoroacetate in plant-animal interactions in Australia. *Oikos* **44,** 55–60.

Mehansho, H., Hagerman, A., Clements, S., Butler, L., Rogler, J., and Carlson, D. M. (1983). Modulation of proline-rich protein biosynthesis in rat parotid glands by sorghums with high tannin levels. *Proc. Natl. Acad. Sci. U.S.A.* **80,** 3948–3952.

Millburn, P. (1978). Biotransformation of xenobiotics by animals. *In* "Biochemical Aspects of Plant and Animal Coevolution" (J. B. Harborne, ed.), pp. 35–73. Academic Press, New York.

Mole, S., and Waterman, P. G. (1985). Tannins and proteolytic enzymes: Enzyme inhibition or substrate deprivation? *J. Chem. Ecol.* **11,** 1323–1332.

Morris, J. G., and Rogers, Q. R. (1982). Nutritionally related metabolic adaptations of carnivores and ruminants. *In* "Adaptations to Terrestial Environments" (N. S. Margaris, M. Arianoutsou-Faraggitaki, and R. J. Reiter, eds.), pp. 165–180. Plenum, New York.

Oh, H. K., Saki, T., Jones, M. B., and Longhurst, W. M. (1967). Effects of various essential oils isolated from Douglas-fir needles, upon sheep and deer rumen microbial activity. *Appl. Microbiol.* **15,** 777–784.

Palmerino, C. C., Rusiniak, K. W., and Garcia, J. (1980). Flavor-illness aversions: The peculiar roles of odor and taste in memory for poison. *Science* **208,** 753–755.

Palo, R. T. (1985). Chemical defense in birch. Inhibition of digestibility in ruminants by phenolic extracts. *Oecologia* **68,** 10–14.

Palo, R. T., Pehrson, Å., and Knutsson, P. G. (1983). Can birch phenolics be of importance in the defense against browsing vertebrates? *Finn. Game Res.* **41,** 75–80.

Partridge, L. (1981). Increased preferences for familiar foods in small mammals. *Anim. Behav.* **29,** 211–216.

Partridge, L., and Maclean, R. (1981). Effects of nutrition and peripheral stimuli on preferences for familiar foods in the bank vole. *Anim. Behav.* **29,** 217–220.

Picman, J., Picman, A. K., and Towers, G. H. N. (1982). Effects of the sesquiterpene lactone, helenin, on feeding rates and survival of the tundra redback vole, *Clethrionomys rutilus*. *Biochem. Syst. Ecol.* **10,** 269–273.

Pienaar, U. de V. (1968). Capture and immobilizing techniques currently employed in South African national parks and reserves. *IBP Handb.* (7), 132–144.

Reichardt, P. B., Bryant, J. P., Clausen, T. P., and Wieland, G. D. (1984). Defense of winter-dormant Alaska paper birch against snowshoe hares. *Oecologia* **65,** 58–69.

Reid, C. S. W. (1973). Limitations to the productivity of the herbage-fed ruminant that arise from the diet. *In* "Chemistry and Biochemistry of Herbage" (G. W. Butler and R. W. Bailey, eds.), Vol. 3, pp. 215–262. Academic Press, New York.

Renwick, A. G., and Williams, R. T. (1972). The fate of cyclamate in man and other species. *Biochem. J.* **129,** 869–879.

Richter, C. P. (1943). Total self regulatory functions in animals and human beings. *Harvey Lect. Ser.* **38,** 63–103.

Robbins, C. T. (1983). "Wildlife Feeding and Nutrition." Academic Press, New York.

Robbins, C. T., Mole, S., Hagerman, A. E., and Hanley, T. A. (1987a). Role of tannins in defending plants against ruminants: Reduction in dry matter digestion? *Ecology* **68,** 1606–1615.

Robbins, C. T., Hanley, T. A., Hagerman, A. E., Hjeljord, O., Baker, D. L., Schwartz, C. C., and Mautz, W. W. (1987b). Role of tannins in defending plants against ruminants: Reduction in protein availability. *Ecology* **68,** 98–107.

Rodman, J. E., and Chew, F. S. (1980). Phytochemical correlates of herbivory in a community of native and naturalized Cruciferae. *Biochem. Syst. Ecol.* **8,** 43–50.

Roze, U. (1985). How to select, climb, and eat a tree. *Nat. Hist.* **94,** 62–69.

Rozin, P. (1967). Specific aversions as a component of specific hungers. *J. Comp. Physiol. Psychol.* **64,** 237–242.

Rozin, P. (1969). Adaptive food sampling patterns in vitamin deficient rats. *J. Comp. Physiol. Psychol.* **69,** 126–132.

Rozin, P., and Kalat, J. W. (1971). Specific hungers and poison avoidance as adaptive specializations of learning. *Psychol. Rev.* **78,** 459–485.

Sanders, E. H., Gardner, P. D., Berger, P. J., and Negus, N. C. (1981). 6-Methoxybenzoxazolinone: A plant derivative that stimulates reproduction in *Microtus montanus*. *Science* **214,** 67–69.

Scheline, R. R. (1978). "Mammalian Metabolism of Plant Xenobiotics." Academic Press, New York.

Schultz, A. M. (1964). The nutrient recovery hypothesis for arctic microtine cycles. II. Ecosystem variables in relation to arctic microtine cycles. *In* "Grazing in Terrestrial and Marine Environments" (D. J. Crisp, ed.), pp. 57–68. Blackwell, Oxford.

Schwartz, C. C., Regelin, W. L., and Nagy, J. G. (1980). Deer preference for juniper forage and volatile oil treated foods. *J. Wildl. Manage.* **44,** 114–120.

Shenk, J. S., Wangsness, P. J., Leach, R. M., Gustine, D. L., Gobble, J. L., and Barnes, R. F. (1976). Relationship between beta-nitropropionic acid content of crown vetch and toxicity in nonruminant animals. *J. Anim. Sci.* **42,** 616–621.

Smith, G. S., Watkins, J. B., Thompson, T. N., Rozman, K., and Klaassen, C. D. (1984). Oxidative and conjugative metabolism of xenobiotics by livers of cattle, sheep, swine and rats. *J. Anim. Sci.* **58,** 386–395.

Smith, R. L., and Williams, R. T. (1966). Implication of the conjugation of drugs and other exogenous compounds. *In* "Glucuronic Acid, Free and Combined" (G. J. Dutton, ed.), pp. 457–491. Academic Press, New York.

Spatz, M., Smith, D. W. E., McDaniel, E. G., and Laqueur, G. L. (1967). Role of intestinal microorganisms in determining cycasin toxicity. *Proc. Soc. Exp. Biol. Med.* **124,** 691–697.

Tagari, H., Henis, Y., Tamir, M., and Volcani, R. (1965). Effect of carob pod extract on cellulolysis, proteolysis, deamination, and protein biosynthesis in an artificial rumen. *Appl. Microbiol.* **13,** 437–442.

Tahvanainen, J., Helle, E., Julkunen-Tiitto, R., and Lavola, A. (1985). Phenolic compounds of willow bark as deterrents against feeding by mountain hare. *Oecologia* **65,** 319–323.

Vaughan, T. A. (1972). "Mammalogy." Saunders, Philadelphia.

Vaughan, T. A., and Czaplewski, N. J. (1985). Reproduction in Stephen's woodrat: The wages of folivory. *J. Mamm.* **66,** 429–443.

Vickery, W. L. (1984). Optimal diet models and rodent food consumption. *Anim. Behav.* **32,** 340–348.

Westoby, M. (1974). An analysis of diet selection by large herbivores. *Am. Nat.* **108,** 290–301.

Zahorik, D. M. (1977). Associative and non-associative factors in learned food preferences. *In* "Learning Mechanisms in Food Selection" (L. M. Barker, M. R. Best, and M. Domjan, eds.), pp. 181–199. Baylor Univ. Press, Waco, Texas.

Zahorik, D. M., and Houpt, K. A. (1977). The concept of nutritional wisdom: Applicability of laboratory learning models to large herbivores. *In* "Learning Mechanisms in Food Selection" (L. M. Barker, M. R. Best, and M. Domjan, eds.), pp. 45–67. Baylor Univ. Press, Waco, Texas.

CHAPTER 15

Exaptation as an Alternative to Coevolution in the Cardenolide-Based Chemical Defense of Monarch Butterflies (*Danaus plexippus* L.) against Avian Predators

LINCOLN P. BROWER
Department of Zoology
University of Florida
Gainesville, Florida

CAROLYN J. NELSON
Pharmacy Department
University of Sydney
Sydney, Australia

JAMES N. SEIBER
Department of Environmental Toxicology
University of California
Davis, California

LINDA S. FINK
Department of Zoology
University of Florida
Gainesville, Florida

CALHOUN BOND
Department of Biology
University of North Carolina
Chapel Hill, North Carolina

I. Introduction: Background of the Study

A. Chemical Defense and the Coevolutionary Arms Race Paradigm

As originally pointed out by Brower (1970), the phenomenon of sequestration of defensive plant chemicals by insects provides an opportunity to investigate coevolutionary interactions among three trophic levels: plants, herbivores, and predators. Thus, many secondary plant chemicals that are toxic are thought to have evolved in response to natural selection exerted by herbivores. Because these chemicals are generally unpalatable (review in Brower, 1984), the plants would suffer less herbivory. Insects that then breached the plants' chemical defenses would have more food and fewer competitors, and consequently become more abundant prey for their own enemies. Since the larvae or nymphs of some of the breachers might not completely break down the ingested poisons in their guts, they could become chemically noxious to their predators. So viewed, this incipient chemical defense arises fortuitously as an evolutionary preadaptation (Simpson, 1953), that is, an exaptation (see Gould and Vrba, 1982). Refinements in the chemical defense and the evolution of aposematic (warning) coloration and behavior could follow. Selection might next increase the predators' ability to avoid the toxic compounds, for example by taste rejection, or result in the predators' insensitivity to their effects following ingestion. Thus prey and predator would enter another round of coevolutionary ploy and counterploy in a never-ending "arms race" (Dawkins and Krebs, 1979).

Coevolution involving species at different trophic levels has been substantially debated with little supporting evidence (Futuyma and Slatkin, 1983a,b). Here we explore possible coevolution of the monarch butterfly

(*Danaus plexippus* L.) as a chemically defended prey vis-à-vis its avian predators. We present new data on the seemingly adaptive distribution of poisonous compounds within the butterfly's body and relate these findings to patterns of experimental and naturally occurring predation.

B. Paradigm Tested: Milkweed Plants, Monarch Butterfly Herbivores, and Avian Predators

1. Milkweeds

In North America 108 species of milkweeds of the genus *Asclepias* (Asclepiadaceae) adaptively radiated in recent geological time (Woodson, 1954). These occur from Mexico northward throughout the United States to southern Canada as a group of morphologically, ecologically, and chemically diverse plants. Most contain steroid glycoside poisons known as cardenolides or cardiac glycosides, so named because of the toxic effects they have on vertebrate hearts (review in Roeske *et al.*, 1976). Within each milkweed species, the cardenolides occur as a specific array of chemically related compounds, which we have called the cardenolide fingerprint pattern (Brower *et al.*, 1984a,b). While the fingerprint of any one milkweed species is relatively invariant, the total amounts of cardenolide vary substantially among individual plants (Seiber *et al.*, 1983).

Certainly one major selective basis for the occurrence of cardenolides is that they provide milkweeds with chemical protection against herbivory. When ingested by vertebrates and some invertebrates, cardenolides affect the heart, elicit emesis, and cause various other sickening responses (Brower, 1984; Berenbaum and Miliczky, 1984). They also are very bitter tasting, a property that evokes universal gustatory rejection in vertebrates (review in Brower, 1984). Evidence that at least two *Asclepias* species that contain large amounts of cardenolide deter mammalian herbivores is adduced from observations that they remain virtually untouched in otherwise heavily grazed fields (Fig. 1).

2. Danaine Butterflies as Ecological Chemists

There are approximately 157 species of butterflies in the subfamily Danainae, and many of the 50 or so whose life histories are known feed on larval foodplants that contain cardenolides and probably other poisonous substances (Ackery and Vane-Wright, 1984). Danaines are "ecological chemists" (Brower and Brower, 1964; Brower, 1969; Brower and Huberth, 1977)—that is, their larvae sequester cardenolides from their foodplants, which make them unpalatable to vertebrate predators. Other purportedly defensive chemicals such as pyrrolizidine alkaloids are sequestered by adult Danaines, and chemically distinct cardiac poisons and highly odorous pyrazines may be synthesized and/or sequestered as well, but will not be considered further here (reviews in Brower, 1984;

Fig. 1. (a,b) A pasture in south Florida contains a lush growth of *Asclepias curassavica* that the cattle are completely avoiding. (Western Dade Co., Florida, December 3, 1981.) (c) An overgrazed pasture in the Coast Range of California in which cattle have similarly avoided an *Asclepias eriocarpa* plant. (Near Santa Ynez, California, June 8, 1974.) (Original Kodachromes by L. P. Brower.)

Ackery and Vane-Wright, 1984; Rothschild, 1985). The larvae of several Danaine species can pass the cardenolides through metamorphosis to the chrysalid, adult, and even to the eggs. Of all species studied, the monarch butterfly is the cardenolide sequesterer par excellence.

Associated with purported general Danaine unpalatability is a suite of aposematic characters in adults that includes contrasting and often bright color patterns and conspicuous behavior. Danaines also have tough integuments, which allow an individual to survive an attack long enough for the predator to sense the unpalatability and release the butterfly unharmed (Swynnerton, 1915, 1919; Carpenter and Ford, 1933; Edmunds, 1974; Huheey, 1984). Even though Danaines are aposematic prey (Ackery and Vane-Wright, 1984), they are within the handling range (Sherry and McDade, 1982) of avian predators as small as chickadees (*Parus sclateri* Kleinschmidt) to as large as Steller's jays (*Cyanocitta stelleri* (Gmelin) (Fink *et al.*, 1983).

3. Bird Predators and the Cardenolide Sorting Hypothesis

When monarch larvae feed on cardenolide-free milkweeds, the butterflies are palatable to blue jays, *Cyanocitta cristata bromia* Oberholser (Brower 1969, 1970; Brower and Glazier, 1975; Brower and Fink, 1985). The typical jay feeding pattern when offered a palatable monarch is to remove and discard the wings and legs and then to swallow the body whole, or to peck apart and eat the thorax, and then to swallow the abdomen. On the supposition that the cardenolides might be differentially distributed as a refined antipredator adaptation, Brower and Glazier (1975) compared gross concentrations and amounts of cardenolides in the wings, thoraces, and abdomens of monarchs reared on the very toxic *Asclepias curassavica* L. Their results, together with those of Parsons (1965) and Nishio (1980), indicated that cardenolide concentrations are highest in the wings, intermediate in the abdomens, and least in the thoraces. By force-feeding measured amounts of each body part to the jays, Brower and Glazier also determined that the cardenolides in the abdomen had about three to five times the emetic potency of those in the thorax or wings. (A cardenolide with a higher emetic potency than another requires a smaller dose to cause emesis.)

Brower and Glazier hypothesized that the higher concentration of cardenolides in the wings has been selected through evolutionary time to facilitate gustatory rejection without harm to the butterfly. They then argued that because the abdomen is the most likely part of the body to be entirely consumed, its intermediate concentration of more emetic cardenolides was an evolved trait to facilitate emesis. Thus if a bird did kill a butterfly, it would reject the least emetic parts by discarding the wings, and then eat the most emetic body part, thereby maximizing the sickening effects. Because of the severe punishment, predators would more rapidly learn to avoid monarchs in subsequent encounters. As discussed elsewhere, the evolution of this trait in monarchs would not be of individual advantage and so would require kin selection (Brower, 1984).

To provide a mechanistic explanation of the emetic potency differences found by Brower and Glazier in the different body parts, we now invoke three lines of evidence. First, monarchs selectively metabolize and store only some of the array of cardenolides present in their foodplants (Reichstein *et al.*, 1968; Roeske *et al.*, 1976; Seiber *et al.*, 1980; Brower *et al.*, 1982, 1984a,b; Marty and Krieger, 1984). Second, preliminary evidence suggested that monarchs store cardenolides of different polarity in their wings, abdomens, and thoraces (Roeske *et al.*, 1976). Third, among a limited number of cardenolides tested, those of greater polarity tend to have lower emetic potencies (review in Duffey, 1977). Thus selective uptake and subsequent sorting of the more polar, less emetic cardenolides into the wings, and of the less polar, more emetic ones into the abdomen would explain the very different emetic potencies of the two body parts. Also, perhaps the polar cardenolides, which are more water-soluble, could be tasted at lower concentrations than the nonpolar cardenolides. This supposition was supported in feeding experiments using blue jays and three purified cardenolides in the digitoxin series (Rothkopf, 1976, as summarized in Brower and Fink, 1985).

4. Birds Break the Cardenolide Chemical Defense Barrier

Monarch butterflies that overwinter along the California coast are largely avoided by birds (Brower and Fink, 1985; Dayton and Bell-Dayton, 1985). In contrast, extensive predation occurs in overwintering areas in Mexico (Calvert *et al.*, 1979). During the 1978–1979 overwintering season in one Mexican colony, birds were estimated to have killed more than 2 million butterflies, about 9% of the colony (Brower and Calvert, 1985). Most predation in Mexico is by the black-headed grosbeak *Pheucticus melanocephalus* (Swainson) and the black-backed oriole *Icterus galbula abeillei* (Lesson), which make daily feeding forays into the colonies in mixed flocks of up to 80 birds. Stomach content analyses indicate monarchs as the major food of both birds (Fink *et al.*, 1983).

Fink and Brower (1981) determined that the grosbeaks exhibit emetic insensitivity to large amounts of cardenolide and eat monarchs indiscriminately. They snap off one or more wings with their forceful bills (Lederer, 1975) and then either swallow the entire body or peck off and swallow the abdomen (Fig. 2, a and b). In contrast, the orioles are sensitive to cardenolides, but, by utilizing taste discrimination, they avoid ingesting emetic dosages. The way in which the orioles feed on monarchs also differs from the grosbeaks: they use their fine bills to pry open the cuticle and eat only the contents of the abdomen (Fig. 2c and Fig. 3a–c) and the thoracic muscles (Fig. 3d–f). If, as in larvae (Thomashow, 1975), cardenolides in the adult monarch cuticle occur in higher concentration than in the body contents, then the orioles' mode of feeding would substantially reduce the amount of cardenolide ingested. Thus the differential distribution of cardenolides within the various body parts of adult monarchs might be even more refined than originally documented by Brower and Glazier (1975).

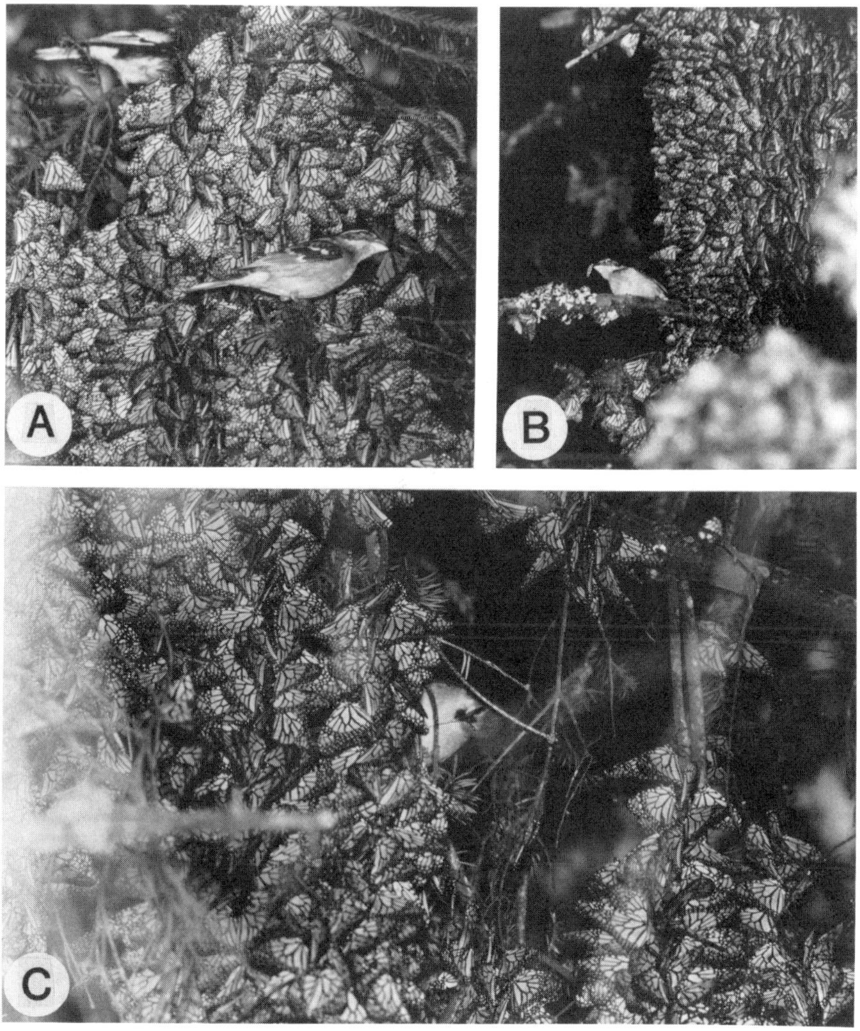

Fig. 2. Methods of eating monarch butterflies by black-headed grosbeaks and black-backed orioles at the Site Alpha overwintering site in Michoacan, Mexico, during mid-January, 1979 and 1980. (a) Part of a large mixed flock, a male oriole (top left) and a female grosbeak (center) feed on the clustering monarchs. The grosbeak has seized a butterfly and will shortly snap off the wings before eating the body. (b) Another female grosbeak has discarded a monarch's wings and thorax and is about to swallow the abdomen. (c) A female oriole has pierced the abdomen and is pulling out and ingesting the soft abdominal contents. (Original Kodachromes by L. P. Brower.)

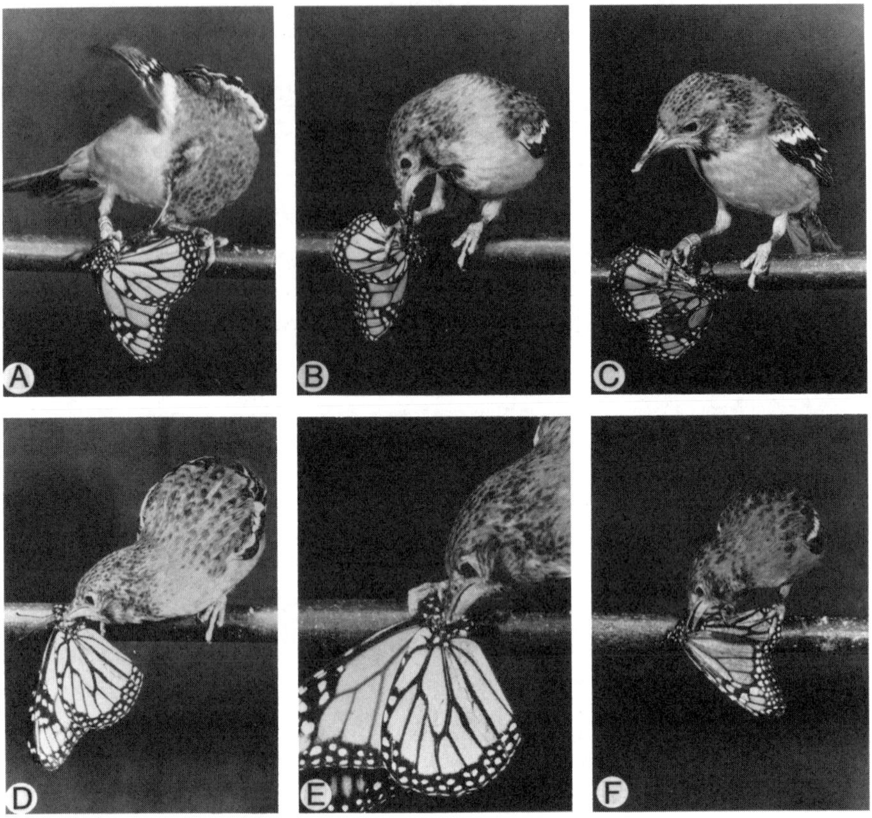

Fig. 3. A laboratory experiment shows how a Northern oriole has adapted its gaping and piercing behavior to feed upon the internal tissues of monarch butterflies. (a) Holding the monarch with its foot, the oriole uses its fine bill to pierce the abdominal cuticle. (b) Having split the abdominal cuticle by forcefully gaping its bill, the bird withdraws the soft abdominal contents with its brush-like tongue. (c) A bit of body contents is manipulated. (d,e) The oriole pierces the thorax and (f) gapes to split the tough cuticle, thereby making the muscle tissue accessible. The oriole was collected during the fall migration in 1979 at Manomet Bird Observatory along the east coast of Massachusetts, and the monarchs were reared on a milkweed lacking cardenolides (*Gonolobus rostratus*) and so were palatable. (Original photos by L. P. Brower.)

II. Purpose of the Present Study

We pose four evolutionary questions based on the above review of the milkweed–monarch–bird paradigm. Has arms race coevolution between the birds and butterflies really been the selective basis for the differential distribution of cardenolides in the monarch's cuticle and body? Are cardenolides distributed in a manner that enhances the probability of taste rejection by birds? Has the oriole's mode of feeding specifically coevolved to breach the cardenolide armaments in the monarch's cuticle? And, finally, is the grosbeak's substantial insensitivity to the emetic effects of cardenolides a direct coevolutionary response to these chemicals?

With respect to the butterflies, an alternative hypothesis is that the high cuticular concentrations of cardenolides and differing emetic potencies of the various body parts did not evolve directly in response to predation or parasitism, but rather are exaptations resulting from biochemical deposition in the development of insect cuticle. With respect to the birds, an alternative hypothesis is that their phylogenetically determined modes of feeding and tolerance to toxins exapted them to prey on chemically defended monarchs.

III. Methods and Materials

A. Rationale of the Cardenolide Analyses

To assure the generality of differential distribution of cardenolides in the cuticle and various body parts of the monarch, we compared butterflies reared on *A. syriaca* L. and *A. curassavica* L. We chose these plants for three reasons. First, *A. syriaca* is the principal foodplant of the eastern North American population that migrates to overwintering sites in Mexico (Seiber *et al.*, 1986, Malcolm *et al.*, 1988), while *A. curassavica* appears the main foodplant for nonmigratory populations throughout the New World tropics (Woodson, 1954; Urquhart, 1960; Brower, 1985). Second, the two plants contain chemically distinct cardenolides. While *A. syriaca* is characterised by generally more polar cardenolides, including aspecioside, syriobioside, and those related to syrioside (Brown *et al.*, 1979; Seiber *et al.*, 1985, 1986), *A. curassavica* contains less polar cardenolides of the calotropin series (review in Seiber *et al.*, 1983). Third, monarchs—because they sequester the respective sets of cardenolides—differ in their emetic potencies to birds. Those caught in the wild in Massachusetts were weakly emetic (Brower and Moffitt, 1974; Fink and Brower, 1981; Brower and Fink, 1985) and undoubtedly had fed as larvae on *A. syriaca* (cf. Fig. 7 in Roeske *et al.*, 1976, with Fig. 4b in this chapter, and Seiber *et al.*, 1986). In contrast, those reared on *A. curassavica* were strongly emetic (Brower *et al.*, 1968; Roeske *et al.*, 1976).

B. Origin of the Butterfly Material

The *A. curassavica*–reared monarchs were the second generation of butterflies obtained from Michoacan, Mexico, on January 20, 1980 (colɔny 14, Calvert and Brower, 1986). They were raised from egg to maturity ın a constant-environment room at 78°F under 16 h light/8 h dark on potted plants originating from Trinidad, W.I. [This stock was infected with the neogregarine protozooan *Ophryocystis elektroscirrha*, a parasite that sporulates between the epidermis and cuticle (McLaughlin and Myers, 1970) and might possibly have affected cardenolide distribution.]

The *A. syriaca*–reared monarchs were collected in the wild as mature larvae in western Massachusetts (Hampshire and Franklin Counties) in September 1980, fed on leaves from the same fields, and were shipped as chrysalids to the University of Florida where they hatched. They were not infected with the neogregarine.

C. Dissection Methodology

Within 24–36 h of emergence, the butterflies were placed individually in glassine envelopes and frozen on dry ice. Without allowing the material to thaw, the heads and legs were clipped off and discarded. The four wings, thorax, and abdomen were then cut from each butterfly and placed in separate aluminum weighing pans on dry ice. In preliminary experiments, the thoracic muscles and abdominal contents were removed by conventional dissection of freshly thawed butterflies with microsurgical instruments (Bond, 1979). Complete separation of the internal tissues from the cuticle proved virtually impossible, and necessitated novel techniques.

The thoraces, immediately after removal from the dry ice chamber, were lyophilized (freeze-dried) in a 500-ml bottle for 20–24 h. They were then placed in a wax-lined petri dish under a dissecting microscope and the dorsal exoskeletal plates were excised with a microsurgical scapel. The dried muscle tissue easily lifted out as large intact blocks with watchmaker's forceps, leaving the dry cuticle behind. The muscles were put in one preweighed aluminum pan and the cuticle in another, and each pan was stored in an evacuated desiccator over silica gel until weighed, extracted, and assayed.

Freeze-drying did not facilitate removal of the abdominal contents because they stuck to the exoskeletal shell. Instead, each abdomen was removed from dry ice to a preweighed aluminum drying pan and slit laterally with a scalpel. As the tissues began to melt, the cuticle was peeled off with forceps. As melting progressed, the pan was put back on the dry ice block to refreeze the material, and the process repeated until all the cuticle was stripped away. The advantage of this method is that lumps of hemolymph and body fluids can be dissected before melting and kept with the rest of the abdominal contents without being lost as flowage across the cuticle. Our dissection methodology improved with time so that cuticle from the abdomens was cleaner in the *A. syriaca* butterflies. The abdominal cuticle and contents, the wings, and undissected whole body parts were dried in a forced draft oven at 60°C and stored in the desiccator until weighed, extracted, and assayed.

D. Cardenolide Analyses

Cardenolide concentrations as micrograms per 0.1 g dry weight of material were determined by spectroassay, and the total amount of cardenolide was calculated by multiplying the concentration times the weight of the material (methods in Brower *et al.*, 1975). We did not extract lipids, and when pooled materials weighed substantially more than 0.2 g, they were extracted in 25-ml rather than 5- or 10-ml volumetric flasks.

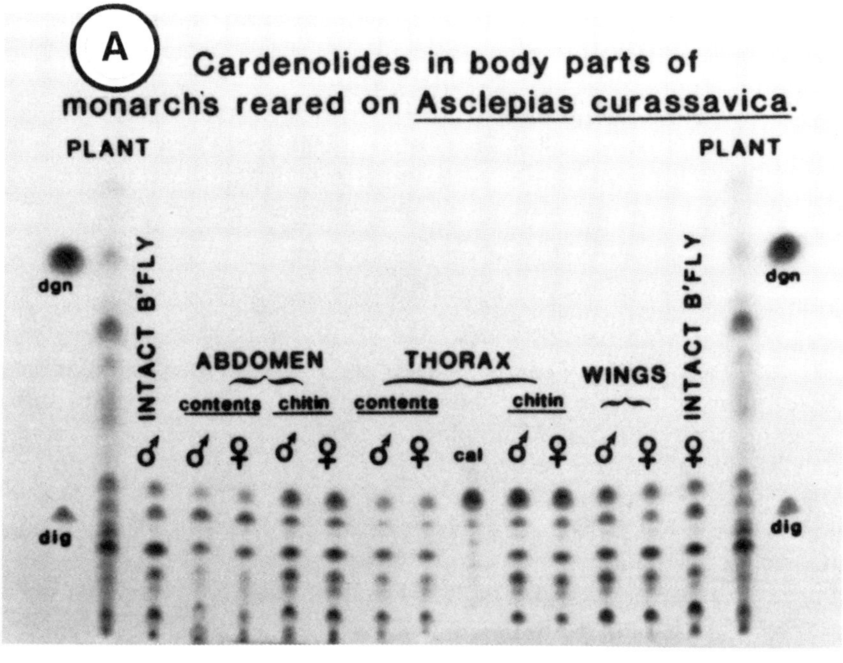

Fig. 4. Thin-layer silica-gel chromatograms showing the array of cardenolides present in the plant leaves, intact butterflies, the cuticle (= "chitin" in figure), the contents of the abdomen, the contents of the thorax, and in the wings of male and female monarch butterflies reared on (a) *Asclepias curassavica* and (b) *A. syriaca*. Cardenolide standards on the chromatograms include 20 μg each of digitoxin ("dig"), digitoxigenin ("dgn"), and an impure mixture of calactin and calotropin ("cal"). The two very different arrays of cardenolides stored by the butterflies correspond to those in the lower R_f range from the respective plants on which they were reared. Note the virtual lack of differential storage of cardenolides in any of the butterfly samples on each plant. See text for chemical and other details.

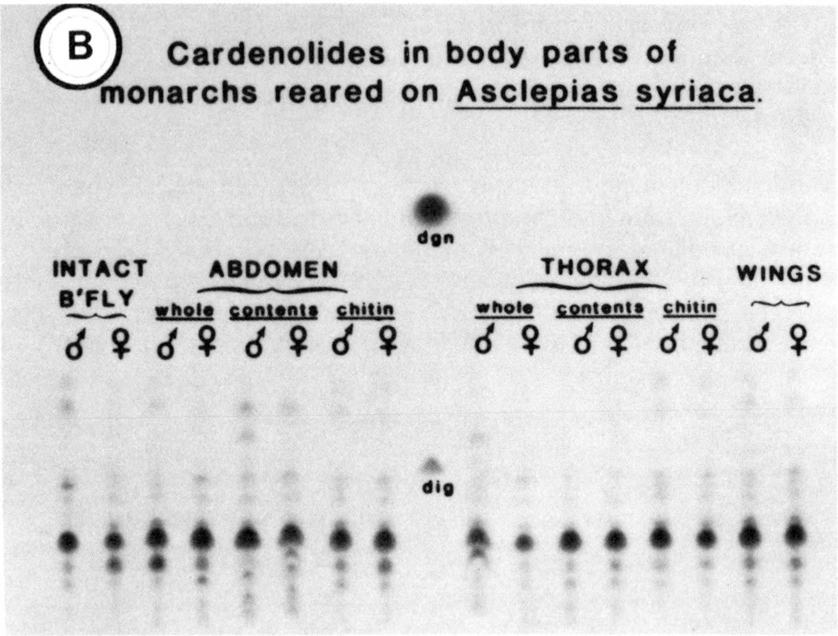

Fig. 4. (*continued*)

We analyzed five body parts: wings, abdominal contents, abdominal cuticle, thoracic muscles, and thoracic cuticle. In order to obtain sufficient cuticular material, cardenolide assays on the *A. curassavica*–reared butterflies were done on pooled material from four individuals, with one pool of females and two pools of males (a total of 15 assays). The *A. syriaca*–reared butterflies store less cardenolide, so we included six butterflies per pool and ran two pools per sex for the same five body parts (a total of 20 assays). We also analyzed intact monarchs reared on each plant, as well as whole thoraces and abdomens for the *A. syriaca*–reared material.

The lower-polarity cardenolides in the residual ethanol extracts of the *A. curassavica* material were prepared for assay by thin-layer chromatography (TLC) as in Brower *et al.* (1982), and the higher-polarity cardenolides in the *A. syriaca* material were run in a modified TLC procedure (Brower *et al.*, 1984b). Fourteen extracts of the pooled material for the *A. curassavica*–reared monarchs were chromatographed along with two extracts of plant leaves as shown in Fig. 4a. Sixteen extracts of the pooled material for the various body parts of the *A. syriaca* butterflies were chromatographed as shown in Fig. 4b.

E. Data and Statistical Analyses

Table I presents the data for each pooled sample for the monarchs reared on thc two milkweed species as dry weights, cardenolide concentrations, and total cardenolide for each of the five body parts. Differences in dry weight, cardenolide concentration, and total cardenolide among the five body parts for butterflies reared on the two plants were statistically analyzed by two-way analysis of variance (ANOVA, GLM) and least-square means (SAS, 1982, 1986). This gave 75 main and interaction contrasts (Table II). We did not analyze the effects of either group or sex. For

Table I. Weights and Cardenolide Contents for the Dissected Body Parts of Adult Monarch Butterflies Reared on Two Milkweeds

Body Part and Group	Dry Weight (mg)		Cardenolide Concentration (μg/0.1g)		Cardenolide Total Amount (μg)	
	cur	syr	cur	syr	cur	syr
All four wings						
Male group 1	39.3	42.3	486	417	191	176
Male group 2	42.1	45.5	409	350	172	159
Female group 1	43.6	46.6	504	297	220	138
Female group 2	—	46.2	—	230	—	106
Abdomen cuticle						
Male group 1	21.3	11.0	348	835	74	92
Male group 2	22.1	12.6	266	668	59	84
Female group 1	20.3	11.2	366	881	74	98
Female group 2	—	11.1	—	815	—	90
Abdomen content						
Male group 1	58.8	67.1	214	59	126	40
Male group 2	66.6	73.0	162	54	108	40
Female group 1	63.9	64.8	219	53	140	34
Female group 2	—	74.8	—	46	—	34
Thorax cuticle						
Male group 1	19.8	22.0	257	203	51	45
Male group 2	21.7	22.2	249	182	54	41
Female group 1	18.7	21.0	290	182	54	38
Female group 2	—	22.0	—	188	—	41
Thorax content						
Male group 1	33.4	42.4	88	106	29	45
Male group 2	32.9	43.6	92	85	30	37
Female group 1	28.0	39.1	79	78	22	31
Female group 2	—	41.1	—	80	—	33

Notes: The value for each group was determined on pooled body parts from four dissected individuals reared on *Asclepias curassavica* (cur) and six individuals reared on *A. syriaca* (syr). The grand means are in Table III.

Table II. Summary of Two-Way Analyses of Variance and Least-Square Mean Contrasts of the Data in Table I

A. Two-Way ANOVA Results

Source	df	SS	MS	*F*	Pr > *F*
Dry weights of the five body parts on the two plants					
Model	9	12,023	1335.8	211.9	0.0001
Error	25	158	6.3		
Corrected total	34	12,180			
Plant	1	53		8.4	0.0078
Body part	4	10,946		434.2	0.0001
Interaction	4	393		15.5	0.0001
Cardenolide concentrations of the five body parts on the two plants					
Model	9	1,719,988	191,110	81.0	0.0001
Error	25	59,001	2360		
Corrected total	34	1,778,989			
Plant	1	4092		1.7	0.1998
Part	4	1,079,071		114.3	0.0001
Interaction	4	460,817		48.8	0.0001
Total cardenolide in the five body parts on the two plants					
Model	9	94,403	10,489	53.7	0.0001
Error	25	4880	195		
Total	34	99,283			
Plant	1	4734		24.3	0.0001
Part	4	78,241		100.2	0.0001
Interaction	4	13,877		17.8	0.0001

B. Least-Square Mean Contrasts

Between Plants	Dry Weight		Concentration		Total Cardenolide	
Wings	0.0813[a]		0.0007		0.0001	
Abdomen cuticle	0.0001		0.0001		0.0498	
Abdomen contents	0.0015		0.0006		0.0001	
Thorax cuticle	0.0375		0.0495		0.2813[a]	
Thorax contents	0.0001		0.9805[a]		0.3818[a]	
Within Plants	**Ac**	**As**	**Ac**	**As**	**Ac**	**As**
Wings versus						
Abd cut[b]	0.0001	0.0001	0.0017	0.0001	0.0001	0.0001
Abd con	0.0001	0.0001	0.0001	0.0001	0.0001	0.0001
Tho cut	0.0001	0.0001	0.0001	0.0006	0.0001	0.0001
Tho con	0.0001	0.0534[a]	0.0001	0.0001	0.0001	0.0001
Abd cut versus						
Abd con	0.0001	0.0001	0.0034	0.0001	0.0001	0.0001
Tho cut	0.5743[a]	0.0001	0.1346[a]	0.0001	0.1730[a]	0.0001
Tho con	0.0001	0.0001	0.0001	0.0001	0.0011	0.0001
Abd con versus						
Tho cut	0.0001	0.0001	0.1036[a]	0.0006	0.0001	0.6707[a]
Tho con	0.0001	0.0001	0.0092	0.3283[a]	0.0001	0.9600[a]
Tho cut versus						
Tho con	0.0001	0.0001	0.0001	0.0067	0.0315	0.6348[a]

[a] Pr values not significantly different.

[b] Abbreviations: cut, cuticle; con, contents; Ac, *A. curassavica;* As, *A. syriaca;* abd, abdomen; tho, thorax.

ease of comparison, the data are summarized as grand means in Table III.

IV. Results

A. Dry Weights, Concentrations, and Amounts of Cardenolide in the Body Parts of Monarchs Reared on *A. syriaca* and *A. curassavica*

Mean values for six intact monarchs reared on A. *syriaca* (three males and three females) were: cardenolide concentration = 187 μg/0.1 g (SD = 63.7); weight = 0.178 g (SD = 0.024); and total cardenolide = 321 μg (SD = 66.1). Mean values for three intact monarchs reared on *A. curassavica* (two males and one female) were: concentration = 298 μg/0.1 g (SD = 10.8); weight = 0.220 g (SD = 0.026); and total cardenolide = 675 μg (SD = 80.7). Thus whole (intact) monarchs reared on *A. syriaca* had dry weights of 81%, cardenolide concentrations of 63%, and total cardenolide amounts of 48% of the respective values of intact monarchs reared on *A. curassavica*. Whereas the weight differences are largely artifacts of different rearing conditions, the cardenolide differences reflect the fact that monarchs seem better able to store cardenolides from *A. curassavica* than from *A. syriaca* (cf. Roeske *et al.*, 1976; Malcolm *et al.*, 1988.

The data for the dissected butterfly parts are in Table I, statistics for the 75 contrasts are in Table II, and the grand mean weights, concentrations, and total cardenolide contents are summarized in Table III. Interplant comparisons by body parts (Table IIA) indicate a lack of significant differences in only four of the 15 contrasts: dry weight of the wings, cardenolide concentrations of the thoracic contents, and total cardenolide in the thoracic cuticle and thoracic contents. The nearly fourfold difference in the concentrations in the abdomens suggests a greater ability of the fat body and/or the hemolymph of monarchs to store cardenolides from *A. curassavica*. The difference in cardenolide concentration in the abdominal cuticle when reared on *A. syriaca* is probably an artifact of the cleaner dissections.

Of the 40 intraplant means for cardenolide concentrations and amounts in the various butterfly parts, only seven contrasts are not significant (Table IIB). These results support and extend the prior Brower and Glazier (1975) findings: with one possible exception, not only the wings, but also the abdominal and thoracic cuticle have higher cardenolide concentrations than the thoracic muscles or abdominal contents. [In the *A. curassavica*–reared butterflies, the cardenolide concentration of the thoracic cuticle is not significantly higher ($p < 0.11$) than that of the abdominal contents.]

Table III. Mean Dry Weights and Cardenolide Contents of the Wings, Abdominal Cuticle and Contents, and Thoracic Cuticle and Contents of Adult Monarch Butterflies Reared on *Asclepias curassavica* and *A. syriaca*

	Dry Weight				Cardenolide					
					Concentration		Total Amount			
	mg		Percent[a]		μg/0.1 g		μg/part		Percent[a]	
Body Part	cur	syr	cur	syr	cur	syr	cur	syr	cur	syr
All four wings	42	45	24%	24%	466	324	194	145	41%	41%
Abdomen cuticle	21	11	12%	6%[b]	327	800[b]	69	91	15%	26%
Abdomen content	63	70	36%	37%	198	53	125	37	27%	10.5%
Thorax cuticle	20	22	11%	11%	265	189	53	41	11%	12%
Thorax content	31	41	17%	22%	86	87	27	37	6%	10.5%
Wings + cuticle	83	78	47%	41%	352	438	316	277	67%	79%
Content abdomen + thorax	94	111	53%	59%	142	70	152	74	33%	21%
Sum all parts	177	189	100%	100%	268	291	468	351	100%	100%

Note: The means are based on Table I.

[a] Percentages are rounded to the nearest whole value.

[b] This difference is probably exaggerated by the dissection methodology; see text.

The overall distribution of cardenolides derived from both plants has interesting implications for predation (Table III). The wings together with the abdominal and thoracic cuticle contain, respectively, 67% and 79% of the total cardenolide in the *A. curassavica*– and *A. syriaca*–reared material. In contrast, the entire body contents contain only 33% and 21% of the total. Thus by discarding the wings, predators avoid ingesting 41% of the cardenolide titer in monarchs reared on both milkweeds. If they also did not ingest the body cuticle, they could avoid 67–79% of the total cardenolide carried in the butterflies, while ingesting 53–59% of the useful biomass.

B. Cardenolide Thin-Layer Chromatography Profiles of the Body Parts of Monarchs Reared on *A. curassavica* and *A. syriaca*

Comparison of Figs. 4a and b indicates large differences in the TLC patterns of stored cardenolides in the butterflies reared on the two plants, in keeping with prior findings (Reichstein *et al.*, 1968; Roeske *et al.*, 1976; Seiber *et al.*, 1983, 1986). The two most nonpolar major cardenolides from *A. curassavica* are calactin and calotropin, which occur at or above the digitoxin reference. At least four other major, more polar cardenolides occur below digitoxin, probably including calotoxin (R_d = 0.66). In contrast, the *A. syriaca*–reared monarchs have a very different TLC pattern: they stored only small amounts of cardenolides above digitoxin, including desglucosyrioside at R_d = 1.3. Below digitoxin, syriobioside at R_d = 0.7 has a low concentration and aspecioside at R_d = 0.6 has the highest concentration. Several other cardenolides in low concentrations occur below aspecioside.

Comparison of the TLC profiles of the butterflies by sex or body parts indicates only very small within-plant caused differences (Fig. 4a and b). Thus while the butterflies sequester the characteristically different cardenolides from the two plant species, from each virtually the same set of cardenolides occurs in both sexes, the wings, cuticle, muscles, and the abdominal contents. The small qualitative differences seem insufficient to cause the various body parts to have different emetic potencies.

V. Discussion

A. Physiological and Biochemical Bases for High Cuticular Concentrations of Cardenolides

The results of the present study substantiate Brower and Glazier's (1975) findings that cardenolide concentrations are highest in the wings compared to the intact body parts and indicate that the differences result from the up to 15-fold higher concentration in the cuticle than in the soft body tissues (Table III).

It is very likely that the dynamic formation, structure, and biochemistry of the exoskeleton in insects (reviews in Neville, 1975; Hepburn, 1985; Blomquist and Dillwith, 1985; Kramer *et al.*, 1985) provide a basis on which exogenous chemicals circulating in the haemolymph can be deposited in the cuticle. During periodic molting, specialized epidermal cells become activated and produce an enzyme-rich moulting fluid, which digests most of the cuticle and its chitin. The fluid is then reabsorbed by the epidermis, leaving only a thin sclerotized outer layer of exocuticle and epicuticle to be shed as the exuvium. During the formation of the new endo-, exo-, and epicuticles, the epidermal cells remobilize both hydrophilic and hydrophobic substances from the hemolymph, and the ensuing sclerotization includes polymerization of lipids, proteins, and chitin. It is known that certain digested secondary plant chemicals present in the hemolymph can also be deposited by the epidermal cells into the newly forming cuticle (Scudder and Meredith, 1982; Bernays and Woodhead, 1982).

Increasing the concentrations of the chemicals in the cuticle above those in the hemolymph appears to occur (Nishio, 1980) and may involve epidermal pumping mechanisms (Wigglesworth, 1972). The selective basis for such pumps may originally have been the deposition of toxic endogenous waste products into the inert cuticle, which is shed at ecdysis. Wigglesworth (1972) referred to this as storage excretion. Once such a mechanism evolved, it could be subject to further natural selection due to efficacies in various other aspects of the insects' lives, that is, it could set the stage for exaptation. One undisputed exapted form of storage excretion involves the yellow and white pteridines that accumulate in the cuticular scales of Pierid butterflies. These pigments are responsible for the insects' wing coloration and play important roles both in thermoregulation (Kingsolver, 1985) and social behavior (Rutowski, 1982).

Exapted storage excretion could also involve digested secondary plant chemicals, which incidentally provided a defensive function, as originally implied by Filshie and Waterhouse (1969). Examples of exogenous defensive substances in insect cuticle that are consistent with exapted storage excretion include an epicuticular substance toxic to fungi (Koidsumi, 1957); high concentrations of cardenolides in the wings and larval, pupal, and adult cuticle of monarch butterflies (Parsons, 1965; Reichstein *et al.*, 1968; Brower and Glazier, 1975; Nishio, 1980; Blum, 1981, 1983, this volume) and in the larval cuticle of the arctiid moth *Cycnia inopinatus* Edwards (Nishio, 1980; Blum, 1983); cannabinol in the larval cuticle of the moth *Arctia caja* L. when fed *Cannabis sativa* L. (Rothschild *et al.*,

1977); and high concentration of pyrrolizidine alkaloids in the adult cuticle of two Ithomiid butterfly species (Brown, 1984).

The varying chemical structures of cardenolides probably serendipitously lend themselves to differential manipulation during development of the adult butterfly. The relatively high concentration in the abdominal cuticle compared to the soft abdominal tissues for the *Asclepias syriaca*-reared monarchs may result from the higher proportion of polar, hydroxylated cardenolides in this plant species compared to those in *A. curassavica*. Seiber *et al.* (1980) and Brower *et al.* (1982) speculated that cardenolide storage is dependent on the presence of free OH groups at C-3′ of the sugar moiety and/or C-3 of the steroid nucleus of *Asclepias* cardenolides. Because this study demonstrates highly concentrated storage predominately in the structural material of the wings, abdomen, and thorax, we can now extend our prior hypothesis to specific adsorption or covalent binding involving OH groups and chitin, protein, or other structural materials. Adsorption would appear to be the more likely mechanism because intact cardenolides can be extracted from the structural material with organic solvents.

We emphasize that our analyses were done on relatively fresh adults (2–3 days old), which might contain mobile cardenolides in soft tissue carried over from metamorphosis. Older adults may show an even more pronounced proportion of cardenolides in the cuticle than we found, although to date there is no evidence addressing this point. Unraveling the nature of the bound state of polar cardenolides in the structural matter of the butterflies is of crucial importance in determining not only why some are stored more effectively than others, but also how and to what extent the bound cardenolides are made less available as defensive agents towards predators.

B. The Avoidance of Self-Poisoning

Cardiac glycosides are cellular toxins because they bind with and inhibit (Na^+,K^+)-ATPase, which is generally responsible for alkali cation transport and the propagation of neuronal action potentials (Skou, 1957; Glynn, 1957). Vaughan and Jungreis (1977) assayed both ATPase and K^+ and Na^+ in the cerebral ganglia of monarchs and found the same mechanism for sustaining action potentials in the central nervous system as in many other animals. They then argued that in order to tolerate cardenolides, monarchs must (1) have insensitive (Na^+, K^+)-ATPase, (2) in some way biochemically protect the ATPase-rich nervous tissue from cardenolides, and (3) store the cardenolides at sites away from the neuronal tissue. They produced evidence for the first two mechanisms: thus monarch (Na^+, K^+)-ATPase not only has reduced sensitivity to cardenolides, but the nervous tissue is bathed in hemolymph containing high titres of K^+ ions, which, by analogy with Glynn's (1957) findings on ouabain (a highly polar cardenolide), they argued, must prevent binding of cardenolide to the nerve cells. Our study provides evidence for the third mechanism: binding of the cardenolides to the larval, pupal, and adult cuticle represents a protected storage site away from the neuronal tissue.

C. Lack of Differential Distribution of Cardenolides of Different Polarity in the Various Body Parts of Adult Monarchs

Our thin-layer chromatographs of monarchs reared on *A. curassavica* and *A. syriaca* do not support the earlier speculation that monarchs differentially sort and store cardenolides in different body parts. This finding is made even more significant because these two milkweeds have arrays of cardenolides differing substantially in their chemical structures (see above) and in the relative distributions of their polarities (cf. Fig. 4a and b). Interestingly, Moore and Scudder (1985) suggested a similar lack of differential tissue storage of cardenolides according to their polarities in *Oncopeltus fasciatus* Dallas.

D. An Alternative to the Cardenolide Polarity-Sorting Hypothesis

If virtually the same proportions of the different kinds of cardenolides occur in the wings, thorax, and abdomen of monarchs, how can we explain the differential emetic potencies of the three body parts? The most likely answer is that the cardenolides are so tightly bound to the cuticle that their bioavailability is reduced (even though they are still easily extracted in organic solvents). The soft body content/cuticle dry weight ratio was 1.71 for the thoraces, and 4.16 for the abdomens (from Table III). Thus, even if the whole thoraces and abdomens had identical concentrations and arrays of cardenolides, a higher proportion of the thoracic cardenolides would be tied up in cuticle, and the overall emeticity of the thorax would therefore be lower.

Clement's (1977) experiments with blue jays provided two further lines of evidence for binding of cardenolides in the cuticle, the first dealing with emetic potency and the second with taste discrimination. She offered seven jays fresh exuviae (shed skins) of fifth-instar monarch larvae that had been reared on *A. curassavica*. The mean dry weight of these skins was 5.86 mg and their cardenolide concentration was 2900 μg/0.1 g, the highest concentration in any monarch butterfly tissue (Thomashow, 1975; Nishio, 1980). Each skin therefore contained an average of 170 μg of cardenolide. The amount of cardenolide necessary to elicit emesis in a blue jay (i.e., the ED_{50}) for monarchs reared on *A. curassavica* is 70–74 μg (Brower and Glazier, 1975; Roeske *et al.*, 1976). Even though each skin contained more than enough poison to make two jays vomit, Clement's seven jays each ate one to two skins without sickening. She also found that the birds showed only weak gustatory discrimination between the exuviae containing these extraordinarily high cardenolide concentrations and the cardenolide-free exuviae (from larvae reared on *Gonolobus rostratus* R. Br.). Rothkopf (1976) (as summarized in Brower and Fink, 1985) determined that blue jays can in fact taste-discriminate among various pure cardenolides applied to bread pieces at concentrations ranging from 1.4 to 96 μg/0.1 g, that is, at least one-thirtieth the concentration of Clement's exuviae. Cuticular binding of the cardenolides is thus again indicated.

Two biological considerations further support the cuticular binding hypothesis. First, lipophilic and hydrophilic substances polymerize with chitin, resulting in the cuticle being notoriously resistant to chemical degradation by almost any biological substances other than chitinases (Hepburn, 1985; Kramer *et al.* 1985). Some birds do have chitinolytic enzymes (Jeuniaux, 1962; Ziswiler and Farner, 1972; LePrince *et al.*, 1979). Second, birds rapidly move food through their digestive systems (Welty, 1962; van der Pijl, 1972; Moermond, 1983). Presumably as a result of these two factors, cuticular insect remains are often sufficiently intact in predator feces to allow detailed taxonomic identification (Herrera, 1981; Jordano, 1981; Redford, 1984). In fact, Fink *et al.* (1983) reported large amounts of monarch and other identifiable arthropod remains in the stomachs of grosbeaks and orioles at the Mexican overwintering site.

We conclude that the cardenolide mixture stored in the monarch cuticle is so tightly bound that much is defecated before the birds can absorb an emetic dose. In contrast, the same cardenolides must be digested rapidly enough from the soft body parts to allow absorbtion of an emetic dosage.

Clement's (1977) data suggesting that the cuticle-bound cardenolides are poorly digested and only weakly tasted at high concentrations are difficult to reconcile with bird beak-mark theory (Brower, 1984). If cardenolides bound in the cuticle are to be effective cues for taste rejection by birds, obviously they must be tasteable. Clearly, the monarch's cardenolide-based defense would be much more effective as a gustatory deterrent if the poisons could be released in an unbound form, such as from glands as occurs in so many chemically defended insects (Eisner and Meinwald, 1966; Blum, 1981). Perhaps the wing scales—which are largely composed of cuticle—have a specialized microstructure or enzyme that is adapted to release the cardenolide upon tasting by vertebrates. The relationships among cardenolide structures, vertebrate taste thresholds, emetic potencies, and cuticular binding merit further investigation.

E. Orioles and Grosbeaks Appear Exapted to Feed on Monarchs

The ability of birds to open their bills forcefully against considerable resistance is called gaping and has been of paramount importance in the adaptive radiation of the Icteridae (Orians, 1985), the family to which orioles belong. Various icterid species utilize gaping to gain access to insects rolled up in leaves, inside stems, and in the ground (Orians, 1985). Northern orioles (*Icterus galbula* L.) pierce fruit with their fine bills, disrupt the pulp by gaping, and then ingest it without eating the skin (Forbush, 1929a,b; Brewster, 1937; Bent, 1965). The black-throated oriole [*Icterus gularis* (Wagler)] behaves similarly and uses a bifurcated and brushy tongue to gain access to the juice (Beecher, 1950). We have found that caged northern orioles perform virtually identical behavior in penetrating the cuticle and eating the soft internal contents of monarch butterflies (Fig. 3). As deduced from Table III, this behavior permits avoidance of 67–79% of the cardenolide. The utilization of gaping to avoid ingesting cardenolide indeed seems a classical demonstration of exaptation and clearly is not a coevolved adaptation.

We also suggest that the grosbeak's ability to ingest large amounts of cardenolide represents a pharmacological exaptation. Thus, earlier in its evolution the taxon to which the grosbeak belonged may well have developed abilities to deal with secondary plant chemicals in its diet, which includes fruits, seeds (including those from *Asclepias*, and *Stillingia* in the notoriously toxic Euphorbiaceae), buds, flowers, and arthropods (Bent, 1968). We speculate that black-headed grosbeaks have upwardly adjusted the amount of toxic secondary chemicals that they can tolerate. Two lines of evidence from our research support this: first, while large quantities of cardenolide did not produce a vomiting response when force-fed to the closely related rose-breasted grosbeak [*P. ludovicianus* (L.)], the birds nevertheless definitely showed symptoms of poisoning (Fink, 1980, Fink and Brower, 1981). Second, we determined that the flocks of grosbeaks and orioles at the overwintering site exhibited a striking 8-day feeding periodicity, which we interpreted as a sickening and recovery cycle (Brower and Calvert, 1985).

Finally, it is worth noting that our exaptation hypotheses are consistent with recent discussions of how various birds have come to feed on toxic food in general. Thus, Herrera (1982, 1985) proposed that the propensities and abilities of various frugivorous birds to eat "aposematic" toxic fruit predisposed them to attack aposematic insects. Moreover, Brown and Neto (1976) suggested that the neotropical tanager *Pipraeidea melanonota* (Vieillot) squeezes out and eats the abdominal contents of Ithomiid butterflies in order to avoid pyrrolizidine alkaloids present in the butterflies' cuticle. This tanager's mode of avoiding the toxins seems likewise to be exapted from the general food mandibulation behavior of the taxon to which it belongs (Moermond, 1983).

VI. Summary and Conclusions

The present study establishes that there is a high concentration of cardenolides in the cuticle of the monarch butterfly. Is this a coevolved response in the butterfly as a direct response to vertebrate predation to facilitate gustatory rejection? The answer seems to be no. High cuticular concentrations probably result because insect cuticle long ago evolved the capacity to serve as a chemical sink allowing excretion of ingested secondary plant chemicals and toxic metabolites via the exuviae during larval molting. When Danaines adapted to milkweeds as a larval food resource, the cardenolides probably were excreted to the cuticular chemical sink and incidentally, via exaptation, led to cardenolide taste rejection by birds.

We also conclude that the oriole's ability to circumvent large amounts of cardenolide by avoiding the cuticle and eating only the body contents did not evolve as a direct response to its feeding on monarch butterflies. Rather, this behavior seems a classical example of exaptation of the bill gaping and piercing behavior that characterizes the family Icteridae to which orioles belong.

Finally, the grosbeak's ability to ingest large amounts of cardenolide without vomiting is unlikely to represent a direct evolutionary response to monarch butterflies as a food resource. Rather, it is explained as a pharmacological exaptation of specializations that allow it to tolerate a variety of ingested toxic secondary plant chemicals.

Thus, what appear to be coevolutionary results of a stepwise arms race may at most represent diffuse coevolution (Futuyma and Slatkin, 1983a,b), or perhaps not coevolution at all. In reviewing community patterns among insects and plants, Strong *et al.* (1984) reached a similar conclusion: that the evidence for coevolution per se is generally weak. We do not deny that the cardenolide defenses and the bird behaviors are current valuable adaptations with respect to each other, but we conclude that their evolution did not result from direct stepwise reciprocal evolutionary responses of the birds to the butterflies, or of the butterflies to the birds.

In future research on the evolution of cardenolide defense, it is important to remember that cardenolide sequestering in Danaine butterflies varies substantially among species. Thus when reared on the same species of milkweeds, the monarch stores higher concentrations ($\sim 1.25\times$) of cardenolides than does the North American queen *Danaus gilippus* C., and about $2.5\times$ higher concentrations than the African queen, *D. chrysippus* L. (Brower *et al.*, 1975; Cohen, 1985; Rothschild *et al.*, 1975; Dixon *et al.*, 1978). Although these three species are members of the subtribe Danaina, the monarch belongs to the subgenus *Danaus* and the two queens are sister species of the related subgenus *Anosia* (Ackery and Vane-Wright, 1984). Other Danaine taxa appear unable to store cardenolides in the adult stage—for example, *Euploea core amymone* Cramer of the subtribe Euploeina—even when reared on the cardenolide-rich *Nerium oleander* L. (Malcolm and Rothschild, 1983). *Danaus affinis* F., *Tirumala hamata* Macleay, *Euploea tulliolus* Fabr., and *E. core* Cramer, captured in the wild during August 1983 in northeastern Australia, all had no or very small amounts of cardenolides (Brower *et al.*, 1988). Carefully selected comparative studies of milkweeds, Danaines, and birds will provide further insight into the evolution of cardenolides and how they have come to be manipulated, utilized, and metabolized in their transfer through the ecological food chain.

Acknowledgments

We are grateful to Tonya Van Hook, Susan C. Glazier, Barbara Tiffany, and Julia Frey for technical help, Drs. George Kidder and Stephen G. Zam for lyophilization, Kathleen Anderson of Manomet Bird Observatory for providing us with fall migrant orioles, and Fred Morrison of Northampton, Massachusetts, for collecting the monarchs on *A. syriaca*. S. B. Malcolm, Barbara Cockrell, James Anderson, Alan Masters, John Glendinning, and Carlos Martinez del Rio offered helpful comments. We thank Amherst College, the University of Florida, and the University of California, Davis, for contributions to our collaborative ecological chemistry program. The research was funded by National Science Foundation grants BSR-8119382 and BSR-8500416 with L. P. Brower as Principal Investigator.

References

Ackery, P. R., and Vane-Wright, R. I. (1984). "Milkweed Butterflies, Their Cladistics and Biology." British Museum (Natural History), London.

Beecher, W. J. (1950). Convergent evolution in the American orioles. *Wilson Bull.* **62,** 51–85.

Bent, A. C. (1965). "Life Histories of North American Blackbirds, Orioles, Tanagers, and Allies." Dover, New York.

Bent, A. C. (1968). "Life Histories of North American Cardinals, Grosbeaks, Buntings, Towhees, Finches, Sparrows, and Allies." Dover, New York.

Berenbaum, M., and Miliczky, E. (1984). Mantids and milkweed bugs: Efficacy of aposematic coloration against invertebrate predators. *Am. Midl. Nat.* **111,** 64–68.

Bernays, E. A., and Woodhead, S. (1982). Incorporation of dietary phenols into the cuticle in the tree locust *Anacridium melanorhodon*. *J. Insect Physiol.* **28,** 601–606.

Blomquist, G. J., and Dillwith, J. W. (1985). Cuticular lipids. *Compr. Insect Physiol. Biochem. Pharmacol.* **3,** 117–154.
Blum, M. S. (1981). "Chemical Defenses of Arthropods." Academic Press, New York.
Blum, M. S. (1983). Detoxification, deactivation, and utilization of plant compounds by insects. *Am. Chem. Soc. Symp. Ser.* **208,** 265–275.
Bond, C., Jr. (1979). Implications of avian predation on monarch butterflies in Mexico. Senior honors thesis, Amherst College, Amherst, Massachusetts.
Brewster, W. (1937). The birds of the Lake Umbagog region of Maine, pt. 3. *Bull Mus. Comp. Zool.* **66,** 408–521.
Brower, L. P. (1969). Ecological chemistry. *Sci. Am.* **220,** 22–29.
Brower, L. P. (1970). Plant poisons in a terrestrial food chain with implications for mimicry theory. *Proc. Annu. Biol. Colloq. Oregon State Univ.* **29,** 69–82.
Brower, L. P. (1984). Chemical defence in butterflies. *Symp. R. Entomol. Soc. London* **11,** 109–134.
Brower, L. P. (1985). New perspectives on the migration biology of the monarch butterfly, *Danaus plexippus* L. *Univ. Tex. Mar. Sci. Inst. Contrib. Mar. Sci.* **27** (Suppl.), 748–785.
Brower, L. P., and Brower, J. V. Z. (1964). Birds, butterflies, and plant poisons: A study in ecological chemistry. *Zoologica* **49,** 137–159.
Brower, L. P., and Calvert, W. H. (1985). Foraging dynamics of bird predators on overwintering monarch butterflies in Mexico. *Evolution* **39,** 852–868.
Brower, L. P., and Fink, L. S. (1985). A natural toxic defense system in butterflies versus birds. *Ann. N.Y. Acad. Sci.* **443,** 171–188.
Brower, L. P., and Glazier, S. C. (1975). Localization of heart poisons in the monarch butterfly. *Science* **188,** 19–25.
Brower, L. P., and Huberth, J. C. (1977). Strategy for survival: Behavioral ecology of the monarch butterfly. (30 minute film). Film 32460, Audiovisual Services, Pennsylvania State University, University Park, Pennsylvania 32460.
Brower, L. P., and Moffitt, C. M. (1974). Palatability dynamics of cardenolides in the monarch butterfly. *Nature (London)* **249,** 280–283.
Brower, L. P., Ryerson, W. N., Coppinger, L. L., and Glazier, S. C. (1968). Ecological chemistry and the palatability spectrum. *Science* **161,** 1349–1351.
Brower, L. P., Edmunds, M., and Moffitt, C. M. (1975). Cardenolide content and palatability of *Danaus chrysippus* butterflies from West Africa. *J. Entomol.* **A49,** 183–196.
Brower, L. P., Seiber, J. N., Nelson, C. J., Lynch, S. P., and Tuskes, P. M. (1982). Plant-determined variation in the cardenolide content, thin-layer chromatography profiles, and emetic potency of monarch butterflies, *Danaus plexippus* reared on the milkweed, *Asclepias eriocarpa* in California. *J. Chem. Ecol.* **8,** 579–633.
Brower, L. P., Seiber, J. N., Nelson, C. J., Lynch, S. P., and Holland, M. M. (1984a). Plant-determined variation in the cardenolide content, thin-layer chromatography profiles, and emetic potency of monarch butterflies, *Danaus plexippus* reared on milkweed plants in California. 2. *Asclepias californica. J. Chem. Ecol.* **10,** 601–639.
Brower, L. P., Seiber, J. N., Nelson, C. J., Lynch, S. P., Hoggard, M. P., and Cohen, J. A. (1984b). Plant-determined variation in the cardenolide content and

thin-layer chromatography profiles of monarch butterflies, *Danaus plexippus*, reared on milkweed plants in California. 3. *Asclepias californica*. *J. Chem. Ecol.* **10,** 1823–1857.

Brower, L. P., Malcolm, S. B., Cockrell, B. J., Edgar, J. S., Nelson, C. S., and Van Hook, T. (1988). In preparation.

Brown, K. S. (1984). Adult-obtained pyrrolizidine alkaloids defend Ithomiine butterflies against a spider predator. *Nature* (*London*) **309,** 707–709.

Brown, K. S. and Neto, J. V. (1976). Predation of aposematic Ithomiine butterflies by tanagers (*Pipraeidea melanonota*). *Biotropica* **8,** 136–141.

Brown, P., van Euw, J., Reichstein, T., Stockel, K., and Watson, T. R. (1979). Cardenolides of *Aslclepias syriaca* L., probable structure of syrioside and syriobioside. *Helv. Chem. Acta* **62,** 414–441.

Calvert, W. H., and Brower, L. P. (1986). The location of monarch butterfly (*Danaus plexippus* L.) overwintering colonies in Mexico in relation to topography and climate. *J. Lepid. Soc.*, **40,** 164–187.

Calvert, W. H., Hedrick, L. W., and Brower, L. P. (1979). Mortality of the monarch butterfly (*Danaus plexippus* L.): Avian predation at five overwintering sites in Mexico. *Science* **204,** 847–851.

Carpenter, G. D. H., and Ford, E. B. (1933). "Mimicry." Methuen, London.

Clement, C. A. (1977). Observations on the reaction of *Cyanocitta cristata bromia* Oberholser to *Danaus plexippus* Linnaeus: The skin flick. Senior honors thesis, Amherst College, Amherst, Massachusetts.

Cohen, J. A. (1985). Differences and similarities in cardenolide contents of queen and monarch butterflies in Florida and their ecological and evolutionary implications. *J. Chem. Ecol.* **11,** 85–103.

Dawkins, R., and Krebs, J. H. (1979). Arms races between and within species. *Proc. R. Soc. London Ser. B* **205,** 489–511.

Dayton, J. J., and Bell-Dayton, E. (1985). Monarch butterfly monitoring. Santa Cruz, California, 1983–1984. Report for Harvey Stanley Associates, Alviso, California.

Dixon, C. A., Erickson, J. M., Kellett, D. N., and Rothschild, M. (1978). Some adaptations between *Danaus plexippus* and its food plant, with notes on *Danaus chrysippus* and *Euploea core* (Insecta: Lepidoptera). *J. Zool. London* **185,** 437–467.

Duffey, S. S. (1977). Arthropod allomones: Chemical effronteries and antagonists. *Proc. Int. Congr. Entomol., 15th, 1976, Washington, D.C.* pp. 323–394.

Edmunds, M. E. (1974). Significance of beak marks on butterfly wings. *Oikos* **25,** 117–118.

Eisner, T., and Meinwald, J. (1966). Defensive secretions in arthropods. *Science* **153,** 1341–1350.

Filshie, B. K., and Waterhouse, D. F. (1969). The structure and development of a surface pattern on the cuticle of the green vegetable bug *Nezara viridula*. *Tissue Cell* **1,** 367–385.

Fink, L. S. (1980). Bird predation on overwintering monarch butterflies. Senior honors thesis, Amherst College, Amherst, Massachusetts.

Fink, L. S., and Brower, L. P. (1981). Birds can overcome the cardenolide defence of monarch butterflies in Mexico. *Nature* (*London*) **291,** 67–70.

Fink, L. S., Brower, L. P., Waide, R. B., and Spitzer, P. R. (1983). Overwintering monarch butterflies as food for insectivorous birds in Mexico. *Biotropica* **15,** 151–153.

Forbush, E. H. (1929a). "Birds of Massachusetts and Other New England States, Part II. Land Birds from Bob-whites to Grackles." Norwood Press, Norwood, Massachusetts.

Forbush, E. H. (1929b). "Birds of Massachusetts and Other New England States, Part III. Land Birds from Sparrows to Thrushes." Norwood Press, Norwood, Massachusetts.

Futuyma, D. J., and Slatkin, M. (1983a). Introduction. *In* "Coevolution" (D. J. Futuyma and M. Slatkin, eds.), pp. 1–13. Sinauer, Sunderland, Massachusetts.

Futuyma, D. J., and Slatkin, M. (1983b). The study of coevolution. *In* "Coevolution" (D. J. Futuyma and M. Slatkin, eds.), pp. 459–464. Sinauer, Sunderland, Massachusetts.

Glynn, I. M. (1957). The action of cardiac glycosides on sodium and potassium movements in human red cells. *J. Physiol.* (*London*) **136,** 148–173.

Gould, S. J., and Vrba, E. S. (1982). Exaptation—A missing term in the science of form. *Paleobiology* **8,** 4–15.

Hepburn, H. R. (1985). Structure of the integument. *Compr. Insect Physiol. Biochem. Pharmacol.* **3,** 1–58.

Herrera, J. C. (1981). Fruit of robins wintering in southern Spanish Mediterranean scrubland. *Bird Study* **28,** 115–122.

Herrera, J. C. (1982). Defense of ripe fruits from pests: Its significance in relation to plant-disperser interactions. *Am. Nat.* **120,** 218–241.

Herrera, J. C. (1985). Aposematic insects as six-legged fruits: Incidental short-circuiting of their defense by frugivorous birds. *Am. Nat.* **126,** 286–293.

Huheey, J. E. (1984). Warning coloration and mimicry. *In* "Chemical Ecology of Insects" (W. J. Bell and R. T. Carde, eds.), pp. 257–297. Sinauer, Sunderland, Massachusetts.

Jeuniaux, C. (1962). Digestion de la chitine chez les oiseaux et les mammifères. *Ann. Soc. Zool. Belg.* **92,** 27–45.

Jordano, P. (1981). Alimentacion y relaciones troficas entre los passeriformes en paso otonal por una localidad de Andalucia central. *Donana Acta Vertebr.* **8,** 103–124.

Kingsolver, J. G. (1985). Butterfly thermoregulation: Organismic mechanisms and population consequences. *J. Lepid. Soc.* **24,** 1–20.

Koidsumi, K. (1957). Antifungal action of cuticular lipids in insects. *J. Insect Physiol.* **1,** 40–51.

Kramer, K. J., Dziadik-Turner, C., and Koga, D. (1985). Chitin metabolism in insects. *Compr. Insect Physiol. Biochem. Pharmacol.* **3,** 75–115.

Lederer, R. J. (1975). Bill size, food size, and jaw force of insectivorous birds. *Auk* **92,** 385–387.

LePrince, P., Dandrifosse, G., and Schoffeniels, E. (1979). The digestive enzymes and acidity of the pellets regurgitated by raptors. *Biochem. Syst. Ecol.* **7,** 223–227.

McLaughlin, R. E., and Myers, J. (1970). *Ophryocystis elektroscirrha* sp. n., a neogregarine pathogen of the monarch butterfly *Danaus plexippus* (L.) and the Florida queen butterfly *Danaus gilippus berenice* Cramer. *J. Protozool.* **17,** 300–305.

Malcolm, S., and Rothschild, M. (1983). A danaid mullerian mimic, *Euploea core amymone* (Cramer) lacking cardenolies in the pupal and adult stages. *Biol. J. Linn. Soc.* **19,** 27–33.
Malcolm, S. B., Cockrell, B. J., and Brower, L. P. (1988). The cardenolide fingerprint of butterflies reared on the common milkweed, *Asclepias syriaca* L. *J. Chem. Ecol.* In press.
Marty, M. A., and Krieger, R. I. (1984). Metabolism of uscharidin, a milkweed cardenolide, by tissue homogenates of monarch butterfly larvae, *Danaus plexippus. J. Chem. Ecol.* **10,** 945–956.
Moermond, T. C. (1983). Suction-drinking in tanagers (Thraupidae) and its relation to frugivory. *Ibis* **25,** 545–549.
Moore, L. V., and Scudder, G. G. E. (1985). Selective sequestration of milkweed *Asclepias* sp. cardenolides in *Oncopeltus fasciatus* (Dallas) (Hemiptera: Lygaeidae). *J. Chem. Ecol.* **11,** 667–687.
Neville, A. C. (1975). "Biology of the Arthropod Cuticle." Springer-Verlag, New York.
Nishio, S. (1980). The fates and adaptive significance of cardenolides sequestered by larvae of *Danaus plexippus* (L.) and *Cycnia inopinatus* (Hy. Edwards). Doctoral dissertation, University of Georgia, Athens, Georgia.
Orians, G. H. (1985). "Blackbirds of the Americas." Univ. of Washington Press, Seattle.
Parsons, J. A. (1965). A digitalis-like toxin in the monarch butterfly, *Danaus plexippus. J. Physiol.* (*London*) **178,** 290–304.
Redford, K. H. (1984). The nutritional value of invertebrates with emphasis on ants and termites as food for mammals. *J. Zool.* (*London*) **203,** 385–395.
Reichstein, T., vón Euw, J., Parsons, J. A., and Rothschild, M. (1968). Heart poisons in the monarch butterfly. *Science* **161,** 861–866.
Roeske, C. N., Seiber, J., Brower, L. P., and Moffitt, C. M. (1976). Milkweed cardenolides and their comparative processing by monarch butterflies (*Danaus plexippus* L.). *Recent Adv. Phytochem.* **10,** 93–167.
Rothkopf, D. M. (1976). The taste rejection and emetic potency of several digitalis series cardenolides in the blue jay, *Cyanocitta cristata bromia* Oberholser. Senior honors thesis, Amherst College, Amherst, Massachusetts.
Rothschild, M. (1985). British aposematic Lepidoptera. *In* "Moths and Butterflies of Great Britain and Ireland." Vol. 2. (J. Heath and A. Maitland Emmett, eds.), pp. 8–62. Harley Books, Colchester.
Rothschild, M., von Euw, T., Reichstein, T., Smith, D. A. S., and Pierre, J. (1975). Cardenolide storage in *Danaus chrysippus* (L.) with additional notes on *D. plexippus* (L.). *Proc. R. Soc. London Ser. B* **190,** 1–31.
Rothschild, M., Rowan, M. G., and Fairbairn, J. W. (1977). Storage of cannabinoids by *Arctia caja* and *Zonocerus elegans* fed on chemically distinct strains of *cannabis sativa. Nature* (*London*) **266,** 650–651.
Rutowski, R. L. (1982). Mate choice and lepidopteran mating behavior. *Fla. Entomol.* **65,** 72–82.
SAS. (1982). "SAS User's Guide: Statistics." SAS Institute, Cary, N.C.
SAS. (1986). "Statistical Program Release 5.16." Northeast Regional Data Center, University of Florida, Gainesville.
Scudder, G. G. E., and Meredith, J. (1982). Morphological basis of cardiac glyco-

side sequestration by *Oncopeltus fasciatus* (Dallas) (Hemiptera: Lygaeidae). *Zoomorphology* **99,** 87–101.

Seiber, J. N., Tuskes, P. M., Brower, L. P., and Nelson, C. J. (1980). Pharmacodynamics of some individual milkweed cardenolides fed to the larvae of the monarch butterfly (*Danaus plexippus*). *J. Chem. Ecol.* **6,** 321–339.

Seiber, J. N., Lee, S. M., and Benson, J. M. (1983). Cardiac glycosides (cardenolides) in species of *Asclepias* (Asclepiadaceae): *In* "Handbook of Natural Toxins 1: Plant and Fungal Toxins." (R. F. Keeler and A. T. Tu, eds.), pp. 43–83. Dekker, Amsterdam.

Seiber, J. N., Lee, S. M., McChesney, M. M., Watson, T. R., Nelson, C. J., and Brower, L. P. (1985). New cardiac glycosides (cardenolides) from *Asclepias* species. *Proc. Aust. U.S.A. Poison. Plants Symp., 2nd,* 1984. pp. 427–437.

Seiber, J. N., Brower, L. P., Lee, S. M., McChesney, M. M., Cheung, H. T. A., Nelson, C. J., and Watson, T. R. (1986). The cardenolide connection between overwintering monarch butterflies from Mexico and their larval foodplant, *Asclepias syriaca. J. Chem. Ecol.* **12,** 1157–1170.

Sherry, T. W., and McDade, L. A. (1982). Prey selection and handling in two neotropical hover-gleaning birds. *Ecology* **63,** 1016–1028.

Simpson, G. G. (1953). "The Major Features of Evolution." Columbia Univ. Press, New York.

Skou, J. C. (1957). The influence of some cations on an adenosine triphosphatase from peripheral nerves. *Biochim. Biophys. Acta* **23,** 394–401.

Strong, D. R., Lawton, J. H., and Southwood, R. (1984). "Insects on Plants: Community Patterns and Mechanisms." Harvard Univ. Press, Cambridge, Massachusetts.

Swynnerton, C. F. M. (1915). Birds in relation to their prey. Experiments on wood-hoopoes, small hornbills, and a babbler. *J. So. Afr. Ornithol. Union* **1915,** 22–108.

Swynnerton, C. F. M. (1919). Experiments and observations bearing on the explanation of form and colouring, 1908–1913, Africa. *Zool. J. Linn. Soc.* **33,** 203–385.

Thomashow, P. M. (1975). The paradox of the cryptic chrysalid. Division III Project, Hampshire College, Amherst, Massachusetts.

Urquhart, F. A. (1960). "The Monarch Butterfly." Univ. of Toronto Press, Toronto.

van der Pijl, L. (1972). "Principles of Seed Dispersal in Higher Plants," 2nd Ed. Springer-Verlag, New York.

Vaughan, G. L., and Jungreis, A. M. (1977). Insensitivity of lepidopteran tissue to ouabain: Physiological mechanisms for protection from cardiac glycosides. *J. Insect Physiol.* **23,** 585–589.

Welty, J. C. (1962). "The Life of Birds." Saunders, Philadelphia.

Wigglesworth, V. B. (1972). "The Principles of Insect Physiology," 7th Ed. Chapman & Hall, New York.

Woodson, R. E., Jr. (1954). The North American species of *Asclepias* L. *Ann. Mo. Bot. Garden* **41,** 1–211.

Ziswiler, V., and Farner, D. S. (1972). Digestion and the digestive system. *In* "Avian Biology" (D. S. Farner and J. R. King, eds.), Vol. 2, pp. 343–430. Academic Press, New York.

CHAPTER 16

Diet Breadth and Insect Chemical Defenses: A Generalist Grasshopper and General Hypotheses

CLIVE G. JONES
Institute of Ecosystem Studies
The New York Botanical Garden
Mary Flagler Cary Arboretum
Millbrook, New York

DOUGLAS W. WHITMAN
USDA-ARS
Insect Biology and Population
Management Research Laboratory
Tifton, Georgia

PETER J. SILK
Pheromone Research Group
New Brunswick Research and
Productivity Council
Fredericton, New Brunswick, Canada

MURRAY S. BLUM
Department of Entomology
University of Georgia
Athens, Georgia

I. Introduction

Many phytophagous insects use chemical defenses against predators. These insects obtain the chemicals in two different ways that have different implications for their host-plant relationships. Insects may make the compounds themselves, or compounds from the diet may be stored in the insect, with or without chemical modification. The evolution of these defenses and the ecological factors affecting quality, quantity, and defensive efficacy of such chemicals are of considerable interest. In this chapter we examine one aspect of chemical coevolution between insect and host plant—that of the relationship between diet breadth and the two different types of chemical defense.

A large number of insect herbivores use autogenous chemical defenses. The compounds range from cycloterpenoids and cardiac glycosides in some chrysomelids, aldehydes and alcohols in Heteroptera, cyanide and carboxylic acids in some Lepidoptera, and phenolics in grasshoppers (for reviews see Duffey, 1976; Blum, 1981; Pasteels *et al.*, 1983a). For these insects the quality, quantity, and efficacy of the defense is not directly related to the chemistry of the host plants. Since the defensive compounds are biosynthesized from dietary precursors, any host-plant effects on chemical defense would occur via the influence of the quality and quantity of dietary precursors, or through the effect of the host plants on the physiological status of the insect. There is little reason to suppose that the chemical defense systems of such insects should necessarily evolve to be tightly coupled to the chemistry of a given host-plant. The question of whether or not diet indirectly affects autogenous defenses has received little attention (Jones *et al.*, 1987). Here we show that experimental dietary specialization has a marked effect on the autogenous defense system of the generalist grasshopper, *Romalea guttata* Houttyun (Orthoptera; Romaleidae; = *R. microptera* Beauvois; Kevan, 1980). We suggest that selection of particular host-plants or mixed diets may play an important constraining role in, and may also be constrained by, the evolution of autogenous defenses in this and other phytophagous insects.

Another group of chemically defended insect herbivores sequesters or stores plant chemicals to defend themselves. Such insects include the monarch butterfly (Brower and Brower, 1964), some crysomelid beetles (Pasteels *et al.*, 1983b), arctiid moths (Rothschild *et al.*, 1970), and lygaeids (Scudder and Duffey, 1972), and involve chemicals as diverse as cannabinoids, allylisothiocyanates, cardenolides, alkaloids, and terpenes (for reviews see Rothschild, 1973; Duffey, 1980; Blum, 1981; Pasteels *et al.*, 1983a). The classic studies of Brower and co-workers (Brower and Brower, 1964) on the monarch butterfly have elucidated many of the chemical relationships between this insect and its host-plants, and led to the theory that some herbivores evolved the special ability to sequester specific molecules from their food plants that they use against predators (Brower *et al.*, 1967). In this theory, adaptation of the insect to a host plant containing these noxious chemicals is a critical part of the coevolutionary process (Ehrlich and Raven, 1964; Brower *et al.*, 1982). Sequestration of chemicals into body tissues or fluids is viewed as an adaptation for putting these compounds out of harm's way (Brower and Brower, 1964; Rothschild, 1973). Sequestered chemicals may confer a selective advantage against predators by decreasing prey palatability (e.g., Brower *et al.*, 1967). This selection reinforces the chemical bond between the insect and its host plant such that the insect may seek out and specialize on hosts with high concentrations of such materials (e.g., Roeske *et al.*, 1976; Seiber *et al.*, 1980, 1986), or particularly bioactive variants of these compounds (e.g., Reichstein *et al.*, 1968). Selection may also result in a more efficient sequestration or selective sequestration of certain components (Aplin *et al.*, 1968; Rothschild and Aplin, 1971; Brower and Glazier, 1975; Brower *et al.*, 1982, 1984a,b). The end result is an insect specialized on a restricted number of host-plants with a chemical defense syndrome that has evolved to be tightly coupled to the host-plant.

We infer this evolutionary process because the many contemporary examples of sequestered chemical defenses are from insects that specialize on, and use, defensive chemicals from a single host species [e.g., *Chrysolina brunsvicensis* Gr. on *Hypericum hirsutum* L. (Rees, 1966, 1969); *Caenocoris nerii* Germar. on *Nerium oleander* L. (von Euw *et al.*, 1971); *Danaus chrysippus* L. on *Asclepias curassavica* L. (Rothschild, 1967)] or a few taxonomically related species with common chemistry [e.g., *Danaus plexippus* L. on Asclepidaceae (Brower *et al.*, 1982, 1984a,b); papilionids on the Aristolochiaceae (Rothschild *et al.*, 1970); *Pieris brassicae* L. on Cruciferae (Rothschild, 1973)]. A few insect species have a fairly wide diet breadth, but seek out plants containing "toxic" chemicals (e.g., some arctiid moths and the pyrgomorphid grasshoppers; Rothschild, 1973). We assume that these insects represent fairly advanced stages or end points in this evolutionary process. Therefore we can only speculate from these extant patterns of chemical defense as to the ecological factors that initiated or constrained the evolution of such a chemical defense syndrome, with the inherent limitations of such circumstantial or inferential evidence. An alternative approach of rearing these specialists on different diets (e.g., Brower *et al.*, 1967) is possible but is difficult because these insects have a limited ability to grow and survive on other hosts, and may be of limited value for understanding the evolutionary process because the insect already has a chemical defense that is adapted to one host.

In the process of examining relationships between diet breadth and autogenous defenses in *R. guttata* we found that experimental dietary specialization resulted in sequestration of host-plant chemicals (Jones *et al.*, 1988). We believe that diet breadth explains the results for *R. guttata* and has important general implications for the initiation of and constraints on the evolution of sequestered defenses in other insects.

In this chapter we discuss the defenses of *R. guttata* and report the effects of manipulating diet breadth on autogenous and sequestered chemical defenses. We propose two simple models to describe the results and hypothesize different mechanisms to explain them. We explore the ecological and evolutionary consequences of our findings for this particular insect. Finally, we generalize from our results and present hypotheses and predictions concerning the relationships between diet breadth, the evolution of autogenous and sequestered defenses, and the extant patterns of chemical defenses in insects.

II. Chemical Defense in *Romalea guttata*

A. Life History and Diet

The lubber grasshopper, *R. guttata,* is a large, flightless, slow-moving grasshopper that lives in early successional or disturbed habitats in the southeastern United States, from Florida to South Carolina, and west to Louisiana. Eggs hatch synchronously in the spring, and larvae aggregate and move in groups, proceeding through five instars before molting to adults in late summer. Following mating, females oviposit in early fall. Within its lifetime, *R. guttata* feeds on a large number of plant species. Over 80 species of herbaceous and woody low-growing plants have been recorded as eaten by this generalist insect in Athens, Georgia. The insect generally consumes plant tissue from one or more species in a meal. Consequently, relatively small amounts of a large number of plants are consumed. Although green leaves dominate the diet, fruits, flowers, stem, bark, fungi, and insect carrion are also eaten, and the insect feeds on all leaf ages as well as senescing and dead tissues.

B. The Defensive Syndrome

Lubber grasshoppers are large, flightless, and slow-moving, are found in aggregations, and have an impressive defensive armory that is well advertised. Larvae have a warning coloration of red, orange, or yellow stripes on a black body. Adults are bright yellow and black and use the wings as a warning display against large predators. The small pink wings with black borders are raised quickly from under the elytra and are opened and closed rapidly in a flashing display. Sharp spines on the hind legs are brought forward and thrust at the predator. Larvae, as in many other grasshoppers, will regurgitate brown, viscous crop contents when attacked. Both larvae and adults are unpalatable to a number of predators that occur in the natural habitat (D. W. Whitman, C. G. Jones, and M. S. Blum, unpublished data). Third-, fourth-, and fifth-instar larvae and adults produce a frothy secretion by expelling air and defensive gland contents through modified metathoracic spiracles (Whitman, Jones, and Blum, unpublished data). The froth has a larger surface area than a droplet of the same volume, has a strong odor, and is produced with an audible hiss (Fig. 1).

C. Chemical Defenses of Wild Insects

Defensive secretions were collected by gently squeezing the insect dorso-ventrally and placing a microcapillary at the spiracular opening. The concentration of selected phenolic constituents was determined by high-pressure liquid chromatography (HPLC) using a C_{18} reverse-phase column, a methanol/water/phosphoric acid solvent gradient, and ultraviolet (UV) detection at 275 nm. The HPLC method was sufficiently sensitive that the secretion components from an individual insect were quantified with up to five replicate analyses (Jones *et al.*, 1986, 1987). Volatiles from the defensive secretion were analyzed by capillary gas chromatography (GC) on a DB-5 column using splitless injection techniques and subambient temperature operation. Again, quantification of secretion components from individual *R. guttata* was feasible, but in most situations there was insufficient secretion for replicate analyses of secretions from individuals (Jones *et al.*, 1988). Structures of compounds in volatile or nonvolatile fractions were determined by GC—mass spectroscopy (GCMS), and confirmed by cochromatography on HPLC or GC, and by GCMS of authentic standards (Jones *et al.*, 1986, 1987, 1988).

To understand the effects of diet on chemical defenses of R. *guttata*, we first examined the defensive chemistry of wild insects. GCMS of secre-

Fig. 1. Frothy metathoracic defense secretion of an adult female *Romalea guttata*.

tions pooled from 20–30 individual lubbers revealed an exceedingly complex composition with a great diversity of chemical structures present (Table I). We selected seven compounds, phenolics, quinones, and an unknown (see Fig. 2), for comparative studies because they were the major components and the major biochemical classes of compounds in the secretion (Table I); because they were easily separated by HPLC; and because such compounds are common constituents of insect defenses being found in as diverse orders as Isoptera, Blattaria, Orthoptera, and Coleoptera (see Blum, 1981, Chap. 7 for review), and are often biologically active as repellents [e.g., bombardier beetle (Eisner *et al.*, 1977), tenebrionids (Tschinkel, 1977)]. The precursors of these compounds are often aromatic amino acids (Blum, 1981, Chap. 16).

We analyzed secretions of wild adult individuals of the same age from a population occurring within a few square meters in the habitat in Georgia. We compared results with the pooled secretions. The phenolic and quinone composition of pooled secretions was markedly different from secretions of individuals. As we analyzed more secretions, it was clear that the secretions of individuals were qualitatively and quantitatively

Table I. A Partial List of Compounds Found in *Romalea Guttata* Defensive Secretions

Group	Reference[a]
Acids	
Acetic acid	1
Fatty Acids[b]	1
Phenols	
Phenol	1, 2
o-Cresol	1, 2
p-Cresol	2
Guaiacol	1, 2
Hydroquinone	1, 2
Catechol	1
Resorcinol	1
2,5-Dichlorophenol	2
Alcohols	
Cyclohexanol	1
A substituted cyclohexanol[b]	1
Cyclopentanols[b]	1
A methyl cyclopentanol[b]	1
A butanediol[b]	1
Aldehydes	
Benzaldehyde	1
3-Methoxybenzaldehyde	1
4-Methoxybenzaldehyde	1
2-Methylbenzaldehyde	1
3-Methylbenzaldehyde	1
4-Methylbenzaldehyde	1
Esters	
Methyl acetate	1
Ketones	
Isophorone	2
Verbenone	2
2,6,6-Trimethyl-Cyclohex-2-ene-1,4-dione	2
Romallenone (allenic sesquiterpene)	1, 3
Hydroxy-2-propanone	1
A thujaplicin (a tropalone)	1
A substituted indene[b]	1
Imidazolidienone[b]	1
Quinones	
p-Benzoquinone	1, 2
Miscellaneous	
Lipids[b]	1
Nitrogen heterocycles[b]	1
Plant pigment degradation products[b]	1
Sulfur volatiles	1

[a] References: 1, Jones *et al.*, 1986, 1988, C. Jones and P. Silk, unpublished data; 2, Eisner *et al.*, 1971; 3, Meinwald *et al.*, 1968.
[b] Compounds or structural groups that have been tentatively identified by GCMS.

very different from each other (Fig. 2). Secretions from some individuals lacked compounds that were present in secretions from other individuals. Figure 3a exemplifies this phenomenon for a wild population of 10 males, where catechol, *p*-benzoquinone, and 4-methoxybenzaldehyde all occurred in less than 100% of the secretions of individuals. Quantitative variation of these compounds in secretions was also extensive. Concentrations varied by up to three orders of magnitude between individuals' secretions. Figure 3b illustrates this variation for wild females, expressed on a log scale with standard errors exceeding half an order of magnitude. Subsequent analyses of secretions from other populations have revealed the same extensive chemical variation, and the causes and ecological and

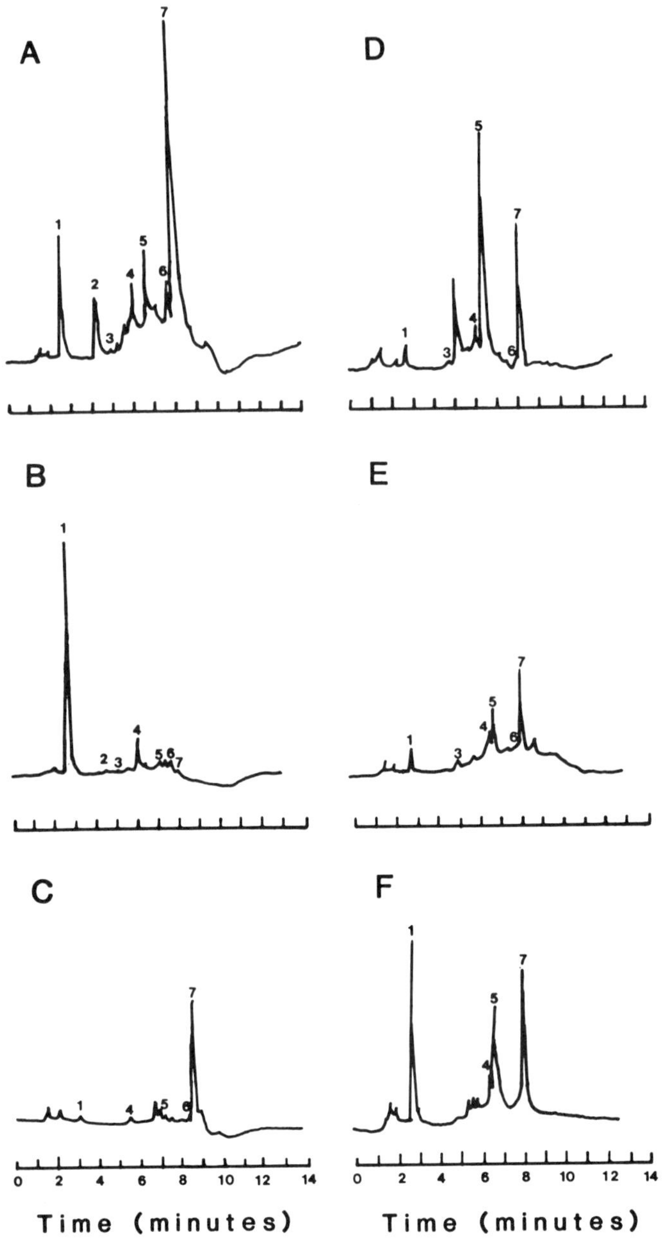

Fig. 2. HPLC traces of secretions of wild adult *Romalea guttata*. The internal standard peak has been deleted for clarity. Missing peak numbers denote compounds absent or below detectable levels. A, Pooled wild females (n = 30); B, C, individual wild females; D, pooled wild males (n = 20); E, F, individual wild males. 1, Hydroquinone (Hyq); 2, catechol (Cat); 3, *p*-benzoquinone (Pbq); 4, phenol (Phe); 5, guaiacol (Gua); 6, 4-methoxybenzaldehyde (Mba); 7, unknown (Unk). (After Jones *et al.*, 1986, Fig. 1).

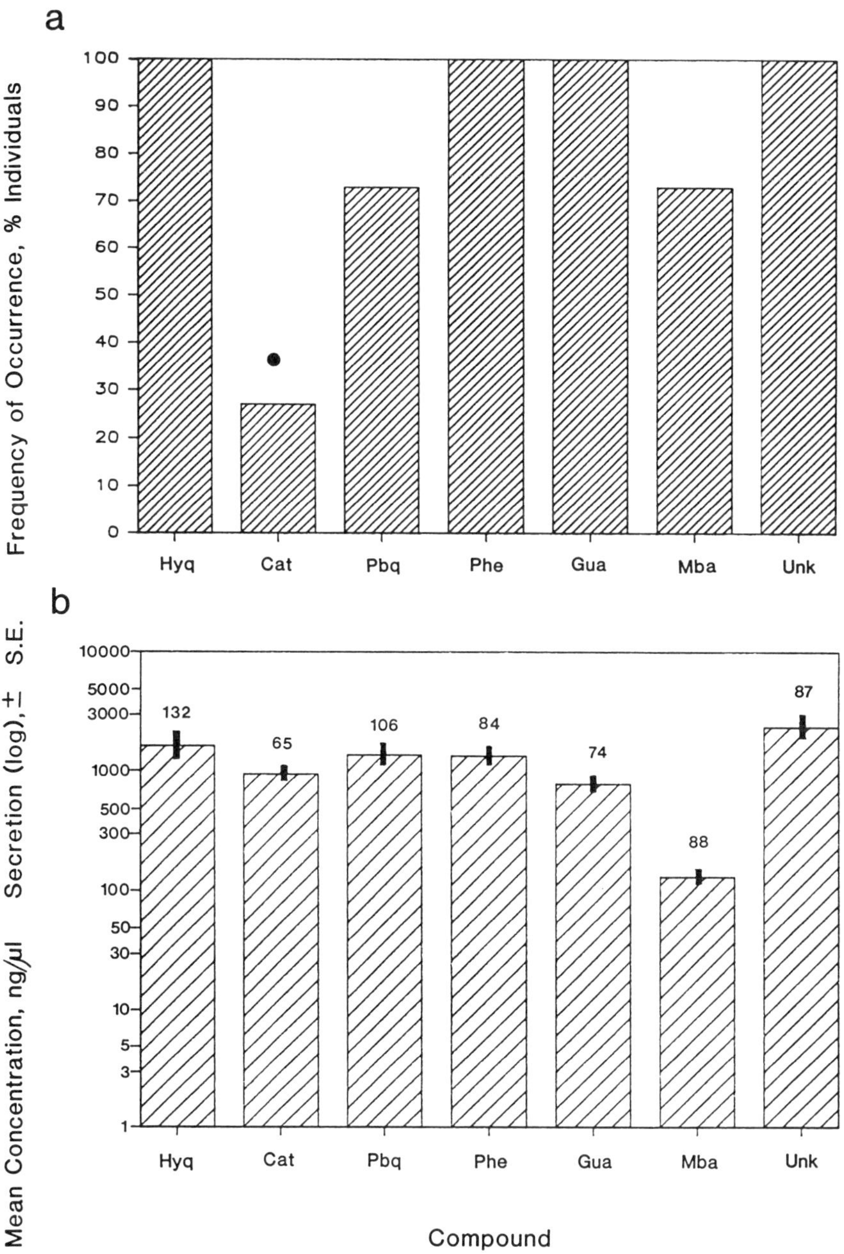

Fig. 3. (a) Frequency of occurrence of autogenous phenolics and quinones in wild adult male *Romalea guttata*. For abbreviations of compounds see Fig. 2. (●) Frequency significantly less than expected, χ^2, $p < 0.05$. (b) Quantitative variation of autogenous phenolics and quinones in wild female adult *Romalea microptera*. Note that concentrations and standard errors ($n = 10$) are on a log scale. For abbreviations of compounds see Fig. 2. Numbers over bars are coefficients of variation (percent).

evolutionary implications have been discussed (Jones *et al.*, 1986). Analysis of the effects of diet on these defensive compounds has to take into account the extensive qualitative and quantitative variation that already exists in wild populations consuming generalist diets.

III. Relationships between Diet Breadth and Chemical Defense in *Romalea guttata*

A. Diet Experiments

Romalea guttata can be reared on diets of different plants. These include diets of different breadth, such as multiple-species versus single-species diets, where the single species also occurs in the multiple-species diet, and diets of the same breadth but different composition, such as different single-species diets. In our experiments we reared lubbers on four types of diet:

1. A "natural" diet of 26 plant species, all of which are eaten in the field, and chosen to approximate the wild, generalist diet.
2. A number of different single-species diets, using species that occurred in the natural diet, and chosen to represent restricted diets. We primarily discuss results of rearing on onion *Allium canadense* L., a preferred host plant of *R. guttata*, and where we have the most complete data.
3. A defined artificial diet (Dadd, 1960) chosen to represent a restricted diet that contained the necessary requirements for growth, survivorship, and reproduction, but lacked any plant defensive chemicals or any of the defensive chemicals made by the insect.
4. Diets of intermediate breadth, such as those containing lettuce, *Lactuca sativa* L., oatmeal, and artificial diet in combination, and a diet of a mixture of six host plants.

B. Comparison with Wild Insects and Origin of Compounds

To determine whether or not insects reared on the natural diet of 26 plant species produced a defensive secretion that was effectively the same as for insects from the wild, we compared quality and quantity of the seven compounds for secretions on these two diets and found them to be the same (Jones *et al.*, 1987).

To establish the origin of the seven phenolic and quinone compounds, we analyzed secretions from insects reared on the defined artificial diet. Secretions of some individuals reared on this diet contained all of the compounds we found in secretions from insects on the natural diet or wild populations. We concluded that these seven compounds were probably biosythesized by the insect from dietary precursors (Jones *et al.*, 1987).

C. Diet Breadth and Autogenous Chemicals

We compared the quality and quantity of phenolic and quinone components of secretions of insects reared on the natural diet of 26 host-plants, and on two restricted diets: the artificial diet and a single-host-plant diet of onion, a species fed on by wild populations and included in the natural diet. Restricting the diet breadth had major effects on the secretion chemistry. First, it reduced the complexity of the secretion. For example, secretions of females contained about 15% and 40% fewer compounds on artificial and onion diets, respectively, than on natural diets (Fig. 4a). This was because individual insects had a reduced probability of containing a given compound when on restricted diets (Fig. 4b), although at least some individuals contained all of the seven compounds. Second, restricting diet breadth reduced the concentrations of all components. Concentrations of some compounds in females, for example, were depressed by two orders

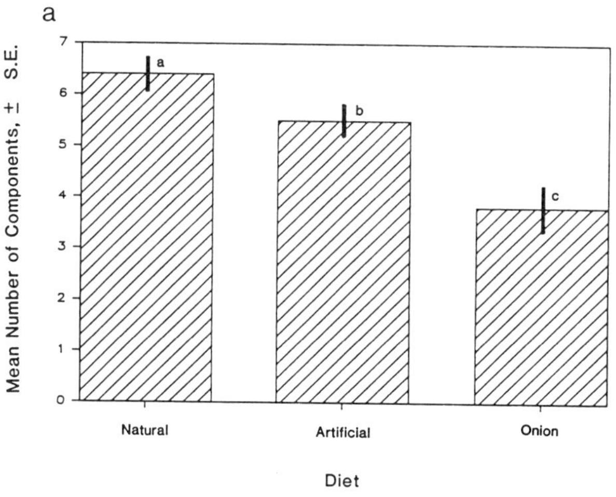

Fig. 4. (a) Mean number of components of autogenous phenolics and quinones in the defensive secretions of adult female *Romalea microptera* reared on three diets. Means with different letters (a, b, c) are significantly different from each other, *t*-tests, $p \leq 0.05$. After Jones *et al.* (1987, Fig. 2). (b) Frequency of occurrence of autogenous phenolics and quinones in adult female *Romalea guttata* reared on natural, artificial, and onion diets. For abbreviations of compounds, see Fig. 2. (●) Frequency significantly less than expected, χ^2, $p \leq 0.05$. After Jones *et al.* (1987, Fig. 1a). (c) Quantitative variation of autogenous phenolics and quinones in adult female *Romalea microptera* reared on natural, artificial, and onion diets [symbols as in (b)]. For abbreviations of compounds, see Fig. 2. Note the log scale. Means with different letters (a, b, c) are significantly different from each other, *t*-tests, $p \leq 0.05$. (After Jones *et al.*, 1987, Fig. 4a).

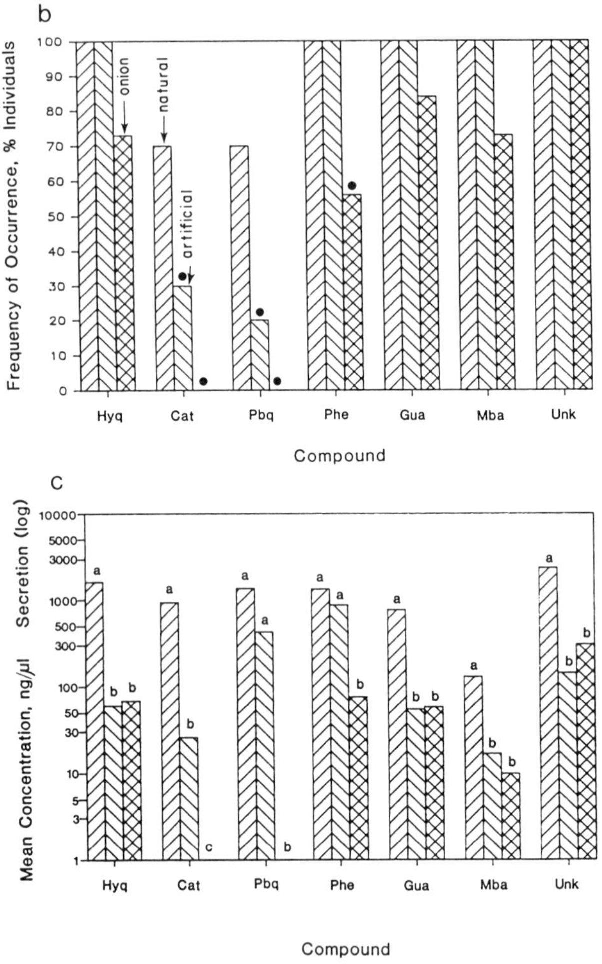

Fig. 4. (*continued*)

of magnitude or more on both restricted diets (Fig. 4c). The effects of restricting diet breadth were greatest for insects on the single-host-plant diet of onion and for females (Jones *et al.*, 1987). Subsequent analyses of individuals on other single-host-plant diets, again using species in the natural diet, have shown the same depression of autogenous compound concentrations. We have also found that diets of intermediate breadth (lettuce plus oatmeal plus artificial diet, or a mixture of six host plants) produce secretions of the same complexity and component quantity as the natural diet and wild populations.

Secretions from insects on the artificial and single-host-plant diets showed the same type of quantitative interindividual variation in the con-

Table II. Sulfur-Containing Volatiles Found by GCMS in Metathoracic Secretions of *Romalea Guttata* Adults Reared on an Onion Diet

1-Propanethiol[a]	Dimethyltrisulfide
1-Pentanethiol	Dipropyltrisulfide
Isopropylsulfide[a]	Alkyl polysulfides
Propylsulfide	Methylthiirane[a]
Methyldisulfide[a]	2,3-Dimethylthiophene
Methylethyldisulfide	2,4-Dimethylthiophene
Methylpropyldisulfide[a]	2,5-Dimethylthiophene[a]
Propyldisulfide[a]	1,3-Dithiane
Isopropyldisulfide[a]	2-Methyl-2-ethylthietane (or isomer)
Isopropylpropyldisulfide	*cis*-Hexahydro-2*H*-cyclopenta(6)thiophene

Sources: Jones *et al.* (1988) and C. Jones and P. Silk (unpublished data).
[a] Compounds quantified by capillary GC.

centrations of these compounds as insects on natural diets and wild populations. So defensive secretion variability per se was not markedly affected by diet (Jones *et al.*, 1987).

D. Diet Breadth and Sequestered Chemicals

We noticed that defensive secretions of insects reared on onion had a distinct onion-like odor as few as 4 days after having been fed on this diet. GCMS of the volatile portion of the defensive secretion of onion-reared insects revealed a large number of sulfur-containing compounds (Table II), including alkyl sulfides and sulfur-containing heterocyclic compounds. Many of these compounds were identical to those isolated and identified from onion (e.g., Carson and Wong, 1961; Brodnitz *et al.*, 1969; Boelens *et al.*, 1971). Other compounds were chemically very similar and were possibly degradation, conjugation, or rearrangement products, or compounds not yet identified from onion. None of these compounds, nor any other sulfur-containing compounds, were detected in GCMS analyses of secretions from insects reared on natural diets that included onion as a component species, or artificial diets, or from wild populations from habitats that contained the preferred host plant of onion. Since we only found onion compounds when the insect was restricted to an onion diet, we concluded that sequestration of these compounds into the defensive secretion was a direct consequence of restriction of diet breadth (Jones *et al.*, 1988). We determined the concentrations of eight of these compounds (see Table II) in adult females using capillary GC at subambient temperatures. Mean concentrations varied from 30 to 115 pg/μl secretion, depending on the compound. Mean concentrations of the eight sulfur volatiles combined was 500 pg/μl, which was about 300-fold less than the combined phenolic and quinone concentrations in secretions of females reared on the same diet (see Fig. 4c).

We did not find any other classes of compounds in the onion diet-reared secretions that were not also present in secretions from insects on the natural diet or from wild populations of insects (i.e., the autogenous phenolics or quinones were present but diminished; see Fig. 4c). This indicates that only the major class of onion natural products, sulfur volatiles, was sequestered in detectable amounts. This may be because they were the only compounds to occur in the plant in sufficient concentrations to be sequestered and detected by GCMS, and/or they were the only compounds of the structural type that has a high probability of being sequestered by this insect. Interestingly, these sulfur volatiles are highly lipophilic, and highly lipophilic compounds are often bioaccumulated in animals (Duffey, 1980).

Subsequent analyses have revealed that sequestration of plant natural products occurs for other single-host-plant restricted diets. For example, insects reared on catnip sequester a derivative of nepitalactone (Blum *et al.*, 1987), but do not contain detectable amounts of this compound when reared on a mixed diet of six species, including catnip. Similarly, insects reared on Johnson grass, *Sorghum halepense* L., contain at least two major HPLC UV_{254}-detectable components (Fig. 5) (as yet unidentified) that are absent in secretions of insects reared on the natural diet that includes *S. halepense* as one of 26 species. It should be noted that GCMS of secretions from insects reared on all single-host-plant diets shows that they contain the autogenous phenolics and quinones, but at reduced concentrations. It is interesting that secretions from insects reared on the artificial diet only contained depressed concentrations of the major GCMS-detectable compounds that were also found in insects reared on the natural and wild diets (i.e., autogenous). The artificial diet (Dadd, 1960) contains no "secondary compounds," and we believe that although this was a restricted diet, sequestration did not take place because there were no compounds to sequester. Our results with onion or catnip predict that inclusion of such compounds in a restricted artificial diet should result in their sequestration into the defensive secretion.

Our data clearly show that restricting the diet breadth of *R. guttata* to monophagous diets containing certain plant natural products results in sequestration of these compounds, and that restriction or specialization is the driving factor. Secretions from grasshoppers fed broad diets containing the same plant species do not contain detectable amounts of these compounds.

E. Diet Breadth and Chemical Defense against Predators

Did the considerable qualitative and quantitative chemical changes caused by restricting diet breadth have any effect on the efficacy of the defense against predators? We bioassayed the secretion for repellent ac-

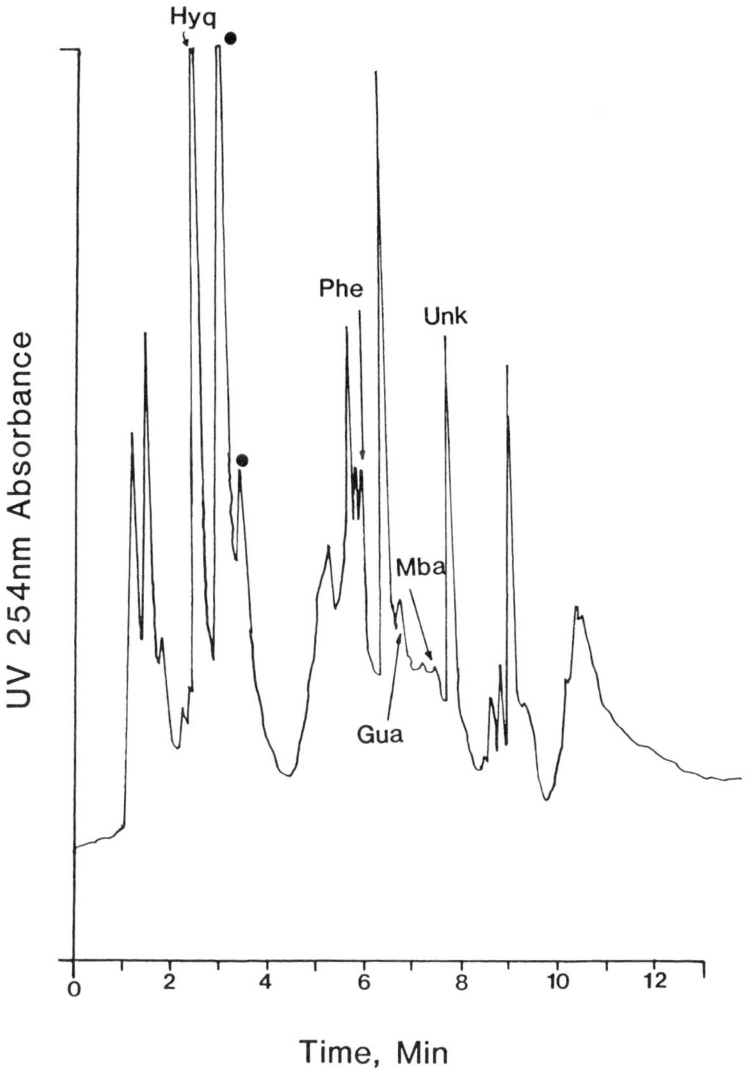

Fig. 5. HPLC trace of secretion from adult female *Romalea guttata* reared on a Johnson grass, *Sorghum halepense* diet, showing peaks absent from secretions of insects on other diets (●). For abbreviations, see Fig. 2.

tivity against a generalist predator, the red imported fire ant, *Solenopsis invicta* Buren. This ant, an introduced species that now occurs in the same habitat as *R. guttata*, preys on *Romalea* in the field (Whitman, Jones, and Blum, unpublished data). To evaluate repellency of the secretion, we compared the number of ants visiting and consuming adjacent baits when the normally attractive sugar-water baits contained equal volumes of pooled defensive secretions of female insects reared on either the artificial diet, onion diet, or natural diet. Ants significantly preferred to visit the artificial-diet secretion bait to the natural-diet secretion bait to the onion-diet secretion bait (Fig. 6) (Jones *et al.*, 1988). Approximately 60%, 25%, and 15% of all foraging ants visited the three baits, respectively. Bait consumption times reflected the visitation rates. The same experiments, repeated with another less aggressive ant species, *Tapinoma* sp., gave similar results with even greater differences between treatments. Recent studies (Whitman, Jones, and Blum, unpublished data) show that the efficacy of the defensive secretion against predators is a good representation of the overall defensive status of the intact, mobile prey.

The differential preference for secretions in baits is entirely compatible with the chemical changes we quantified. Thus the artificial diet secretion and bait that had a reduced phenolic and quinone concentration and that did not contain any sequestered compounds was the most preferred. The natural diet secretion and bait with its high phenolic and quinone concentration and lack of sequestered components was of intermediate prefer-

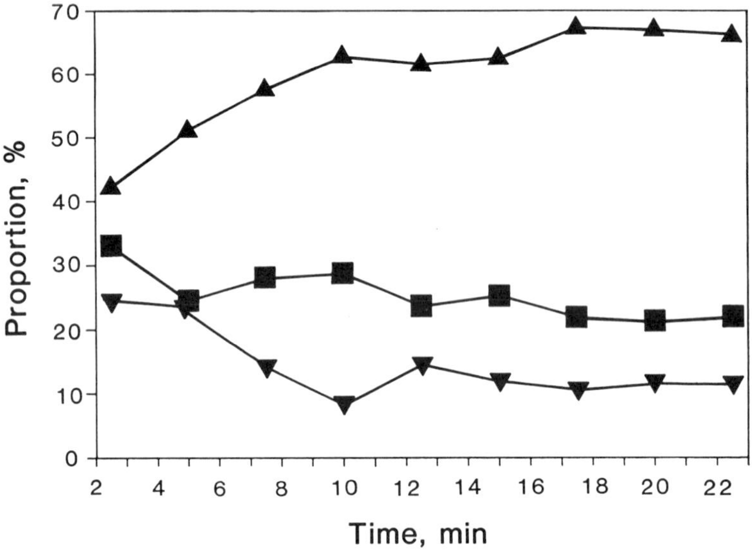

Fig. 6. Proportion of *Solenopsis invicta* ants visiting 0.1 *M* sucrose baits containing secretions, with time. Data are from five colonies and three replicate trials. ▲, Artificial diet secretion plus bait; ■, natural diet secretion plus bait; ▼, onion diet secretion plus bait. Proportions are significantly different over the entire trial (Kruskal–Wallis rank sums, $p < 0.0001$). (After Jones *et al.*, 1988).

ence. The single-host-plant onion diet secretion and bait had very low quinone and phenolic concentrations and, as such, should have been highly preferred, similar to the artificial diet secretion. However, this secretion contained sequestered sulfur volatiles that were absent from other secretions, and did not contain any other compounds that were not also in the artificial diet or natural diet secretions. Therefore, the decreased preference was due to the sequestered sulfur volatiles (Jones *et al.*, 1988).

Subsequent experiments with *S. invicta* and secretions of R. *guttata* from other single-host-plant diets reveal that the above phenomena with onion is by no means unique. Of a total of eight different single-host-plant diets bioassayed in comparison to secretions from natural diets, one was more preferred, three were not significantly different from the natural diet, despite reductions in autogenous compounds, and four were less preferred. Clearly, then, restricting diet breadth of *R. guttata* has a high probability of producing a defensive secretion of lower or equal preference (seven of eight) than the autogenous defenses to this predator.

F. A Summary of Dietary Effects on Chemical Defense

Our experiments show that diet breadth has a major effect on the quality, quantity, and efficacy of chemical defense for this generalist insect herbivore. Restricting the diet breadth from diets of 26 host plants to single plant species experimentally mimicked the process of specialization, and had the following overall effects on the defensive secretion. First, it reduced the complexity, consistency, and concentrations of autogenous compounds. Second, it resulted in the sequestration of host-plant compounds. Third, it changed the defensive efficacy of the secretion in a manner entirely consistent with the chemical changes. A reduction in autogenous phenolic concentrations led to increased predator preference, while sequestration led to decreased preference in many cases. On the onion diet, for example, preference decreased fourfold, despite the fact that the sequestered active ingredients in the secretion were at 300-fold lesser concentrations than the autogenous phenolics and quinones.

IV. Mechanisms, Models, and Consequences to *Romalea guttata*

What mechanisms could explain the effects of diet breadth, and how can these effects be modelled? Our studies showed two main effects on the chemical defense of restricting diet breadth: first, a reduction in the production of autogenous compounds, and second, sequestration of plant compounds. Since we obtained a reduction in autogenous compounds without sequestration for insects on the artificial diet, it seems reasonable that these two phenomena can be treated separately, although, of course, they are related because they both involve diet breadth as a major factor.

A. Reduction of Autogenous Defenses

Why should restricting diet breadth reduce secretion chemical complexity and secretion component concentrations? We propose two possible mechanisms that are not mutually exclusive. The first possibility is that biosynthesis of autogenous compounds is dependent upon the quality and quantity of precursors in the diet. We will assume that precursors do not occur at the same concentrations in all plant species. If so, then the probability of feeding on a diet that does not contain adequate precursors is likely to increase with increasing specialization, because a single host-plant species randomly drawn from the mixed diet has a lower probability of containing the appropriate precursors than a mixed diet. This mechanism does not exclude the possibility that some single plant species might have adequate or even very high concentrations of precursors. If so, we would predict that feeding lubbers on some of our single-plant-species diets would result in concentrations of autogenous compounds similar to, or perhaps higher than, those for insects on natural diets. Although there are differences in the concentration of autogenous compounds for insects reared on different single-plant-species diets, in all cases so far examined, autogenous compounds are substantially (greater than one order of magnitude of concentration) reduced on all single species diets. To rigorously test this hypothesis we have to identify the precursors for the biosynthesis of these compounds and manipulate their concentrations while keeping all other dietary factors constant, and relate concentration of precursors in single plant diets to the concentration of autogenous compounds in the secretion.

An alternative hypothesis proposes that the effect of diet breadth on autogenous compound production is indirect. In contrast to the previous hypothesis, we assume that precursors are more or less ubiquitous in all plants and occur at relatively uniform concentrations—or at least, that variation in precursor concentrations occurs above a minimum threshold concentration of precursors necessary for biosynthesis of autogenous compounds. We also assume that single-plant-species diets are usually of poorer nutritional quality to lubbers than mixed-species diets. We do have evidence to support this latter assumption—growth and survivorship of *R. guttata* on most (but not all) single-plant-species diets is significantly reduced in comparison to natural diets (Whitman, Jones, and Blum, unpublished data). This hypothesis states that under conditions where the diet is of poor nutritional quality, the insect becomes physiologically stressed. Resources (such as amino acids, for example) or energy that would normally be allocated to biosynthesis of defenses become limiting, and the insect reallocates these resources to more basic requirements of maintenance, growth, and reproduction. This results in a reduction in the biosynthesis of defensive compounds (Jones *et al.*, 1987). Testing this hypothesis would require identification of precursors, then feeding lubbers a constant concentration of precursors while manipulating the nutritional suitability of the diet, and relating growth, survivorship, and reproduction of lubbers on single- and multiple-host diets to the quality and quantity of defensive secretion compounds. We do know that defensive secretion volume decreases on single-host-plant diets (Whitman, Jones, and Blum, unpublished data), and that female lubbers, which have a larger demand for resources allocated to reproduction compared to males, also show a larger decrease in autogenous compound production when they are fed on restricted diets (Jones *et al.*, 1987).

Irrespective of whether one or both hypotheses are involved, we propose a model to describe the effects of diet breadth on autogenous defenses that helps interpret the phenomenon and aids in the prediction of consequences. The model is based on some of the data. We know that the single-plant diet of onion depressed autogenous compound concentrations by 98%. Artificial diet depressed concentrations by 82% (Jones *et al.*, 1987). Other single-species diets decrease compound concentrations to a similar degree. However, diets of lettuce, oatmeal, and artificial diet combined, diets of six host-plant species combined, natural diets of 26 species, and wild diets all have similar mean concentrations of autogenous compounds. We have modeled these changes as a threshold response curve, with concentration of defensive secretions as the ordinate, and diet breadth as the abscissa. Figure 7 shows this proposed curve for *R. guttata*. There is an abrupt change in the amount of compound produced, with relatively small changes in diet breadth, rather than a linear relationship. Although the exact position of the inflection on the threshold curve will depend on the suitability of the diet consumed (with either or both of the hypothesized mechanisms), it appears that it should occur between one and three host plants—the point of considerable dietary specialization. The shape of the curve implies that the autogenous defensive system is resistant to the influence of specialization up to a certain point, but thereafter further specialization causes this defense to become effectively nonexistent. As we shall show later, the shape of the curve has important implications for understanding the evolutionary consequences of diet breadth effects for *R. guttata*.

B. Sequestration of Plant Chemicals

Why should restricting diet breadth result in sequestration of plant compounds? Our study clearly shows that onion volatiles occur in secre-

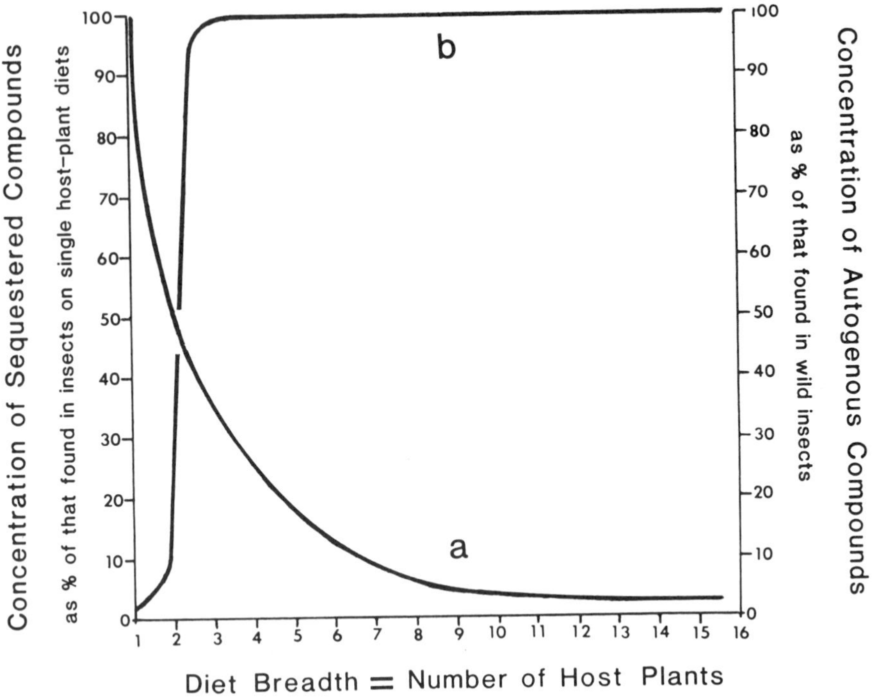

Fig. 7. Proposed models of effects of diet breadth on chemical defense in *Romalea guttata*. Curve a, hyperbolic dilution curve for sequestered defenses; curve b, threshold response curve for production of autogenous defenses.

tions only when onion is the sole diet, and not when it is included in a mixed diet of 26 species (Jones *et al.*, 1988); this was also the case with Johnson grass. Similarly, a derivative of catnip nepitalactone is found in secretions of insects on the single-species diet, but is absent from secretions of insects on a mixed six-species diet (Blum *et al.*, 1987). ("Absent" means below the level of detection by GC, HPLC, or GCMS.) We hypothesize that these results can best be explained on the basis of how diet breadth affects the concentration of compounds ingested by the insect. Our argument is as follows. We know that these compounds are present in both the mixed and single-species diets, and that the insect is capable of sequestering at least these compounds into its defensive secretion. We propose that this phenomenon is casual sequestration and could apply to other compounds ingested in the diet of *Romalea*. We argue that the concentration of chemicals sequestered will relate to the dose ingested. We assume that lubbers will eat more or less equal amounts of all dietary components in a mixed diet over the course of our experiments. This assumption is supported by data from feeding trials and field observations, showing that lubbers continually switch host plants and mix their diets on a daily basis (Jones, Whitman, and Blum, unpublished data). Then, when lubbers feed on a single species containing a given compound, they will ingest a higher concentration of this compound than when they feed on the mixed diet including this species. This increased concentration ingested would be reflected in an increased concentration sequestered. So specialization results in sequestration of these compounds simply because the insect ingests high enough concentrations of these compounds for them to be sequestered into the defensive secretion in detectable amounts. On the other hand, generalist diets dilute the concentrations of chemicals contributed by any one species in the diet by all the other plant species. The hypothesis still applies if all of the species in the mixed diet had different chemicals that could be sequestered on single-species diets. Each one of these chemicals would be at too low a concentration in a mixed diet for them to be sequestered in detectable amounts.

We justify our assumption of casual sequestration because there is no evidence to suggest that lubbers have evolved to specifically sequester onion volatiles or nepitalactone derivatives, for example. It is more reasonable to suppose that these phenomena are examples of fortuitous or casual sequestration (*sensu* Duffey, 1980). Casual sequestration of liphophilic chemicals and some hydrophilic compounds is very common in animals and plants. Examples include lignins and cardenolides in mistletoes on their hosts (Boonsong and Wright, 1961; Freudenberg, 1965), carotenoids in many vertebrates (Fox, 1976), alkaloids in bird tissues and mammalian livers (van Veen, 1973), and the well-known examples of bioaccumulation and bioconcentration of pesticides such as DDT in fatty tissues of many animals (O'Brien, 1967; Matsumara, 1975; Wilkinson, 1976). Duffey (1980) cites over 20 examples of casual sequestration that are exclusive of insects. Casual sequestration is also common in insects. Examples include carotenoid pigments in many insects (Needham, 1968), pesticides in tissues (Eto, 1974), and a diversity of plant natural products in sucking insects (Patwardhan and White, 1973) (for review see Duffey, 1980, Table 1). There are also a number of examples of sequestration of plant chemicals that are putatively or demonstrably used as pheromones or in growth and development. These compounds include terpenes (Hughes, 1974; Hedin, 1977; Birch, 1978), amino acids (Harrewijn, 1978), and alcohols (Howard *et al.*, 1976) (for review see Duffey, 1980, Table 2). Finally there is a substantial body of literature reviewing casual and/or specific sequestration by insects of chemicals as diverse as anthraquinones, alkaloids, cardenolides, flavanoids, terpenes, allylisothiocyanates, and polyacetylenes for—in most cases—putative defensive purposes (see reviews by Rothschild, 1973; Duffey, 1976, 1980; Blum, 1981; Pasteels *et al.*, 1983a). In some insects (e.g., the milkweed bug and the monarch) there is good evidence of specific adaptations for sequestration, but in most cases the process, the adaptations, and the functions of the sequestration are unknown. Suffice it to say that sequestration of plant chemicals is common and many examples of casual sequestration are known. It has also been suggested that fortuitous sequestration of compounds in small doses may lead to regional differences in prey palatability (Duffey, 1976).

Our argument concerning concentration sequestered versus dose ingested is as follows. A large number of factors determine the exact concentration of these compounds in body tissues—such as the fat content of the animal and the potential for degradation or excretion of these compounds (O'Brien, 1967; Matsumara, 1975; Wilkinson, 1976), their structure (Seiber *et al.*, 1980; Jones *et al.*, 1983), and their lipophilicity (Duffey, 1980). However, there is, at simplest, a fundamental dose relationship between the concentration in the body tissues or fluids and environmental dose and consumption rate (per unit body weight). Such a relationship holds for such diverse compounds as bioaccumulated pesticides (O'Brien, 1967; Matsumara, 1975; Wilkinson, 1976), and sequestered plant compounds used in defense. For example, Duffey *et al.* (1978) showed that the concentration of cardenolides in the dorso-lateral spaces of the milkweed bug is directly related to the dose administered. In the monarch butterfly, such a relationship was shown to be true for four cardenolides at doses of less than about 25×10^{-7} mol consumed (Seiber *et al.*, 1980, Table 1). At higher concentrations (100–1000 μg/0.1 g plant dry weight) this linear relationship plateaus and the concentration stored no longer relates strongly to the concentration ingested (Brower *et al.*, 1982). Nevertheless, for lower concentrations in this insect and in other cases studied, compounds seem to bioaccumulate in a reasonably dose-dependent manner.

Our hypothesis derives from results with some plant species, and therefore has certain caveats. Lubbers are almost certainly not capable of sequestering all compounds from all species that we have recorded as hosts—although they have shown the ability to sequester a remarkable diversity of structures. Certain compounds may well have a much higher probability of sequestration than others. Some compounds are probably degraded or excreted and some compounds may be modified by the insect following sequestration. The rate of sequestration is probably not the same for all compounds. The slope and intercept of the proposed dose relationship (concentration in secretion related to concentration in diet) is probably not the same for each compound. Nevertheless, we argue that the concentration of each compound that is sequestered will be affected by diet breadth in the same general way.

Testing this hypothesis requires varying the concentration of sequestrable compounds by adding them to a fixed breadth diet in different amounts. Here the concentration in the defense secretion should vary in the same way. By analogy, experimentally holding the concentration of compounds constant per unit food ingested (by adjusting dose) and varying diet breadth should result in no change in concentration of sequestered compounds in the defense secretion.

We propose a simple dilution model to help interpret the sequestration phenomenon in *R. guttata*. The concentration of a compound from one plant becomes diluted as other plant species that do not contain the same compound are added to the diet; the simplest manner in which this could occur is a hyperbolic dilution curve. Figure 7 shows such a curve where the ordinate is the concentration of the sequestered compound in the insect as a percent of that found when feeding only on the single host-plant from which the compound comes. The abscissa is diet breadth, expressed as the number of host-plants in the diet. For every doubling of diet breadth, there would be a halving of concentration. Thus, if the insect feeds only on species A, then the concentration of compound A from species A will be the maximum that can be sequestered (= 100%). If the insect were to double its diet breadth by feeding on species A and species B, then the concentration of compound A will be halved. Similarly, the concentration of compound B from species B will also be half of its maximum concentration. The hyperbolic shape of the curve means that the largest changes in concentration occur in the initial increase in diet breadth from one to two and two to four host-plant species. These reduce concentrations by 50% and 25%, respectively. Thereafter, progressive doubling of diet breadth produces progressively smaller reductions in concentration. Consequently, sequestered chemicals that are present in biologically active concentrations when the insect is feeding on a single plant species would be rapidly diluted with increasing diet breadth. Dilution to concentrations below biological activity against predators would seem very likely.

Our hypothesis predicts that if we had GC, HPLC, or GCMS instrumentation with sufficiently sensitive detectors, we would find trace amounts of the same compounds that are sequestered on single-species diets in secretions of insects on mixed diets, unless there is a minimum threshold concentration for any sequestration to occur. We were not able to detect onion compounds in secretions from insects fed a mixed diet of 26 plant species including onions. Using the proposed hyperbolic dilution model and the known concentration range of 25–125 pg/μl for each onion volatile in secretions of lubbers fed onion only, we would expect about 4% (1–5 pg/μl) in lubbers fed a 26-species diet. This concentration would be below our detection limits, and we do not know whether or not these compounds are found in trace amounts, and cannot test our hypothesis from these data.

There are other data that bear on this issue. First, we have found trace amounts of a tremendous diversity of compounds in the defensive secretion of wild insects (Jones, Silk, Whitman, and Blum, unpublished data; see also Table I). A similar diversity was found by Eisner *et al.* (1971). Most of these compounds remain unidentified, because they are present in such low amounts and, of course, these trace compounds may be autogenous. However, some of these compounds appear chemically very similar to some plant compounds such as pigments and may be plant-derived. Second, Eisner *et al.* (1971) reported 2,5-dichlorophenol in the defensive secretion of *R. guttata* collected in an area that had been treated with the herbicide 2,4-D. The secretions contained the autogenous phenolics and quinones and the halogenated phenol. Grasshoppers from an adjacent, untreated area contained only the phenolics and quinones. So here we had a situation in which secretions contained a sequestered compound in concentrations much greater than trace amounts, even though *R. guttata* was feeding on a generalist diet of host plants. We suggest that the uniform treatment of all the plant species with herbicide resulted in the ingestion of high concentrations of halogenated phenol and this is why *Romalea* sequestered the compound while on a generalist diet. From the perspective of the phenol alone, we argue that this situation is analagous to consuming a single-species diet containing the compound.

C. Evolutionary Constraints and Consequences

Our experiments do not represent the field situation for *R. guttata*. In the wild, the lubber is an extreme generalist, with individuals perhaps having a wider diet breadth within their lifetimes than most generalists. Nevertheless, lubbers have the potential to show changes in the chemical quality, quantity, and origin of the defense and its efficacy against predators, if diet breadth were to change. We also know that these changes can occur very rapidly—adult insects may stop producing significant amounts of phenolics and quinones and contain detectable concentrations of sequestered compounds as few as 4–5 days after restriction to a single-plant-species diet. It seems reasonable to ask why the phenomenon does not occur in the field. Why don't lubbers specialize on one host-plant species and sequester the chemicals for defense? Understanding the ecological factors constraining lubber defenses may help us understand evolution of chemical defense in other insects.

We consider that there are three major factors that should cause selection to favor maintenance of generalism:

1. Lubber growth, survivorship and reproduction are generally best on mixed diets (Whitman, Jones, and Blum, unpublished data). Although lubbers can survive to reproduce on some single-species diets, and can even grow, survive, and reproduce as well on some single-species diets as on a mixed diet (e.g., onion), performance on many single-species diets is poor.

2. Lubbers have low vagility and occur in habitats of high plant species diversity. Large monocultures of one plant species tend not to occur and the body size and aggregative behavior of lubbers suggests that small patches of one plant species, or single plants, may not be an adequate food supply for a population to reach recruitment.
3. Lubber feeding behavior enhances the probability of diet mixing, despite the fact that lubbers show distinct preferences for certain plant species, such as onion (Jones, Whitman, and Blum, unpublished data).

These three factors should result in selection for generalist diets and, as we have shown, mixed diets produce the highest concentration of autogenous defenses. The autogenous defense is effective against a number of predators—particularly birds and reptiles—but not all predators (Whitman, Jones, and Blum, unpublished data).

It is possible to envision situations in which these selection pressures could change and lead to a decrease in diet breadth. For example, this might happen if feeding behavior were to change to preference for one or a few species on which lubbers could grow, survive, and reproduce (e.g., onion), or if a lubber population became isolated in a large monoculture of these species (e.g., an onion field) or a habitat of reduced plant diversity containing these species. If this were to occur, then our experiments predict a reduction in autogenous defenses together with the potential for sequestration to occur. In order for these changes to result in a sequestered defense syndrome in evolutionary time, a number of criteria would have to be fulfilled:

1. Compounds would have to be sequestered initially. Even given all the caveats in Section IV,C, this seems a likely outcome for some restricted diets.
2. These compounds should not be autotoxic and should have no direct adverse effect on growth, survivorship or reproduction. We have no evidence to suggest that sequestration on onion diets, for example, has costs. Cost-free sequestration is known for other insects [e.g., *Oncopeltus fasciatus* Dallas (Isman, 1977; Jones *et al.*, 1983); monarch (Seiber *et al.*, 1980; Brower *et al.*, 1982)], although we would never claim that all sequestration in *Romalea* would be cost-free.
3. Production of autogenous compounds should be costly. Our hypothesis involving stress-induced reduction in autogenous compound production implies such a cost, and costs are known in insects such as *O. fasciatus* (Duffey *et al.*, 1978).
4. The defensive secretion containing sequestered compounds must be effective against predators, despite its reduced autogenous compound concentration. Based on our data, this is quite likely.

If all these criteria are fulfilled, as well as the initial conditions changing diet breadth, then we would expect selection for the sequestered defensive syndrome. Based on the two proposed models, intermediate diet breadth may be evolutionarily unstable, since an insect feeding on two to four hosts might initially have both an ineffective autogenous defense and too low a concentration of sequestered chemicals for effective defense. We have shown that reduction in autogenous defenses, sequestration, and a change in defensive efficacy occur rapidly and simultaneously. This, together with the preexisting armory of defenses such as spines, wing flashing, and bright colors that would act to reinforce defense against predators, could lead to rapid selection.

It is important to note that while all the previous criteria are reasons why evolution of sequestered defenses may or may not occur in lubbers, in these scenarios selection cannot commence until diet breadth is changed, for reasons that do not necessarily have to do with chemical defense. While there could well be subsequent feedback reinforcing specialism/sequestration once it has occurred in lubbers in a manner analogous to coevolution in monarchs (Brower *et al.*, 1982), changes in diet breadth are necessary initially. Thus a reduction in diet breadth initiates the sequestered defensive syndrome. Furthermore, since efficacious autogenous defenses are most likely dependent on the maintenance of a generalist diet, diet breadth is an important constraint on the autogenous defensive syndrome.

V. Diet Breadth and Insect Chemical Defenses: General Hypotheses

We consider that the hypothesized mechanisms and proposed models are simple, parsimonious explanations of our experiments with *R. guttata*, and have relevance to the ecology and evolution of chemical defense in this insect. Can any of our findings be extended to other insects? What are the ecological and evolutionary implications of diet breadth and chemical defense in insects? Although diet breadth affects autogenous defenses and sequestration simultaneously in *Romalea*, different mechanisms are involved. Therefore, we examine the evolutionary implications of diet breadth to autogenous chemical defenses in insects separately from diet breadth and sequestration. We also examine the evolutionary implications of feedback between defensive efficacy and diet breadth. Our hypotheses derive directly from the *Romalea* study and use most of the same arguments and evidence. We present hypotheses and proposed mechanisms, make predictions where possible, and provide a few examples where available. We discuss assumptions and alternatives, and suggest tests of these hypotheses. We do not attempt to test these hypotheses or review patterns of chemical defense in insects in this chapter.

A. Constraints, Autogenous Defenses, and Generalists

Diet breadth can be an important constraint on, and may also be constrained by, autogenous chemical defenses of generalist herbivorous insects. We suggest that decreases in diet breadth can adversely affect the defense because restricted diets physiologically stress the insect and/or reduce the availability of precursors for biosynthesis of the defense. One consequence is that generalists that specialize, for some other reason, may temporarily or permanently lose their autogenous defenses. The hypothesis predicts that there may be insects of intermediate diet preference, or specialists that have no chemical defenses, but were once autogenously defended generalists. Perhaps such insects have anatomical remnants of these defensive systems. The hypothesis also suggests that if predation is an important selection pressure favoring maintenance of effective autogenous defenses, this in turn should select for the maintenance of a generalist diet. So although the interaction between diet breadth and autogenous defenses does not lead to a classical chemical coevolutionary situation—where some chemicals of the host plant are the defense, and defense reinforces specialization on that host—it may produce an analogous evolutionary situation where selection for effective autogenous defenses and broad diets reinforce each other.

This hypothesis assumes that the consequences of restricting diet breadth are adverse. It is possible that some single-host or intermediate-breadth diets do not lead to stress and/or have adequate precursors, and in these cases changes in defensive status would not necessarily occur. We also assume that the stress/precursor mechanisms we proposed for *Romalea* hold good for other autogenously defended generalists. While this does not seem unreasonable, other data on diet effects on generalist autogenous defenses are lacking. Our hypothesis also assumes that there are selection pressures on generalists that lead to specialization. While there are clearly selection pressures for the maintenance of generalism (e.g., see Section IV,C), there is also good evidence to suggest selection for specialization. For example, it appears that the bulk of insects attempting and succeeding in colonizing new host plants are generalists, which then subsequently specialize. This is true for such plant species as soybean (Turnipseed and Kogan, 1976), grape (Bournier, 1977), maize (Chiang, 1978), sugarcane (Strong *et al.*, 1977), parsnip (Berenbaum 1981a), thistles (Goeden and Ricker, 1968; Goeden, 1971, 1974), wintercress (Root and Tahvanainen, 1969), and vetch (Wheeler, 1974) (see Strong *et al.*, 1984, for review).

It is possible to test this hypothesis by experimentally manipulating diets of autogenously defended generalists, by altering either precursor concentrations or nutritional suitability and measuring chemistry and efficacy of defense. We predict that reducing diet breadth would produce the same adverse effects as feeding the insect on a stressful and/or precursor-deficient diet.

B. Constraints, Autogenous Defenses, and Specialists

Diet breadth may constrain and be constrained by autogenous chemical defenses of specialist herbivorous insects. Increases in diet breadth or switching of hosts may lead to a decrease in efficacy of autogenous defenses because stress on a new host or diet is likely to be greater, and/or the new host(s) may be deficient in precursors. Again this does not always have to occur—some hosts may be equally good or better. Again we predict loss of autogenous defenses in some specialists. As with generalists, effective autogenous defenses may require maintenance of an appropriate diet that does not result in stress and/or reduction in precursors—although here, such a diet might be a single host, rather than a mixed diet. In such a case selection for effective autogenous defense and specialist diets may reinforce each other in a coevolutionary manner.

This hypothesis first assumes that we can extend the stress/precursor argument out of the specific context of diet breadth to a more general context of diet suitability—that is, dietary stress or dietary deficiency in precursors, arising for any reason, may lead to a reduction in autogenous defenses. Evidence for effects of diet on autogenous defenses in specialists is lacking, although assumptions that host precursor concentrations are important are implicit in the studies of Pasteels and Daloze (1977; Daloze and Pasteels, 1979) for autogenous cardenolides of chrysomelids, and of Nahrstedt and Davis (1983) and Wray *et al.* (1983) for autogenous cyanogens of Heliconiini. Second, we assume that there are selection pressures leading to changes in host plants for specialists. Although this appears less frequent for specialists than in generalists (Strong *et al.*, 1984), it does occur, and is reported, for example, for Lepidoptera on *Pinus contorta* Dougl. ex Loud. and *Picea* sitchensis (Bong.) Carr. (Winter, 1974), some insects on *Buddleia davidii* Franch. in Britain (Owen and Whiteway, 1980), mirids on many Norwegian plants (Taksdal, 1965), *Pieris brassicae* and *P. rapae* L. on *Tropaeolum majus* L., and swallowtails on some hosts (Berenbaum, 1981b) (for review, see Strong *et al.*, 1984).

Testing this hypothesis requires the same type of experiments that we propose for generalists. For example, one could measure quality, quantity, and efficacy of autogenous defenses for specialists reared on diets deficient in precursors or that are nutritionally marginal or inadequate. It might also be possible to compare defenses of insects that feed on a few plant species in different parts of their range, or even to rear such insects on mixed diets of two species and compare chemistry and efficacy of their autogenous defenses.

C. Initiation of Sequestered Defenses in Generalists

A reduction in diet breadth is an important initiator of evolution of sequestered defenses in generalist herbivorous insects. We hypothesize that initiation occurs because on restricted diets insects will ingest higher concentrations of chemicals from a given plant than on mixed diets. Higher concentrations ingested can result in bioaccumulation of a sufficient concentration of chemicals in body tissues and fluids for there to be biological activity against predators. This hypothesis is a direct generalization from our work with *Romalea* and has the same basic assumptions concerning the nature of bioaccumulation processes, the basic relationship between concentration of chemicals in the diet and their concentration in body tissues and fluids, and also has the same caveats regarding sequestration—namely, not all compounds have an equal probability of being sequestered; some compounds may well be subsequently degraded or excreted; some may be chemically modified; and sequestration rates and the relationship between concentration of chemicals in the diet and in the insect will not be the same for all compounds (see Section IV,B). Each of these factors will markedly influence the type of chemical and its concentration in the insect, and thereby influence the potential for activity against predators. However, our hypothesis states that irrespective of these provisos, the potential for defense is unlikely to be realized unless a reduction in diet breadth occurs, because this acts as the ecological equivalent of concentration.

Our hypothesis makes some specific predictions about defense in contemporary insects, which we illustrate with a few examples. We do not present these examples as a comprehensive or critical evaluation of our hypothesis.

1. Sequestered defenses should be more common in specialists than in generalists. Most of the classic examples of sequestration such as the monarch (Brower and Brower, 1964) and milkweed bug (Duffey *et al.*, 1978) are from specialists. Of the approximately 40 species reported as sequestering plant compounds by Rothschild (1973), over 75% are specialists, although this tabulation is not an unbiased sample by any means.
2. Sequestered defenses should be rare in generalists. The same table (Rothschild, 1973) lists only two species of arctiid moths [*Arctia caja* L. and *Cycnia* (=*Diaphora*) *mendica* Clerk, a lygaeid (*Spilostethus pandurus* Scop.)], and possibly three species of pyrgomorphid grasshoppers that can reasonably be called generalists. One other example is the grasshopper *Zonocerus variegatus* L. (Bernays *et al.*, 1977).

3. Sequestered defenses are more likely to be found in generalists that feed on many plants across their range but locally specialize on one or very few plants (*sensu* Fox and Morrow, 1981), or in generalists that specialize temporarily during one or more feeding stages. This is because these are the conditions under which a generalist will ingest high concentrations of plant chemicals. Interestingly, both the arctiid moth, *A. caja* (Rothschild and Aplin, 1971), and *Z. variegatus* L. (Bernays *et al.*, 1977) are examples of generalists that can sequester a variety of "toxins" and are local specialists.
4. Insects of intermediate diet breadth (i.e., oligophages) that use sequestered defenses will tend to use hosts that have the same defensive chemicals, irrespective of whether or not these plants are taxonomically related. Again, this is because feeding on one host that has defensive chemicals, and on two to three other hosts that do not, is likely to reduce the concentration of chemicals. By way of example, the cerambycids *Tetraopes oregonensis* Le Conte and *T. tetraophthalmus* Furst have hosts in two different families (Asclepidaceae and Apocynaceae), but both contain cardenolides (Duffey, 1970; Duffey and Scudder, 1972). The fact that insects sequester the same compounds from unrelated hosts has been used as an argument for specialization on toxic plants (Rothschild, 1971).
5. Some specialists may sequester chemicals even though they are not used for defensive purposes (i.e., have no activity against predators). This is because our hypothesis says that bioaccumulation can occur irrespective of whether or not that compound turns out to be effective against predators. This phenomenon appears to be quite common with the flavanoids and carotenoids that are found in many insects but do not appear to serve a defensive function (e.g., Feltwell and Valadon, 1970, 1972; Rothschild, 1971; Wilson, 1985).

This hypothesis can be tested in a number of ways. Generalists that have no chemical defense could be reared on mixed and restricted diets of various plants; the body tissues and fluids could be analyzed for trace-plant natural products; growth, survivorship, and reproduction could be measured; and the insects could be bioassayed against predators. Similar experiments could be done using artificial diets to which different doses of chemicals could be added. Alternatively, it should be possible to manipulate the diet breadth and dose of chemicals for generalist insects such as Arctiids and *Zonocerus* that have sequestered chemical defenses and locally or temporarily specialize. Similarly, oligophages that have sequestered defenses and hosts with common chemistry could be reared on mixtures of varieties of their hosts that have low or high concentrations of these compounds. It would also seem worthwhile to look for more cases where compounds are sequestered but do not appear to serve a defensive or other obvious function. Finally, it should be possible to examine the relationships between diet breadth and the frequency of occurrence of sequestered defenses in insects, perhaps focusing on well-studied taxa that contain specialists and generalists, such as the Lepidoptera and Orthoptera, and avoiding the problems of inherent biases of reporting that exist in the literature (e.g., aposematic insects may be overreported).

D. Constraints, Sequestered Defenses, and Specialists

Diet breadth is an important constraint on and may be constrained by sequestered defenses in specialist herbivorous insects. This hypothesis offers a mechanistic explanation of part of the evolutionary relationship between a specialist with a sequestered defense syndrome and its host-plant. We argue that an increase in diet breadth, for some other reason, could result in dilution of the concentration of defensive chemicals and potentially reduce defensive efficacy. In this sense the defense is constrained by diet breadth. On the other hand, if predation selects for effective defenses, there would be selection for maintenance of the existing reduced diet breadth. This hypothesis predicts that any increase in diet breadth could potentially lead to the loss of the sequestered chemical defense syndrome, unless the new hosts also contain these chemicals. This might be an explanation for oligophages that have sequestered defenses and hosts of common chemistry, similar to oligophages arising from a reduction in diet breadth of generalists (see Section V,C).

This hypothesis assumes the same type of relationship between concentration ingested and concentration in defense (see Section IV,B), and also that there are selection pressures that can lead to an increase in diet breadth (see Section V,B). Testing the hypothesis would follow the same basic ideas as those proposed for generalists (see Section V,C). Specialists could be reared on hosts with varying concentrations of compounds, or varieties with and without such chemicals, and the concentration and efficacy of the defense measured. Analogous studies have been done by Brower *et al.* with the monarch (1982; Seiber *et al.*, 1980). An alternative would be to carry out dosing experiments for insects reared on natural or artificial diets, augmenting the diet with the natural defensive chemicals or with synthetic analogs. Finally, it may be possible to study natural populations of such specialists in areas where host range has expanded and compare the defensive chemistry and efficacy of populations on the single and multiple hosts.

E. Conclusion

Chemical defense in insects is cited as an example of diversity and eclecticism in natural products (Blum, 1981). Attempts to discern and understand patterns in this diversity seem worthwhile. We would never claim that considerations of diet breadth will explain all of the patterns of chemical defense in insects. However, we do consider that diet breadth is a significant factor that can initiate and constrain the evolution of chemical defenses. Effects of diet breadth on sequestration can be relatively simply modeled as the ecological equivalent of concentration and dilution, and effects can be applied to both autogenous and sequestered defenses in specialists and generalists. Further consideration of the role of diet breadth may help us understand the evolutionary and ecological patterns and processes that underlie the eclectic diversity of chemical defenses in insects.

Acknowledgments

We thank Tim Hess, Beth Price, Steve Compton, Patrick Cusack, and S. H. Tan for analytical help in the project, and Jim Coleman, Alcinda Lewis, Gary Lovett, Jacques Pasteels, and Kevin Spencer for critical comment. As usual, reviewers improved the manuscript. Contribution to the program was from the Institute of Ecosystem Studies, The New York Botanical Garden. Financial support was provided by the Mary Flagler Cary Trust and National Science Foundation (DEB-8117943 to C. G. Jones and DEB-8117999 to M. S. Blum).

References

Aplin, R. T., Benn, M. H., and Rothschild, M. (1968). Poisonous alkaloids in the body tissues of the cinnabar moth (*Callimorpha jacobeae* L.). *Nature (London)* **219,** 747–748.

Berenbaum, M. (1981a). Patterns of furanocoumarin production and insect herbivory in a population of wild parsnip (*Pastinaca sativa* L.). *Oecologia* **49,** 236–244.

Berenbaum, M. (1981b). Effects of linear furanocoumarins on an adapted specialist insect (*Papilio polyxenes*). *Ecol. Entomol.* **6,** 345–351.

Bernays, E. A., Edgar, J. A., and Rothschild, M. (1977). Pyrrolizidine alkaloids sequestered and stored by the aposematic grasshopper, *Zonocerus variegatus*. *J. Zool.* **182,** 85–87.

Birch, M. C. (1978). Chemical communication in pine bark beetles. *Amer. Sci.* **66,** 409–419.

Blum, M. S. (1981). "Chemical Defenses of Arthropods." Academic Press, New York.

Blum, M. S., Whitman, D. W. Arrendale, R. F., and Severson, R. F. (1987). Herbivores and toxic plants: Evolution of a menu of options for processing allelochemicals. *Insect Sci. Applic.* **8,** 459–463.

Boelens, M., Valois, P. J. D., Wobben, H. J., and Gen, A. V. D. (1971). Volatile flavor compounds from onion. *J. Agric. Food Chem.* **19,** 984–991.

Boonsong, C., and Wright, S. E. (1961). The cardiac glycosides present in mistletoes growing on *Nerium oleander*. *Aust. J. Chem.* **14,** 449–457.

Bournier, A. (1977). Grape insects. *Annu. Rev. Entomol.* **22,** 355–376.

Brodnitz, M. H., Pollock, C. L., and Vallon, P. P. (1969). Flavor components of onion oil. *J. Agric. Food Chem.* **17,** 760–763.

Brower, L. P., and Brower, J. V. Z. (1964). Birds, butterflies and plant poisons: A study in ecological chemistry. *Zoologica* **49,** 137–159.
Brower, L. P., and Glazier, S. C. (1975). Localization of heart poisons in the monarch butterfly. *Science* **188,** 19–25.
Brower, L. P., Brower, J. V. Z., and Corvino, J. M. (1967). Plant poisons in a terrestrial food chain. *Proc. Natl. Acad. Sci. U.S.A.* **57,** 893–898.
Brower, L. P., Seiber, J. N., Nelson, C. J., Lynch, S. P., and Tuskes, P. M. (1982). Plant-determined variation in the cardenolide content, thin-layer chromatography profiles, and emetic potency of monarch butterflies, *Danaus plexippus,* reared on the milkweed *Asclepias eriocarpa* in California. *J. Chem. Ecol.* **8,** 579–633.
Brower, L. P., Seiber, J. N., Nelson, C. J., Lynch, S. P., Hoggard, M. P., and Cohen, J. A. (1984a). Plant-determined variation in cardenolide content and thin-layer chromatography profiles of monarch butterflies, *Danaus plexippus* L. reared on milkweed plants in California. 3. *Asclepias californica. J. Chem. Ecol.* **10,** 1823–1857.
Brower, L. P., Seiber, J. N., Nelson, C. J., Lynch, S. P., and Holland, M. M. (1984b). Plant-determined variation in the cardenolide content, thin-layer chromatography profiles, and emetic potency of monarch butterflies, *Danaus plexippus* L. reared on milkweed plants in California. 2. *Asclepias speciosa. J. Chem. Ecol.* **10,** 601–639.
Carson, J. F., and Wong, F. F. (1961). The volatile flavor components of onions. *J. Agric. Food Chem.* **9,** 140–143.
Chiang, H. C. (1978). Pest management in corn. *Annu. Rev. Entomol.* **23,** 101–123.
Dadd, R. H. (1960). The nutritional requirements of locusts—1. Development of synthetic diets and lipid requirements. *J. Insect Physiol.* **4,** 319–347.
Daloze, D., and Pasteels, J. M. (1979). Production of cardiac glycosides by chrysomelid beetles and larvae. *J. Chem. Ecol.* **5,** 63–77.
Duffey, S. S. (1970). Cardiac glycosides and distastefulness: Some observations on the palatability spectrum of butterflies. *Science* **169,** 78–79.
Duffey, S. S. (1976). Arthropod allomones: Chemical effronteries and antagonists. *Proc. Int. Congr. Entomol.* **15,** 323–394.
Duffey, S. S. (1980). Sequestration of plant natural products by insects. *Annu. Rev. Entomol.* **25,** 447–477.
Duffey, S. S., and Scudder, G. G. E. (1972). Cardiac glycosides in North American Asclepiadaceae, a chemical basis for unpalatability in brightly coloured Hemiptera and Coleoptera. *J. Insect Physiol.* **18,** 63–78.
Duffey, S. S., Blum, M. S., Isman, M. B., and Scudder, G. G. E. (1978). Cardiac glycosides: A physical system for their sequestration by the milkweed bug. *J. Insect Physiol.* **24,** 639–645.
Ehrlich, P. R., and Raven, P. H. (1964). Butterflies and plants: A study in coevolution. *Evolution* **18,** 586–608.
Eisner, T., Hendry, L. B., Peakall, D. B., and Meinwald, J. (1971). 2,5-Dichlorophenol (from ingested herbicide?) in defensive secretion of grasshopper. *Science* **172,** 277–278.
Eisner, T., Jones, T. H., Aneshansley, D. J., Tschinkel, W. R., Silberglied, R. F.,

and Meinwald, J. (1977). Chemistry of defensive secretions of bombardier beetles (Brachinini, Metriini, Ozaenini, Paussini). *J. Insect Physiol.* **23,** 1383–1386.

Eto, M. (1974). "Organophosphorous Pesticides: Organic and Biological Chemistry." CRC, Cleveland.

Feltwell, J. S. E., and Valadon, L. R. G. (1970). Plant pigments identified in the common blue butterfly. Nature, (*London*) **255,** 969.

Feltwell, J. S. E., and Valadon, L. R. G. (1972). Carotenoids of *Pieris brassicae* and of its food plant. *J. Insect Physiol.* **18,** 2203–2215.

Fox, D. L. (1976). "Animal Biochromes and Structural Colors." Univ. of California Press, Berkeley.

Fox, L. R., and Morrow, P. A. (1981). Specialization: Species property or local phenomenon. *Science* **211,** 887–893.

Freudenberg, K. (1965). Lignin: Its contribution and formation from *p*-hydroxycinnamyl alcohols. *Science* **148,** 595–600.

Goeden, R. D. (1971). The phytophagous insect fauna of milk thistle in Southern California. *J. Econ. Entomol.* **64,** 1101–1104.

Goeden, R. D. (1974). Comparative survey of the phytophagous insect faunas of Italian thistle, *Carduus pycnocephalus*, in southern California and southern Europe relative to biological weed control. *Environ. Entomol.* **3,** 464–474.

Goeden, R. D., and Ricker, D. W. (1968). The phytophagous insect fauna of Russian thistle (*Salsola kali* var. *tenuifolia*) in southern California. *Ann. Entomol. Soc. Am.* **61,** 67–72.

Harrewijn, P. (1978). The role of plant substances in polymorphism of the aphid *Myzus persicae*. *Entomol. Exp. Appl.* **24,** 198–214.

Hedin, P. A. (1977). A study of factors that control the compounds which comprise the boll weevil pheromone. *J. Chem. Ecol.* **3,** 279–289.

Howard, R., Matsumara, F., and Coppel, H. C. (1976). Trail-following pheromones of the rhinotermitidae: Approaches to their authentication and specificity. *J. Chem. Ecol.* **2,** 147–166.

Hughes, P. R. (1974). Myrcene: A precursor of pheromones in *Ips* beetles. *J. Insect Physiol.* **20,** 1271–1275.

Isman, M. B. (1977). Dietary influences of cardenolides on larval growth and development of the milkweed bug *Oncopeltus fasciatus*. *J. Insect Physiol.* **23,** 1183–1187.

Jones, C. G., Hoggard, M. P., and Blum, M. S. (1983). Is sequestration structure-specific in the milkweed bug, *Oncopeltus fasciatus*? *Comp. Bioch. Physiol.* **76C,** 283–284.

Jones, C. G., Hess, T. A., Whitman, D. W., Silk, P. J., and Blum, M. S. (1986). Idiosyncratic variation in chemical defenses between individual generalist grasshoppers. *J. Chem. Ecol.* **12,** 749–761.

Jones, C. G., Hess, T. A., Whitman, D. W., Silk, P. J., and Blum, M. S. (1987). Effects of diet breadth on autogenous chemical defense of a generalist grasshopper. *J. Chem. Ecol.* **13,** 283–297.

Jones, C. G., Whitman, D. W., Compton, S. J., Silk, P. J., and Blum, M. S. (1988). Diet breadth determines the type and efficacy of chemical defense in a generalist grasshopper. *J. Chem. Ecol.* (submitted).

Kevan, D. K. M. (1980). *Romalea guttata* (Houttyun), name change for well-known "Eastern Lubber Grasshopper" (Orthoptera: Romaleidae). *Entomol. News* **91,** 139–140.

Matsumara, F. (1975). "Toxicology of Insecticides." Plenum, New York.

Meinwald, J., Erickson, K., Hartshorn, M., Meinwald, Y. C., and Eisner, T. (1968). Defensive mechanisms of arthropods. XXIII. An allenic sesquiterpenoid from the grasshopper *Romalea microptera. Tetrahedron Lett.* **25,** 2959–2962.

Nahrstedt, A., and Davis, R. H. (1983). Occurrence, variation and biosynthesis of the cyanogenic glucosides linamarin and lotaustralin in species of the Heliconiini (Insecta: Lepidoptera). *Comp. Biochem. Physiol.* **75B,** 65–73.

Needham, A. E. (1968). Insect biochromes: Their chemistry and role. *In*, (M. Rockstein, ed.), pp. 233–305. "Biochemistry of Insects" Academic Press, New York.

O'Brien, R. D. (1967). "Insecticides, Action and Metabolism." Academic Press, New York.

Owen, D. F., and Whiteway, W. R. (1980). *Buddleia davidii* in Britain: History and development of an associated fauna. *Biol. Conserv.* **17,** 149–155.

Pasteels, J. M., and Daloze, D. (1977). Cardiac glycosides in the defensive secretion of chrysomelid beetles: Evidence for their production in insects. *Science* **197,** 70–72.

Pasteels, J. M., Gregoire, J.-C., and Rowell-Rahier, M. (1983a). The chemical ecology of defense in arthropods. *Annu. Rev. Entomol.* **28,** 263–289.

Pasteels, J. M., Rowell-Rahier, M., Braekman, J. C., and Dupont, A. (1983b). Salicin from host plant as a precursor of salicylaldehyde in defensive secretion of chrysomeline larvae. *Physiol. Entomol.* **8,** 307–314.

Patwardhan, V. N., and White, J. W., Jr. (1973). Problems associated with particular foods. *In* "Toxicants Occurring Naturally in Foods" (Comm. Food Protection F.N.B., N.R.C., eds.), pp. 495–507. Natl. Acad. Sci., Washington, D.C.

Rees, C. J. C. (1966). Ph.D. thesis, Oxford University.

Rees, C. J. C. (1969). Chemoreceptor specificity associated with choice of feeding site by the beetle *Chrysolina brunsvicencis* on its food plant *Hypericum hirsutum. Entomol. Exp. Appl.* **12,** 565–583.

Reichstein, T., von Euw, J., Parsons, J. A., and Rothschild, M. (1968). Heart poisons in the monarch butterfly. *Science* **161,** 861–866.

Roeske, C. N., Seiber, J. N., Brower, L. P., and Moffitt, C. M. (1976). Milkweed cardenolides and their comparative processing by monarch butterflies (*Danaus plexippus* L.) *Recent Adv. Phytochem.* **10,** 93–167.

Root, R. B., and Tahvanainen, J. O. (1969). Role of winter cress, *Barbarea vulgaris*, as a temporal host in the seasonal development of the crucifer fauna. *Ann. Entomol. Soc. Am.* **62,** 852–855.

Rothschild, M. (1967). Mimicry: The deceptive way of life. *Nat. Hist.* **76,** 44–51.

Rothschild, M. (1971). Speculations about mimicry with Henry Ford. *In* (R. Creed, ed.), pp. 202–223. "Ecological Genetics and Evolution" Blackwell, Oxford.

Rothschild, M. (1973). Secondary plant substances and warning colouration in insects. *Symp. R. Entomol. Soc. London* **6,** 59–83.

Rothschild, M., and Aplin, R. T. (1971). Toxins in tiger moths (Arctiidae: Lepidoptera). *Pestic. Chem.* **3,** 177–182.

Rothschild, M., Reichstein, T., von Euw, J., Aplin, R. T., and Harman, R. R. M. (1970). Toxic Lepidoptera. *Toxicon* **8,** 293–299.

Scudder, G. G. E., and Duffey, S. S. (1972). Cardiac glycosides in the Lygaeinae. *Can. J. Zool.* **50,** 35–52.

Seiber, J. N., Tuskes, P. M., Brower, L. P., and Nelson, C. J. (1980). Pharmacodynamics of some individual cardenolides fed to larvae of the monarch butterfly (*Danaus plexippus* L.). *J. Chem. Ecol.* **6,** 321–339.

Seiber, J. N., Brower, L. P., Lee, S. M., McChesney, M. M., Cheung, H. T. A., Nelson, C. J., and Watson, T. R. (1986). Cardenolide connection between overwintering monarch butterflies from Mexico and their larval food plant *Asclepias syriaca. J. Chem. Ecol.* **12,** 1157–1170.

Strong, D. R., Jr., McCoy, E. D., and Rey, J. R. (1977). Time and the number of herbivore species: The pests of sugarcane. *Ecology* **58,** 167–175.

Strong, D. R., Lawton, J. H., and Southwood, T. R. E. (1984). "Insects on Plants. Community Patterns and Mechanisms." Blackwell, Oxford.

Taksdal, G. (1965). Hemiptera (Heteroptera) collected on ornamental trees and shrubs at the agricultural college of Norway. *Ås. Norsk Entomol Tidsskr.* **13,** 5–10.

Tschinkel, W. R. (1977). A comparative study of the chemical defensive system of tenebrionid beetles: Chemistry of the secretions. *J. Insect Physiol.* **21,** 753–783.

Turnipseed, S. G., and Kogan, M. (1976). Soybean entomology. *Annu. Rev. Entomol.* **21,** 247–282.

van Veen, A. G. (1973). Toxic properties of certain unusual foods. *In* "Toxicants Occurring Naturally in Foods" (Comm. Food Protection F.N.B., N.R.C. (eds.), pp. 448–476. Natl. Acad. Sci. Washington, D.C.

von Euw, J., Reichstein, T., and Rothschild, M. (1971). Heart poisons (cardiac glycosides) in the lygaeid bugs *Caenocoris nerii* and *Spilostethus pandurus. Insect Biochem.* **1,** 373–384.

Wheeler, A. G., Jr. (1974). Phytophagous arthropod fauna of crown vetch in Pennsylvania. *Can. Entomol.* **106,** 897–908.

Wilkinson, C. F., ed. (1976). "Insecticide Biochemistry and Physiology." Plenum, New York.

Wilson, A. (1985). Flavonoid pigments in marbled white butterfly (*Melanargia galathea*) are dependent on flavonoid content of larval diet. *J. Chem. Ecol.* **11,** 1161–1179.

Winter, T. G. (1974). New host plant records of Lepidoptera associated with conifer afforestation in Britain. *Entomol. Gaz.* **25,** 247–258.

Wray, V., Davis, R. H., and Nahrstedt, A. (1983). Biosynthesis of cyanogenic glycosides in butterflies and moths: Incorporation of valine and isoleucine into linamarin and lotaustralin by *Zygaena* and *Heliconius* species (Lepidoptera). *Z. Naturforsch.* **38C,** 583–588.

CHAPTER 17

Chemical Mimicry

MARK KWASNY STOWE*
Department of Organismic
and Evolutionary Biology
Harvard University
Boston, Massachusetts

* Address correspondence to Mark K. Stowe, c/o Dr. J. Reiskind, Department of Zoology, University of Florida, Gainesville, Florida 32611.

A WIDE VARIETY OF ORGANISMS mimic chemical signals. However, such mimicry is very poorly understood in comparison to visual mimicry. Moreover, it is likely that we are aware of a disproportionately small fraction of chemical mimicry systems; the chemical senses, poorly developed in humans, are dominant for many, if not most, organisms. In this chapter I (1) briefly consider some unusual aspects of chemical signals, (2) review all chemical mimicry systems known to me, and (3) consider, some ways that such systems might evolve. (Related phenomena are treated in Appendix B.)

I. Terminology

I use "signal" to mean any potential or actual transmission of information. This conflicts with Lloyd's (1977, 1983) terminology, where "signal" is restricted to interorganism information transmissions that benefit both emitter and receiver. For general discussions of information flow, it is necessary to have a term for any information that is actively or passively released by a living or nonliving source (a term that is appropriate to vision, hearing, contact chemoreception, and other modalities). For discussions of early stages in the evolution of interorganism signals, it is necessary to have a term that avoids assumptions concerning harm or benefit. Lloyd's (1977, 1983) otherwise useful system does not include such a term. No English word is completely appropriate. "Signal" as it is used in the engineering literature comes closest to the intended meaning ("signal to noise ratio," "the signal from a pulsar").

A "mimic" organism profits from the response of a "dupe" organism to a signal from the mimic. The response is the same as the one that the dupe normally makes to a particular signal of interest from a (living or nonliving) entity in the dupe's environment. The dupe perceives the mimic signal to be an instance of this "model" signal (definition/terminology after Pasteur, 1982). A "mimicry system" consists of a mimic species, the model entities, and the set of all sympatric species that act as dupes. I make minimal use of "mimicry categories" such as Batesian mimicry. (See Wickler (1968), Vane-Wright (1976), Wiens (1978), and Pasteur (1982) for various viewpoints, and Lloyd (1984) for a delightful review emphasizing mimicry systems that defy categorization. These works are hereafter referred to as the "mimicry reviews.")

II. Unusual Aspects of Chemical Signals

Signals can provide two types of information about their sources: identity and location. To identify the signal source, organisms frequently use the signal's small-scale temporal and spatial details. For example, to identify a bird species from its song, we often use the temporal patterning of the song's changing pitch. To identify a butterfly species from its visual appearance, we often use the spatial patterning of the butterfly's varied colors. Do organisms ever use temporal and spatial details to identify the sources of chemical signals?

Under most natural conditions, any patterning that is present at the source of an air- or water-borne chemical signal is quickly overwhelmed by the turbulence and flow of the fluid media. However, temporal patterning in fluid-borne, long-distance chemical communication is possible under restrictive conditions, (refs. in Elkinton and Carde, 1984; Mayer and Mankin, 1984), and is important in some soil organisms (e.g., Bonner, 1983). Moreover, there is no reason that surface chemical signals could not vary spatially or temporally. Despite these possibilities, there are relatively few examples in the literature of organisms using details of small-scale temporal or spatial patterning in chemical signals for identification. [In many moth species, females pulse their sex pheromones (review in Conner *et al.,* 1980). However, it is doubtful that the pheromone plume from a pulsed source is effectively different from one produced by continuous release (Baker *et al.,* 1985). There is no direct evidence for Conner *et al.*'s 1980 suggestion that pulsing helps males locate females (see Cardé, 1986, for a review, and other likely advantages of pulsing).]

In many nonchemical mimicry systems, the mimic must reproduce, to a greater or lesser extent, the temporal and spatial details of the model signal; this is probably not the case in most chemical mimicry systems. The relative "simplicity" of chemical signals is one reason for suspecting that chemical mimicry may operate somewhat differently from most other forms of mimicry.

Visual and auditory signals possess many elemental properties (e.g., tone, color) that are "scalable" in a simple way. If one draws pairs of tones from a set, there is an obvious scale (wavelength) that one can use to judge relative similarity of tone pairs. In contrast, if one is given a set of different chemical compounds, there is no single, obvious scale for judging relative similarity. There are in fact many conceivable ways of decoding the chemical environment. Not surprisingly, organisms have evolved diverse chemical decoding schemes involving different arrays of variously shaped chemical receptors. Even though our understanding of these schemes is very poor, one could argue that the chemical sense is intrinsically the sense that is "least shared" among species. Therefore, in chemical mimicry systems, the perception of mimic signals may be very different from dupe species to dupe species, and it will be particularly dangerous to generalize from one system to others. [See Theimer (1982) for further discussion of differences between the chemical and nonchemical senses.]

III. A Review of Known Systems

Unfortunately, space permits only a very brief description of each system. Although it pains me to do so, I omit descriptions of the elaborate entrapping mechanisms and other incredible behavior and morphology that accompanies these systems. These are some of the most exciting natural history stories I know of, and I urge any reader unfamiliar with them to seek out the references. [For detailed descriptions and photographs see, in particular the reviews in Slack (1979) and Meeuse and Morris (1984).] Throughout, taxonomy will follow cited literature.

A. The Various Levels of Evidence for Mimicry

Chemical mimicry systems are poorly known for several reasons. (1) Many occur in low-light environments where chemical communication is particularly prevalent but fieldwork is difficult: the organisms live in soil, or are strictly nocturnal. I know of no reported marine chemical mimicry systems, but as Pasteur (1982) notes, this is surely an artifact of our sensory limitations, and the difficulty of fieldwork. (2) In many of the chemical mimicry systems, at least one of the species involved, usually the mimic, is quite rare. (3) Many of the organisms have proven resistant to study in the laboratory and must be studied in the field, often in locations that are remote from the instrumentation necessary for analyzing the chemical signals.

The level of evidence for the existence of mimicry varies considerably among the systems described in the literature. In many cases, mimicry must be inferred from natural history observations alone. Where the effect of the mimic signal on the dupe has been demonstrated, it is often impossible to exclude the possibility of nonchemical signals imperceptable to the experimenters. In a few cases, the mimic signal has been analyzed and the effects of the signal on the dupe have been reproduced with blends of the appropriate synthetic compounds. However, even these most advanced studies of chemical mimicry systems suffer in comparison to the best work with other types of chemical communication. (1) The techniques employed until now allow only the vaguest estimation of component release rates and ratios in the original and the synthetic signal. (Release rates and ratios are often critical in chemical communication.) (2) The composition of the signal is usually inferred from whole-gland extracts rather than from collections of volatiles actually released into the environment. (Frequently the volatiles a gland gives off are quite different from what it stores.) (3) The signal analyzed is usually a combination of material from many individuals, each of which may be producing different signals. If sensitive new techniques for accurate measurement and reproduction of chemical signals are more widely applied, our knowledge of the chemical mimicry systems should improve dramatically (see techniques and discussion in Heath *et al.*, 1986a,b, 1988; Heath and Tumlinson, 1986; Stowe *et al.*, 1987).

A few of the reports I review demonstrate that dupes prefer air bearing the odor of mimic material to "unscented" air in two-choice laboratory tests (or the reports provide analogous test results for soil organisms). However, such tests provide only very preliminary evidence that mimic odors are able to attract dupes from a distance in the field. (For more comments on methodology, see Müllerian mimicry section below.)

For most systems, I do not describe the evidence for mimicry. (A few strictly suggestive cases are identified as such.) Instead, I concentrate on describing (1) the mimic and model signals [always including the chemistry in the (rare) instances where this is possible] and (2) what is known of the mimic's effect on the dupe's fitness. For many chemical prey lure systems, it is not clear what the lure (mis)represents. I include these cases on the assumption that the mimic must imitate some signal of interest to the dupe. My principal goal is to illustrate the widespread nature of this interesting phenomenon, which desperately needs further investigation.

B. Deception in the Social Communication of Bees, Ants, Termites, and Aphids

Many associates of social insects appear to chemically manipulate the behavior of their hosts through airborne odors, cuticular contact, and "feeding" secretions. Associates elicit from their hosts attraction, regurgitation, appeasement, transport into the nest, etc. In some cases, this manipulation apparently involves mimicry of host pheromones (reviews in Wilson, 1971; Hölldobler, 1977; Cottrell, 1983). I review below some of the better-understood systems (see also the prey-lure and pollination sections).

Some nest parasites of social insects (Fig. 1) are apparently protected from host attacks because their cuticle chemistry is very similar to that of their hosts. The complex cuticle hydrocarbon blends of parasite and host are virtually identical, and the hosts apparently cannot (chemotactilely) distinguish parasites from conspecifics. Nest parasites in the staphylinid beetle genus *Trichopsenius* biosynthesize their hydrocarbon mixture; the pattern is specific to the beetle species and matches the species-specific mixture of their termite hosts (Howard *et al.*, 1980, 1982). In contrast, the ant associate *Myrmecaphodius excavatacollis*, a scarabaeid beetle, acquires its cuticle hydrocarbons through contact with the host. The ability to acquire hydrocarbons apparently allows the beetles to successively invade the nests of several species of hosts with different hydrocarbon blends (Vander Meer and Wojcik, 1982). Howard *et al.* (1988) have recently found that hydrocarbon blend matching also occurs in syrphid flies (*Microdon* species) that live in ant nests.

A related phenomenon is the "wolf-in-sheep's-clothing strategy" (Eisner *et al.*, 1978) used by at least two groups of predators. Larvae of the neuropteran *Chrysopa slossonae* (Chrysopidae) cover themselves with wax plucked from their wax-coated aphid prey (*Prociphilus tesselatus*). The ants that tend these aphids and defend them from most other predators fail to recognize the neuropterans because of their covering and do not attack them (Eisner *et al.*, 1978, includes chemistry). Similarly, two species of spider that prey on ants, *Strophius nigricans* (Thomisidae)

Fig. 1. Two "nest parasite" beetles that avoid eliciting aggression from their social insect hosts by mimicking host cuticular hydrocarbon patterns. (Both photos are of U.S. species and were taken in artificial nests.) (a) *Trichopsenius frosti* beetles (Staphylinidae) live by soliciting meals from their termite hosts. The beetles aggregate (photo) and mate only on the bodies of host queens (*Reticulitermes flavipes*). These host-specific termitophiles biosynthesize their cuticular hydrocarbons (see Howard *et al.*, 1980, 1982, and refs. therein). (Courtesy of R. W. Howard.) (b) *Myrmecaphodius excavaticollis* beetles (Scarabaeidae) feed on ant host larvae (photo) and solicited meals. These beetles acquire cuticular hydrocarbons from their hosts and are able to integrate themselves into the nests of several ant species (in photo: *Solenopsis invicta* (see Vander Meer and Wojcik, 1982). (Courtesy of D. P. Wojcik.)

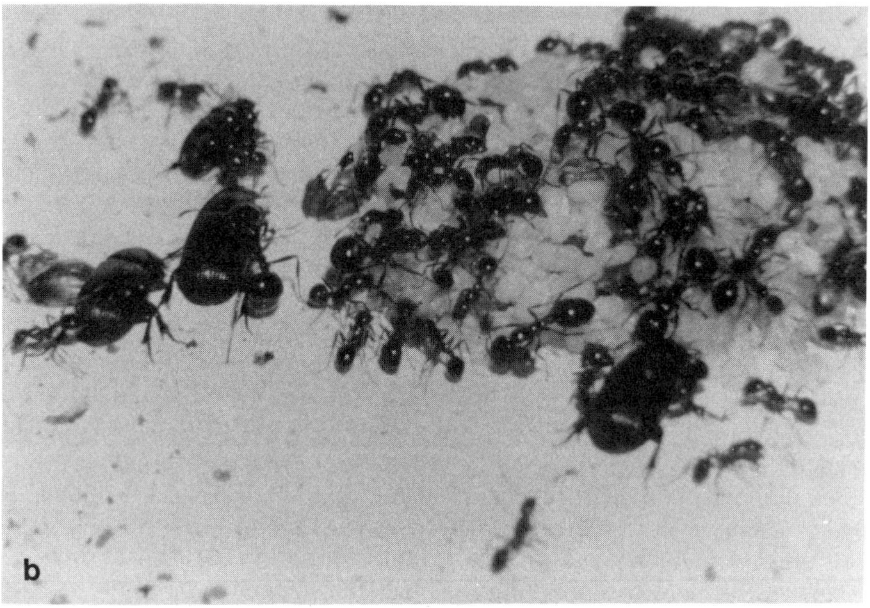

Fig. 1. (*continued*)

and *Aphantochilus rogersi* (Aphantochilidae), are able to mingle with their prey by holding a dead ant aloft and using it as a pacifying "shield" against approaching ants (Oliveira and Sazima, 1985). In both spider and lacewing systems, it is likely that the chemical signals provided by the "sheep's clothing" are critical to the success of the strategy.

A number of arthropods may repel attacking ants by giving off odors that mimic ant alarm pheromones. Kistner and Blum (1971) suggest that some staphylinid beetles associated with ant nests use alarm pheromone mimicry to ward off attacking hosts (but see Hölldobler *et al.*, 1981). The defensive secretions of a wide array of arthropods are known to contain compounds identical to the principal components of common ant alarm pheromones; many mutillid wasps, opilionids, cockroaches, and beetles produce C_6 to C_9 alkanones and various beetles produce *n*-undecane and formic acid (reviews in Fales *et al.*, 1980; Blum, 1980, 1984). Blum (1980, 1984) and Fales *et al.* (1980) suggest that the use of these compounds has evolved repeatedly in arthropods, like those just mentioned, that dwell on the ground and frequently contact ants.

Tests are needed to confirm that these arthropods actually repel ants in the same manner as alarm pheromones, and to show that the suspect compounds (in the concentrations actually found in the defensive sprays) are responsible. Particularly at night, these compounds might also "frighten" nonformicid predators that associate ant alarm pheromones with distasteful or dangerous ants. Protective (Batesian) mimicry is an additional, untested possibility. (This assumes that the predator can eat the mimic and is not deterred by the noxious nature of the secretion. Mimicry systems that involve defensive odors must be interpreted with caution—see Müllerian mimicry section below.)

An alarm pheromone for many aphids, $(E)\beta$-farnesene, is a major component of the volatiles emitted by the leaves of the wild potato *Solanum berthaultii* (Gibson and Pickett, 1983). These leaves repel aphids. Air from around leaves and air with similar concentrations of $(E)\beta$-farnesene produce a dispersal response in aphids similar to that produced by their own alarm pheromone. The evidence suggests that the effect is mimetic, but a role for other volatile components cannot be excluded. Because the plants are covered with potentially lethal, sticky, glandular hairs (Gibson and Pickett, 1983), aphids might actually benefit from being repelled.

A number of social insects specialize in plundering nests belonging to other social insect species. Some of these invaders release strong odors during raids, which profoundly affect the behavior of the victims such that they are incapable of mounting a normal coordinated defense of the nest. There are at least three (not necessarily exclusive) hypotheses concerning the efficacy of these "raiding allomones" (reviews and discussion in Blum, 1966; Blum and Brand, 1972). (1) The odors mimic the alarm pheromones of their victims (the chemistry is similar in some cases). (2) The compounds may be intrinsically irritating and/or highly excitatory to olfactory neurons (possibly because of some similarity to the alarm pheromone). Released in large quantities (often in enclosed spaces), they may simply overwhelm the sensory input of the victims, making chemical communication amongst the victims impossible. (3) The compounds may be "omens of impending disaster" (Blum and Brand, 1972) signaling the presence of an unbeatable enemy; selection may have favored those nests with a submissive behavioral response that minimizes casualties. (This assumes that the nests are left weakened but with some chance of reproduction.)

Raiding allomones have been identified for some bees that raid bees and some ants that raid ants. Bees in this category include (a) *Lestrimelitta limao,* which uses citral (Blum, 1966; Blum *et al.*,1970; see Weaver *et al.*, 1975b), and (b) *Trigona tataira,* which uses highly reactive C_7 and C_9 conjugated alkendiones (Bian *et al.*, 1984; see Weaver *et al.* 1975b, for other possibilities). Ants include (a) *Formica* species in the *sanguinea* group, which use decyl-, dodecyl-, and tetradecylacetates (Regnier and Wilson, 1971; see also Blum *et al.*, 1970; Alloway, 1979), and (b) *Solenopsis fugax,* which uses *trans*-2-butyl-5-heptylpyrrolidine (Blum *et al.*, 1980). (*Monomorium pharaonis,* which produces 2,5-dialkylpyrrolidines, might also be a raider.) *Lestrimelitta limao* and *F. sanguinea* are perhaps the only two groups where the mimetic effect has a reasonable chance of being important (but see Müllerian and dupes-benefit mimicry sections below). In both systems, the raiders specialize on a small number of victim species. In *L. limao* there is indirect evidence that the alarm pheromone chemistry of the victim species is important; citral (the raiding allomone) is used by victim species but not by congeners that are never raided (Blum *et al.*, 1970; see Smith and Roubik,1983).

Female bees in the Anthophorid genus *Nomada* are brood parasites of a number of Andrenid and Melittid solitary bee species. *Nomada* males emit and probably spray females with cephalic secretions during copulation. Tengö and Bergström (1976a, 1977a) propose that the composition of the male spray is such that it gives females a lingering odor that is similar to that of their hosts. This similarity presumably reduces the disruption of host nest-building activity caused by the presence of *Nomada* females. Tengö and Bergström base their proposal (in part) on the close correspondence between male parasite and female host chemistry in six of the nine parasite/host systems they studied. In these six systems the set of principal components of the parasite male cephalic secretion is identical to the set of principal components of the host female Dufour's gland secretion [components: one or more of octadecyl acetate, octadecyl butanoate, geranyl octanoate, farnesyl hexanoate (Tengö and Bergström, 1976a, 1977a)]. Unfortunately, there is at present only indirect observational evidence for the proposal (the parasites have proven very difficult to work with; J. Tengö, personal communication). Such a system may have started with males attracting their mates by mimicking the scents that females use to locate hosts. [Chemical information is available for other genera and families where future work might reveal similar phenomena (Hefetz *et al.*, 1982; Borg-Karlson and Tengö, 1980).]

Phorid flies in the genus *Diplonevra* are termite parasitoids with an unusual method of obtaining hosts (Disney, 1986). A female fly approaches its victim (always a worker of a Nasutitermitinae species), touches its abdomen and then runs away from the colony. The termite follows quickly behind the fly and is led some distance away—presumably to a suitable final resting place. The victim acts as though it is strongly attracted by the fly's odor. (Perhaps the odor mimics that of a host pheromone.)

C. Sexual Ruses

In a wide variety of organisms, males have been observed to mimic females under conditions where this gives them an advantage in courtship. In red-sided garter snakes (*Thamnophis sirtalis parietalis*), some males apparently produce the odor of the female pheromone; solvent extracts of the skin of "she-males" are as exciting to males as extracts from females (Mason and Crews, 1985). She-males confuse fellow males, and are able to sneak matings with real females in highly competitive, male-biased "mating balls." (However, this mimicry may not represent an alternative mating strategy so much as it does a developmental abnormality; see Mason, 1986, 1987).

In a number of insect species, males have been reported to pheromonally elicit courtship behavior (and in some cases long-distance attraction) from other males. Homosexual courtship has been observed in *Drosophila melanogaster* (numerous refs. in Peschke, 1985; Hall, 1986a), many solitary bees (Borg-Karlson, 1987; Cane and Tengö, 1981, and refs. therein, includes chemistry; J. Tengö, personal communication), staphylinid beetles in the genera *Leistotrophus* (Forsyth and Alcock, 1988) and *Aleochara* (*A. curtula*; Peschke, 1985, 1987), and in a few other insects (refs. in Peschke, 1985). This phenomenon might be more widespread; it could be easily overlooked in species where it is only newly eclosed males that elicit courtship.

This male–male courtship need not involve mimicry. Consider species where (1) characteristic odors are associated with eclosion (e.g., modified cuticle lipids), (2) newly eclosed males and females do not differ in any way that makes their odors different, and (3) newly eclosed females are capable of mating. If receptive females are scarce, then males may increase their reproductive fitness by seeking out and courting any newly eclosed individual whose odor they detect, even if there is a significant chance that the individual is a male.

Only if males derive some benefit from attracting courting males is such a system mimetic. Hall (1986a) suggests that (perhaps) in the *Drosophila* system, young males that attract older males gain by learning the local dialect of the courtship song. Cane and Tengö (1981) note that newly eclosed male solitary bees, which must dig their way up to the soil surface, derive some benefit from the attracted, sexually excited males; the older males dig down toward the emerging bees. However, they believe that this benefit is probably more than offset by the delay and by occasional injury associated with homosexual mauling. In the two staphylinid beetle systems, young males that mimic females clearly benefit; they avoid eliciting aggression from older (stronger) males (Peschke, 1985; Forsyth and Alcock, 1988). (In this system, young males cannot court females. Older males cannot mimic females and at the same time successfully court real females.) Similarity in female and young male cuticle hydrocarbon patterns suggests that cuticle chemistry mediates the behaviors (Peschke, 1985). Numerous subtleties not treated here complicate the treatment of these and other intraspecific mimicry systems—(see Markl, 1979; Hall, 1986b; Peschke, 1987, and refs. therein).

In many parthenogenetic species females solicit copulation from males of a closely related species. Chemical signals almost certainly play a role (see in particular Woodruffe, 1958). Males lose time and sperm. The females gain because they need stimulation (but not sperm) from males to trigger development. (Solicited males typically belong to the species that is thought to have given rise to the parthenogenetic species—e.g., refs. in McKay, 1971.) Lloyd (1984) suggests that, in some species, females may also gain nutritive rewards during insemination. Chemical signals probably play a role in any system where males or females falsely represent themselves to be receptive females [see dupes-benefit section below, and Lloyd (1984) and Mitter and Klun (1987)— includes chemistry—for more intraspecific examples, and see pollination and prey lure sections below for more interspecific examples].

D. Lures for Carriers of Pollen and Pycniospores

An enormous number of plants mimic odors of interest to their pollinators. Yet surprisingly little experimental and chemical work has been done [see refs. in Dafni (1984) and Yeo (1972, p. 53) for studies into the cues used by pollinators]. The vast literature in this area (mostly superb natural history) is covered in a number of reviews (the "pollination reviews": van der Pijl and Dodson, 1966, pp. 129–142; Wickler, 1968; Yeo, 1972; Wiens, 1978; Pasteur, 1982; Dafni, 1984; Meeuse and Morris, 1984; see also Kevan and Baker, 1983; Faegri and van der Pijl, 1979, pp. 102–106). Plants usually mimic visual and/or tactile signals in addition to the chemical signals. However, in most systems where tests have been conducted, chemical signals alone are sufficient for attracting the pollinator. I review the principal model categories paying special attention to (1) those not covered in past reviews and (2) those where chemical work has been done.

1. Oviposition-Site/Food Mimicry

Many flowers emit fetid odors that attract "coprophilous insects" (i.e., insects—mostly flies and beetles—that oviposit on, or feed at, carrion or dung) (reviews by Meeuse cited in Dafni, 1984; see Ponchet *et al.*, 1982, for more scent chemistry). The flowers of various Araceae produce ammonia, indole, skatole, trimethylamine, and other amines. The first four compounds are major components of the odor of urine and feces (see Table I). For various coprophilous insects, the first three compounds are used to (1) locate oviposition sites, and/or (2) stimulate oviposition (refs. in Table I).

Many of these flowers also produce excess heat and carbon dioxide (refs. in Dafni, 1984). The heat presumably increases the emission of the volatiles mentioned above, and may be valuable to the plant for other

Table I. Organisms That Mimic Fecal/Carrion Odors and the Chemistry of Mimic Volatiles

mimic organism	dupe role	hydrogen sulfide	methylmercaptan	two- to six-carbon acids	one- to six-carbon aldehydes	phenol	2-phenylethanol	methoxybenzene compounds	benzoic acid	benzaldehyde	phenylacetic acid	phenylacetaldehyde†	ammonia*	indole*	skatole*	other amines	references
Phallus impudicus	spore vectors			x		x	x	x		x		x					Kikuchi et al 1984
		x	x	x	x						x	x					List & Freund 1966-1968
Splachnum species	spore vectors			x					x		x			?		?	Pyysalo et al 1978, 1983
various Araceae	pollen vectors												x	x	x	x	refs. in Dafni 1984
various Uredinales	spore vectors																Savile 1976
Chamaeleo species	prey																Ogilvie 1966

Notes: Mimics exploit coprophilous insects for various purposes (the "dupe role" column). For the upper set of species, mimic volatiles have been analyzed. Numerous studies have identified volatiles from feces and decaying material (mammalian feces/urine: Miner, 1977; Yasuhara *et al.*, 1984; Spoelstra, 1980, and refs. therein; avian feces: Yasuhara, 1987, and refs. therein; rotting meat: review by Dainty, 1982; refs. in Gill, 1986, pp. 52–56; rotting algae: Yasuhara and Fuwa, 1981). In the table, mimic volatile compounds identical to fecal volatile compounds are indicated. Different mimics may imitate different fecal/carrion models. In both mosses and mushrooms, odors vary considerably between species (Pyysalo *et al.*, 1978, 1983; W. R. Burk, personal communication). This may be one reason why the mimics attract different sets of coprophilous insects. Question marks indicate compounds possibly present in mimics (see text). Asterisks indicate those compounds that have been reported to attract and/or stimulate oviposition by coprophilous insects (refs. in Dafni, 1984, pp. 264–265; see also Kaib, 1974, and refs. therein; Vale *et al.*, 1986, and refs. therein). Daggers indicate compounds not presently known from fecal odors, but similar to known components. In life, all that stinks is not feces.

reasons. CO_2 is an inevitable by-product of heat production, but it might also serve to increase the amount of time pollinators stay at the flower; there is (tentative) evidence that CO_2 helps to stimulate carrion fly oviposition (refs. in Dafni, 1984). [Similarly, CO_2 may or may not play a role in the attraction of the blood-sucking insects that pollinate *Arum conophalloides* (see also Faegri and van der Pijl, 1979, pp. 102–106)]

A large number of flowers attract insects (mostly flies) that oviposit on mushrooms (Fig. 2) (pollination refs., Vogel, 1978a). While there is a developing literature on the odor of fungi themselves (Kikuchi *et al.*, 1984; Camazine, 1985, and refs. therein), the mushroom-like odors of the flowers have, unfortunately, never been analyzed.

A third group of brood/food-site mimics imitate the scent of fresh or decaying fruit, and attract beetles and flies (Gottsberger and Amaral, 1984; review in Gottsberger, 1977; Dafni, 1984). Methyl esters, which are

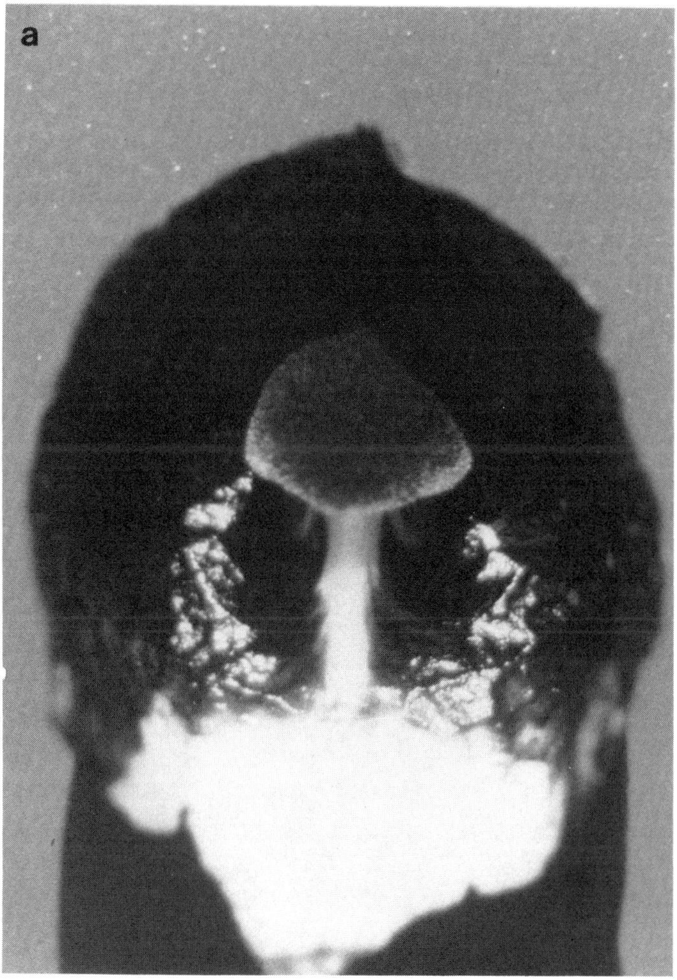

Fig. 2. A mushroom-like flower, and a flower-like mushroom; two poorly known organisms that might be both chemical and visual mimics. (Both photographs are lateral views of uprooted structures.) (a) Flowers in several families are pollinated by fungivorous flies that are apparently attracted by the mushroom-like odor of the flowers. Vogel (1978a) predicts that the poorly known Central American *Aristolochia arborea* (photo) is a member of this pollination syndrome. In many of the flower species in this syndrome, structures resembling mushrooms (photo) or parts of mushrooms are present. The gill-like involutions of these structures (not visible in photo) may stimulate fly exploration behavior and facilitate pollination. (Female flies frequently lay eggs in the flowers.) (Courtesy of W. Barthlott.)

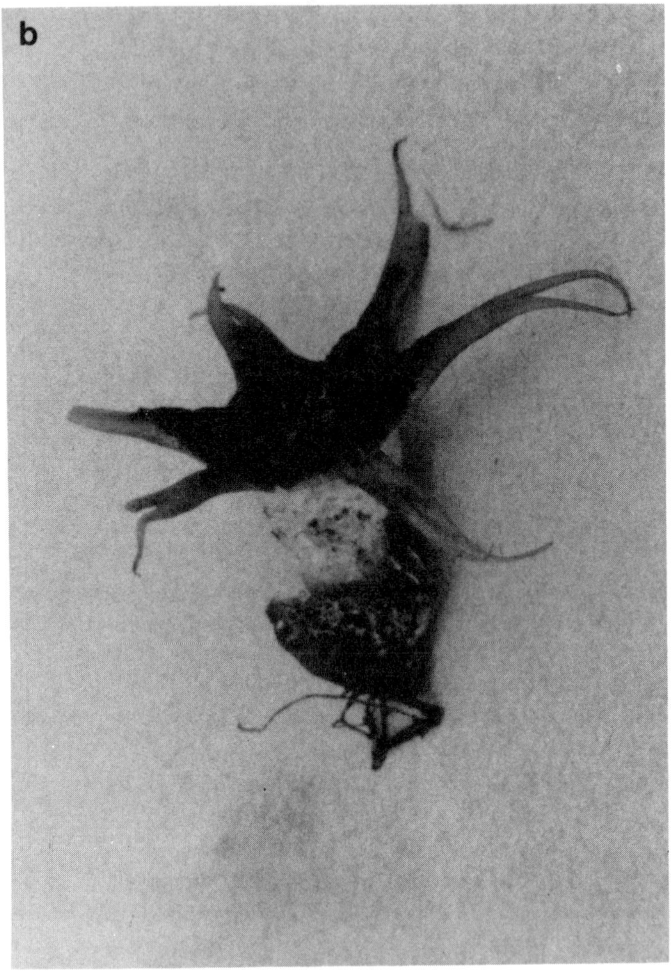

(b) The "stinkhorn mushrooms" in the order Phallales attract fly spore dispersers with a fetid odor (review in Ingold, 1953, pp. 120–125). Little is known concerning the biology of *Aseroë rubra* (photo), a species that is found in warm areas in many parts of the world. The petal-like structures are usually red, but may be pink, orange, or white (Dring, 1980). The resemblance to a flower may be purely fortuitous. (Many stinkhorn mushrooms are oddly-shaped and brightly colored.) On the other hand, the (unreported) spore dispersers might be attracted to flowers as well as carrion/dung. (Courtesy of R. D. Goos.)

present in many fruit odors and in the scents of some of these flowers (refs. in Gottsberger, 1977), may play a key role.

The effect on the fitness of the pollinators varies enormously among the different brood/food-site mimicry systems just mentioned (review in Dafni, 1984). Many (perhaps most) of the flowers confine their pollinators for extended periods—usually for a day, but in some systems for as long as several days. This is a significant fraction of the adult lifespan of many short-lived pollinators such as flies. In many of the brood-site mimicry systems, female insects lay eggs inside the flowers where larvae have no chance of survival (see Dafni, 1984; Vogel, 1978a; Meeuse and Morris, 1984). In a few systems, female or female-stage flowers trap and kill the pollinators. On the other hand, many of the flowers that confine pollinators provide a small amount of nectar. Some brood/food-site mimics actually provide significant nutritional reward (Gottsberger and Amaral, 1984; refs. in Dafni, 1984), and a few provide sites where larvae may develop.

Some orchids may mimic aphids to attract the aphid-seeking insects that are the plants' principal pollinators. Orchids in two distantly related genera apparently lure female aphidophagous syrphid flies. The orchid *Epipactis consimilis* is pollinated by at least eight fly species, principally two *Sphaerophoria* species (Ivri and Dafni, 1977), *E. gigantea* by two *Syrphis* species (T. C. Boyden, unpublished), and *Paphiopedilum rothschildianum* by the fly *Dideopsis aegrota* (Atwood, 1985). Female flies are attracted to aphids because they normally lay their eggs near aphid infestations (fly larvae eat aphids). Flies are apparently also attracted to orchids and frequently lay eggs while exploring the flowers. (Exploration and oviposition are concentrated in areas of the flower where flies are likely to activate the pollination mechanism.) Flies might be attracted by flower visual cues and, in the case of *Epipactis* species, nectar. However, the fact that olfactory cues from aphids are important for stimulating egg-laying in related aphidophagous syrphids (refs. in Ivri and Dafni, 1977) suggests that the scent of the flowers may be important to the mimicry. More observations are necessary to confirm that (1) the syrphid larvae do not usually survive and (2) the syrphid females do not belong to species that frequently lay eggs on uninfested plants that are likely to be infested later (Chandler, 1968). Faegri and van der Pijl (1979) suggest that aphids might also be mimicked by the ant-pollinated orchid *Oberonia thwaitesii*.

2. Sex-Pheromone Mimicry

A large number of nectarless flowers (mostly orchids) attract male insect pollinators by mimicking female insects (Fig. 3) (see pollination reviews). In many of these systems, part of the flower visually resembles the female insect; for a few systems males might be attracted by visual cues alone (see systems involving tachinid flies in van der Pijl and Dodson, 1966, pp. 129–142). However, for all the other systems, there is evidence from observations of anemotactic flight to hidden flowers, and from various experiments, that the flower odor mimics the female sex pheromone. Many of the flowers are also tactile mimics of the female insects (e.g., Ågren *et al.*, 1984). In most (but not all) of these systems, males attempt to mate with the flower (“pseudocopulation”). However,

Fig. 3. Orchids that mimic hymenopteran sex pheromones and attract male hymenopteran pollinators (see Table IIa, text). Pollinators frequently “pseudocopulate” with the flowers. The two orchids shown here almost certainly evolved sex-pheromone mimicry independently (refs. in Dafni, 1984). (a) A European orchid, *Ophrys insectifera,* and its male sphecid wasp pollinator, *Argogorytes fargei.* (Courtesy of B. Kullenberg.) (b) An Australian orchid, *Caladenia lobata,* and its male tiphiid wasp pollinator, *Thynnoides bidens.* (Courtesy of W. P. Stoutamire.)

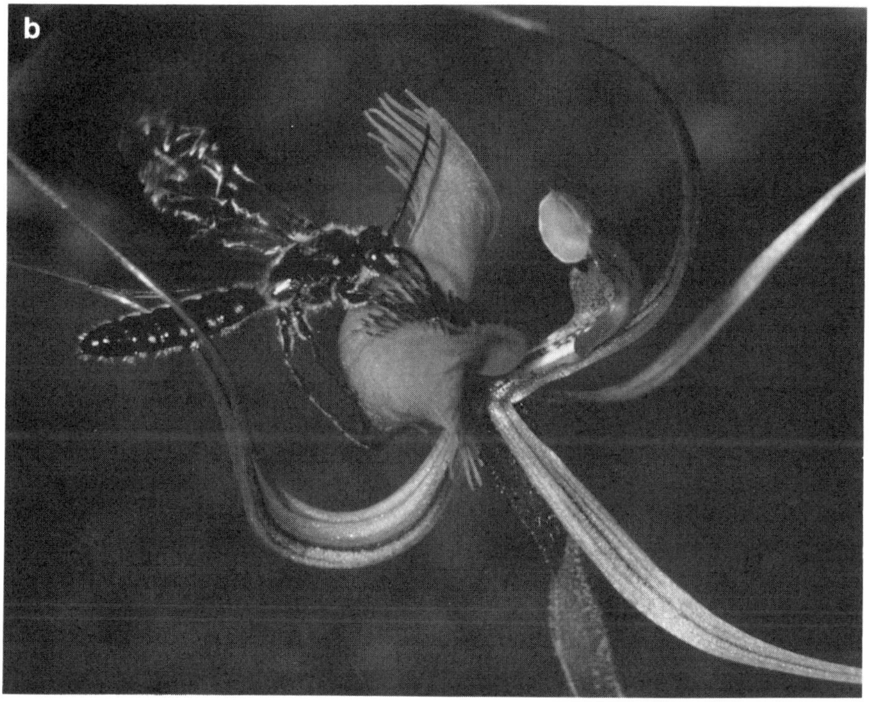

Fig. 3. (*continued*)

ejaculation is apparently very rare (refs. in Dafni, 1984). Wasted time and energy are the only obvious costs to most of the males involved.

The mimicry of sex pheromones by the orchids in the European genus *Ophrys* is better understood than any other odor mimicry system ("the *Ophrys* refs.": Kullenberg, 1961, 1973; Kullenberg and Bergström, 1976; Tengö, 1979; Borg-Karlson, 1987, 1988; Borg-Karlson and Tengö, 1986; Borg-Karlson and Groth, 1988; Borg-Karlson *et al.*, 1985; 1987; summary and refs. in Borg-Karlson, 1985; see also Bergström *et al.*, 1980, for techniques and terminology). There are approximately 30 *Ophrys* species. All studied species sexually attract male insects: principally bees and wasps (at least 11 genera in six families—see Table IIa), but also, in some areas, beetles (Scarabaeidae and possibly Elateridae).

For many of the *Ophrys* systems, flower and pollinator material has been chemically analyzed and bioassayed (*loc. cit.*). In field tests, odor sources that release *Ophrys* flower material excite known pollinators of the flower species; attraction and copulation attempts have been elicited from most of the hymenopteran genera as well as the scarab beetle *Phyllopertha horticola*. (In these tests, odor sources were paired with controls, and were constructed to minimize visual cues.) For many *Ophrys* species, it has also been possible to collect and analyze volatiles emitted by living flowers, and to attract pollinators with (synthetic) mixtures of some of the compounds present in flower volatiles.

Unfortunately, the model insect scents are not as well understood. [In this discussion of models I am elaborating ideas presented in Tengö and Bergström (1976b; 1977b) and Bergstrom (1978).] *Ophrys* blends might mimic one or more of the following models: (1) scents produced by eclosing virgin females, (2) scents produced by virgin females intent on attracting males, (3) scents produced by mated females as part of nest-making and other activities (males are attracted to mated females that resist—perhaps not always successfully?—male copulation attempts), (4) scents produced by *males* during courtship (which may attract other males intent on usurping copulation), and (5) scents used by males to mark points in the environment. Different *Ophrys* species may mimic different models. Some species may employ blends that mimic more than one model at the same time. Presumably no *Ophrys* blends mimic male marking scents—model (5)—alone; these scents attract bees of both sexes, whereas flowers attract only males.

Various insect glands (in particular the Dufour's gland and mandibular gland), are suspected as the sources of model scents. The best evidence to date for the involvement of these glands is that gland solvent extracts attract males in field tests. This evidence is available for only a few pollinator/gland combinations. The attraction of males with volatiles collected from living pollinators has not been achieved. Current knowledge of pollinator signal chemistry is based almost entirely on solvent extracts of suspect glands (*Ophrys* refs.).

The mixtures of compounds found by chemical analyses of flower and pollinator material may differ in several ways from the actual mimic and model odor signals.

1. Analyzed material may contain compounds not present in the signal. This is particularly likely for pollinator gland extracts, which may include (a) compounds that are never released; and (b) stored compounds used for several different signals. Although less likely, *Ophrys* volatile collections might also contain some extraneous compounds. The scents of some flowers vary depending on the time of day (review in Vogel, 1983, p. 583). If this is true for *Ophrys,* then material from collections made over long periods (most exceeded 12 h) might include compounds that are only emitted at times of the day when flowers are not attracting pollinators.
2. For a few *Ophrys* species, and many of the Hymenoptera, analyses may have missed a few important signal compounds that are present in very small amounts. (However, the set of volatile compounds produced by the flowers is very thoroughly established for most *Ophrys* species.)
3. The ratios of important compounds may be very different from those analyzed. This is inevitable for gland extracts (see Heath *et al.*, 1988b). It may also be important for *Ophrys* volatile collections. For at least two *Ophrys*

species, there is evidence of large variation in scents among individuals. Most analyses combine material from many individuals, and therefore may not reflect the ratios put out at any one time by a single individual. (These considerations apply to most chemical mimicry studies, and reflect shortcomings of techniques available until present.)

Despite these potential problems, chemical analyses have successfully revealed (sometimes extensive) chemical similarities in pollinator and flower material in most of the *Ophrys* systems examined. The volatiles of *O. lutea* and *O. fusca* share many compounds with the mandibular glands of their various *Andrena* pollinators. (Mandibular gland extracts are attractive to males in field tests.) Among the C_6 to C_{17} alkanes, alkanols, alkanals, and alkanones found in flower volatiles are compounds that (a) are found in the glands of many pollinator species, and (b) are attractive to males in field tests, notably 6-methyl-5-hepten-2-one and C_6 to C_{14} alkan-1-ols and alkan-2-ols. Among the mono- and sesquiterpenes found in the flower volatiles, several oxygenated terpenes are found in the bees and are attractive to males, particularly geraniol/al, nerol/al, and (*E,E*)-farnesol. Of the other flower compounds—hydrocarbons, acids, esters and aromatics (mostly minor components)—only esters are known to be shared with the bees. *Ophrys iricolor* extracts contain an oxygenated terpene, linalool, which is found in, and known to be attractive to the *Colletes* pollinators. *Ophrys* species in the sections *Araneiferae* and *Arachnitiformes* show similar chemical overlap with their *Andrena* and *Colletes* pollinators. Some of these orchids also produce farnesyl octanoate, a compound found in *Andrena* female Dufour's glands. *Ophrys insectifera* volatiles are dominated by C_{11} to C_{19} (particularly C_{15}) saturated and unsaturated alkanes. This spectrum of compounds is very similar to that of the Dufour's gland extracts of its *Argogorytes* pollinator species. However, field tests with these compounds have failed so far to attract males. Flower volatiles also include sesquiterpenes. *Ophrys* species in the sections *Bombyliflorae* and *Fuciflorae* produce oxygenated terpenes, aliphatic alcohols and acetates. There is no apparent similarity with the chemistry of material from the *Eucera* bee pollinators. However, because of the limitations mentioned above, similarity in the chemistry of the actual mimic and model signals cannot be ruled out.

Sex-pheromone mimicry occurs in at least 11 Australian orchid genera (see refs. in Table IIb; see also Graham, 1983). These genera almost certainly evolved such mimicry independently of *Ophrys* (see Table IIb; refs. in Dafni, 1984). All studied species of the genus *Cryptostylis* attract the same pollinator, the ichneumonid wasp *Lissopimpla excelsa*. *Chiloglottis, Drakaea, Caladenia, Arthrochilus, Paracaleana,* and *Spiculaea* species attract male tiphiid wasps of the subfamily Thynninae. *Calochilus* species are pollinated by scoliid wasps, *Caleana* species by pergid sawflies, and some *Diuris* species by halictid bees. The orchid *Leporella fimbriata* attracts male alate ants (*Myrmecia* sp.). Some species show geographical variation in their pollinators. While certain *Caladenia* species mimic sex pheromones, others mimic food sources. (At first glance most of these genera would appear to have a much less diverse set of pollinators than *Ophrys*. However, any comparisons of pollinator specificity with that genus are probably premature given its enormous man-hours-of-observation advantage.)

Some other flowers (poorly known) also apparently mimic sex pheromones. The orchid *Orchis galilaea* attracts male halictid bees (*Halictus marginatus*) (Bino *et al.*, 1982; Borg-Karlson, 1985, p. 10). *Guiera senegalensis* (Combretaceae), which appears to attract males of the sphecid wasp *Tachysphex,* is the oniy reported nonorchid sex-pheromone mimic (Kullenberg, 1961, p. 266). (For other possible examples see van der Pijl and Dodson, 1966, pp. 129–142, refs. in Dafni, 1984; Nilsson, 1979.)

3. Poorly Understood Mimicry of Various Resources

The orchid *Epidendrum anceps* is pollinated at night by moths that are attracted by the flower's scent (Goss and Adams 1976). The moths are almost exclusively males, primarily of the ctenuchid species *Lymire edwardsii*. To the human nose (N. Williams, personal communication, the smell of the flower is similar to that of plants that harbor high concentrations of pyrrolizidine alkaloids (PAs). In many moth and butterfly species (including many ctenuchid species), males synthesize their sex pheromones from PAs that they collect from plants (refs. in Pliske, 1975). Moth pollinators of *E. anceps* may be attracted to the flower because they identify its scent with that of an alkaloid source. It is not known whether the moths actually obtain the appropriate alkaloids from the flower (see Pliske, 1975). Even if they do not, they may obtain some benefit from the small amount of nectar present in the flower. [See Goss and Adams (1976) and Pliske (1975) for other plant species that may attract PA-seeking pollinators.]

The "bladder flower plant" *Araujia sericofera* is pollinated by a wide variety of moths (refs. in Cantelo and Jacobson, 1979a). The flowers' pollinia attach themselves to the moths' proboscises on the first flower visit. In the process of pollination during a subsequent visit, the pollinia may become wedged inside the flower. Flowers are frequently observed with trapped moths, which eventually die (refs. in Cantelo and Jacobson 1979a). Cantelo and Jacobson (1979a) present preliminary evidence that moths are attracted by flowering plants, and by phenylacetaldehyde, a major volatile constituent of flower extracts. Interpretation of this possibly mimetic system is difficult at present. First, all reported observations and studies have been made in areas where the plant is an exotic. Where the plant is native, it may only attract pollinators that (a) are not trapped and (b) benefit from the nectar that the plant offers. (Even in nonnative areas, it is not clear that a high percentage of visiting moths are trapped.) Second, if the flower odor mimics or coincidentally resembles an odor of interest to moths, it is not clear what that odor is (but see Cantelo and Jacobson, 1979b). The plant appears to be at best a weak and very nonspecific attractor of moths. It may have a negligible impact on any one moth species. Third, the sex of visiting moths has not been reported.

Euglossine bee males collect scents from miscellaneous natural sources (e.g., fruit, decaying wood) and from orchids that depend on the males for their pollination (reviews in Feinsinger, 1983; refs. in Schemske and Lande, 1984). Bee species pollinate characteristic sets of orchid species, and the relationship is apparently governed by the chemistry of the scents that the orchids offer as the sole reward to their visitors. One hypothesis is that male bees use collected scents (or derivatives) to attract females, and the flowers mimic scents that the bees would otherwise collect from other sources (reviews in Feinsinger, 1983; Schemske and Lande, 1984). However, since both bees and flowers benefit, such a system could start (or eventually end up with) the production by the flower of an odor perceptually similar to, but chemically different from, a nonorchid odor used by the bee, complicating interpretation of this mimicry system (see discussion in dupes-benefit section below). (Unfortunately, while a great deal is known about orchid chemistry and orchid/bee interaction, almost nothing is known about the courtship behavior and the sex-pheromone chemistry of the bees themselves.)

4. Mimicry of Rewarding Flowers

Chemical signals are likely to be important in at least a few of the systems where flowers that do not reward their pollinators mimic those that do. (Numerous systems have been cited as examples. However there has been very little experimental work, and none to show that floral scent is important.) These systems fall into several (overlapping) categories (for each see the pollination reviews, Dafni, 1984, 1986). (1) Within a single species, nonrewarding female flowers mimic rewarding male flowers (refs. and chemical information in Thien *et al.*, 1985). (2) Within a species, a nonrewarding population of flowers mimics a rewarding population (see Mainero and Martinez del Rio, 1985, pp. 198–199). (3) A nonrewarding species mimics a specific rewarding species. The mimic often flowers toward the end of the model's flowering period, taking advantage of pollinator experience with the model (see Mainero and Martinez del Rio, 1985). (4) Some nonrewarding species apparently do not mimic any particular species; the model is "virtual" (Pasteur, 1982). These mimics often flower early in the flying season of the pollinator to take advantage of pollinator inexperience (see Dafni, 1984). [Dafni (1984) argues that flowers of this type that produce no scent will benefit because pollinators will be forced to discriminate solely on the basis of visual cues. However, some are scented—see chemical information in Nilsson (1980) and refs. in Burns-Balogh *et al.* (1985).]

5. Lures Employed by Rusts

The pycnia of many rust fungi produce odors attractive to the insect "pollinators" that carry the pycniospores to rusts of the opposite mating type. Some genera (found on conifers) produce carrion odors, others (found on flowering plants) floral fragrances (D. Savile, 1976, personal communication). Despite the economic importance of many rust species, almost nothing is known about their "pollination ecology." There are almost no surveys of insects visitors to pycnia. Rust pycnia produce a "nectar," but its value to "pollinators" is unknown. Ingold (1953), Savile (1976) and Stoffolano (1988) summarize our very meager knowledge in this area.

Many more odor-mimetic pollination systems probably exist than are discussed here (see Wiens, 1978, p. 388). The phorid fly (*Megaselia* sp.) pollination (Cooke, 1984) of an orchid—*Rhizanthela gardneri*–that blooms underground is one of many systems where the existence of chemical mimicry would not be surprising.

E. Lures for Dispersers of Spores

In many mosses of the family Splachnaceae spores are dispersed by insects (review in Koponen and Koponen, 1978). The sporophytes emit strong fetid odors, which attract coprophilous flies (Koponen and Koponen, 1978; Troilo and Cameron, 1981; Cameron and Troilo, 1982; Cameron and Wyatt, 1986). Flies pick up spores as they explore the mosses (Cameron and Troilo, 1982; Cameron and Wyatt, 1986). These insects are particularly likely to carry the spores to new patches of dung or carrion—the substrates on which these mosses grow. Flies that visit mosses sample surfaces with their proboscises. However, there is apparently no food reward; flies never initiate true feeding behavior, and soon leave (Troilo and Cameron, 1981).

Volatile compounds from the mosses (Pyysalo *et al.*, 1978, 1983) include numerous octane derivatives (alcohols, ketones, and aldehydes, including several known from mushroom odors), and various volatile organic acids, including the acetic-through-hexanoic acid series, phenylacetic acid, and benzoic acid; all known from fecal odors (see Table I). Very early chemical studies (refs. in Koponen and Koponen, 1978) indicated that trimethylamine and chemical relatives of indole are also present (these compounds are also known from fecal odors). The volatiles are apparently synthesized in the spore-releasing structures (Pyysalo *et al.*, 1983). [The substrate does not contribute to the odor; by the time mosses have grown to the spore-producing stage, the original odor of the carrion or dung has long since disappeared (A. Koponen, personal communication).]

Many mushrooms in the order Phallales also attract coprophilous fly spore dispersers with a fetid odor (reviews in Ingold, 1953, pp. 120–125; refs. in Stoffolano *et al.*, 1988) (see Fig. 2b). Fly visitors have been identified for a number of species (Cobb, 1906, 1909; Nielsen, 1968; refs. in Burk, 1980; review in Stoffolano *et al.*, 1988). Flies attracted to *Phallus impudicus* eat the spore-containing slime ("gleba") on the outside of the mushroom; viable spores are later deposited in the fly feces (Schremmer, 1963). The gleba is probably not a true reward; its nutritional value is very low in the single mushroom species (*Mutinus caninus*) where measurements have been made (Stoffolano *et al.*, 1988).

The odor of *Phallus impudicus* chemically resembles fecal volatiles (refs. in Table I, Stijve, 1965; note differences in reported chemistry). Major components include acetic acid (a major fecal odor component) and phenylacetaldehyde (similar to phenylacetic acid, another major fecal odor component). [See Ingold (1953, pp. 120–125, 145–147) for other possible fungal systems.]

F. Prey Lures

1. Plants and Fungi that Lure Prey

A diverse assemblage of plants and fungi trap, kill, and digest animals (reviews in Givnish *et al.*, 1984; Lloyd 1942; and Slack, 1979; other refs. in Duddington and Wyborn, 1972; Zuckerman and Jansson, 1984; Simons, 1981, reviews unstudied, potentially carnivorous plants). Many (perhaps most) of the species in this assemblage chemically attract their prey. Unlike most of the animal predators mentioned below, the majority of these plants and fungi (1) can survive without carnivory and (2) capture a significant percentage of their prey by chance. However, various evidence suggests that the ability to lure prey is important to these organisms.

Fungi that capture and digest nematodes are ubiquitous. Numerous studies have demonstrated that the nematodes are attracted to the fungi, and can be attracted to filtrates and other preparations of fungal material (review in Zuckerman and Jansson, 1984). Possibly the fungi mimic nematode feeding cues. Different fungus species attract different sets of nematode species (most attract fungal-feeding and plant parasitic nematodes). Attraction of prey is also possible in a second, less-studied class of fungi, whose members capture rotifers after they bite specialized hyphal knobs (Lloyd, 1942; Duddington and Wyborn, 1972).

Barber and Page (1976; review in Barber, 1978) suggest that the mucilaginous seeds of many plants (including many common crucifers such as *Capsella bursa-pastoris*) (1) attract and entrap soil organisms such as nematodes and (2) make use of nutrients from the captured animals. (Currently, the evidence for this proposition is indirect.)

In many familiar carnivorous plants, chemical mimicry probably aids in the capture of insects (Lloyd, 1942; Slack, 1979). Unfortunately, although biologists and abundant enthusiasts outside academia have long been fascinated by these plants, there have been remarkably few ecological studies. For most species, even the common ones, descriptions of prey more specific than "insects" or "small flies" are unavailable. Prey have been recorded for the tank bromeliad *Brocchinia reducta* (Givnish *et al.*, 1984), *Dionaea muscipula* (the Venus fly trap) (Lichtner and Williams, 1977), and some *Sarracenia* (pitcher plant) species (the "*Sarracenia* refs.": Wray and Brimley, 1943; Fish, 1976; Rymal and Folkerts, 1982; Folkerts, 1982; Folkerts and Crossley, 1988; Hermann *et al.*, 1988). The Hermann *et al.* (1988) study is the first long-term field experiment to demonstrate that captured insects significantly increase carnivorous plant growth rates.

Nectar glands help position prey for capture in many carnivorous plants including *Dionaea* and the pitcher plants (all five genera in all three families—Sarraceniaceae, Nepenthaceae, Cephalotaceae). Various observations suggest that many species also attract prey from a distance with odors (and visual cues). (However, experiments to test this hypothesis have never been conducted.) Sweet "nectar-like" scents are produced by many pitcher plant species, and by the tank fluid in the bromeliad *B. reducta*. The latter is the only species in its genus known to be carnivorous. Tank fluid odor is unusual (possibly unique) among tank bromeliads, and could be one of the more important adaptations of the plant for carnivory (Givnish *et al.*, 1984). *Sarracenia* species with a floral odor tend to capture larger numbers of flower-visiting insects than species without a floral odor. Other trends in prey capture might also be odor mediated (see *Sarracenia* refs.).

Almost nothing is known concerning the chemistry of carnivorous-plant volatiles. In the only study that I know of, terpenes, amines, and a large number of other compounds were identified from extracts of *Sarracenia flava* (Miles *et al.*, 1975; Mody *et al.*, 1976). The role of these compounds in interactions with prey was not determined (contra the claims of Mody *et al.*, 1976).

Among those carnivorous plants that capture their prey with sticky droplets, there is less indication of an important role for scent. The possible attraction of some insects to the large droplets of *Drosera* and other sundews may be purely visual. However, some *Drosophyllum* species have a "honey" odor (Lloyd, 1942). Also, many *Pinguicula* species give off a "fungus" odor, and "small flies" are a major prey item (Lloyd, 1942, pp. 2, 106; Slack, 1979, p. 107); are some of these flies attracted fungus gnats?

Prey attraction has been suggested for some of the carnivorous plants that capture prey underwater (Lloyd, 1942; Slack, 1979). The closely related genera *Utricularia* and *Polypompholyx* grow in aquatic, moist-soil, or rain forest epiphyte environments. They capture prey that trigger their small, underwater (or water-film) suction traps (review in Wallace, 1978). Very little is known about the prey of most species, particularly those that live in highly specialized habitats. In at least one species, filaments probably guide prey to the trap; many of the prey (particularly chydorid crustaceans) normally travel along algal filaments eating epiphytes (Meyers and Strickler, 1979). However, many workers have suggested that "odors" may also attract prey to the traps, which bear secre-

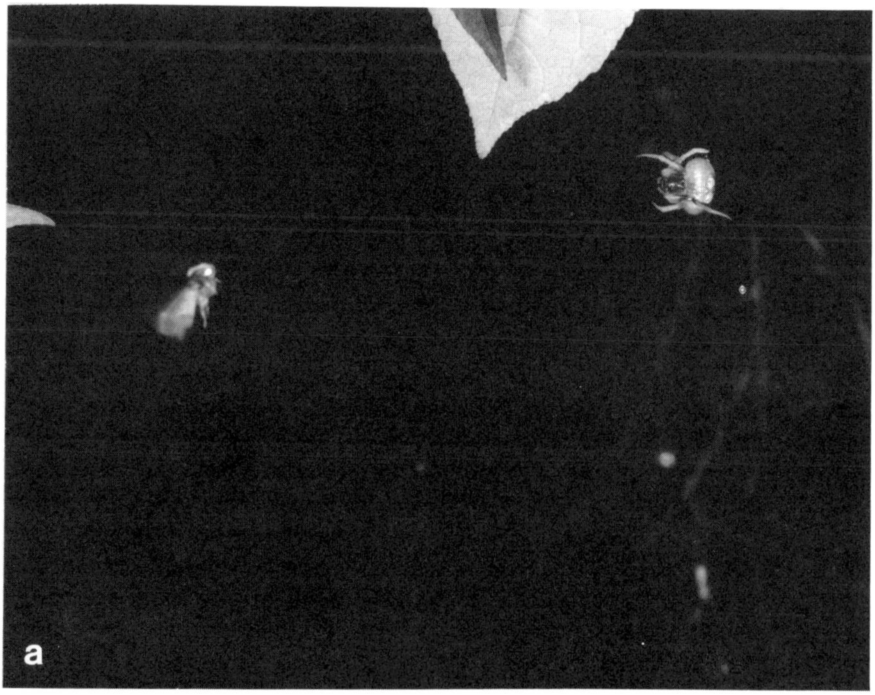

Fig. 4. Spiders that mimic moth sex pheromones and attract male moth prey (see Table IIa, text). Differences in behavior, morphology, and internal anatomy make it almost certain that sex-pheromone mimicry evolved independently in the two spiders shown (see Stowe, 1986, 1988b). Both photos were taken in Gainesville, Florida. (a) A tineid male moth (*Acrolophus plumifrontinellus*) approaches a hunting bolas spider *Mastophora bisaccata*. The spider has detected the wing vibrations of the moth and "cocked" the "bolas" (a line ending in a drop of glue). When the moth is close enough, the spider will flick the bolas at the moth. (b) Hunting *Kaira alba* individuals snatch moth prey from the air.

Fig. 4. (*continued*)

tory trichomes (refs. in Lloyd, 1942; and in R. L. Wallace, 1978, and personal communication). Chemical lures would not be surprising in two unrelated aquatic genera, *Genlisea* and *Aldrovanda*; to be captured, prey must pull themselves through narrow slits. In *Genlisea* these slits are associated with mucilage-secreting glands (refs. in Lloyd, 1942).

2. Animals that Lure Prey

A large number of spiders, representing at least two independent evolutionary lines, apparently attract their male moth prey by mimicking the odor of female moth sex pheromones (Eberhard 1977; Stowe and Icenogle, 1988; Stowe *et al.*, 1987; Yeargan, 1988; review in Stowe, 1986, 1988b). The spiders capture the attracted moths with a minimal web or snatch them from the air (Fig. 4). This is a widespread phenomenon involving spiders and moths in tropical and temperate habitats worldwide. At least nine (possibly 15) araneoid genera of spiders attract moths belonging to at least seven families.

Table IIa presents the relationship of the moths and the few spider species whose prey have been recorded (see Appendix A). For many spider species only one to three prey species are known (e.g., Eberhard, 1977, 1980). However, none of these species has been the subject of long-term studies. One species that *has* been the subject of a long-term (7-year) study, *Mastophora cornigera*, captures at least 19 moth prey species (Stowe and Icenogle, 1988). [Very preliminary observational evidence suggests that nematocerous dipteran prey are attracted by very young individuals of some of these spiders, and by some other araneoid spiders (Eberhard, 1981; Stowe, 1988a; refs. in Stowe, 1986).]

Unfortunately, the spider/moth systems are difficult to study; spiders (1) are hard to obtain in large numbers (they are cryptic, strictly nocturnal and very rarely seen) and (2) readily engage in hunting behavior only in the field. However, a recently developed volatile-collection technique has overcome some of the difficulties of working with these animals, and made examination of spider signal chemistry possible for the first time (Stowe *et al.*, 1987). Volatiles from hunting *M. cornigera* include the following known moth sex pheromones (not found in simultaneous control collections): (*Z*)-9-tetradecenyl acetate (Z9–14:Ac), (*Z*)-9-tetradecenal (Z9–14:Al), and (*Z*)-11-hexadecenal (Z11–16:Al). There was considerable variation in the composition of the samples, which strongly

Table II. The Diversity of Mimics and Dupes in Sex-Pheromone Mimicry Systems. (a) Spiders and Moths; (b) Orchids and Hymenoptera

a

Lepidoptera [Hodges et al 1983]

Araneae: Araneidae [Levi and Coddington 1983]		Tineoidea: Tineidae: *Acrolophus*	Gelechioidea: Gelechiidae: Gelechiinae	Yponomeutoidea: Plutellidae	Tortricoidea: Tortricidae: Olethreutinae	Pyraloidea: Pyralidae: ?	Pyraloidea: Pyralidae: Pyraustinae	Pyraloidea: Pyralidae: Crambinae	Noctuoidea: Geometridae: Larentiinae	Noctuoidea: Noctuidae: Herminiinae	Catocalinae	Acronictinae	Amphipyrinae	Hadeniinae	Noctuinae	Heliothinae	References
Araneinae	*Kaira alba*	3					1				1						A
Celaeneae [Simon 1892]	*Celaenia kinbergi*												1	1	1	1	A
	Taczanowskia sp.					1											A
Mastophorinae [Gertsch 1955]	*Dicrostichus magnificus*										1		?	?			A,B
	Ordgarius sexspinosus											?					E
	Mastophora phrynosoma								1								A
	M. bisaccata (cornigera)	2									3						A
	M. cornigera	1	1	1				1					1	8	5	1	A,C
	M. dizzydeani				1								1	1			A
	M. hutchinsoni							1		1				2			D

(*continued*)

Table II. (*continued*)

b

Hymenoptera [Brothers 1975]: Symphyta — Tenthredinoidea — Pergidae; Apocrita — Ichneumonoidea — Ichneumonidae; Aculeata — Vespoidea — Vespiformes — Tiphiidae, Formicidae, Scoliidae; Sphecoidea — Spheciformes — Sphecidae; Apiformes — Colletidae, Halictidae, Andrenidae, Megachilidae, Anthophoridae

Orchidaceae [Dressler 1981]: Orchidoideae — Orchideae — Orchidinae; Diurideae — Diuridinae, Caladeniinae; Spiranthoideae — Cranichideae — Cryptostylidinae

Orchidaceae [Dressler 1981]		Perg.	Ichn.	Tiph.	Form.	Scol.	Sph.	Coll.	Hali.	Andr.	Mega.	Anth.	References[a]
Orchidoideae – Orchideae – Orchidinae	(Orientales)												
	Ophrys reinholdii											x	C,L
	O. cretica											2	L,G
	(Fuciflorae)												
	O. fuciflora s. str.										?	x	A
	O. f. candica											1	A
	O. scolopax s. str.											6	A,L
	O. tenthredinifera											9	A
†*	*O. apifera*											3	A
	(Bombyliflorae)												
	O. bombyliflora											10	A
	(Araneiferae)												
	O. sphecodes s. str.						?	?		x			B
	O. s. provincialis							1		1			B
	O. s. litigiosa									?	1		B
	O. s. mammosa									1			B
	O. s. atrata					?				x		?	B,C,L
*	*O. tarentina*										1		B
	O. ferrum-equinum										x	?	L,G
	O. bertoloniiformis					?					x	1	A,C,L
	(Arachnitiformes)												
*	*O. arachnitiformis*							1		2			B,F
*	*O. splendida*									1	1	?	B,C
	(Fusci-Luteae)												
*	*O. fusca* s. str.									10		1	B,L
*	*O. f.* "grossblütig"							1		1			B,L
*	*O. lutea* s. str.									9			B,L
*	*O. atlantica*										1		B,L
*	*O. dyris* s. str.											2	B,L
*	*O. d. kelleri*					?				x			B,J
*	*O. omegaifera*											x	B,D,L
*	*O. iricolor*									2	1		B,L
‡*	*O. iricolor*											x	D
	(Ophrys)												
	O. insectifera						2						C
	O. speculum					1							C,E
§	*Orchis galilaea*								1				H
Diurideae – Diuridinae †	*Calochilus*					x							I
§	*Diuris*								?				M
Caladeniinae	*Leporella*				x								N
§	*Caladenia*			x									I
†	*Chiloglottis*			x									I
	Drakaea			x									I
	Spiculaea			x									I
	Arthrochilus			x									I
	Paracaleana			x									I
	Caleana	x											I
Spiranthoideae · Cranichideae – Cryptostylidinae	*Cryptostylis*		1										I

(*continued*)

Table II. (*continued*)

Notes: Entries are conservative estimates of the number of known dupe species in each category (see text, Appendix A). See text for possible model scents other than female sex pheromones. Taxonomic unit [citation]: I use the cited study for my definition of the taxonomic unit and (unless otherwise indicated) all the unit's subdivisions. (Section): Names indicate the species group or genus section. See references for the authors of the species and species groups. Several of the subdivisions of *Ophrys* are controversial (see review in Borg-Karlson, 1985). Solid horizontal lines separate mimic groups that almost certainly evolved sex pheromone mimicry independently (see Stowe, 1986; Dafni, 1984; Dressler, 1981). This division is conservative and line-delimited groups of taxa are not necessarily monophyletic. Mimic genera not broken down into species: for all studied members of the genus (1) there is a single dupe species or (2) dupes belong to genera that are poorly known taxonomically such that it is difficult to determine the number of species. Shaded columns: These dupe (= model) families or subfamilies have been assigned to a sex pheromone chemistry category (Table IIa only, see Tamaki, 1984; Wong *et al.*, 1985; Mayer and McLaughlin, 1988). Solid lines border columns for groups where this assignment is based on information from a relatively large number of species within the group (or the chemistry of the prey species in the group is known). Dashed lines border columns for groups where the evidence is preliminary. Columns are light grey for those dupe (= model) taxa that use C_{10} to C_{18} terminal functional group type pheromones, medium grey for those dupe (= model) taxa that use C_{19} to C_{21} hydrocarbon type pheromones. X in place of a number: There are one or more known species in this category. However, assigning an exact number is either difficult or inappropriate (see references). Question mark in place of a number: One or more species in this category may act as dupes but the evidence is indirect or observations are scanty (see references). Asterisk: The species's taxonomic position in its genus is uncertain. Dagger: The mimics are not completely dependent on dupes; they frequently autopollinate (orchids), or make a large web and catch many prey by chance (spiders) (no examples included, but see Horton, 1979 . Section symbol: Only some members of the genus are sex-pheromone mimics. Double dagger: There is some doubt that the published reports refer to the same mimic species. Therefore, there is more than one entry for this species.

[a] References: Most of these are reviews that cite the original studies. b: A, Kullenberg *et al.* (1984b); B, Warncke and Kullenberg (1984); C, Kullenberg (1973); D, Vogel (1976); E, Paulus and Gack (1981); F, Godfery (1921, 1922); G, Kullenberg *et al.* (unpub.); H, Bino *et al.* (1982); I, Stoutamire (1974, 1975, 1983, 1985); J, Kullenberg (1961); K, Paulus and Gack (1983); L, Paulus and Gack (1986); M, Coleman (1932, 1933); N, Bates (1979); Peakall (1984); a: A, Stowe (1986, 1988b; B, D. Clyne (unpublished); C, Stowe and Icenogle (1988); D, Yeargan (1988); E, Shinkai (1988).

suggests that there is inter- or intraindividual variation in the blends of compounds the spiders produce.

Sex-pheromone information is available for eight moth species captured by *M. cornigera*. Male moth–attracting blends can be made for each of these prey species by mixing compounds from the set Z5–10:Ac, Z9–14:Ac, Z9–14:Al, Z11–16:Ac, Z11–16:Al (refs. in Stowe *et al.*, 1987). These five compounds are part of the same biosynthetic series in female moths (Bjostad and Roelofs 1983); (Z)-[$CH_3(CH_2)_3CH{=}CH(CH_2)_{n-7}R$] ($n = 10, 12, 14, 16, 18$; R = CHO, CH_2OCOCH_3). Our study demonstrates that spiders produce at least three of these compounds.

In future work we hope to test the hypotheses, suggested by this study and field evidence, that (1) spiders attract male moths with blends of moth sex pheromone compounds and (2) spiders produce different blends to attract different sets of prey species. We suspect that the spiders are capable of producing compounds beyond those found in this preliminary study. To characterize distinct attractant blends, it is probably necessary to work with brief collections from individual spiders with known moth-capturing histories. (Our study analyzed mostly pooled material.)

An unusual system of prey attraction occurs in the social spider *Mallos gregalis* (Dyctinidae). Spider colonies make a large communal web and catch primarily muscoid flies. Unlike most social spiders, *M. gregalis* individuals do not remove consumed prey from the web. Tietjen *et al.* (1987) have shown that feeding by these spiders alters the environment for microflora in fly carcasses such that yeasts outcompete bacteria (in contrast to dead flies that have not been fed upon, or flies fed upon by other spider species). They have also shown that these yeasts produce a sweet odor that attracts more flies to the web, which suggests a spider/yeast mutualism.

Adults of the staphylinid beetles species *Leistotrophus versicolor* prey on flies (Forsyth and Alcock, 1988; R. Silberglied, unpublished; M. Thayer, personal communication). When dung is available, beetles hunt in territories and catch flies that are attracted to the dung. Away from dung, beetles hunt by waving their abdomens in the air. Various flies (principally phorids) fly directly to the beetles and are seized. The prey appear to be attracted by odors emanating from the glandular tip of the beetle's abdomen. [The model scent in this system is not clear. See Dettner (1983) for chemical analyses of the abdominal glands of other staphylinids.]

Ogilvie (1966) reports preliminary evidence for a similar phenomenon in certain chameleon species (genus *Chamaeleo,* especially *C. bitaeniatus* and *C. hoehnelii*) that possess a "temporal pouch"—a gland that opens at the angle of the jaw. This gland produces a secretion with a smell similar to decaying meat. On numerous occasions, he observed chameleons wipe their jaws on branches, back away a short distance, and capture flies apparently attracted to the "jaw-wiped" spot (jaw-wiped branches give off the decayed meat odor). He showed that this odor attracts flies in two-choice tests. The odor may also have a role in social communication. [Cowgill (1966) suggests that the primate *Perodicticus potto* might also attract insect prey. However, I find this evidence unconvincing.]

Adults of the chigger mite *Womersia strandtmanni* appear to attract their collembolan prey (Huber 1978). The Collembola wander up to the mites, imbibe secretions from an area on the mite hysterosoma, and are quickly paralyzed. Perhaps the mites mimic the odor of a food source. A very similar hunting technique is employed by the reduviid bug *Ptilocerus ochraceus*. The secretion of the abdominal trichomes of these bugs attracts *Hypoclinea bituberculata* ants, which feed on the secretion and are quickly paralyzed and subsequently eaten (Jacobson, 1911; summary and illustrations in Wilson, 1971). Perhaps the bugs mimic honey-dew-secreting aphids. Weaver *et al.* (1975a) report tentative indications that another reduviid bug (*Apiomerus pictipes*) attracts bee (*Trigona fulviventris*) prey with a scent (possibly by taking advantage of nest-defense behavior).

G. Chemical Crypsis

In mimicry the model is some entity in the environment that is of interest to the dupe. In crypsis the "model" is the background, or an entity in the background. Distinguishing between the two related phenomena is sometimes difficult (see Wiens, 1978). Crypsis involves (1) reproducing background signals, as in many cases of visual crypsis and/or (2) eliminating signals that stand out from the background. Chemical (like auditory) crypsis presumably takes the latter form in most cases (but see below). Stopping chemical "leakage" is not necessarily trivial and may be very important to organisms trying to escape kairomone-guided predators. There are probably many biochemical and structural adaptations for this purpose that we will never come to recognize.

1. Masking

Some organisms alter the chemical signals they would otherwise emit by *adding* chemical signals from other sources. Such "masking" may make it less likely that they are correctly identified by predators or prey. I consider three ways that organisms might employ masking. (The examples include cryptic and mimetic systems.)

1. Organisms might try to alter their chemical signals to more closely resemble background chemical signals. Fishlyn and Philips (1980) show that the snail *Notoacmea paleacea* incorporates into its shell compounds from the surfgrass (*Phyllospadix*) that it feeds on. Because starfish (*Leptasterias*) that contact these snails usually fail to attack these potential prey, Fishlyn and Philips (1980) suggest that the presence of the surfgrass compounds results

in the inability of the starfish to distinguish the snail from the plant background.

2. Organisms could conceivably mimic the odor of strong-smelling, neutral objects in the environment such as carcasses or feces. A diverse array of mammalian carnivores engage in "scent-rubbing behavior"; animals roll in and rub themselves with material from carcasses and a wide variety of other strong scent sources (review in Reiger, 1979). The animals might benefit (while hunting) by smelling like something other than a carnivore. However, this is only one of a number of untested hypotheses for the function of this behavior. [Organisms might also use fetid-odor mimicry to *escape* predators—see Swennen (1968), Franq (1969), and Eisner *et al.* (1967) for possible examples.]
3. Organisms might mimic the odor of specific dangerous or inedible organisms (see Camazine, 1985). Strangely, despite the abundance of Batesian visual mimicry systems, I know of only one reported suggestion of Batesian chemical mimicry. [For other possible examples see ant alarm pheromone mimicry, above, and Wiens (1978, p. 378).] Siberian chipmunks (*Eutamias sibiricus asiaticus*) that find a snake carcass gnaw off pieces of skin, chew the pieces, and apply the "extraction" to their fur (Kobayashi and Watanabe 1986). They also rub themselves with snake urine and feces. Tests with material from numerous animals showed that this behavior is elicited exclusively by material from snakes. Snake odors frighten many animals that are potential snake prey (including many carnivores that may prey on the chipmunks) (refs. in Kobayashi and Watanabe, 1986). Kobayashi and Watanabe (1986) suggest that the chipmunks' acquired odor deters predators, particularly in situations where the predator enters the chipmunk's burrow. (This would make snake-scent application behavior a "sheep in wolf's clothing" strategy.)

2. Müllerian Mimicry

In Müllerian mimicry, the aposematic signals used by two or more dangerous or unpalatable organisms converge (or evolve in parallel). These organisms benefit from the consequent enhancement of learning by predators to avoid such prey. Because it is difficult to designate one of the organisms as the model, some authors believe that the term mimicry should not be applied to this phenomenon (see mimicry reviews). (The evolutionary discussion below is only partly applicable to these systems.)

There are few reports of Müllerian chemical mimicry. Rothschild (1961, 1964) described possible convergent patterns in insect "warning odors" as perceived by the human nose. She predicted that improving chemical analytical techniques would confirm such patterns. Blum (1984, p. 211) also notes that Müllerian odor mimicry could be commonplace, giving as a possible example the use of 4-methyl-3-heptanone as an alarm pheromone by many sympatric ant species that possess powerful stings or other chemical defenses. Other researchers have proposed that the common occurrence of pyrazines in diverse insect warning odors may be in part a product of Müllerian odor mimicry (see Guilford *et al.*, 1987).

Two types of (as yet unavailable) evidence are needed to convincingly demonstrate the existence of Müllerian odor mimicry. (1) Experiments with relevant predators must be conducted to verify that the chemical similarity is effective in enhancing predator learning. (2) More evidence of *mimetic* chemical similarity is needed. The demonstration that an assemblage of species uses the same compound as the principal component in their warning odors is generally not sufficient. Such similarity could be the product of chance convergence on a small set of (functionally similar) compounds. This is particularly likely in the case of hymenopteran "warning odors"; most are dominated by a relatively restricted set of widely distributed compounds.

What sort of evidence might be found that would favor mimicry over chance convergence? (1) Chemical studies might uncover convergent patterns in intraspecific geographical variation of warning odors. (2) Species assemblages might be found that produce warning odors composed of nearly identical complex blends. (3) Phylogenetic studies of one or more assemblages might indicate that, in numerous independent evolutionary lines, chemical convergence is a derived state.

However, it will probably be difficult (perhaps impossible) to find good evidence for Müllerian chemical mimicry. Few odors are strictly warning odors; most are defensive (noxious) by themselves. The evolution of warning odor chemistry has probably been dominated by the selective advantages of maximizing noxiousness. In social insects, defensive odor evolution is further complicated by the fact that the odors serve in intraspecific communication. (These observations apply to all forms of mimicry involving defensive odors.)

Many workers have suggested that a phenomenon analogous to Müllerian mimicry occurs in flowers (review in Dafni, 1984, 1986). They suggest that certain assemblages of sympatric, pollinator-rewarding flowers benefit by being similar in appearance and increasing their apparent abundance to pollinators. Convergent odor patterns might be found in some of these species. In nature, Müllerian and Batesian mimicry systems (both floral and defensive) form a continuum (see mimicry reviews).

IV. Possible Evolutionary Patterns

Our present knowledge of chemical mimicry systems in terms of chemistry, neurophysiology, and ecology is primitive. Nonetheless, it is valuable to consider some preliminary hypotheses concerning the evolution of these systems. [See also Bergström (1988), a related review that was not available to me. Some of the numerous assumptions I make in this section are discussed in Appendix C.]

For the study of evolutionary questions, two major mimicry complexes— the spider/moth and orchid/pollinator (sex-pheromone mimicry) complexes—should prove particularly useful (see Appendix A). (1) Mimics are dependent on a relatively well delimited set of dupe species; strong interaction and some degree of coevolution seem likely. (2) Large numbers of species of mimics and dupes are involved. (3) Many of the taxa in both systems have been the subject of intensive systematic study, including numerous relevant moth groups (see Stowe, 1986; Hodges *et al.*, 1983), the orchid genus *Ophrys*, and many relevant bee and wasp taxa (refs. in Borg-Karlson 1985). (4) In both systems, there are at least two independent lines of evolution leading to sex pheromone mimicry. (5) Sex pheromone–related chemistry, neurophysiology, and behavior is better understood in moths than in any other group of organisms (see Tamaki, 1984; Mayer and Mankin, 1984; Cardé and Baker, 1984), and the chemical work on *Ophrys* orchids and their pollinators is extensive (*Ophrys* refs.).

My views concerning the evolution of the orchid/pollinator sex-pheromone mimicry systems conflict with some of the views expressed by several of the principal *Ophrys* workers. In my opinion, the nectarless orchids harm their pollinators; males that are attracted lose time, and, on the average, must achieve slightly lower mating success (see Dafni, 1987, p. 86). In contrast, Kullenberg *et al.* (1984a) and Burns-Balogh *et al.* (1985, pp. 52–53) feel that *Ophrys* flowers are likely to benefit their pollinators in at least three ways. (1) The flowers attract males to, and keep them concentrated in, appropriate habitats. (2) The sexual stimulation from the flowers increases the amount of time males spend flying in search of females. (Factors 1 and 2 increase the chances of males and females meeting.) (3) The presence of the flowers means that males spend more time being attracted to flowers and less time attempting to mate with mated females; females therefore have more time for nest-building. I consider these group selection-oriented proposals implausible. Even if encounters with mimics provide some indirect benefits, I suspect that visiting flowers is much less profitable than spending the same amount of time searching for females (cf. Cameron and Wyatt, 1986). I expect that selection will favor males that can distinguish between flowers and females, and that avoid flowers.

In the discussion below, I illustrate most points with examples from sex-pheromone mimicry (usually spider/moth) systems (see Appendix C). This emphasis has a serious drawback. In sex-pheromone mimicry systems, the dupes and the sources of model signals belong to the same species. The reader must resist the subconscious temptation to equate model and dupe; the points I make are meant to apply to mimicry systems in general.

A. Preadaptations and the Original Mimic Signal

Evolving the ability to biosynthesize relatively unique compounds may be crucial for the development of some mimicry systems. However, most of the compounds that have been identified so far from mimic signals are known to occur in a variety of animals and plants, not just in the mimic, model, and closely related taxa. Three preadaptations for chemical mimicry that are probably more important than the biosynthesis of unique compounds are (1) the production of an effective combination of compounds, (2) the release of relevant compounds in appropriate relative concentrations, and (3) the production of the signal in the correct ecological context.

In my evolutionary scenarios, I assume that mimic signals begin as "blends" of "components" consisting of (1) some "incidental" compounds that do not play a role in the mimicry and (2) a few "key" compounds that mimic the effects of key compounds in the model signal. A mimic key compound may not be identical to its model key compound. If some of the conformations of the mimic and model compounds are similar, the two compounds may interact with and stimulate the dupe organism's receptors in a similar way. For example, in many moth species, males are attracted in experiments to compounds that are not present in the female moth sex-pheromone blend. Male-attracting blends can often be created by replacing a major key component of the female blend with a compound possessing a slightly different functional group or double-bond position (e.g., Tumlinson *et al.*, 1986; see also Steck *et al.*, 1982; Cardé and Baker, 1984, and refs. therein).

I now consider the evolution of the mimic signal and the dupe's response in three mimicry system categories: (1) mimicry systems where mimics decrease average dupe fitness, (2) systems where mimics have no effect on dupe fitness, and (3) systems where mimics increase dupe fitness. Of course, these three categories exist only on paper; mimicry systems in nature presumably form a continuous spectrum.

In each category, I use reported mimicry systems in my arguments. However, I do not pretend that any of these systems can be categorized with certainty. Consider the spider/moth systems. Dupes are clearly harmed by encounters with mimics. However, spiders might have a negligible effect on average dupe fitness; spiders might be much rarer than virgin female moths, or be restricted to only a small part of the moth's geographic range. With luck, future work will provide some of the critical information; it may be possible for example to (1) quantify the rates at which dupes encounter mimics versus models (cf. Lloyd, 1984, p. 64) or (2) test for differences in dupe behavior in areas with and without mimics.

B. Dupes Are Harmed: Arms Races

In systems where mimics lower average dupe fitness, mimic and dupe may engage in an evolutionary arms race of increasing signal discrimination on the part of the dupe, and increasing signal refinement on the part of the mimic. This is most likely to happen in systems where (1) the species interact for a long time and (2) the original mimic blend is significantly different from the model. I will consider two ways that dupes might detect mimic signals (but see Appendix C).

1. Detection of Alien Components

A dupe might detect a mimic signal by sensing those compounds that are never present in the model signal. These "alien components" might be (1) incidental compounds or (2) key compounds that differ from their models. Selection should favor dupe individuals that (1) possess receptors capable of detecting alien components, and (2) do not respond to relevant signals if they contain alien components.

In theory, it is always possible to discriminate between a model compound and a nonidentical mimic. The only prerequisite is a receptor sensitive to one or more of those conformations that the two compounds do not share. In practice, it appears that most organisms readily evolve the ability to discriminate among, and respond appropriately to, compounds that are important in their ecology. Sensitivity to predator/prey/host kairomones is ubiquitous (e.g., review in Haynes and Birch, 1984). Inhibition of behaviors due to detection of alien components is common in the sex-pheromone communication of moths (and many other groups—review in Cardé and Baker, 1984). Researchers find that in many moth species, the addition of even small amounts of certain compounds to male-attracting blends makes them unattractive. It is generally thought that these compounds are not present in conspecific female blends, but are present in otherwise very similar blends produced by the females of certain sympatric species; selection favors males that are inhibited by these compounds and therefore fail to fly to females of the wrong species. In some cases, inhibition may have evolved in response to the presence of the compounds in blends of sympatric spider species.

If dupes evolve the ability to detect alien components, then mimics should (1) replace alien key components with model key components and (2) decrease alien incidental components to below detectable levels (but see below). Failing this, the mimetic relationship will end.

In most of the systems where some signal chemistry is known, mimic key compounds are apparently identical to the corresponding model key compounds. However, we know very little. Furthermore, researchers are more likely to notice and report systems where mimic and model compounds are identical. [Many venom and toxin components that "mimic" hormones and neuropeptides are very different from their "models"—e.g., "mimic" opium alkaloids and "model" endorphin peptides (see Appendix B).]

An alien incidental component *can* persist in an arms race if it is rendered undetectable by environmental chemical noise. Thus mimic signals from plants might "safely" include common plant odors not present in the model signal. (In some mimicry systems that involve surface chemical signal sources, the environment may contribute very little chemical noise, and this argument may not be applicable.)

2. Detection of Ratios Outside the Range Found in Models

A second way that a dupe might detect a mimic signal is by sensing ratios of two or more components that are outside the range of those found in the model. Many male moths are very sensitive to ratios of components in blends of female sex-pheromone compounds. This is apparently common among sympatric moth species that produce blends that only differ in the ratio of the relevant components (refs. in Cardé and Baker, 1984). Sensitivity to ratios is also thought to be important in caste and nest-mate recognition in social insects (see Howard *et al.*, 1980, 1982). There is some experimental evidence that selection can change ratio preferences (Hoffman *et al.*, 1984; but see Roelofs *et al.*, 1986, discussed below). For most of the discussion on ratios, analogous arguments apply to absolute concentration.

Detection of mimics by means of ratio discrimination may be complicated by the (often considerable) variation in the model. A few studies have measured individual variation in female moth sex-pheromone blends (Löfstedt *et al.*, 1985; Haynes, *et al.*, 1984; Barrer *et al.*, 1987, and refs. therein). These studies suggest that (1) most females produce blend ratios close to some "nominal blend" ratio, but (2) there is a wide absolute range of ratios that are at least occasionally produced by females. In one study of volatiles collected from individual calling female moths (*Pectinophora gossypiella*), the fraction of (*Z,Z*)-7,11-hexadecadienyl acetate in the volatile sample averaged 61% (± 5% SD), but ranged from 39% to 74% (Haynes *et al.*, 1984; see also Linn and Roelofs, 1985).

Dupe use of ratios to detect mimics may be further complicated by inconsistency in ratio assessment. Cardé *et al.* (1976) demonstrated variation within individual male moths in the response to female sex-pheromone blends with various ratios of pheromone compounds. Their findings suggest that whenever an individual male moth encounters a blend with a ratio very different from that in the nominal female blend, there is always a small chance that it will be attracted. Selection may not completely eliminate this variation in male response.

Clearly, the range of blend ratios that, at least occasionally, will be perceived by dupes as falling within the absolute ratio range of the model could be quite broad (see also below). Mimics in an arms race must produce blends that fall somewhere within this range or the mimetic relationship will end.

One mimic blend distribution that may persist in an arms race is a distribution that broadly overlaps that of the model. Let us suppose that, for each range of blend ratios, mimic signals constitute a constant fraction of all signals in the dupe's environment. In such an environment all blend ratios are equally risky and the dupe should respond to all of them in the same way. If the mimic signal includes no alien components, then no improvement of chemical acuity will allow the dupe to discriminate between model and mimic.

But mimics may not be able to reproduce the nominal model blend. Consider a system where (1) the mimic cannot produce one of the components in the model blend and (2) a small fraction of model blends is missing that component. The mimic is able to produce a signal in the absolute ratio range of the model, but the signal is a "rare ratio" signal (one that corresponds to a small fraction of model signals). Mimic signals are also rare ratio signals in any system where (1) the mimic is incapable of producing a ratio of components that is as extreme as that in the nominal model blend, but (2) some model signals are within the range that the mimic is capable of producing. [Extreme ratios are apparently important in the sex-pheromone communication of many moth species where bioassays indicate that male attraction peaks at ratios as extreme as 500 : 1 (e.g., Steck *et al.*, 1980; but see also Heath *et al.*, 1988a).]

Consider a dupe that encounters a rare-ratio signal in such a system. The signal has a higher chance of coming from a mimic than do other signals in the model range. Nonetheless, the dupe individual should respond to the signal if, for that individual at that time, the potential benefits of finding a model outweigh the increased risk of finding a mimic. It is more likely, on average, that dupes will benefit by responding, if one or more of the following is true: (1) responding to a model greatly increases reproductive fitness, (2) responding to a mimic decreases reproductive fitness only slightly, (3) there are very few mimics relative to models, (4) there is a small chance of encountering in the future a signal that is more likely to come from a model. In most of the mimicry systems reviewed in this chapter, the model signal represents a particularly limiting resource for the dupe (condition 1 is true). Many moths and bees mate only once and there are very few sexually active females compared to sexually active males. Dung is an ephemeral resource and most coprophilous flies are short-lived, *r*-selected organisms (condition 4 is also true). In such systems, it seems likely that a rare-ratio mimic signal could persist in an arms race.

In rare-ratio systems, selection may produce a conditional dupe response such that only a small fraction of the dupe population responds to signals that might come from mimics (see below). Mimicry will persist if there are enough responding dupes to enable mimics to survive and reproduce.

3. Signal Passage through the Environment

Chemical signals may change in at least three ways as they move away from their source. (The following arguments do not apply to the surface chemical systems.)

1. Components of signals may be differentially adsorbed and re-released by solid surfaces that they contact as they pass through the environment. There is some evidence that plant and soil surfaces adsorb a significant fraction of the compounds in airborne odor plumes. However, there have been too few measurements to know how generally important this phenomenon is, or whether it leads to significant changes in the ratios of odor components (review in Perry and Wall, 1986). Although I am not aware of any measurements, adsorption and re-release seems likely to alter ratios in signals released in soil.
2. A chemical signal source may release into the medium a mixture of two compounds with different diffusion rates. Whether diffusion causes the ratio of the two compounds to change in an airborne signal as it moves farther away from the source has been the subject of much theoretical dispute and few measurements (review in Elkinton and Cardé, 1984). Diffusion is more likely to alter component ratios in signals released in soil (where there is no turbulence; see Elkinton and Cardé 1984).
3. As the signal moves farther away from the source, the absolute concentration of each signal component decreases, until it eventually falls below the threshold of perception of the receiver. For some signals, only a subset of the signal's components (those that are more abundant or more easily perceived) will be perceptable at intermediate distances from the source (see suggestions in Kullenberg and Bergström 1976).

Mimics may be able to employ blends with ratios that are equivalent to the altered ratios sometimes found in model signals far from the model source. For example, a mimic that cannot make a minor component of a model blend might successfully employ an "incomplete" mimic blend if the component is not perceptable to dupes in distant model signals (but see Linn *et al.*, 1986, discussed below). In an arms race, mimics employing such blends might also have to reproduce the lower absolute concentrations of model signals distant from the source. Such mimic blends may still fail in an arms race, if dupes (1) can use cues other than concentration to determine that they are close to a signal source and (2) use this information to discriminate against altered ratio mimic signals. [However, dupes may often be limited in their ability to assess absolute concentrations, and distances from signal sources (see Elkinton and Cardé, 1984). See Kullenberg *et al.* (1984a, p. 93) for another possible environmental influence.]

4. Highly Variable Models

In some systems the model signal may be so variable that selection does not favor dupes that are inhibited by alien incidental components or skewed ratios. This might apply to dupes in systems that involve mimicry of carrion/fecal odors or generalized floral odors.

C. Dupes Are Not Affected: Signal Refinement

In this section, I assume (1) that mimics have no effect on average dupe fitness and therefore (2) that dupes' responses to possible chemical stimuli remain fixed. How might mimic blends evolve under these conditions?

Consider the set of variant blends that might be produced by occasional mimic individuals. These blends might differ from the original mimic signal because the ratio of components has changed, or because they include a precursor or other relative of one of the original compounds. If some of these "related" blends elicit a stronger response from dupes and a stronger response benefits the mimic, then the mimic blend should change, until mimics are producing that relative (of a relative of a relative . . .) of the original blend that evokes the maximum response from the dupe.

This "refined" blend will often be quite different from the model blend, as in the following five examples.

1. A refined blend might include a large number of alien incidental compounds that were present in the original signal.
2. A mimic that originally produced a two-key compound blend identical to a two-key compound model blend might end up making a blend with a highly skewed ratio. Consider a hypothetical system involving a fecal-odor mimic and a coprophilous insect dupe. For this dupe, higher hydrogen sulfide ratios indicate higher nutritional-quality feces, and the dupe prefers high ratios of hydrogen sulfide to ammonia. Such a dupe might be optimally attracted by volatiles with a higher ratio of the two compounds than occurs in nature.
3. A mimic may produce compounds at much higher absolute concentrations than occur in nature. This might be the case for some spider mimics that produce incomplete moth sex pheromone blends—in many (perhaps most) moths, such blends are only effective if the concentration of the compounds is much higher than in the natural blend (Linn *et al.*, 1986).
4. The original mimic blend may have included an alien key component that is too biosynthetically "distant" from the model for the mimic to "bridge the gap." If the alien key component and its relatives are not as effective as the

model, then the mimic may end up with an alien key component blend that is inferior to the model.

5. It is conceivable that in some systems, the most effective mimic blend for the dupe is a blend that mimics two or more different models at the same time (see the discussion of *Ophrys* models in the pollination section above). In such a system, a mimic might end up producing a blend that is a mixture of a few of the compounds from each model (those that it can biosynthesize).

The refined mimic blend is only likely to be the same as the model blend if all of the following statements are true: (1) The original blend was related to the model blend. (2) There are no "super-normal" related blends that elicit a stronger response from the dupe than the model. (3) There were no alien incidental compounds in the original mimic blend. (4) There is no advantage to mimicking more than one model blend.

It is possible that in most mimicry systems, mimics have a negligible average effect on dupes, and arms races do not take place. If this is the case, then the composition of most mimic signals will be governed by refinement alone.

D. Dupes Benefit: Drift

In systems where both mimic and dupe benefit, selection will favor dupes that respond to mimic signals. There may be no selective advantage for mimics that improve on the (possibly vague) resemblance of the original mimic signal to the model. Moreover, in cases where the mimic signal slowly changes, the dupe should continue to respond even if the signal drifts far from the model.

Consider a flower that produces a scent that attracts pollinators by mimicking the odor of the pollinators' oviposition site. Pollinators eat part of the flower and therefore are rewarded. [Many authors have suggested that all floral odors started with food- or oviposition site–odor mimicry in primitive angiosperms (e.g., Pellmyr and Thien, 1986).] Later in the evolution of the plant, the flower attracts pollinators with a nonmimetic scent and provides them with a nutritional reward from specialized structures. At what point did the system cease to be mimetic? [See Otte (1974, p. 398) for analogous objections to fish tail-fin spots visually "mimicking" fish eggs.]

Dupe-benefit mimicry systems involving the chemical senses are likely to be particularly difficult to study (as mimicry systems). It will be necessary in many cases to demonstrate that mimic and model signals of different chemical composition are perceived similarly by the dupe.

Consider certain changes in cuticle chemistry in females of at least some fly species. Some of the dominant cuticle hydrocarbons in mated females are compounds that are virtually absent in virgin females but plentiful in males (e.g., Scott, 1986). The presence of these compounds in mated females inhibits male courtship. This is true despite the persistence in female cuticle of other hydrocarbons that are characteristic of females and by themselves stimulate courtship. Are mated females mimicking males? Is it likely that males are really deceived when female compounds are also present? If both males and females benefit [my assumption—see Thornhill and Alcock (1983, pp. 336–338)], then it is simpler to consider this phenomenon as nonmimetic communication. [These systems are complex; male-inhibiting compounds include both female-produced compounds, and compounds transferred from males to females during mating (see Scott, 1986).]

The most obvious examples of systems where both mimic and dupe benefit are (some of the) systems where mimic and model are (1) the same species or (2) a pair of mutualist species. Dupes may benefit in the Euglossine bee/orchid systems and rust/rust-"pollinator" systems mentioned above. They might also benefit in the potato/aphid (alarm pheromone mimicry) system and the raiding allomone systems. (Even though raiders harm victims, victims may suffer even greater harm if they fail to desist in response to the raiding allomones.)

E. Systems Where There Is More Than One Dupe Species

In numerous chemical mimicry systems, individual mimic species interact with a large number of dupe species. I first consider four overlapping categories of "multiple-dupe" systems.

1. One Mimic Signal/One Popular Model

In "popular model" systems, there is a single model that is important to many dupe species. For example, the mushroom *Dictyophora ravinelii* mimics fecal odors and attracts at least 10 coprophilous fly species (Stoffolano *et al.*, 1988). The remaining categories are all "multiple-model" systems.

2. One Mimic Signal/Multiple Equivalent Models

In the sex-pheromone mimicry systems a single mimic species may attract as many as 19 dupe species (see Tables IIa and IIb). In some of these systems mimics might employ a single blend that mimics a set of equivalent models; all the attracted males (dupes) belong to a set of species where the females (models) use essentially identical sex-pheromone blends (model signals). In the dupe taxa the diversity of sex pheromone chemistry is modest in comparison to the enormous diversity of dupe species. For example, large numbers of moth species in unrelated groups use identical pheromone compounds (cf. Tamaki, 1984; Mayer and McLaughlin, 1988). Many related moth species use very similar pheromone blends (refs. in Cardé and Baker, 1984). Such species are typically sexually active in different times and places (presumably due in part to selection for reproductive isolation). Some mimics may interact with a large set of dupe species merely because the mimics are "sexually active" over longer periods of time and/or wider geographic areas than the dupes. Spiders hunt at different times of night and over most of the year, while most moths are sexually active during only part of the year and during only a few hours at characteristic times of the night.

3. One Mimic Signal/Multiple Nonequivalent Models

In theory at least, mimics could also use a "multiple-interpretation" blend that elicits responses from more than one dupe species even though (1) the nominal model blends are different for the various dupe species and (2) each dupe species normally responds to only one of the model blends. In systems where there is no arms race, the model blends need not be similar, and many kinds of multiple-interpretation blend are conceivable (for two possibilities see Appendix A). I will discuss a hypothetical system where the mimic employs a multiple-interpretation blend that could persist in an arms race.

Consider a spider hunting in an area where three moth species are sexually active at the same time. Assume that the sex-pheromone blends of the female moths all have the same two compounds as the two major blend components. Males discriminate among females of different species on the basis of (1) differences in the ratio of the two compounds and (2) the presence or absence of a third minor constituent. Spiders might attract males of all three species with a blend of just the two compounds in a ratio that represents an average of the different ratios of the nominal female moth blends.

Dupes may be able to discriminate between this multiple-interpretation blend and the nominal model blends. This multiple-interpretation blend might persist in an arms race nonetheless if (1) at least a few of the model blends match the multiple-interpretation blend and (2) at least some fraction of the dupe population of each species benefits on average in responding to what might be a mimic or model signal.

This multiple-interpretation blend might have low "activity"—i.e., it might elicit a response from only a small fraction of the population of each dupe species. Nonetheless, such a blend may be preferable to blends that have high activity for just one dupe species. Consider a spider hunting at a time of year when the only potential moth prey species that are sexually active are species whose actual availability (the number of males flying in the vicinity of the spider) is unpredictable and varies independently for each moth species (see Stowe and Icenogle, 1988). In such situations, a multiple-interpretation blend has an advantage over a high-activity, high-specificity blend because the latter will attract nothing if the relevant moth species is not available. The multiple-interpretation blend may attract a relatively small percentage of the flying males and yet have no disadvantage over the high-activity, high-specificity blend, because spiders can only make use of a very small percentage of the available male moths: spiders can only eat three or four individual moths in one night. Although selection presumably favors high specificity and high activity in the blends produced by female moths, this is not necessarily the case for spider blends.

4. Multiple Mimic Signals/Multiple Nonequivalent Models

Producing many distinct mimic signals is another way that individuals of a single species may mimic different models (or sets of models). There is preliminary evidence for intraspecific variation in blends in both the orchid/pollinator and spider/moth systems. (It is not clear at present whether this is intra- or strictly interindividual variation.) In the spider *Mastophora cornigera*, some blend variation is almost certainly necessary to account for the fact that compounds that are necessary for attraction of males of some prey species are strongly inhibitory to other species (Stowe and Icenogle, 1988; Stowe *et al.*, 1987, and refs. therein).

At present, intraspecific variation in mimic signals is well established only in visual mimicry systems. Different individuals mimic different models in the butterfly *Papilio dardanus* (models: different unpalatable butterflies) and the brood-parasitic European cuckoo *Cuculus canorus* (models for the cuckoo eggs: the eggs of different host birds); both systems are reviewed in Wickler (1968). Single individuals facultatively mimic different models at different times in the mimicry systems involving certain species of the firefly genus *Photuris*. *Photuris* females lure male firefly prey of various species by producing the flash responses of females of the appropriate species (review in Lloyd, 1984). [In a number of bird species, individuals "mimic" different heterospecific bird songs at different times (e.g., Howard, 1974). However, it is not clear what any of the species involved gains or loses.]

5. Flux in the Set of Dupe Species

For some mimic species, there might be a steady turnover in the set of dupe species. For example, changes in the distribution of a mimic species and a large number of potential dupe species might frequently lead to new mimic–dupe interactions. For mimics in some systems, the number of potential dupe species is very large; mimic–dupe interaction (1) is affected relatively little by the size and morphology of the dupe and/or (2) depends on just a few widely shared features of the dupe's normal response to the model. For example, mimic spiders will (1) accept any palatable prey in a wide size range and (2) only require that prey be attracted close enough for capture [in contrast to many orchid sex-pheromone mimics that depend on more species-specific aspects of sexual behavior (see, for example, Stoutamire, 1974)].

Another potential source of flux is the appearance of new mimic blends that elicit responses from new dupe species. Because chemical mimic signals are presumably controlled by very few genes (see Roelofs *et al.*, 1986), signal repertoires have the potential of evolving quite rapidly in chemical mimicry systems. A minor genetic change in the mimic might lead to (1) a mimic blend with a new ratio of compounds or (2) a blend containing a new compound: a precursor or other biosynthetic relative of a compound in the old blend. Numerous orchids, including many *Ophrys* species, occasionally hybridize. These hybrids may emit novel mimic blends (see Dafni, 1987, pp. 93–94). How likely are these new blends to be similar enough to extant models that they elicit appropriate responses from new dupe species?

Consider some properties of female moth sex pheromones, the models in spider/moth mimicry systems. (1) The blends in most species of the major subfamilies of the largest moth family, the Noctuidae, obey the "one-step rule": each blend component is only one enzyme step removed from one of the other components (Steck *et al.*, 1982; see Bjostad and Roelofs, 1983). The one-step rule may apply to blends in other families as well. (2) Many sex-pheromone "communication channels" are tightly packed with species—any mixture of two common sex-pheromone compounds is likely to fall in the ratio range of some species (Greenfield, 1983; refs. in Cardé and Baker, 1984). (3) Moths use many different combinations of compounds in their blends. Field screening of moth sex-pheromone compounds has been successful largely because a significant fraction of compound combinations (and ratios) attract one or more moth species. For example, in one study where 441 three-compound mixtures were tested in crop fields in Canada, 11% of the mixtures attracted at least one moth species (Steck *et al.*, 1979; cf. compounds and moth species in Stowe *et al.*, 1987).

Therefore, at least in moth/spider systems, new mimic blends have a relatively good chance of being useful; new blend compounds are likely to be identical to compounds found in some model blends, and new combinations and new ratios of compounds are likely to match combinations and ratios found in some models. In sex-pheromone mimicry systems, the production of a new mimic compound may be instantly profitable for a mimic. In contrast, the evolution of new female pheromone compounds in dupe (= model) species must be accompanied by changes in male olfaction. It is possible that mimic compound chemistry generally diversifies faster than dupe (= model) sex-pheromone chemistry. This might be one reason why, in the sex-pheromone mimicry systems, a few genera of mimics attract many families of dupes.

In a mimicry system where flux in the set of dupe species is possible, the mimic may frequently lose arms races with individual dupe species and yet mimicry will persist because new dupe species become available. This condition has been proposed for (visual) egg mimicry by the European cuckoo *Cuculus canorus* (Dawkins and Krebs, 1979; Wickler, 1968; see also Mason and Rothstein, 1986). Cuckoos regularly parasitize 12 host species, but cuckoo eggs have been reported at least occasionally in nests of birds belonging to 180 different species. The resemblance of cuckoo eggs to host eggs varies from host species to host species. The hosts in turn appear to vary in their ability to recognize eggs and thwart brood parasitism. The differences in cuckoo/host interactions may reflect a slow but steady turnover in the set of host species.

6. A Variety of Relationships within a Single System

In any system where a mimic interacts with many dupe species, the relationship between mimic and dupe may be very different for each dupe. Consider an extreme, hypothetical example involving one moth-attracting spider species and five moth prey species A, B, C, D, and E, caught at different times of the year. (1) The range of moth A has only recently overlapped that of the spider species, and an arms race has not developed. Spiders attract males of this species with an alien component blend where one key component is one enzyme step different from the model key component in the female moth blend but is attractive to male moths nonetheless. (2) Spiders attract moths of species B and C with a blend of two compounds in a ratio that is the average of the ratios of the two compounds in the two female moth blends. The blend is inefficient; it attracts only a small fraction of flying male moths that detect it. However, one of the moths is usually abundant and spiders feed to satiation. (3) Moth species D has been involved in a long-running arms race with the spider species. The spiders efficiently attract males of this moth with a blend of five compounds that closely resembles the female moth blend. (4) The spiders attract moth species E with a blend of components that matches the blend produced by *male* moths during courtship of females. This blend attracts other males, which hope to steal a copulation. [The production of rival-male-attracting blends of "female" sex-pheromone compounds by males during courtship has been demonstrated in a number of noctuid moths and is conceivably widespread (Heath *et al.*, 1988c, and refs. therein).]

7. The Take-Home Message

Species interactions in chemical mimicry systems are potentially complex. Researchers working with these systems need to keep the following subtleties in mind. (1) Selection pressures that mold mimic signals are not the same as those that mold model signals; numerous kinds of mimic signal are possible. (2) Mimic signals may or may not closely resemble model signals, and may or may not elicit a response from the dupes that is as strong as a response to the model. Therefore, chemical analyses or bioassays alone may be misleading. (3) Within a single multiple-dupe system, or within a set of systems involving related mimics, general rules governing the relationship of mimic and model signals, (if they exist at all), will be discerned only after many different mimic/dupe relationships have been examined.

F. Systems Where Dupes Learn

Mimicry may be more likely to develop from species interactions where the incipient dupe is incapable of learning. If, as a normal part of their behavior, dupes learn to discriminate between model signals and similar signals present in the environment, then dupes may learn to avoid incipient mimics so quickly that mimics rarely benefit, and mimicry fails to develop. Vogel (1975, 1978b) suggests that this is why mimetic pollination systems involving nonrewarding flowers include relatively few examples of flowers that mimic food sources of bees (but see Dafni, 1984).

Dupes may learn to avoid mimics in some orchid/pollinator systems. *Ophrys* flowers are apparently less attractive to some bee pollinators as the season progresses; a smaller and smaller percentage of those bee individuals that fly in the vicinity of the flower shows the full response (e.g., Kullenberg *et al.*, 1984a). In at least one pollinator species, initial experiments indicate that bees learn to avoid flowers after a small number of encounters (Paulus *et al.*, 1983). [Casual observation suggests that a similar phenomenon occurs in *Caladenia* pollinators (W. P. Stoutamire, personal communication).] Whether avoidance is based on visual or odor cues is not known.

It would be interesting to compare the "learning skill" of bees (preferably different populations of the same species) in areas with and without orchids. Naive bees in areas with orchids might learn to avoid flowers faster than naive bees in areas without orchids. If this pattern was repeated for several bee species, then this would be good evidence that the bees' learning capacity is, in part, an evolutionary response to the harmful effects of orchids (Paulus *et al.*, 1983).

In an arms race, dupes might enhance their capacity to learn to avoid mimic odors rather than develop "hard-wired" (alien component/ratio) inhibition responses (see Hall, 1986b, for a pertinent, counterintuitive example). Learning might be less expensive in terms of neurons, and a more general solution, particularly for dupes that must deal with (possibly variable) signals from numerous mimic species. I will not attempt to treat here the numerous ways that dupes could learn to avoid mimics.

G. Systems Where Dupe Response Changes over Time

Arms races may produce systems where the dupe's response to mimic signals changes over time. For example, it would not be surprising if dupes are more likely to respond to rare-ratio mimic signals as they near the end of their lifespan. Some of the prey (various *Acrolophus* species) of the moth-attracting spiders lack mouthparts and do not feed as adults. These moths apparently live for only 3 or 4 days and are sexually active for less than 1 h each evening (Stowe, unpublished). If an old male is unlikely to stay alive as long as the average search time required to find a virgin female, it may increase the male's chances of reproduction on average if it approaches a female-like odor source—even when the odor is relatively unlikely to come from a female. Although it is of general interest, no one, to the best of my knowledge, has ever measured the age of dupes responding to mimics in any aggressive mimicry system.

In some systems, dupes may change their response to rare-ratio signals in a way that reflects changing mimic and model phenologies. The dupes in orchid/pollinator systems represent many evolutionary lines. Some pollinator groups may not have evolved the ability to learn to avoid flowers, and it may be that, in some systems, seasonal dropoff of dupe visits to flowers occurs independently of experience. Such a changing dupe response could evolve if mimic and dupe (= model) populations predictably change in such a way that it is profitable to respond to rare-ratio signals early in the season but not late in the season. (There are other possible explanations for seasonal dropoff of pollinator response. It is conceivable that in some orchid species, the flowers change their scent as the season progresses to match changing populations of available pollinator species. If enough pollinator species are examined, some species may be found whose response to the flowers' scent increases as the season progresses.)

H. Systems Where Dupes and Models Are the Same Species

There are unique possibilities for systems where the dupe is the same species as the source of the model signal, and some genes influence both the production of, and the response to the model signal. In these systems, selection on the dupe may result in a change in the model signal. Roelofs *et al.* (1986) showed that in at least one moth species, males that prefer blends with an unusually high ratio of compounds father females whose blends have unusually high ratios of compounds. Paradoxically, however, they also found that after several successive generations of selection there was very little change from the original blend ratio. If, contrary to these results, spider predation were to affect sex-pheromone ratios in successive generations of moths, the ratio of blend components in both spider and female moth signals might increase until an extreme ratio representing physiological limits is reached. Even if the average ratio of the moths is more extreme than that of the spiders, the spiders will still be able to catch moths if there is enough remaining variation (independent of selection) in moth signal distributions. However, moth blends might not change, for two reasons. (1) If there is a common sympatric moth species with a more extreme blend ratio, selection on both sexes to avoid interspecific attraction may be stronger than selection on males to avoid uncommon spiders (see related arguments in West-Eberhard, 1983, p. 159). (2) Moth gene flow from areas without spiders may overwhelm any effects in the areas with spiders. (See Haynes *et al.*, 1984, for analogous arguments concerning the possibility of the development of 'resistance' to pheromone-based control of moths in agriculture.)

V. Summary

Chemical mimicry has evolved in a wide variety of taxa. The following three sets of mimicry systems are particularly noteworthy because (a) they involve many species, (b) the mimic and model signals have been chemically analyzed, and (c) the ecology of (some of) the mimics and dupes is relatively well understood.

1. Diverse organisms, including rusts, mushrooms, mosses, flowering plants, and possibly some vertebrates, mimic the odor of feces/carrion and exploit coprophilous insects. Known fecal odor components are produced by some of the mimics (Table I).
2. Various orchids sexually excite and attract their male insect pollinators (primarily bees and wasps—Table IIb).
3. Numerous spiders mimic moth sex pheromones and attract male moth prey from a number of families (Table IIa).

For both sets 2 and 3, similarity in the chemistry of mimic volatiles and model sex pheromones has been demonstrated. (See text for various model possibilities other than female sex pheromones.)

Even for the well-studied systems, our understanding of signal chemistry and dupe perception is primitive. Using information from better understood forms of chemical communication, I speculate about the evolution of chemical mimicry systems. Many of my arguments are pertinent to aggressive mimicry systems in general.

In many systems, mimics may have little effect on average dupe fitness, and mimic signals may differ from models in a number of ways. An arms race of increasing signal discrimination on the part of the dupes and increasing signal refinement on the part of the mimics may take place in systems where mimics lower average dupe fitness. Under some conditions, this will result in a system where mimic signals closely resemble model signals qualitatively and quantitatively. Under other conditions, various types of nonidentical mimic signals may persist.

If variance in model signals is large, mimics may elicit sufficient dupe response with a "rare-ratio" signal—a blend of compounds, in a ratio that is present in only a small percentage of the model signals. Mimics might use such a blend if (1) they are incapable of reproducing the nominal model blend or (2) the rare-ratio blend is advantageous because it elicits a response from more than one dupe species. In such mimicry systems, the relative costs and benefits of responding to a rare-ratio blend may change over the season or during the lifetime of the dupe as the population sizes of mimics, models, and dupes change. In some systems, therefore, dupe response may vary over the season and/or lifetime of the dupe.

In a large number of chemical mimicry systems, the mimic species interacts with many dupe species. In a multiple-dupe system, a single mimic signal may imitate more than one model signal, and elicit responses from more than one dupe species, and/or mimics may be capable of producing more than one mimic signal.

Even though we are probably aware of only a small fraction of extant systems, chemical mimicry is apparently widespread. Mimics may have been an important influence on the evolution of chemical communication in many dupe taxa.

Acknowledgments

B. J. Ploger provided invaluable and exceptionally generous editorial assistance at every stage of manuscript preparation. For criticism of early drafts I also thank B. Kullenberg, G. Bergström, J. Tengö, A. K. Borg-Karlson, R. Cardé, L. Brower, H. W. Levi, D. Bowers, J. H. Tumlinson, J. Sivinski, J. E. Lloyd, M. Obin, and an anonymous reviewer. Short sections of the manuscript were criticized by J. Koach, S. Hermann, A. Koponen, and D. Savile. I am grateful to R. W. Howard, E. Shinkai, W. J. Tietjen, A. Forsyth, A. K. Borg-Karlson, J. Tengö, K. V. Yeargan, R. Cameron, J. G. Stoffolano, D. Clyne, and T. C. Boyden for sharing unpublished information, and B. Kullenberg for very generous help in constructing Table IIa. Rare, excellent photographs (which converted to black and white with varying success) and permission to use them were kindly provided by B. Kullenberg, W. P. Stoutamire, R. W. Howard, D. P. Wojcik, R. Dressler, W. Barthlott, H. F. Neubauer, B. J. D. Meeuse, W. R. Burk, and R. D. Goos. Helpful information and discussion were supplied by J. E. Lloyd, J. Sivinski, G. Gottsberger, J. Tengö, N. Williams, B. J. D. Meeuse, A. Forsyth, W. R. Burk, W. P. Stoutamire, and D. Griffin. I thank K. Spencer for the original symposium invitation.

Research by the author on spider/moth mimicry systems was partially supported by Harvard University, the National Geographic Society (grant 2506-82), the U.S. National Science Foundation (predoctoral fellowship), and a cooperative agreement with the U.S. Department of Agriculture Insect Attractants Laboratory (USDAIAL). J. Reiskind and the University of Florida Zoology Department, and J. H. Tumlinson, P. D. Greany and the USDAIAL provided generous use of their facilities during preparation of the manuscript. I apologize to the numerous persons not mentioned here who have helped me, and to those authors whom I have slighted by citing reviews rather than research papers. Finally, I thank in advance those who contribute to future reviews on related topics by informing me of relevant studies.

References

Ågren, L., Kullenberg, B., and Sensenbaugh, T. (1984). Congruences in the pilosity between three species of *Orphrys* (Orchidaceae) and their hymenopteran pollinators. *Nova Acta Reg. Soc. Sci. Ups. Ser. V:C* **3,** 15–25.

Alcock, J., and Forsyth, A. (1988). Female mimicry in the rove beetle *Leistotrophus versicolor.* In preparation.

Alloway, T. M. (1979). Raiding behavior of two species of slave-making ants, *Harpagoxenus americanus* (Emery) and *Leptothorax duloticus* Wesson (Hymenoptera: Formicidae). *Anim. Behav.* **27,** 202–210.

Atwood, J. T. (1985). Pollination of *Paphiopedilum rothschildianum:* Brood-site deception. *Nat. Geogr. Res.* **Spring,** 247–254.

Baker, T. C., Willis, M. A., Haynes, K. F., and Phelan, P. L. (1985). A pulsed cloud of pheromone elicits upwind flight in male moths. *Physiol. Entomol.* **10,** 257–265.

Barber, J. T. (1978). *Capsella bursa-pastoris* seeds. Are they carnivorous? *Carniv. Plant Newslett.* **7,** 39–42.

Barber, J. T., and Page, C. R. (1976). Mucilaginous seed pellicles. *What's New Plant Physiol.* **8,** 1–5.

Barrer, P. M., Lacey, M. J., and Shani, A. (1987). Variation in the relative quantities of airborne sex pheromone components from individual female *Ephestia cautella* (Lepidoptera: Pyralidae). *J. Chem. Ecol.* **13,** 639–653.

Bates, R. (1979). *Leporella fimbriata* and its ant pollinators. *J. Native Orchid Soc. So. Aust.* **3,** 9–10.

Beckage, N. E. (1985). Endocrine interactions between endoparasitic insects and their hosts. *Annu. Rev. Entomol.* **30,** 371–413.

Bergström, G. (1978). Role of volatile chemicals in *Ophrys*-pollinator interactions. *In* "Biochemical Aspects of Plant and Animal Coevolution" (J. B. Harbourne, ed.), pp. 207–232. Academic Press, London.

Bergström, G. (1988). Transition from chemical resemblance to chemical mimetism. *Colloq. Mimicry, Soc. Entomol. Fr. Paris 16–19 Dec. 1981.* In press.

Bergström, G., Appelgren, M., Borg-Karlson, A. K., Groth, I., Strömberg, S., and Strömberg, S. (1980). Studies on natural odoriferous compounds XXI. Techniques for the isolation/enrichment of plant volatiles in the analyses of *Ophrys* orchids (Orchidaceae). *Chem. Scripta* **16,** 173–180.

Bian, Z., Fales, H. M., Blum, M. S., Jones, T. H., Rinderer, T. E., and Howard, D. F. (1984). Chemistry of cephalic secretion of the fire bee *Trigona (Oxytrigona) tataira*. *J. Chem. Ecol.* **10,** 451–461.

Bino, R. J., Dafni, A., and Meeuse, A. D. J. (1982). The pollination of *Orchis galilaea* (Bornm. et Schulze) Schltr. (Orchidaceae). *New Phytol.* **90,** 315–319.

Bjostad, L. B., and Roelofs, W. L. (1983). Sex pheromone biosynthesis in *Trichoplusia ni:* Key steps involve delta-11 desaturation and chain-shortening. *Science* **220,** 1387–1389.

Blum, M. S. (1966). Chemical releasers of social behavior. VIII. Citral in the mandibular gland secretion of *Lestrimelitta limao* (Hymenoptera: Apoidea: Melittidae). *Ann. Entomol. Soc. Am.* **59,** 962–964.

Blum, M. S. (1980). Arthropods and ecomones: Better fitness through ecological chemistry. *In* "Animals and Environmental Fitness" (R. Gilles, ed.), pp. 207–222. Pergamon, Oxford.

Blum, M. S. (1984). Alarm pheromones. *In* "Comprehensive Insect Physiology, Biochemistry and Pharmacology" (G. A. Kerkut and L. I. Gilbert, eds.), pp. 193–223. Pergamon, Oxford.

Blum, M. S., and Brand, J. M. (1972). Social insect pheromones: Their chemistry and function. *Am. Zool.* **12,** 553–576.

Blum, M. S., Crewe, R. M., Kerr, W. E., Keith, L. H., Garrison, A. W., and Walker, W. M. (1970). Citral in stingless bees: Isolation and functions in trail-laying and robbing. *J. Insect Physiol.* **16,** 1637–1648.

Blum, M. S., Jones, T. H., Hölldobler, B., Fales, H. M., and Jaouni, T. (1980). Alkaloidal venom mace: Offensive use by a thief ant. *Naturwissenschaften* **67,** 144–145.

Bonner, J. T. (1983). Chemical signals of social amoebae. *Sci. Am.* **248,** 114–120.

Borg-Karlson, A. K. (1985). Chemical and behavioral studies of pollination in the genus *Ophrys* L. (Orchidaceae). Ph.D. Thesis, *R. Inst. Technol. Stockholm.*

Borg-Karlson, A. K. (1987). Chemical basis for the relationship between *Ophrys* orchids and their pollinators III: Volatile compounds of species in the *Ophrys* sections *Fuciflorae* and *Bombyliforae* as insect mimetic attractants/excitants. *Chem. Scripta* **27,** 313–320.

Borg-Karlson, A. K. (1988). Attraction of *Phyllopertha horticola* (Coleoptera: Scarabeidae) males to *Ophrys* flower fragrances in the section *Fuciflorae.* In preparation.

Borg-Karlson, A. K., and Groth, I. (1986). Volatile compounds from flowers of four species in the sections *Arachnitiformes* and *Araneiferae* of the genus *Ophrys* as insect mimetic attractants/excitants. *Phytochemistry* **25,** 1297–1299.

Borg-Karlson, A. K., and Tengö, J. (1980). Pyrazines as marking volatiles in philanthine and nyssonine wasps. (Hymenoptera: Sphecidae). *J. Chem. Ecol.* **6,** 827–835.

Borg-Karlson, A. K., and Tengö, J. (1986). Chemical mimetism? Key substances in the *Ophrys lutea-Andrena* relationship. *J. Chem. Ecol.* **12,** 1927–1941.

Borg-Karlson, A. K., Bergström, G., and Groth, I. (1985). Chemical basis for the relationship between *Ophrys* orchids and their pollinators I: Volatile compounds of *Ophrys lutea* and *O. fusca* as insect mimetic attractants/excitants. *Chem. Scripta* **25,** 283–294.

Borg-Karlson, A. K., Bergström, G., and Kullenberg, B. (1987). Chemical basis for the relationship between *Ophrys* orchids and their pollinators II: Volatile compounds of *Ophrys insectifera* and *O. speculum* as insect mimetic attractants/excitants. *Chem. Spripta,* **27,** 303–312.

Brothers, D. J. (1975). Phylogeny and classification of the aculeate Hymenoptera, with special reference to Mutillidae. *Univ. Kansas Sci. Bull.* **50,** 483–648.

Burk, W. R. (1980). A bibliography of North American Gasteromycetes: I. Phallales. *Bibl. Mycol.* **73.**

Burns-Balogh, P., Borg-Karlson, A. K., and Kullenberg, B. (1985). Evolution of the monandraceous Orchidaceae VI. Evolution and pollination mechanisms in the subfamily Orchidoidea. *Can. Orchid J.* **3,** 29–57.

Camazine, S. (1985). Olfactory aposematism: Association of food toxicity with a naturally occurring odor. *J. Chem. Ecol.* **11,** 1289–1296.

Cameron, R. G., and Troilo, D. B. (1982). Fly-mediated spore dispersal in *Splachnum ampullaceum. Mich. Bot.* **21,** 59–65.

Cameron, R. G., and Wyatt, R. (1986). Substrate restriction in entomophilous Splachnaceae: Role of spore dispersal. *Bryologist* **89,** 279–284.

Cane, J. H., and Tengö, J. (1981). Pheromonal cues direct mate-seeking behavior of male *Colletes cunicularius* (Hymenoptera: Colletidae). *J. Chem. Ecol.* **7,** 427–436.

Cantelo, W. W., and Jacobson, M. (1979a). Phenylacetaldehyde attracts moths to bladder flower and to blacklight traps. *Environ. Entomol.* **8,** 444–447.

Cantelo, W. W., and Jacobson, M. (1979b). Corn silk volatiles attract many pest species of moths. *J. Environ. Sci. Health* **A14,** 695–707.

Cardé, R. T. (1986). Epilogue: Behavioral mechanisms. *In* "Mechanisms in Insect Olfaction" (T. L. Payne, M. C. Birch, and C. E. J. Kennedy, eds.), pp. 175–181. Clarendon, Oxford.

Cardé, R. T., and Baker, T. C. (1984). Sexual communication with pheromones. *In* "Chemical Ecology of Insects" (W. J. Bell and R. T. Cardé, eds.), pp. 355–383. Sinauer Associates, Sunderland, Massachusetts.

Cardé, R. T., Baker, T. C., and Roelofs, W. L. (1976). Sex attractant responses of male Oriental fruit moths to a range of component ratios: Pheromone polymorphism? *Experientia* **32,** 1406–1407.

Chandler, A. E. F. (1968). Some host-plant factors affecting oviposition by aphidophagous Syrphidae (Diptera). *Ann. Appl. Biol.* **61,** 415–423.

Cobb, N. A. (1906). Fungus maladies of the sugar cane. *Hawai. Sugar Plant. Assoc. Exp. Sta. Div. Pathol. Physiol. Bull.* **5,** 1–254.

Cobb, N. A. (1909). Fungus maladies of the sugar cane. *Hawai. Sugar Plant. Assoc. Exp. Sta. Div. Pathol. Physiol. Bull.* **6,** 5–30.

Coleman, E. (1932). Pollination of *Diuris pedunculata R. Br. Victorian Nat.* **49,** 179–186.

Coleman, E. (1933). Pollination of *Diuris sulphurea R. Br. Victorian Nat.* **50,** 3–8.

Conner, W. E., Eisner, T., Vander Meer, R. K., Guerrero, A., Ghiringelli, D., and Meinwald, J. (1980). Sex attractant of an Arctiid moth (*Utetheisa ornatrix*): a pulsed chemical signal. *Behav. Ecol. Sociobiol.* **7,** 55–63.

Cooke, J. (1984). Hidden assets. *Nat. Hist.* **93,** 74–77.

Cottrell, C. B. (1983). Aphytophagy in butterflies: Its relationship to myrmecophily. *Zool. J. Linn. Soc.* **80,** 1–57.

Cowgill, U. (1966). *Perodicticus potto* and some insects. *J. Mammal.* **47,** 156–157.

Dafni, A. (1984). Mimicry and deception in pollination. *Annu. Rev. Ecol. Syst.* **15,** 259–278.

Dafni, A. (1986). Floral mimicry—Mutualism and unidirectional exploitation of insects by plants. *In* "Insects and the Plant Surface" (B. Juniper and R. Southwood, eds.), pp. 81–90. Arnold, Baltimore.

Dafni, A. (1987). Pollination in *Orchis* and related genera: Evolution from reward to deception. *In* "Orchid Biology: Reviews and Perspectives." Vol. IV (J. Arditti, ed.), pp. 81–104. Cornell University, Ithaca, New York.

Dainty, R. H. (1982). Biochemistry of undesirable effects attributed to microbial growth on proteinaceous foods stored at chill temperatures. Food Chem. **9,** 103–113.

Damian, R. T. (1979). Molecular mimicry in biological adaptation. *In* "Host-Parasite Interfaces: At Population, Individual, and Molecular Levels" (R. B. Nichol, ed.), pp. 103–126. Academic Press, New York.

Damian, R. T. (1984). Immunity in schistosomiasis: A holistic view. *In* "Contemporary Topics in Immunobiology." (J. Marchalonis, ed.), pp. 359–420. Plenum, New York.

Dawkins, R., and Krebs, J. R. (1979). Arms races between and within species. *Proc. R. Acad. London Ser. B* **205,** 489–511.

Dettner, K. (1983). Vergleichende Untersuchungen zur Wehrchemi und Drüsenmorphologie abdomenaler Aberwehrdrüsen von Kurzflügen aus dem Subtribus Philonthina (Coleoptera: Staphylinidae). *Z. Naturforsch. C* **38,** 319–328.

Disney, R. H. L. (1986). Two remarkable new species of scuttle-fly (Diptera: Phoridae) that parasitize termites (Isoptera) in Sulawesi. *Syst. Entomol.* **11,** 413–422.

Dressler, R. (1981). "The Orchids, Natural History and Classification." Harvard Univ. Press, Cambridge, Massachusetts.

Dring, D. M. (1980). Contributions toward a rational arrangement of the Clathraceae. *Kew Bull.* **35,** 1–96.

Duddington, C. L., and Wyborn, C. H. E. (1972). Recent research on the nematophagous hyphomycetes. *Bot. Rev.* **38,** 545–565.

Eberhard, W. G. (1977). Aggressive chemical mimicry by a bolas spider. *Science* **198,** 1173–1175.

Eberhard, W. G. (1980). The natural history and behavior of the bolas spider *Mastophora dizzydeani* sp. n. (Araneae). *Psyche* **87,** 143–169.

Eberhard, W. G. (1981). The single line web of *Phoroncidia studo* Levi (Araneae, Theridiidae) a prey attractant? *J. Arachnol.* **9,** 229–232.

Eisner, T. E., and Halpern, B. P. (1971). Taste distortion and plant palatability. *Science* **172,** 1362.

Eisner, T., van Tassell, E., and Carrell, J. E. (1967). Defensive use of a "fecal shield" by a beetle larva. *Science* **158,** 1471–1473.

Eisner, T., Hicks, K., Eisner, M., and Robson, D. S. (1978). "Wolf-in-sheep's-clothing" strategy of a predaceous insect larva. *Science* **199,** 790–794.

Elkinton, J. S., and Cardé, R. T. (1984). Odor dispersion. *In* "Chemical Ecology of Insects" (W. J. Bell and R. T. Cardé, eds.), pp. 73–91. Sinauer, Sunderland, Massachusetts.

Elstner, E. F. (1983). Hormones and metabolic regulation in disease. "Biochemical Plant Pathology" (J. A. Callow, ed.), pp. 415–431. Wiley, Chichester.

Faegri, K., and van der Pijl, L. (1979). "The Principles of Pollination Ecology," pp. 102–106. Pergamon, Oxford.

Fales, H. M., Jaouni, T. M., Schmidt, J. O., and Blum, M. S. (1980). Mandibular gland allomones of *Dasymutilla occidentalis* and other mutillid wasps. *J. Chem. Ecol.* **6,** 895–903.

Feinsinger, P. (1983). Coevolution and pollination. *In* "Coevolution" (D. J. Futuyma and M. Slatkin, eds.), pp. 282–310. Sinauer, Sunderland, Massachusetts.

Fish, D. (1976). Insect-plant relationships of the insectivorous pitcher plant *Sarracenia minor. Fla. Entomol.* **59,** 199–203.

Fishlyn, D. A., and Phillips, D. W. (1980). Gastropod defenses against predators. *Biol. Bull.* **158,** 34–48.

Folkerts, G. W. (1982). The Gulf Coast pitcher plant bogs. *Am. Sci.* **70,** 260–267.

Folkerts, D. and Crossley, D. H. (1988). Resource partitioning in Sarracenia. *Oecologia.* In preparation.

Forsyth, A., and Alcock, J. (1988). Alternative foraging tactics and sexual selection in the rove beetle *Leisotrophus veriscolor. Behav. Ecol. Sociobiol.,* in preparation.

Franq, E. N. (1969). Behavioral aspects of feigned death in the opossum *Didelphus marsupialis. Am. Midl. Nat.* **81,** 556–568.

Gertsch, W. S. (1955). The North American bolas spiders of the genera *Mastophora* and *Agatostichus. Bull. Am. Mus. Nat. Hist.* **4,** 221–254.

Gibson, R. W., and Pickett, J. A. (1983). Wild potato repels aphids by release of aphid alarm pheromone. *Nature (London)* **302,** 608–609.

Gill, C. O. (1986). The control of microbial spoilage in fresh meats. *In* "Advances in Meat Research" (A. M. Pearson and T. R. Dutson, eds.), pp. 49–88. Avi Publ., Westport, Connecticut.

Givnish, T. J., Burkhardt, E. L., Happel, R. E., and Weintraub, J. D. (1984). Carnivory in the bromeliad *Brocchinia reducta,* with a cost/benefit model for the general restriction of carnivorous plants to sunny, moist, nutrient-poor habitats. *Am. Nat.* **124,** 479–497.

Godfery, M. J. (1921). The fertilisation of *Ophrys apifera. J. Bot. (London)* **59,** 285–287.

Godfery, M. J. (1922). Notes on the fertilisation of orchids. *J. Bot. (London)* **60,** 359–361.

Goss, G. J., and Adams, R. M. (1976). The reproductive biology of the epiphytic orchids of Florida IV. Sexually selective attraction of moths to the floral fragrance of *Epidendrum anceps* Jacquin. *Am. Orchid Soc. Bull.* **45,** 997–1001.

Gottsberger, G. (1977). Some aspects of beetle pollination in the evolution of flowering plants. *Plant Syst. Evol. Suppl.* **1,** 211–226.

Gottsberger, G., and Amaral, A. (1984). Pollination strategies in Brazilian *Philodendron* species. *Ber. Dtsch. Bot. Ges.* **97,** 391–410.

Graham, D. K. F. (1983). Interaction between an ichneumonid wasp and the Australian tongue orchid in New Zealand. *Rec. Auckland Inst. Mus.* **20,** 217–222.

Greenfield, M. D. (1981). Moth sex pheromones: An evolutionary perspective. *Fla. Entomol.* **64,** 4–17.

Greenfield, M. D. (1983). Reproductive isolation in clearwing moths (Lepidoptera: Sesiidae), a tropical-temperate comparison. *Ecology* **64,** 362–375.

Guilford, T., Nicol, C., Rothschild, M., and Moore, B. P. (1988). The biological roles of pyrazines: Evidence for a warning odour function. *Biol. J. Linn. Soc.,* **31,** 113–128.

Hall, J. C. (1986a). Mutants of biological rhythms and conditioned behavior in *Drosophila* courtship. *In* "Evolutionary Genetics of Invertebrate Behavior: Progress and Prospects" (M. D. Huettel, ed.), pp. 237–246. Plenum, New York.

Hall, J. C. (1986b). Learning and rhythms in courting, mutant *Drosophila. Trends Neuro Sci.* **9,** 414–418.

Harbourne, J. B. (1977). "Introduction to Ecological Biochemistry." Academic Press, London.

Haynes, K. F., and Birch, M. C. (1984). The role of other pheromones, allomones, and kairomones in the behavioral responses of insects. *In* "Comprehensive Insect Physiology, Biochemistry and Pharmacology" (G. A. Kerkut, G. A. and L. I. Gilbert, eds.), pp. 225–255. Pergamon, Oxford.

Haynes, K. F., Gaston, L. K., Pope, M. M., and Baker, T. C. (1984). Potential for evolution of resistance to pheromones: Interindividual and interpopulational variation in chemical communication system of the pink bollworm moth. *J. Chem. Ecol.* **10,** 1151–1165.

Heath, R. R., and Tumlinson, J. H. (1986). Correlation of retention times on liquid crystal capillary column with reported vapor pressures and half lives of compounds used in pheromone formulations. *J. Chem. Ecol.* **12,** 2081–2088.

Heath, R. R., Teal, P. E. A., Tumlinson, J. H., and Mengelkoch, L. J. (1986). Prediction of release ratios of multi-component pheromones from rubber septa. *J. Chem. Ecol.* **12,** 2133–2143.

Heath, R. R., Dueben, B. D., and Leppla, N. (1988a). Dial-a-blend: A system for formulating multicomponent pheromones. *J. Chem. Ecol.,* submitted.

Heath, R. R., Landolt, P. J., Tumlinson, J. H., and Calkins, C. O. (1988b). A system to collect volatile chemicals emitted by insects. *J. Chem. Ecol.* In press.

Heath, R. R., Landolt, P. J., Leppla, N., and Dueben, B. D. (1988c). Identification of a male-produced pheromone of *Anticarsia gemmatilis* (Hubner) (Lepidoptera: Noctuidae) attractive to conspecific males. *J. Chem. Ecol.* **14,** 1121–1130.

Hefetz, A., Eickwort, G. C., Blum, M. S., Cane, J., and Bohart, G. E. (1982). A comparative study of the exocrine products of cleptoparasitic bees (*Holcopasites*) and their hosts (*Calliopsis*) (Hymenoptera: Anthophoridae, Andrenidae). *J. Chem. Ecol.* **8,** 1389–1397.

Hermann, S., Platt, W. J., and von Ende, C. (1988). The importance of ancillary resources in the growth and reproduction of the pitcher plant *Sarracenia flava.* In preparation.

Hirai, K., Shiwaku, K., Tsuboi, T., Torii, M., Nishida, H., and Yamane, Y. (1983). Biological effects of *Spirometra erinacei* Pl. on several species of rodents. *Z. Parasitenk.* **69,** 489–495.

Hodges, R. W. *et al.* (1983). "Checklist of the Lepidoptera of North America North of Mexico." Classey, and Wedge Entomological Foundation, London.

Hoffmann, A. A., Parsons, P. A., and Nielsen, K. M. (1984). Habitat selection: Olfactory response of *Drosophila melanogaster* depends on resources. *Heredity* **53,** 139–143.

Hölldobler, B. (1977). Communication in the social Hymenoptera. *In* "How Animals Communicate" (T. A. Sebeok, ed.), pp. 418–471. Indiana Univ. Press, Bloomington.

Hölldobler, B., Möglich, M., and Maschwitz, U. (1981). Myrmecophilic relationship of *Pella* (Coleoptera: Staphylinidae) to *Lasius fuliginosus* (Hymenoptera: Formicidae). *Psyche*. **88,** 347–74.

Horton, C. C. (1979). Apparent attraction of moths by webs of araneid spiders. *J. Arachnol.* **7,** 88.

Howard, R. D. (1974). The influence of sexual selection and interspecific competition on mockingbird song (*Mimus polyglottos*). *Evolution* **28,** 428–438.

Howard, R. W., McDaniel, C. A., and Blomquist, G. J. (1980). Chemical mimicry as an integrating mechanism: Cuticular hydrocarbons of a termitophile and its host. *Science* **210,** 431–433.

Howard, R. W., McDaniel, C. A., Nelson, D. R., Blomquist, G. J., Gelbaum, L. T., and Zalkow, L. (1982). Cuticular hydrocarbons of *Reticulotermes virginicus* (Banks) and their role as potential species- and caste-recognition cues. *J. Chem. Ecol.* **8,** 1227–1239.

Howard, R. W., Akre, R. D., and Garnett, W. B. (1988). Chemical mimicry in obligate predators of carpenter ants. *Ann. Entomol. Soc. Am.* In preparation.

Huber, I. (1978). Prey attraction and immobilization by allomone from nymphs of *Womersia strandtmanni* (Acarina: Trombiculidae). *Acarologia* **20,** 112–115.

Inglett, G. E. (1977). Artificial sweeteners: Natural and man-made. *J. Toxicol. Environ. Health* **2,** 207–214.

Ingold, C. T. (1953). "Dispersal in Fungi." Clarendon, Oxford.

Ivri, Y., and Dafni, A. (1977). The pollination ecology of *Epipactis consimilis* Don (Orchidaceae) in Israel. *New Phytol.* **79,** 173–177.

Jacobson, E. (1911). Biological notes on the Hemipteran *Ptilocerus ochraceus*. *Tidschr. Entomol.* **54,** 175–179.

Kaib, M. (1974). Die Fleisch- und Blumenduftrezeptoren auf der Antenne der Schmeissfliege *Calliphora vicina*. *J. Comp. Physiol.* **95,** 105–121.

Kennedy, L. M., Saul, L., Sefeuka, R., and Stevens, D. A. (1988). Hovecins: Selective sweetness-reducing principle from *Hovenia dulcis* leaves. *Chem. Senses*. In press.

Kevan, P. G., and Baker, H. G. (1983). Insects as flower visitors and pollinators. *Annu. Rev. Entomol.* **28,** 407–453.

Kikuchi, T., Kadota, S., Tanaka, K., and Nishi, A. (1984). Odorous metabolites of an acellular slime mold, *Physarum polycephalum* Schw., and a basidiomycete, *Phallus impudicus* Pers. *Chem. Pharmacol. Bull.* **32,** 797–800.

Kistner, D. H., and Blum, M. S. (1971). Alarm pheromone of *Lasius* (*Dendrolasius*) *spathepus* (Hymenoptera: Formicidae) and its possible mimicry by two species of *Pella* (Coleoptera: Staphylinidae). *Ann. Entomol. Soc. Am.* **64,** 589–594.

Kobayashi, T., and Watanabe, M. (1986). An analysis of snake-scent application behavior in Siberian chipmunks (*Eutamias sibiricus asiaticus*). *Ethology* **72,** 40–52.

Koponen, A., and Koponen, T. (1978). Evidence of entomophily in Splachnaceae (Bryophyta). *Bryophytorum Bib.* **13,** 569–577.

Kullenberg, B. (1961). Studies in *Ophrys* pollination. *Zool. Bidr. Uppsala* **34,** 1–330.

Kullenberg, B. (1973). New observations on the pollination of *Ophrys* L. (Orchidaceae). *Zoon Suppl.* **1,** 9–14.

Kullenberg, B., and Bergström, G. (1976). Hymenoptera Aculeata males as pollinators of *Ophrys*. *Zool. Scripta* **5,** 13–23.

Kullenberg, B., Borg-Karlson, A. K., and Kullenberg, A. L. (1984a). Field studies on the behavior of the *Eucera nigrilabris* male in the odour flow from the flower labellum extract of *Ophrys tenthredinifera*. *Nova Acta Reg. Soc. Sci. Ups. Ser. V:C* **3,** 79–110.

Kullenberg, B., Buel, H., and Tkalcu, B. (1984b). Übersicht von Beobachtungen über Besuche von *Eucera*- und *Tetralonia* Männchen auf *Ophrys*—Blüten (Orchidaceae). *Nova Acta Reg. Soc. Sci. Ups. Ser. V:C* **3,** 27–40.

Levi, H. W., and Coddington, J. (1983). Progress report on the phylogeny of the orb-weaving family Araneidae and the superfamily Araneoidea (Arachnida: Araneae) (Abstract). *Verh. Naturwiss. Ver. Hamburg* **26,** 151–154.

Lichtner, F. T., and Williams, S. E. (1977). Prey capture and factors controlling trap narrowing in *Dionaea* (Droseraceae). *Am. J. Bot.* **64,** 881–886.

Linn, C. E., and Roelofs, W. L. (1985). Response specificity of male pink bollworm moths to different blends and dosages of sex pheromone. *J. Chem. Ecol.* **11,** 1583–1590.

Linn, C. E., Campbell, M. G., and Roelofs, W. L. (1986). Male moth sensitivity to multicomponent pheromones: Critical role of female-released blend in determining the functional role of components and active space of the pheromone. *J. Chem. Ecol.* **12,** 659–668.

List, P. H., and Freund, B. (1966). Phenylcrotonaldehyde und andere Aldehyde als Geruchstoffe der Stinkmorchel *Phallus impudicus* L. *Naturwissenschaften* **53,** 585.

List, P. H., and Freund, B. (1967). 2-Phenyl-buten(2)-al(1), Reichstoffe der Stinkmorchel. *Naturwissenschaften* **54,** 368.

List, P. H., and Freund, B. (1968). Geruchstoffe der Stinkmorchel *Phallus impudicus* L. 18. Mitteilung über Pilzinhaltsstoffe. *Plant Med. Suppl.* 123–132.

Lloyd, F. E. (1942). "The Carnivorous Plants." Chronica Botanica, Waltham, Massachusetts. (Republished in 1976 by Dover, New York.)

Lloyd, J. E. (1977). Bioluminescence and communication. *In* "How Animals Communicate" (T. A. Sebeok, ed.), pp. 164–183. Indiana Univ. Press, Bloomington.

Lloyd, J. E. (1983). Bioluminescence and communication in insects. *Annu. Rev. Entomol.* **28,** 131–160.

Lloyd, J. E. (1984). On deception, a way of all flesh, and firefly signalling and systematics. *In* "Oxford Survey of Evolutionary Biology" (R. Dawkins and M. Ridley, eds.), pp. 48–84. Oxford Univ. Press, Oxford.

Löfstedt, C., Lanne, B. S., Löfqvist, J., Appelgren, M., and Bergström, G.

(1985). Individual variation in the pheromone of the turnip moth *Agrotis segetum*. *J. Chem. Ecol.* **11,** 1181–1196.

McKay, F. E. (1971). Behavioral aspects of population dynamics in unisexual-bisexual *Poeciliopsis* (Pisces: Poeciliidae). *Ecology* **52,** 778–790.

Mainero, J. S., and Martinez del Rio, C. (1985). Cheating and taking advantage in mutualistic associations. *In* "The Biology of Mutualism" (D. H. Boucher, ed.), pp. 192–216. Croom Helm, Beckenham.

Markl, H. (1979). Manipulation, modulation, information, cognition: Some of the riddles of communication. *Fortschr. Zool.* **31,** 163–194.

Mason, R. T. (1986). Pheromone mimicry in garter snakes. *In* "Chemical Signals in Vertebrates IV" (D. Duvall, D. Müller-Schwarze, and R. M. Silverstein, eds.), pp. 279–283. Plenum, New York.

Mason, R. T. (1987). Sex pheromones and the mediation of reproduction in the red-sided garter snake *Thamnophis sirtalis parietalis*. Ph.D. Thesis, Univ. of Texas at Austin, Austin.

Mason, R. T., and Crews, D. (1985). Female mimicry in garter snakes. *Nature (London)* **316,** 59–60.

Mason, P., and Rothstein, S. I. (1986). Coevolution and avian brood parasitism: Cowbird eggs show evolutionary response to host discrimination. *Evolution* **40,** 1207–1214.

Mayer, M. S., and Mankin, R. W. (1984). Neurobiology of pheromone perception. *In* "Comprehensive Insect Physiology, Biochemistry and Pharmacology" (G. A. Kerkut and L. I. Gilbert, eds.), pp. 95–143. Pergamon, Oxford.

Mayer, M. S., and McLaughlin, J. R. (1988). "Handbook of Insect Sex Pheromones." CRC Press, Cleveland, Ohio, in press.

Meeuse, B. J. D., and Morris, S. (1984). "The Sex Life of Flowers." Facts on File, New York.

Meyers, D. G., and Strickler, J. D. (1979). Capture enhancement in a carnivorous plant: Function of antennae and bristles in *Utricularia vulgaris*. *Science* **203,** 1022–1025.

Miles, D. H., Kokpol, U., Mody, N. V., and Hedin, P. A. (1975). Volatiles in *Sarracenia flava*. *Phytochem. Rep.* **14,** 845–846.

Miner, J. R. (1977). Characterization of odors and other volatile emissions. *Agric. Environ.* **3,** 129–137.

Mitter, C., and Brooks, D. R. (1983). Phylogenetic aspects of coevolution. *In* "Coevolution" (D. J. Futuyma and M. Slatkin, eds.), pp. 65–98. Sinauer, Sunderland, Massachusetts.

Mitter, C., and Klun, J. A. (1987). Evidence of pheromonal constancy among sexual and asexual females in a population of the fall cankerworm *Alsophila pometaria* (Lepidoptera: Geometridae). *J. Chem. Ecol.* **13,** 1823–1830.

Mody, N. V., Henson, R., Hedin, P. A., Kokpol, U., and Miles, D. H. (1976). Isolation of the insect paralyzing agent coniine from *Sarracenia flava*. *Experientia* **32,** 829–830.

Mueller, J. F. (1980). A growth factor produced by a larval tapeworm and its biological activity. *In* "Growth and Growth Factors" (K. Shizume and K. Takano, eds.), pp. 193–201. Univ. of Tokyo Press, Tokyo.

Nielsen, B. O. (1968). Studies on the dipterous fauna of the stinkhorn (*Phallus impudicus* Pers.). *Nat. Jutl.* **14,** 89–98.

Nilsson, L. A. (1979). Anthecological studies of the lady's slipper, *Cypripedium calceolus* (Orchidaceae). *Bot. Notr* **132,** 329–347.
Nilsson, L. A. (1980). The pollination ecology of *Dactylorhiza sambucina* (Orchidaceae). *Bot. Notiser* **133,** 367–385.
Ogilvie, P. W. (1966). An anatomical and behavioral investigation of a previously undescribed pouch found in certain species of the genus *Chamaeleo*. Ph.D. thesis, Univ. of Oklahoma, Norman.
Oliveira, P. S., and Sazima, I. (1985). Ant-hunting behavior in spiders with emphasis on *Strophius nigricans* (Thomisidae). *Bull. B. Arachnol. Soc.* **6,** 309–312.
Otte, D. (1974). Effects and functions in the evolution of signalling systems. *Annu. Rev. Ecol. Syst.* **5,** 385–417.
Pasteur, G. (1982). A classificatory review of mimicry systems. *Annu. Rev. Ecol. Syst.* **13,** 169–199.
Paulus, H. F., and Gack, C. (1981). Neue Beobachtungen zur Bestäubung von *Ophrys* (Orchidaceae) in Südspanien, mit besonderer Berücksichtigung des Formenkreises *Ophrys fusca* agg. *Plant Syst. Evol.* **137,** 241–258.
Paulus, H. F., and Gack, C. (1983). Untersuchungen zur Bestäubung des *Ophrys fusca*-Formenkreises: Ein Beitrag zum Biospecies-Konzept der Gattung *Ophrys*. *Jahresber. Naturwiss. Ver. Wuppertal* (= Die Orchidee-Sonderheft) **36,** 65–72.
Paulus, H. F., and Gack, C. (1986). Neue Befunde zur Pseudokopulation und Bestäuberspezifität in der Orchideengattung *Ophrys*-Untersuchungen in Kreta, Süditalien und Israel. *Jahresber. Naturwiss. Ver. Wuppertal* (= Die Orchidee-Sonderheft) **39,** 48–86.
Paulus, H. F., Gack, C., and Maddocks, R. (1983). Beobachtungen und Experimente zum Pseudokopulations-verhalten an *Ophrys*—Das Lernverhalten von *Eucera barbiventris* Männchen an *Ophrys scolopax* in Südspanien. *Jahresber. Naturwiss. Ver. Wuppertal* (= Die Orchidee-Sonderheft) **36,** 73–79.
Peakall, R. (1984). Observations on the pollination of *Leporella fimbriata* (Lindl.) A. S. George. *Orchadian* **8,** 44–45.
Pellmyr, O., and Thien, L. B. (1986). Insect reproduction and floral fragrances; Keys to the evolution of the angiosperms? *Taxon* **35,** 76–85.
Perry, J. N., and Wall, C. (1986). The effect of habitat on the flight of moths orienting to pheromone sources. *In* "Mechanisms in Insect Olfaction" (T. L. Payne, M. C. Birch, and C. E. J. Kennedy, eds.), pp. 90–96. Clarendon, Oxford.
Peschke, K. (1985). Immature males of *Aleochara curtula* avoid intrasexual aggressions by producing the female sex pheromone. *Naturwissenschaften* **72,** 274–275.
Peschke, K. (1987). Cuticular hydrocarbons regulate mate recognition, male aggression and female choice of the rove beetle, *Aleochara curtula*. *J. Chem. Ecol.* **13,** 1993–2008.
Phelan, P. L., and Baker, T. C. (1987). Evolution of male pheromones in moths: Reproductive isolation through sexual selection? *Science* **235,** 205–207.
Pliske, T. E. (1975). Pollination of pyrrolizidine alkaloid-containing plants by male Lepidoptera. *Environ. Entomol.* **4,** 474–479.
Ponchet, M., Martin-Tanguy, J., Marais, A., and Martin, C. (1982). Hydroxycin-

namoyl acid amides and aromatic amines in the inflorescences of some Araceae species. *Phytochemistry* **21,** 2865–2869.

Pyysalo, H., Koponen, A., and Koponen, T. (1978). Studies on entomophily in Splachnaceae (Musci). I. Volatile compounds in the sporophyte. *Ann. Bot. Fenn.* **15,** 293–296.

Pyysalo, H., Koponen, A., and Koponen, T. (1983). Studies on entomophily in Splachnaceae (Musci). II. Volatile compounds in the hypophysis. *Ann. Bot. Fenn.* **20,** 335–338.

Regnier, F. E., and Wilson, E. O. (1971). Chemical communication and "propaganda" in slave-maker ants. *Science* **172,** 267–269.

Reiger, I. (1979). Scent-rubbing in carnivores. *Carnivore* **2,** 17–25.

Riddiford, L. M. (1975). Host hormones and insect parasites. *In* "Invertebrate Immunity" (K. Maramorosch and R. E. Shope, eds.), pp. 339–353. Academic Press, New York.

Roelofs, W. L., Du, J. W., Linn, C., Glover, T. J., and Bjostad, L. B. (1986). The potential for genetic manipulation of the redbanded leafroller moth sex pheromone blend. *In* "Evolutionary Genetics in Invertebrate Behavior: Progress and Prospects" (M. D. Huettel, ed.), pp. 263–272. Plenum, New York.

Rothschild, M. (1961). Defensive odours and Müllerian mimicry among insects. *Trans. R. Entomol. Soc. London* **113,** 101–121.

Rothschild, M. (1964). A note on the evolution of defensive and repellant odors of insects. *Entomologist* **97,** 276–280.

Rymal, D. E., and Folkerts, G. W. (1982). Insects associated with pitcher plants (*Sarracenia:* Sarraceniaceae), and their relationship to pitcher plant conservation: A review. *J. Ala. Acad. Sci.* **53,** 131–151.

Savile, D. (1976). Evolution of the rust fungi (Uredinales) as reflected by their ecological problems. *Evol. Biol.* **9,** 137–207.

Schemske, D. W., and Lande, R. (1984). Fragrance collection and territorial display by male orchid bees. *Anim. Behav.* **32,** 935–937.

Schmidt, J. O. (1982). The biochemistry of insect venoms. *Annu. Rev. Entomol.* **27,** 339–368.

Schremmer, F. (1963). Wechselbeziehungen zwischen Pilzen und Insekten. *Osterr. Bot. Z.* **110,** 380–400.

Scott, D. (1986). Sexual mimicry regulates the attractiveness of mated *Drosophila melanogaster* females. *Proc. Natl. Acad. Sci. U.S.A.* **83,** 8429–8433.

Sequeira, L. (1973). Hormone metabolism in diseased plants. *Annu. Rev. Plant Physiol.* **24,** 353–380.

Shinkai, E. (1988). Some notes on the biology of *Ordgarius sexspinosus*. In preparation.

Simon, E. (1892). "Histoire naturelle des Araignées," Vol. 1. Librairie Encylcopédique de Roret., Paris.

Simons, P. (1981). How exclusive are carnivorous plants? *Carniv. Plant Newslett.* **10,** 65–80.

Slack, A. (1979). "Carnivorous Plants." MIT Press, Cambridge, Massachusetts.

Smith, B. H., and Roubik, D. W. (1983). Mandibular glands of stingless bees (Hymenoptera: Apidae) chemical analysis of their contents and biological function in two species of *Melipona*. *J. Chem. Ecol.* **9,** 1465–1472.

Smith, V. V., and Halpern, B. P. (1983). Selective suppression of judged sweetness by ziziphins. *Physiol. Behav.* **30,** 867–874.

Spoelstra, S. F. (1980). Origin of objectionable odorous components in piggery wastes and the possibility of applying indicator components for studying odour development. *Agric. Environ.* **5,** 241–260.

Steck, W., Chisholm, M. D., Bailey, B. K., and Underhill, E. W. (1979). Moth sex attractants found by systematic field testing of 3-component acetate-aldehyde candidate lures. *Can. Entomol.* **111,** 1263–1269.

Steck, W., Underhill, E. W., and Chisholm, M. D. (1980). Trace components in lepidopterous sex attractants. *Environ. Entomol.* **9,** 583–585.

Steck, W., Underhill, E. W., and Chisholm, M. D. (1982). Structure-activity relationships in sex-attractants for North American noctuid moths. *J. Chem. Ecol.* **8,** 731–754.

Stijve, T. (1965). Een Chemisch onderzoek van de grote Stinkzwam (*Phallus impudicus* Pers.). *Coolia* **11,** 40–41.

Stoffolano, J. G. (1988). The role of insects in the transport of rust pycniospores. In preparation.

Stoffolano, J. G., Yin, C., and Wu, T. (1988). Nutritive deception of flies by the mushroom *Phallus impudicus*. In preparation.

Stoutamire, W. P. (1974). Australian terrestrial orchids, thynnid wasps and pseudocopulation. *Am. Orchid Soc. Bull.* **43,** 13–18.

Stoutamire, W. P. (1975). Pseudocopulation in Australian terrestrial orchids. *Am. Orchid Soc. Bull.* **44,** 226–233.

Stoutamire, W. P. (1983). Wasp-pollinated species in *Caladenia* (Orchidaceae) in south-western Australia. *Aust. J. Bot.* **31,** 383–394.

Stoutamire, W. P. (1985). *Spiculaea ciliata* Lindl., the Australian Elbow Orchid. *Can. Orchid J.* **3,** 19–23.

Stowe, M. K. (1986). Prey specialization in the Araneidae. *In* "Spiders Webs, Behavior and Evolution" (W. Shear, ed.), pp. 101–131. Stanford Univ. Press, Stanford, California.

Stowe, M. K. (1988a). The natural history of *Phoroncidia americana* (Araneae: Theridiidae). *Bull. Mus. Comp. Zool.*, in preparation.

Stowe, M. K. (1988b). The behavior and ecology of moth-attracting spiders in two araneid genera, *Kaira* and *Mastophora*, that evolved sex pheromone mimicry independently. *Behav. Ecol. Sociobiol.*, in preparation.

Stowe, M. K., and Icenogle, W. (1988). Patterns in the phenology of prey capture by moth-attracting spiders. I. *Mastophora cornigera* (Araneae:Araneidae). *Oecologia*. In preparation.

Stowe, M. K., Tumlinson, J. H., and Heath, R. R. (1987). Chemical mimicry: Bolas spiders which attract male moth prey emit components of prey sex pheromone blends. *Science* **236,** 964–967.

Swennen, C. (1968). Nest protection of eiderducks and shovelers by means of faeces. Ardea **56,** 248–258.

Tamaki, Y. (1984). Sex pheromones. *In* "Comprehensive Insect Physiology, Biochemistry and Pharmacology" (G. A. Kerkut, and L. I. Gilbert, eds.), pp. 145–191. Pergamon, Oxford.

Tengö, J. (1979). Odour released behavior of *Andrena* male bees. *Zoon* **7,** 15–48.

Tengö, J., and Bergström, G. (1976a). Odor correspondence between *Melitta* females and males of their nest parasite *Nomada flavopicta* K. (Hymenoptera: Apoidea). *J. Chem. Ecol.* **2,** 57–65.

Tengö, J., and Bergström, G. (1976b). Comparative analysis of lemonsmelling secretions from heads of *Andrena* F. (Hymenoptera, Apoidea) bees. *Comp. Biochem. Physiol.* **55B,** 179–188.

Tengö, J., and Bergström, G. (1977a). Cleptoparasitism and odor mimetism in bees: Do *Nomada* males imitate the odor of *Andrena* females? *Science* **196,** 1117–1119.

Tengö, J., and Bergström, G. (1977b). Comparative analysis of complex secretions from heads of *Andrena* bees (Hym., Apoidea). *Comp. Biochem. Physiol.* **57B,** 197–202.

Theimer, E. T., ed. (1982). "Fragrance Chemistry: The Science of the Sense of Smell." Academic Press, New York.

Thien, L. B., Bernhardt, P., Gibbs, G. W., Pellmyr, O., Bergström, G., Groth, I., and McPherson, G. (1985). The pollination of *Zygogynum* (Winteraceae) by a moth *Sabatinca* (Micropterigidae), an ancient association? *Science* **227,** 540–544.

Thornhill, R., and Alcock, J. (1983). "The Evolution of Insect Mating Systems." Harvard Univ. Press, Cambridge, Massachusetts.

Tietjen, W. J., Ayyagari, L., and Uetz, G. W. (1987). Symbiosis between social spiders and yeast: The role in prey attraction. *Psyche,* **94,** 151–158.

Troilo, D. B., and Cameron, R. G. (1981). Comparative behavior of *Pyrellia cyanicolor* (Diptera: Muscidae) on the moss *Splachnum ampullaceum* and on substrates of nutritional value. *Great Lakes Entomol.* **14,** 191–195.

Tumlinson, J. H., Mitchell, E. R., Teal, P. E. A., Heath, R. R., and Mengelkoch, L. J. (1986). Sex pheromone of the fall armyworm, *Spodoptera frugiperda* (J. E. Smith): Identification of components critical to attraction in the field. *J. Chem. Ecol.* **12,** 1909–1926.

Vale, G. A., Flint, S., and Hall, D. R. (1986). The field responses of tsetse flies *Glossina* spp. (Diptera: Glossinidae) to odours of host residues. *Bull. Entomol. Res.* **76,** 685–693.

Vander Meer, R. K., and Wojcik, D. B. (1982). Chemical mimicry in the myrmecophilous beetle *Myrmecaphodius excavatacollis. Science* **218,** 806–808.

van der Pijl, L., and Dodson, C. H. (1966). "Orchid Flowers: Their Pollination and Evolution," pp. 129–142. Univ. of Miami Press, Coral Gables, FL.

Vane-Wright, R. I. (1976). A unified classification of mimetic resemblances. *Biol. J. Linn. Soc.* **8,** 25–56.

Vogel, S. (1975). Mutualismus und Parasitismus in der Nutzung von Pollenträgern. *Ver. Dtsch. Zool. Ges.* 102–110.

Vogel, S. (1976). Zur *Ophrys*-Bestäubung auf Kreta. *Jahresber. Naturwiss. Ver. Wuppertal* (= Die Orchidee) **29,** 131–139.

Vogel, S. (1978a). Pilzmückenblumen als Pilzmimeten. I and II. *Flora (Jena)* **167,** 329–366; 367–398.

Vogel, S. (1978b). Evolutionary shifts from reward to deception in pollen flowers. *In* "The Pollination of Flowers by Insects" (A. Richards, ed.), pp. 89–96. Academic Press, London.

Vogel, S. (1983). Ecophysiology of zoophilic pollination. *In* "The Encyclopedia of Plant Physiology." Vol. 12c (O. L. Lange *et al.*, eds.), pp. 559–624. Springer-Verlag, Berlin.

Wallace, R. L. (1978). Substrate selection by larvae of the sessile rotifer *Ptygura beauchampi*. *Ecology* **59,** 221–227.

Warncke, K., and Kullenberg, B. (1984). Übersicht von Beobachtungen über Besuche von *Andrena*- und *Colletes cunicularius*-Männchen auf *Ophrys*-Blüten (Orchidaceae). *Nova Acta Reg. Soc. Sci. Ups. Ser. V:C* **3,** 41–55.

Weaver, E. C., Clarke, E. T., and Weaver, N. (1975a). Attractiveness of an assassin bug to stingless bees. *J. Kansas Entomol. Soc.* **48,** 17–18.

Weaver, N., Weaver, E. C., and Clarke, E. T. (1975b). Reactions of five species of stingless bees to some volatile chemicals and to other species of bees. *J. Insect Physiol.* **21,** 479–494.

West-Eberhard, M. J. (1983). Sexual selection, social competition, and speciation. *Q. Rev. Biol.* **58,** 155–183.

Wickler, W. (1968). "Mimicry in Plants and Animals." Wiedenfeld & Nicholson, London.

Wiens, D. (1978). Mimicry in plants. *Evol. Biol.* **11,** 365–403.

Wilson, E. O. (1971). "The Insect Societies." Belknap, Cambridge, Massachusetts.

Wong, J. W., Underhill, E. W., MacKenzie, S. L., and Chisholm, M. D. (1985). Sex attractants for geometrid and noctuid moths. *J. Chem. Ecol.* **11,** 727–756.

Woodroffe, G. E. (1958). The mode of reproduction of *Ptinus clavipes panzer* from *mobilis* Moore (= *P. latro* auct.)(Coleoptera: Ptinidae). *Proc. R. Entomol. Soc. London Ser. A* **33,** 25–30.

Wray, D. L., and Brimley, C. S. (1943). The insect inquilines and victims of pitcher plants in North Carolina. *Ann. Entomol. Soc. Am.* **36,** 128–137.

Yasuhara, A. (1987). Identification of volatile compounds in poultry manure by gas chromatography–mass spectrometry. *J. Chromatogr.* **387,** 371–378.

Yasuhara, A., and Fuwa, K. (1981). Characterization of odorous compounds in rotten blue-green algae. *Agric. Biol. Chem.* **46,** 1761–1766.

Yasuhara, A., Fuwa, K., and Jimbu, M. (1984). Identification of odorous compounds in fresh and rotten swine manure. *Agric. Biol. Chem.* **48,** 3001–3010.

Yeargan, K. V. (1988). Ecology of a bolas spider, *Mastophora hutchinsoni*: Phenology, hunting tactics, and evidence of aggressive chemical mimicry. *Oecologia* **74,** 524–530.

Yeo, P. F. (1972). Floral allurements for pollinating insects. *In* "Insect-Plant Relationships." (H. F. van Emden, ed.), pp. 51–57. Blackwell, Oxford.

Zuckerman, B. M., and Jansson, H. (1984). Nematode chemotaxis and possible mechanisms of host/prey recognition. *Annu. Rev. Phytopathol.* **22,** 95–113.

APPENDIX A: COMMENTS ON TABLES IIa AND IIb

Tables IIa and IIb have two purposes.

1. They are intended to indicate the number of dupe species that each mimic species interacts with. However, the use of species totals has several drawbacks. (a) The tables combine information from studies done in many areas. While some spiders interact with many moth species in a single area, some

orchids probably interact with only a few Hymenoptera species in any one area, but many species over the entire range. (b) As the tables indicate, some mimic species interact with many dupe species. However, others probably interact with very few. For example, it is relatively well established that several orchids (e.g., *O. speculum* and the *Cryptostylis* species) are dependent on a single pollinator (refs. in table). In the tables, it is impossible to distinguish these specialized mimics from mimics that are merely poorly known (the latter are probably more numerous). (c) The tables do not indicate how much evidence there is for mimicry in particular mimic/dupe interactions. However, I have counted only relatively well-established dupe species (i.e., the higher two evidence levels in Kullenberg *et al.*, 1984b; Warncke and Kullenberg, 1984; Stowe and Icenogle, 1988). For controlled studies demonstrating attraction of some of the dupe species see the *Ophrys* refs., Eberhard (1977), Stowe and Icenogle (1988), Stowe (1986, 1988b), and Yeargan (1988). (d) All well-established dupe species are treated as equal, obscuring the fact that some dupe species interact with mimics much less frequently than others.

2. The tables are intended to show the taxonomic diversity of mimic and dupe species and to provide a hint at the diversity of model signal chemistry. Female sex pheromones have been identified for a large number of moth species (see large databases in Tamaki, 1984; Mayer and McLaughlin, 1988). Most moth sex-pheromone compounds fall into one of two "chemical categories": the C_{17} to C_{21} hydrocarbon pheromones or the C_{10} to C_{18} terminal functional group pheromones (see Wong *et al.*, 1985; Mayer and McLauglin, 1988). As indicated in the table, all the moth subfamilies that contain spider prey fall into one of the following three "chemotaxonomic categories": (1) the chemistry is not known for any members of the subfamily; (2) all species for which there is information use C_{17} to C_{21} hydrocarbon pheromones; and (3) all species for which there is information use C_{10} to C_{18} terminal functional group pheromones.

As the table indicates, some spider species catch moth species from both of the latter categories. There are at least five possible explanations for their ability to do so. (1) The prey species in one of the chemotaxonomic categories use compounds from the chemical category "opposite" to the one employed by all previously studied species in the subfamily. Spiders are able to attract all the moths with compounds from one chemical category. (2) The prey species in one chemotaxonomic category use the "expected" chemical category of compounds, but they are attracted to some compounds in the 'opposite' chemical category. Spiders are able to attract all the moths with compounds from one chemical category. [I am unaware of any reports as of yet of field tests and other assays of synthetic compounds where moths in one chemotaxonomic category are attracted by compounds of the 'opposite' category; see Wong *et al.*, (1985), and refs. therein.] (3) Spiders sometimes produce compounds of one category, sometimes compounds of the 'opposite' category. (4) Spiders use mixtures of the two types of compounds in blends that attract both types of moths. (5) Spiders use compounds that (a) do not belong to either chemical category, but (b) are attractive to moths in both chemotaxonomic categories. (In both cases 4 and 5, spiders would be employing multiple-interpretation blends that contain alien components for both chemotaxonomic categories of dupes.)

Unfortunately, virtually no sex-pheromone chemistry is known for hymenopteran taxa outside of *Ophrys* pollinator species (Mayer and McLaughlin, 1988). Although there are probably some chemotaxonomic trends among *Ophrys* pollinator species (at the level of genera and species groups; see Borg-Karlson, 1985), I have not attempted to assign pollinators to sex pheromone categories.

In the future, tables similar to these may suggest how important the diversification of mimic signal chemistry has been to the development of new mimic/dupe interactions. [See Mitter and Brooks's (1983) discussion of 'colonization' patterns in the phylogenetic analysis of coevolving organisms.]

Difficulties with *Ophrys*

In many sections of this orchid genus it is difficult to separate hybrids, species and subspecies. Therefore, I never combine information from different subspecies. I do not present pollinator totals for every form (species or subspecies) where information is available. If closely related forms all attract pollinators from the same set of families ("pollinator-family set"), I usually present information for only one of the forms. For each section of the genus, I have selected forms so as to include the forms with the highest number of pollinator species known for the section, and those that have the lowest number. I have also included enough forms so that there is at least one form for each pollinator-family set known from the section.

Conflicts with Previous Reports

Stoutamire (1974) mistakenly indicated that thynnids pollinate *Calochilus* orchids (W. P. Stoutamire, personal communication). Apparent conflicts within the *Ophrys* pollinator references are the result of shifting opinions concerning the identification of *Ophrys* forms.

APPENDIX B: RELATED PHENOMENA

A few of the arguments in this review might be relevant to five phenomena that share some properties with chemical sense mimicry. (1) A number of plants contain taste-modifying substances (refs. in Smith and Halpern, 1983; review in Inglett, 1977; Kennedy *et al.*, 1988) that in theory could function to deceive herbivores and seed dispersers (Eisner and Halpern, 1971). However, no ecological work has been done. (2) Many toxins in plants (e.g., review in Harbourne, 1977, pp. 83–101) and (3) components of venoms (see review in Schmidt, 1982) "mimic" the effects of hormones, neuropeptides, etc. (4) Many parasites and pathogens "mimic" host hormones (reviews by Sequeira, 1973; Elstner, 1983; Riddiford, 1975; Beckage, 1985; Mueller, 1980; see also Hirai *et al.*, 1983). (5) Many parasites apparently are protected by antigens resembling host antigens, and by other immunomodulatory compounds (reviews in Damian, 1979, 1984; the author terms the phenomenon "molecular mimicry").

All of these phenomena are similar to chemical sense mimicry in that components of a "mimic signal" bind to a "dupe" protein receptor in a manner similar to some "model" compounds of interest, and a response is elicited that benefits the "mimic." In chemical sense mimicry, the dupe receptor proteins are part of an integrated network of receptors that form a system capable of discriminating between many (often subtly) different compounds, and mixtures of compounds. This is also true of phenomenon 1, and to some extent phenomenon 5 above. It is probably not true for most cases of 2–4.

"Mimicry" in the context of 1–4 (particularly 2 and 3) is not easily delimited. "Mimic" compounds range from those that are identical to "model" compounds, to those that are very different and bind to receptors in a completely different way (see in particular refs. in Inglett, 1977). Compounds in high concentrations may achieve the effects of very dissimilar compounds at lower concentrations. The "response" of the "dupe" to the "mimic signal" may be similar to that elicited by the "model signal," or it may be completely different (see Mueller, 1980). At present, it appears that these statements apply less often to chemical sense mimicry systems. However, water is the "signal" medium in 1–4, and we know nothing about the chemistry of waterborne mimic signals in aquatic and within-soil chemical mimicry systems. (Water is capable of transporting a far greater variety of compounds than air.)

In many instances of phenomena 3 and 4, "mimics" harm "dupes," but there is very little selective advantage for dupes with better signal "discrimination." By the time "dupes" have received the "signal," they have lost nearly all chances of reproducing.

APPENDIX C: ASSUMPTIONS AND OVERSIMPLIFICATIONS

I use spider/moth examples in my evolutionary discussion despite the fact that our knowledge of the mimics' signal chemistry is limited to one preliminary study (Stowe *et al.*, 1988). I assume from very little evidence that many moth-attracting spider species use moth sex-pheromone compounds to attract prey, and that they biosynthesize the compounds. I also assume that the findings in the cited moth studies are applicable to prey species of spiders. Spider/moth systems may not behave as I have indicated. However, I do expect that some of my predictions apply to some chemical mimicry systems.

I make many simplifying assumptions. I discuss alien component and ratio detection separately even though they are certainly not independent processes in real organisms. I ignore the fact that absolute concentration affects the range of ratios that are acceptable in moths (see Cardé and Baker, 1984; Linn and Roelofs, 1985, and refs. therein). I assume that there is a one-to-one correspondence between mimic key compounds and model key compounds. Although, in many systems, mimics employ nonchemical as well as chemical mimic signals, I do not consider either the numerous ways that the two signal types might interact, or the effects of such interactions on evolutionary changes.

I state that inhibition due to detection of alien components is common in sex-pheromone communication. Actually, it is difficult to evaluate how many reports of inhibitory compounds represent inhibition by an alien compound. Many important minor components of female blends may be present in such small quantities that they will usually not be detected. A reported inhibitor may appear to be absent from a female blend, when in fact (1) it is present in the female blend and (2) it increases blend attractivity in very small amounts (Steck *et al.*, 1980). A true alien compound inhibitor would be inhibitory in any detectable concentration.

I assume that the nominal model blend elicits stronger responses from dupes than blends that are representative of only a small percentage of model signals. This ignores the "rare male advantage" in *Drosophila*, and other cases of organisms preferring rare signals over common signals (see review in West-Eberhard, 1983). However, most of these cases involve signals produced by the "nonchoosing sex" (see West-Eberhard, 1983). I am not aware of any chemical mimicry systems where such signals are the model signals. In this review, I have not considered the effects of sexual selection and a number of other factors that have probably been very important in the evolution of pheromone communication (see Greenfield, 1981; Thornhill and Alcock, 1983, pp. 159–162, 432–444; Phelan and Baker, 1987, and refs. therein.)

CHAPTER 18

The Chemistry of Coevolution

KEVIN C. SPENCER
Department of Medicinal Chemistry and Pharmacognosy
University of Illinois
Chicago, Illinois

IT IS NOT MY PURPOSE here to review critically the chapters preceding, which would be both presumptuous and unfair to the depth each has explored its subject, but rather to provide an update on our collective thoughts on the subject of coevolution. Specifically, we have vigorously considered the role of chemistry in the establishment and control of coevolution, have reevaluated current concepts of coevolution in light of this new approach, and have separately generated some new ideas on what coevolution is and how it got to be that way. I will try to point out some of the many interesting directions in thought taken by the contributors to this volume, as well as the sometimes controversial and contradictory interim conclusions drawn in regards to the role of chemistry in mediating coevolution.

I. Coevolution as a Chemical Process

A general feeling expressed by many authors in this volume is that coevolution is not all that common (Chapters 2–4, 8, 11, 14, 15). Where the process is doubted, the presence or selective function of chemicals is usually not known. In those systems where coevolution is considered to be acting (5–7, 10, 13, 17), plant chemistry is determined to be an essential component of its establishment and development, and in comparative studies of other systems (2, 9, 12, 16) the same conclusion is reached.

Evidence for coevolution is weak, generally, but strongest in plant–herbivore associations (1). Evidence for plant chemical control in the establishment and maintenance of coevolution is stronger yet.

In every system of plants and herbivores where plant chemistry has diversified extensively, we find at least some evidence for coevolution. In every well-described coevolutionary system of plants and herbivores we find a diversified chemistry. We are left with the inescapable conclusion that, in plant–herbivore specialization, coevolution is a function of plant chemistry. We can extend this observation cautiously to higher trophic levels in interactions involving sequestration-based defenses, in herbivores interacting with mimics, and in interacting guilds of herbivores. It seems likely that predators and parasitoids will also respond to plant chemistry through herbivores, but these interactions are rarely described as coevolutionary.

Coevolution then is a process of interaction that represents a special case of the larger process of specialization, and specialization versus generalization is a much more important force in evolution than is that process that has been called coevolution.

Specialization and generalization in herbivores are processes certainly affected by plant chemistry, and possibly entirely determined by changes in plant chemistry over time. Generalist herbivores represent species that either have not yet specialized, or are no longer specializing, and their mode of ecological and evolutionary interaction with plants is also chemical. Generalization is thus a special case of specialization. The potential exists for coevolution to be a result in any host–plant–specialist interaction, and that process of coevolution will be a product of plant chemistry.

Chemical communication between herbivores and plants is the primary driving force behind what we recognize as coevolution, and that process is better represented as an accelerated process of genetic change facilitated by the efficiency of chemical information exchange. Coevolution is no different from evolution, except in being faster, and directed. We do not have data herein to bring to bear on other forms of coevolution, especially mutualisms, and I restrict my discussion to herbivores and to trophic levels immediately dependent on herbivores. Yet, it seems possible that facilitated information exchange is an important factor in the establishment of these systems as well.

II. Coevolved Systems and the Environment

Having decided on the importance of chemical events in the establishment, maintenance, and dissolution of coevolved interactions, it behooves us to consider external factors that directly affect evolution of chemistries or phenotypic expression. Several authors point out the void in our knowledge of the effects of environmental factors and nutrition in directly limiting expression of plant chemistry (3–5, 7, 9–14). If indeed coevolution is a process consisting of an interplay between the rapid development of chemistries by plants and their detection and subsequent toleration by herbivores, then it is clear that any environmental control of expression of plant chemistry represents environmental control of the coevolutionary process. That such a direct cause and effect might exist between the abiotic environment and systems characterized by a very high degree of biological information exchange is fascinating. We can only postulate at this time that regulatory mechanisms exist in plants that can modify these abiotic constraints in ways that may be selected to be responsive to coevolutionary events. Might similar but usurptive regulatory mechanisms exist in herbivores? It appears that evolutionary biologists will have to work more closely with plant and insect physiologists in order to address such questions.

III. Chemicals as Agents of Selection

Biochemical expression and regulation in plant chemistry is referred to by most authors herein, but a concomitant discussion of insect biochemical and physiological parameters may be found to be deficient. We simply do not know what happens to most plant chemicals when they are ingested, nor usually do we know what happens to the herbivore that eats them, nor to its subsequent predator. More is known about toxicological effects of plant chemicals to mammals than any other class of organisms (14), but most studies of chemical coevolution involve insects. The small amount of data available on actual biochemical mechanisms of action of plant chemicals suggests most directly that toxification is a complex, specific, and regulated process (5–7, 14). Most authors agree that the actual fitness and selection effects of these compounds are rarely known and simply cannot be assumed. Measures of selection effects and potentials must be made in the laboratory (5–7) and in the field before the effects of ecological factors (3, 4, 7, 8), chemical variation (10–13), or genetic contribution (2, 5) upon coevolution can be quantified.

IV. Chemical Variation

Variation in expressed chemistry is viewed as a selected deterrence to specialized herbivory spatially (10), biogeographically (8, 15), seasonally (10–13), temporally (10), and through evolutionary time (5–7, 10).

This variation may occur within a chemical class (6, 10), or may involve balances between several biosynthetic classes (5, 7, 9) including primary metabolites (8, 10, 12). This variation can also be expressed as a deterrent to generalist herbivores when inducible (12, 13) or when the chemicals involved can reduce the fitness of an average generalized herbivore (4, 7, 11–14).

All authors feel that resource predictability, or plant apparency, is a factor of great importance in controlling specialization and coevolution when the mechanisms of detection and defense are chemical, and specific chemical arrays are present that contain both antispecialist and anti-generalist compounds (7). Most plant chemicals have not been bioassayed for their range of activity against specialists or generalists in their environment (4, 5, 7, 11, 14). Many chemicals are assumed to be antispecialist compounds expressed through coevolutionary events, and this has rarely been tested in the field (3–5, 7, 11). Toxicity is a relative term; there is no such thing as a toxic compound (7, 14) or array (4, 5) when considered independent of its biological target.

V. Higher-Order Interactions

Plant chemicals may influence the success and coevolution of species in trophic levels above that of herbivory. Sequestration of chemicals by a generalist to generate defenses may depend directly on the breadth of plant chemicals available in the diet (16). As in plants, nutrient limitations may alter the defensive array, and variation of chemicals presented in the defense seems to be as important as quantity (16). The sequestration process, while poorly understood, may be either specialized (6, 15) or generalized (16). This process may be under evolutionary control and may be different in specialists than in generalists. Exaptations for sequestration may explain many apparently coevolved systems (15, 17), and we may wish to reinterpret these as less complex chemical relationships. Measurements of selection are needed—the effect of mimics upon models' fitness remains unknown.

The observation that insects can detect individual chemical structures (5–7, 15) whereas predators sometimes cannot (15) leads to the question: might the different sensory abilities toward particular chemicals between herbivores and predators in coevolved systems represent a measure of the difference in intensity of coevolution between the herbivore/plant and predator/herbivore pairs? Does this have a measurable biochemical or genetic component?

The control of sequestration, either through diet or biochemical regulation, may lead ultimately to the presence in individuals of variable chemical arrays (17), which change through time during selection. Convergence in chemistry of aposematic forms may result in chemical arms races between mimics and models very similar to the coevolutionary processes described for plants and herbivores. Phylogenetic analyses of higher trophic level interactions can yield the same kinds of evidence for coevolution (1, 15, 17) as found for other systems. It may be possible to quantify the changes that have occurred between chemical characters of each mimic, model, and plant species in a large and complex multitrophic coevolved system, and to use these data to estimate the genetic interdependence of various trophic levels in coevolution. The covariance in chemistry between species pairs should relate to the variance in the chemistry throughout the system as a quantitative function of this interdependence.

VI. Coadaptation and Cospeciation

The events usually taken as proof of the coevolutionary process are cophylogenetic correlation (1, 2). The building of these phylogenies involves the increasing organization of species through association, selective interaction, generation of specific adaptations, fixation of adaptation, coadaptation, and cospeciation.

We can detect cophylogenetic correlation only after all of these process have occurred. Sequential or parallel associations may mislead us, and the lack of cophylogenetic correlation in any case cannot be taken as evidence against coevolution (2). Active avoidance or dissolution of coevolved systems may be selected for (1, 2, 7), and it may be common for systems to coevolve only for a short time—we may thus fail to observe them.

Several authors suggest that coevolution can be established quite rapidly (2, 6, 7). We may fail to recognize the process due to our overly simplistic expectations of its result (5, 13). Most authors conclude that it is very clear that species may not be as important as local populations in establishment of responsorial and reciprocal interactions that lead ultimately to observed coevolved lineage pairs. Chemistry is correlated with reproductive barriers in several systems (2, 4, 5, 7–10, 12, 17), and this correlation may be present but unobserved in many others.

We will probably be frustrated in our attempts to define and describe coevolution as a cophylogenetic phenomenon until we are able to adequately document the regulatory processes and selective pressures that control it. Plant chemistry is a factor that increases the rate of genetic exchange to an extent sufficient to alter the rate and direction of coevolutionary events. In chemically mediated systems, evolution probably proceeds by leaps and bounds of rate coupled with constraint in direction. This process is far better described as a punctuated equilibrium phenomenon than as a linear and gradual coevolution.

VII. Questions and Directions

Is coevolution a universal end-result of all species interactions? Does coevolution lead to committed genome fractions so large that genetic and ecological variation becomes extremely limited, communities become fixed, and only stochastic or catastrophic events can break these commitments and reintroduce system-wide variation?

Is there selection to coevolve as a general response to the availability of biotic energy sources and the high pay-off in being able to efficiently or exclusively exploit them, or is the impetus to coevolve a passive and selection-neutral manifestation of association, shared habitat, or coexistence? Either way, would there then be selection to avoid coevolutionary entrapment (and perhaps to maintain genetic variability)?

Do specialist herbivores and their coevolved host plants have reduced genetic variation, and if so, will this serve as a quantitative measure of the age of the interaction and the degree of coevolution? Do generalists have higher genetic variability?

One hundred million years of plant–herbivore association means several times 100 million generations, depending on the number of generations per year for a given species, but when the multiple associations of a species of plant or insect are considered together with the actual numbers of individuals in many lineages that have existed in such contact, the number of generations involved is nearly (but not quite) infinite. How do coevolution and chemical diversification depend on generation time and length of association?

Mammals are generalists different from other herbivores in terms of learning abilities, metabolism, detoxification abilities, motility, and in having extensively developed microbial gut floras, but they do respond to plant chemistry. Have mammals evolved as generalists following the same chemical selective pressures generated in systems of insects and plants? If learning and detoxification are genetically disjunct processes, do plants respond by generating chemicals that separate these effects?

How much more do we need to know about how different plant secondary chemicals affect herbivores?

Nutritive metabolism and detoxification are separately selectable processes. To what degree has plant primary metabolism constrained the digestive systems of insects in their ability to evolve responses to secondary metabolism?

Palatability and chemical perception are processes distinct from detoxification and tolerance. What is the degree of linkage disequilibrium between behavioral and sensory modalities and biochemical capabilities?

Correlation studies are not adequate to detect and describe the coevolution of a system. Experimental and manipulative multivariate studies addressing chemistry, biochemical regulation, and detoxification in both plants and herbivores, edaphic factors, primary metabolic requirements, genetics, ecological factors, and variation of these over space and through time are needed to ultimately define the process of coevolution.

SUBJECT INDEX

Biosystematic Index